Vegetable Diseases

A Color Handbook

Steven T. Koike

Plant Pathology Farm Advisor, Monterey County, University of California Cooperative Extension

Peter Gladders

Plant Pathologist, ADAS Boxworth, Cambridge, UK

Albert O. Paulus

Plant Pathology Department, University of California, Riverside

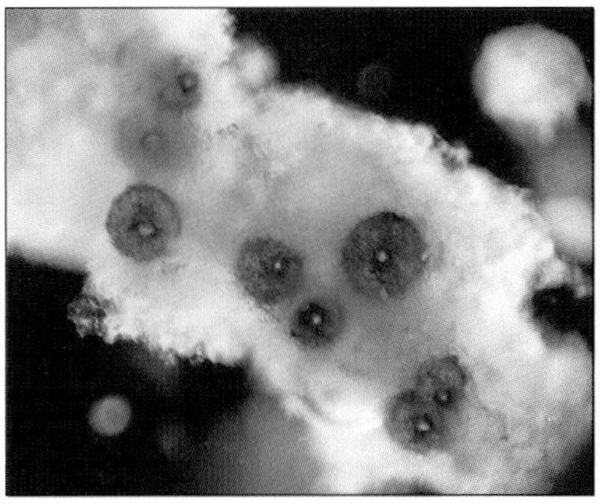

ELSEVIER

BOSTON • SAN DIEGO

Academic Press is an imprint of Elsevier

Steven Koike dedicates this book to Margaret, Evan, and Andrew Koike

First published in the United States of America in 2007 by
Academic Press, an imprint of Elsevier
30 Corporate Drive, Suite 400, Burlington, MA 01803, USA
525 B Street, Suite 1900, San Diego, CA 92101-4495, USA

ISBN-10: 0-12-373675-7
ISBN-13: 978-0-12-373675-8

Library of Congress Cataloging-in-Publication Data
A catalog record for this book is available from the Library of Congress

For information on all Academic Press publications
visit our website at books.elsevier.com

Commissioning editor: Jill Northcott
Project manager: Ayala Kingsley
Copy editor: Derek Hall
Proof-reader: Lynne Wycherley
Indexer: Jill Dormon
Book design and layout: Ayala Kingsley
Colour reproduction: Tenon & Polert Colour Scanning Ltd, Hong Kong
Printed by: Grafos SA, Spain

Plant Protection Handbooks Series

Alford: *Pests of Fruit Crops – A Color Handbook*

Biddle/Cattlin: *Pests, Diseases and Disorders of Peas and Bean – A Color Handbook*

Blancard/Lot/Maisonneuve: *Diseases of Lettuce and Related Salad Crops – A Color Atlas*

Bridge/Starr: *Plant Nematodes of Agricultural Importance – A Color Handbook*

Fletcher/Gaze: *Mushroom Pest and Disease Control – A Color Handbook*

Koike/Gladders/Paulus: *Vegetable Diseases – A Color Handbook*

Wale/Platt/Cattlin: *Pests and Diseases of Potatoes – A Color Handbook*

Contents

Preface

FROM THE LAND that is clean and good, by the will of its Cherisher, springs up produce, (rich) after its kind: but from the land that is bad, springs up nothing but that which is niggardly: thus do we explain the signs by various (symbols) to those who are grateful.

7 (Al-Araf): Verse 58 , The Quran

MAY THE WINDS blow sweetly,
May the rivers flow sweetly,
May plants and herbs be sweet to us,
May days and nights be sweet to us. . . .

X. 48–50, Yajur Veda. Taitt. Aranyaka

. . . IF THERE IS FAMINE in the land, if there is pestilence,
if there is blight or mildew, locust or grasshopper. . . .

I Kings 8:37, The Bible

IF THERE IS BLIGHT OR MILDEW on one's crops, what is one to do? The topic of diseases that damage and kill plants is hardly a new subject. For as long as humans have foraged for, grown, traded, or eaten edible plants, disease-causing organisms have been present and exacted their toll on quality, yields, and consumer satisfaction of these commodities. Impacts of plant diseases on the lives of people range from the nuisance of losing a few plants in one's garden, to significant economic losses to a farmer, and finally to widespread famine due to extensive crop losses in a region. Throughout human history, devastating crop losses have sometimes resulted in subsequent loss of human life and disruptive migrations of the inhabitants. Humanity's dependence on healthy crops and reliable sources of food, therefore, transcends all barriers of culture, nation, and time. We all need to eat and to feed our children.

Vegetables are an essential and increasingly popular component of human diets today. Collectively, vegetable crops are a major part of agricultural commerce. The vegetable industry produces large volumes of high quality commodities that are intensely marketed and can be delivered locally and regionally or shipped internationally. Consumer standards and market requirements mandate excellent quality produce. The diseases that affect vegetables compromise such quality and therefore are of great importance to grower, shipper, marketer, and consumer. Vegetable production and marketing in the 21st Century has been fashioned by technology and developments that are unique to our times, including molecular biology, globalization of international trade, awareness of the benefits and dangers of synthetic pesticides, and insights into specific health benefits of vegetable foods.

This book is written to address the broad topic of diseases that affect vegetables. Part 1 offers a brief introduction to vegetable crops, descriptions of the disease-causing agents, suggested strategies for identifying and diagnosing vegetable diseases, and general principles in controlling them. In this book we describe diseases that are primarily caused by pathogens (biotic diseases). Problems caused by nutritional and physiological disorders and environmental and cultural factors (abiotic problems) are mostly not covered.

The rest of the book (Part 2) is divided into chapters on the principal crop groups (and further subdivided if different plants within the group suffer from distinct sets of diseases) and describes the major diseases that affect those vegetables. The diseases are, for the most part, organized first by pathogen type and then by pathogen name. (We should point out that each crop chapter does not include all possible diseases and that the disease list is therefore not exhaustive.) Of special note are chapters devoted to spinach – an increasingly popular vegetable – and to specialty crops and herbs. Each disease entry includes a brief introduction to the disease, detailed description of symptoms, information on the pathogen and disease development, and suggestions on how to manage the problem. For pathogens that affect several crops, full details are presented in only one chapter in order to reduce unnecessary repetition; for other crops that are subject to the same disease, reference will be made to the more complete chapter. Selected references are included that will allow interested readers to further research the subject.

Our collective experience in applied research, extension education, and working closely with farmers and industry members has shaped our approach. Our aim is to increase recognition and diagnosis of vegetable diseases and to provide information on biology and control of the problems. A particular feature of this book are the many high-quality color photographs that illustrate most of these vegetable diseases and which will assist the reader in identifying and understanding them.The glossary at the end lists much of the terminology used in plant pathology and related fields.

In an effort to keep this book timely and reduce the amount of information that rapidly becomes outdated, we have not included specific information on vegetable crop cultivars, pesticide product recommendations, and seed treatments. Such information can change from year to year and also varies greatly between regions, countries, and continents. Seed treatments, in particular, can be implemented in many ways depending upon the practitioner, the nature and location of the seed treatment facility, and so on. For up-to-date and area-appropriate recommendations on vegetable cultivars and disease control chemicals and treatments, consult local extension agents, agricultural consultants, or other professionals who are familiar with the location.

We have written this book with a very broad and diverse audience in mind. We hope this effort will help and be of interest to the following persons: research and extension plant pathologists; diagnosticians and plant lab personnel; teachers of agriculture and related subjects; university students in agriculture and related fields; commercial farmers, vegetable producers, and farm managers; agriculturalists in the fields of seed production, vegetable breeding, agrichemicals, pest control, marketing, and other subjects; government and regulatory persons dealing with agriculture; home gardeners and hobbyists.

ACKNOWLEDGEMENTS

We have many collaborators with the University of California, ADAS, and other agencies who have assisted us with research and extension projects on vegetable diseases. We acknowledge the following persons for such contributions: DM Ann, NJ Bradshaw, CT Bull, EE Butler, WS Clark, DA Cooksey, JML Davies, RM Davis, Dorothy M Derbyshire, BW Falk, RL Gilbertson, TR Gordon, AS Greathead, Kim Green, GW Griffin, MJ Griffin, S Hammond, P Headley, D Henderson, DR Jones, OW Jones, K Kammeijer, BC Knight, FN Martin, GM McPherson, the late SC Melville, R Michelmore, the late JM Ogawa, TM O'Neill, GS Saenz, RF Smith,KV Subbarao, HJ Wilcox, DJ Yarham.

We thank David Yarham, John Fletcher, and our two reviewers for their efforts in evaluating the book and suggesting improvements.

We greatly thank the following persons for contributing photographs to this book: Mike Asher, Lindsey du Toit, John T Fletcher, Philip Hamm, Mary Hausbeck, Nikol Havranek, Gerald Holmes, David R Jones, AP Keinath, Frank Laemmlen, David Langston, Robert McMillan, Gene Miyao, Krishna Mohan, Tim O'Neill, Ken Pernezny, Melodie Putnam, Richard Smith, Mike Stanghellini, Tom Turini, William Wintermantel.

We thank Manson Publishing for providing us the opportunity to present this work. We especially thank Jill Northcott for assistance, guidance, and encouragement with this project, and Ayala Kingsley and Derek Hall for their careful work on our manuscript.

Steven Koike thanks the many farmers, pest control advisors, and agriculturalists in California who taught him so much about vegetable crops.

PART 1

Introduction to Vegetable Crops and Diseases

- VEGETABLE PRODUCTION
  The increasing importance of vegetables; vegetable commodities; technological advances; challenges in vegetable production; the challenge of plant diseases
- CAUSES OF DISEASE
  Diseases caused by biotic agents (bacteria, viruses, fungi, fungus-like pathogens); disorders caused by abiotic factors and physiological conditions
- DIAGNOSING DISEASE
  The importance of accurate, timely diagnosis; overall strategy; asking the right questions; conducting the examination; incorporating laboratory tests; diagnosing virus diseases; digitally assisted diagnosis; compiling information and drawing a conclusion
- CONTROLLING DISEASE
  Criteria for commercial control of vegetable diseases; disease control options; other aspects of disease management

Vegetable production

Vegetables are a critically important part of the human diet. Along with grains and fruits, vegetables are one of the three major food groups of plant origin. The highly perishable nature of most vegetable commodities, along with the market demand for a steady supply of high-quality, disease-free produce, makes vegetable production a challenge for the farmers of the world.

The increasing importance of vegetables

The role of vegetables in the human diet has increased for several reasons. Nutritional and medical research elucidates the value and role of these foods in keeping us healthy. Vegetables provide essential carbohydrates, proteins, fibers, vitamins, and minerals. Researchers are indicating that components found in vegetables provide additional health benefits such as the following: reduced cancer risk associated with lycopene in tomato and watermelon fruit, beta-carotene from carrots and squash, lutein from broccoli and peas, and glucosinolates from crucifers; lowered blood pressure and other benefits for the circulatory system associated with anthocyanins in beets, red cabbage, and kidney beans; assistance with depression associated with B vitamin folate in legumes and spinach; prevention of age-related macular degeneration of the eye attained from the antioxidant lutein in spinach leaves; reduced heart disease and cancer risk associated with allicin in onion and garlic. These and other examples highlight specific roles that vegetables have in possibly addressing human diseases.

In addition to nutritional and health issues, vegetables bring a welcome aesthetic value to the table. Because there are so many different kinds and varieties of vegetables, these commodities are an important source of diversity, color, taste, and texture in cooking. Specialty vegetables, baby leaf and baby vegetable products, and organically produced commodities provide additional choices to the people preparing and eating vegetables. New vegetable types, vegetable products, and the changing tastes and demands of the consumer have further encouraged growth in the vegetable producing industry. For example, ready-to-eat bagged vegetable products have greatly increased in popularity and availability. These products consist of diverse vegetable and salad components that are washed, chopped into consumable portions, mixed together, and sealed into plastic bags that allow for many days of shelf-life.

Finally, the modern world market situation is responsible for allowing significant increases in vegetable production to take place. The current global economy and intense export/import businesses make it possible for vegetables produced in one part of the world to be quickly shipped and made available to markets and consumers anywhere else in the world. Transportation systems and postharvest handling have improved over the years, resulting in high-quality produce with excellent shelf-life being available year-round to the consumer.

The emergence of large corporate, national, and international marketing chains and companies has also affected the way vegetables are produced. Such huge conglomerates in large measure are able to exert control over commodities such as vegetables. Vegetable growers and suppliers must agree to provide steady supplies of high-quality produce at fixed prices in order to sell to such chains. These business agreements place pressure on growers who cannot allow disease and other production problems to interfere with harvest schedules and projections.

Total world vegetable production has increased almost three-fold over the last 30 years according to FAO reports (***Table 1***). This phenomenal growth applies to many of the major vegetable crops, including tomato, onion, eggplant, pepper, carrot, cucurbits, spinach, and lettuce. Increases in corn, pea, and bean production have doubled. In comparison to the growth of vegetable production, over the same 30-year period the human population has increased approximately 1.6 times, from 3.84 to 6.22 billion people.

TABLE 1 Thirty-year trends for world production of major vegetable crops

Vegetable crop	1972	1982	1992	2002
Artichoke	1.40	1.18	1.24	1.26
Asparagus	1.16	1.42	2.24	5.08
Bean	2.68	3.18	3.66	5.65
Bean, string	1.02	1.12	1.32	1.72
Broad bean	0.66	0.87	1.00	1.02
Cabbage	27.39	39.39	39.78	62.47
Carrot	7.96	11.24	13.99	21.02
Cauliflower	4.48	6.57	10.24	15.05
Corn	4.84	5.77	7.26	8.40
Cucumber, gherkin	10.63	14.72	18.61	36.40
Eggplant (aubergine)	6.59	9.14	11.47	29.93
Garlic	2.98	4.76	7.34	12.11
Leek and other alliums	0.79	0.80	1.41	1.52
Lettuce	6.89	9.00	12.42	18.75
Mushroom	0.79	1.14	1.94	3.07
Okra	1.82	2.61	3.61	4.90
Onion and shallot, fresh	1.60	2.53	3.25	4.36
Onion, dry	16.94	24.57	32.35	51.91
Pea	4.86	5.98	6.75	9.06
Pepper	6.19	8.23	11.33	22.17
Pumpkin, squash, gourd	6.23	8.60	10.84	16.91
Spinach	2.34	3.40	4.42	10.31
Tomato	38.49	57.67	74.76	108.50
Total fresh vegetables	**158.73**	**223.89**	**281.23**	**429.40**

Units = million metric tonnes [million MT]
Listed vegetables are for fresh market and do not include crops grown for dried or processed products. Source: FAO (http://apps.fao.org)

Vegetable commodities

What exactly is a vegetable? Like many words, the term itself has different meanings and degrees of ambiguity depending upon the context of its use. A formal dictionary definition is the following: *vegetable* = usually an herbaceous plant grown for the edible part that is eaten as part of a meal; also, such an edible part (Merriam-Webster's Collegiate Dictionary). This definition, however, is broader than the agricultural/horticultural understanding of 'vegetable' which usually excludes traditional fruit commodities such as peach, pear, apple, cherry, grape, blackberry, strawberry, and many others. Botanically, a vegetable is usually defined as those parts of a plant derived from a vegetative organ (root, bulb, crown, stem, rhizome, petiole, leaf) and so would not include reproductive structures like tomato and pepper fruit, cauliflower and artichoke flower buds, and bean and pea pods and seeds. The conventional usage of 'vegetable' is a combination of all these meanings. Vegetables traditionally cover all crops derived from vegetative plant parts, and also include selected flower, fruit, and seed commodities that over time were considered vegetables through common use and inclusion in non-dessert foods.

In the USA, the issue of 'what is a vegetable' needed to be addressed by the country's highest court. In 1893, the US Supreme Court ruled on whether tomato was a vegetable or fruit. This authoritative legal body decided that for Americans the tomato would be considered a vegetable. This ruling was necessary because businesses and regulatory agencies did not know whether tomatoes were subject to commerce laws for fruits or vegetables. Interestingly, ketchup derived from processed tomatoes was quite popular many years before this legal discussion; ketchup was declared America's National Condiment in 1830.

Like all other plants, each vegetable crop plant is given scientific genus and species names. Each vegetable is then grouped taxonomically into plant families (*Table 2*). Knowledge of these plant families and the relatedness between certain vegetable commodities is often useful in understanding plant pathogen dynamics and devising disease management strategies. While some pathogens are able to infect hundreds of plants in many different plant families, most of these organisms have much smaller host ranges. In many cases, the pathogens will tend to infect only crops within these plant families. Rust (*Puccinia allii*) of garlic infects onion and chives, but not non-alliums. The black rot pathogen of crucifers (*Xanthomonas campestris* pv. *campestris*) attacks many plants in the Brassicaceae, but not plants in other families. The extensive research information available for sugar beet diseases can in large measure be used to help understand diseases of the closely related table beet and Swiss chard.

TABLE 2 Plant families and selected commercial vegetable crops

Plant family	Examples of vegetable crops
Alliaceae	Onion, garlic, leek, chives, shallot
Amaranthaceae	Beet, Swiss chard, spinach
Apiaceae	Carrot, celery, celeriac, parsley, parsnip, cilantro, fennel, dill, chervil
Asparagaceae	Asparagus
Asteraceae	Lettuce, endive, radicchio, artichoke, Jerusalem artichoke, salsify
Brassicaceae	Broccoli, Brussels sprout, cabbage, cauliflower, mustard, arugula, radish, broccoli raab, watercress
Cucurbitaceae	Cucumber, squash, melon, pumpkin, watermelon
Fabaceae	Bean, broad/faba bean, pea
Lamiaceae	Mint, basil, oregano, marjoram, sage, catnip
Poaceae	Corn, lemongrass
Polygonaceae	Rhubarb
Solanaceae	Pepper, tomato, eggplant, potato, tomatillo
Valerianaceae	Corn salad

Technological advances

There have been major changes and improvements in the production of vegetables. To begin with, there is now a great awareness of, and appreciation for, the importance of seedborne pathogens. The seed producing industry now invests a great amount of research and effort into providing high-quality, relatively pathogen-free seed to growers. Such seed results from the following steps: establishment of clean mother seed stock sources, placement of seed increase plantings in arid regions, vigilant monitoring of seed fields, roguing of fields having symptomatic and off-type plants, development and use of sensitive pathogen detection methods, and treatment of seed to reduce what pathogens might be present.

Advances in seed physiology, such as seed priming and coating treatments, also contribute to improvements in seed quality, seedling vigor, and reduction in disease problems. Seed coat treatment is a highly developed, sophisticated science. Seeds are coated with various substances so that they can be more easily handled during planting, germination can be modified and controlled, and protection against pests and diseases can be ensured. Seed priming is a relatively recent development. Seed priming treatments are applied before seeds leave the packaging facility, and consist of processes that invigorate seeds and actually initiate some physiological steps in germination. When primed seeds are placed in the soil or rooting medium, seeds will germinate more rapidly and uniformly than non-primed seeds.

Vegetable transplants to start new plantings are less frequently grown in the ground, but instead now are often produced in sanitized transplant trays, using soilless rooting media, and placed in greenhouses where conditions can be strictly controlled. In the USA, the great majority of celery, cauliflower, tomato, pepper, lettuce, and melon transplants are grown under this system.

Vegetable crop breeding programs continue to produce new, vigorous, productive cultivars. The creation of improved hybrids brings about increases in yield and in some cases resistance to diseases and pests. While not universally accepted at the consumer level, the fields of molecular biology and genetic engineering potentially can further advance vegetable plant breeding.

Postharvest handling methods have improved greatly. Postharvest research has informed the industry about the optimum ways of harvesting, transporting, and storing vegetable commodities. Improvements have therefore been made in the ways these vegetables are handled once they leave the field. These improvements have increased the overall quality of marketed vegetables, improved their shelf-life, and made possible the long-range transport and sale of vegetables to distant overseas destinations.

A revolutionary postharvest advancement for vegetables was the rapid development and marketing of bagged, ready-to-eat vegetable products. Developed on a large scale within the past 10 years, a wide variety of vegetable commodities are harvested and taken to processing facilities where they are washed, cut and prepared, and placed in bags made of specialized, semipermeable materials. These specially designed plastics control the concentrations of oxygen and carbon dioxide inside the bags and allow these products to have significant shelf-life. Properly refrigerated, these cleaned and chopped products may last perhaps as long as 7 to 10 days after they go on the market. Products are intended to be ready-to-eat and do not usually need further preparation. These 'value added' or 'lightly processed' products have greatly changed the vegetable market by providing novel products that are convenient for the consumer. Vegetables included in these products include lettuce, carrot, celery, cauliflower, broccoli, cabbage, mustards, arugula, endive, radicchio, spinach, beet leaves, Swiss chard, and others.

Research on plant pathogens has also aided the vegetable industry. More sensitive, highly accurate pathogen detection methods are now available for many seedborne pathogens of vegetables. In-depth studies on disease epidemiology help explain the biology of vegetable pathogens and enable better disease control strategies to be devised. Extensive breeding efforts develop resistant cultivars. Safer fungicides that are less disruptive to the environment are being used on vegetables. Compared to older generation pesticides, such chemicals are applied at very small rates and enable growers to control diseases while adding less fungicide to the environment.

Challenges in vegetable production

Despite advances in vegetable production and disease management, many challenges face growers of vegetables. Because of the nature of today's worldwide market, there are extremely high expectations for growers to provide ample supplies of high-quality, disease-free produce that has extended shelf-life. Markets are looking to suppliers to provide such produce over many months each year. Competition to maintain a market share in this global vegetable market is intense.

In addition to requiring high quality, the world market is demanding food that is safe to consume. Vegetables for both local and regional markets and for use overseas must be free of human, foodborne pathogens and contaminants such as *Escherichia coli*, *Salmonella*, other enteric bacteria, and various other microorganisms. Mistakes in the handling chain from field to table can result in consumer exposure to such pathogens. This requirement for cleanliness and careful handling adds another pressure in the vegetable producing business.

Another challenge is the need to produce high-quality vegetables while reducing the use of synthetic pesticides. Consumers worldwide want fewer chemicals on their food products, but still demand high quality and long shelf-life. Vegetable growers must grapple with both demands. As time passes, there are fewer fungicides remaining to control diseases, fewer insecticides to control insect vectors of pathogens, and fewer herbicides to control weeds that harbor both insect vectors and pathogens. Reduction in pesticide use is a food safety and health issue as well as an environmental one. Individuals and groups are seeking to better preserve the wildlife and environment around us, and hence seek reduced pesticide use. The example of the soil fumigant methyl bromide and the international movement to ban the agricultural use of this chemical is a prominent case study that illustrates these trends and issues.

The challenge of plant diseases

For as long as humans have foraged for, grown, traded, sold and bought, or eaten vegetables and other edible plants, disease-causing organisms have reduced quality, yields, shelf-life, and consumer satisfaction of these commodities (see the references cited in this chapter that give examples of plant disease concerns throughout history). Therefore, managing these diseases is an essential task of all producers. If a particular grower cannot provide vegetables having these high standards, then other farmers in the region or the world will take over that market share. Therefore, disease control has become an extremely critical aspect in the drive to produce excellent quality vegetables. Ironically, the same extensive international system that makes possible the selling, buying, and transporting of harvested vegetables across the globe also can create new disease problems for growers. Pathogens that occur in one part of the world can be moved on contaminated or diseased seed and plant material to other parts of the world where that pathogen never before had occurred.

The cost of producing vegetables continues to climb, placing ever-increasing economic pressure on growers. Rising land use, water, seed, equipment, fuel, fertilizer, pesticide, and labor expenses force growers to maximize yields and seek economical ways to control diseases. A final contemporary challenge is the worldwide trend of reducing the use of certain pesticides on food commodities. For environmental and human health reasons, some pesticides are no longer available for use on vegetables and others will either be eliminated in the future or their use will be severely restricted. Therefore, a disease control program that relies on fungicides requires revision and development of new and integrated strategies. Because of the challenges posed by plant pathogens, it will be imperative for growers and field personnel to be familiar with the vegetable diseases they may face.

References on vegetables and general production

Anonymous. 2003. *Basic Horticultural Statistics for the United Kingdom Calendar and Crop Years 1992/93–2002/3*. Defra Publications, London.

Anonymous. 1992. *Basic Horticultural Statistics for the United Kingdom Calendar and Crop Years 1981–1990*. Ministry of Agriculture, Fisheries and Food Publications, London.

Bailey, L. H. 1949. *Manual of Cultivated Plants*. Macmillan Publishing Company, New York.

Bailey, L. H. and Bailey, E. Z. 1976. *Hortus Third: A Concise Dictionary of Plants Cultivated in the United States and Canada*. Macmillan Publishing Company.

Bosland, P. W. and Votava, E. J. 2000. *Peppers: Vegetable and spice capsicums. Crop Production Science in Horticulture 12*. CABI Publishing.

Bown, D. 1995. *Encyclopedia of Herbs and Their Uses*. Dorling Kindersley, London.

Brewster, J. L. *Onions and Other Vegetable Alliums. Crop Production Science in Horticulture Series, No. 3*. CAB International.

Davidson, A. 2002. *The Penguin Companion to Food*. Penguin Books.

Duke, J. A. 2002. *Handbook of Medicinal Herbs*. Second edition. CRC Press.

Duke, J. A. and duCellier, J. L. 1993. *CRC Handbook of Alternative Cash Crops*. CRC Press.

Ellison, D. 1999. *Cultivated Plants of the World*. New Holland Publishers.

Fahey, J. W. and Stephenson, K. K. 1999. Cancer chemoprotective effects of cruciferous vegetables. *HortScience* 34:1159–1163.

Garbutt, N. 2000. Meeting consumer demands for food safety: European retailer protocols for crop protection programmes. *Proceedings of the BCPC Conference – Pests & Diseases* 2000 1: 129–132.

George, R. A. T. 1999. *Vegetable Seed Production*. 2nd edition. CABI Publishing.

Janick, J. 1981. *Plant Science: An Introduction to World Crops. Third edition*. W. H. Freeman and Company.

Jones, J. B. 1999. *Tomato Plant Culture: In the Field, Greenhouse, and Home Garden*. CRC Press.

Larkcom, J. 1991. *Oriental Vegetables*. Kodansha International.

Maynard, D. N. and Hochmuth, G. J. 1997. *Knott's Handbook for Vegetable Growers*. Fourth edition. John Wiley & Sons, Inc.

Mitton, P. J. 2000. The implication of food assurance schemes for pesticide manufacturers. *Proceedings of the BCPC Conference – Pests & Diseases 2000* 1: 143–148.

Peterson, J. 1998. *Vegetables*. William Morrow and Company.

Robinson, R. W. and Decker-Walters, D. S. 1996. *Cucurbits. Crop Production Science in Horticulture Series, No. 6*. CAB International.

Rubatzky, V. E., Quiros, C. F., and Simon, P. W. 1999. *Carrots and Related Vegetable Umbelliferae. Crop Production Science in Horticulture Series No. 10*. CABI Publishing.

Rubatzky, V. E. and Yamaguchi, M. 1997. *World Vegetables: Principles, Production, and Nutritive Values*. Second Edition. Chapman & Hall.

Ryder, E. J. 1999. *Lettuce, Endive, and Chicory. Crop Production Science in Horticulture Series, No.9*. CABI Publishing.

Savona, N. 2003. *The Kitchen Shrink*. Duncan Baird Publishers, London, 144 pp.

Schneider, E. 2001. *The Essential Reference: Vegetables From Amaranth to Zucchini*. William Morrow.

University of California. 1998. *Specialty and Minor Crops Handbook. Second edition*. DANR Publication 3346.

Wallwork, C. 2000. Working within food assurance protocols: current farm practice in the UK. *Proceedings of the BCPC Conference – Pests & Diseases* 2000 1: 139–142.

Wargovich, M. J. 2000. Anticancer properties of fruits and vegetables. *HortScience* 35:573–575.

Wien, H. C. 1997. *The Physiology of Vegetable Crops*. CAB International.

Wiersema, J. H. and Leon, B. 1999. *World Economic Plants: A Standard Reference*. CRC Press.

References on history of plant disease concerns

Ainsworth, G. C. 1969. History of plant pathology in Great Britain. *Annual Review of Phytopathology* 7:13–30.

Akai, S. 1974. History of plant pathology in Japan. *Annual Review of Phytopathology* 12:13–26.

Campbell, C. L., Peterson, P. D., and Griffith, C. S. 1999. *The Formative Years of Plant Pathology in the United States*. American Phytopathological Society Press.

Fish, S. 1970. The history of plant pathology in Australia. *Annual Review of Phytopathology* 8:13–36.

Kiraly, Z. 1972. Main trends in the development of plant pathology in Hungary. *Annual Review of Phytopathology* 10:9–20.

Klinkowski, M. 1970. Catastrophic plant diseases. *Annual Review of Phytopathology* 8:37–60.

Large, E. C. 1962. *The Advance of the Fungi*. Dover Publications. Reprinted by American Phytopathological Society Press. 2003.

Nevo, D. 1995. Some diseases of agricultural crops and their control in the land of Israel during biblical, mishnaic, and talmudic times. *Phytoparasitica* 23:7–17.

Nolla, J. A. B. and Fernandez Valiela, M. V. 1976. Contributions to the history of plant pathology in South America, Central America, and Mexico. *Annual Review of Phytopathology* 14:11–29.

Orlob, G. B. 1971. History of plant pathology in the middle ages. *Annual Review of Phytopathology* 9:7–20.

Padmanabhan, S. Y. 1973. The great Bengal famine. *Annual Review of Phytopathology* 11:11–26.

Raychaudhuri, S. P., Verma, J. P., Nariani, T. K., and Sen, B. 1972. The history of plant pathology in India. *Annual Review of Phytopathology* 10:21–36.

Thurston, H. D. 1973. Threatening plant diseases. *Annual Review of Phytopathology* 11:27–52.

Walker, J. C. 1975. Some highlights in plant pathology in the United States. *Annual Review of Phytopathology* 13:15–29.

White, N. H. 1992. A case for the antiquity of fungal parasitism in plants. *Advances in Plant Pathology* 8:31–37.

Causes of disease

A MULTITUDE of microscopic and sub-microscopic pathogens threaten vegetable crops and can cause significant damage: principal among them are bacteria, fungi, and viruses. Knowledge of the names, characteristics, and biology of these pathogens is critical in the effort to identify and control them.

Diseases caused by biotic agents

Some understanding of the taxonomy of plant pathogenic organisms is required to identify them accurately, understand their biology, and ultimately to best control them. Taxonomy is the science that classifies, categorizes, and names organisms. The taxonomy of plant pathogens, like that of all organisms, continues to change and develop. Ongoing research, discovery of additional species, and modern molecular and biochemical methods allow researchers to more precisely define these classifications, resulting in occasional but significant changes in previous taxonomic schemes.

The major groups of plant pathogens are the fungi and fungus-like organisms, bacteria and other prokaryotes, and viruses and viroids. These major groups cover the vast majority of the pathogens that affect plants, though a few other miscellaneous pathogens exist. A few plant pathogens are higher vascular plants such as dodder (*Cuscuta* species), broomrape (*Orobanche* species), witchweed (*Striga* species), and mistletoes (species of *Arceuthobium*, *Phoradendron*, and *Viscum*). In very humid or tropical areas there are a few other pathogens that are classified in other microorganism groups. The alga *Cephaleuros virescens* is pathogenic on citrus and other plants in warmer regions of the world. Flagellate protozoa, or flagellates, are pathogenic on plants in tropical areas. Examples of flagellates are *Phytomonas* species on coffee, cassava, and coconut palm. Microscopic unsegmented worms, or nematodes, are important parasites on plants but are usually considered parasitic pests and not pathogens. ***Tables 3, 4*** (page 21), and ***5*** (page 23) list the primary bacterial, fungal, and viral pathogens that affect vegetable crops. ***Table 7*** (page 26) lists the vegetable diseases that are discussed and illustrated in this book.

BACTERIA

Bacteria are microorganisms placed in the kingdom Prokaryota and are typically single-celled organisms, lack a membrane-bound nucleus, and usually have an enclosing cell wall. Most plant pathogenic bacteria are rod-shaped, Gram-negative species in the genera *Acidovorax, Agrobacterium, Erwinia, Pseudomonas, Ralstonia, Serratia*, and *Xanthomonas. Rhizomononas* bacteria are thus far only pathogenic on lettuce, are Gram-negative, and live in the soil as soil inhabitants. Pathogenic *Clavibacter* bacteria, previously in the genus *Corynebacterium*, are rod-shaped Gram-positive species. *Streptomyces* species are Gram-positive actinomycete bacteria that are found in soils, produce an aerial mycelium and chains of spores, and form filamentous, branching colonies. Xylem-inhabiting, pathogenic bacteria in the *Xylella* group are not pathogenic on vegetables but cause diseases of woody plants and trees.

There are many vegetable crop pathogens in the *Erwinia, Pseudomonas*, and *Xanthomonas* genera. In most cases these pathogens are readily isolated onto standard microbiological agar media. Presumptive identification to genus, and sometimes species, level requires matching strain features with known characteristics such as colony morphology and color on certain agar media, reactions to a number of biochemical tests, serological reactions, chemical composition, fatty acid and other physiological profiles, and molecular profiles.

Species in the *Pseudomonas* and *Xanthomonas* genera are often further divided into sub-species categories based on host ranges. For example, the *P. syringae* pathogen that infects tomato is primarily restricted to tomato. Therefore, pathologists go further in their classification scheme for *P. syringae* and designate the tomato pathogen as a specific pathovar (pv.). This organism is therefore named *Pseudomonas syringae* pv. *tomato*. In official taxonomic systems, the pathovar designation is not a taxonomic term, but is a useful means for pathologists and agriculturalists to identify host–pathogen relationships and distinguish between closely related but distinct bacterial pathogens.

Mollicutes are prokaryotes that differ from other bacteria in important ways. Mollicutes notably lack cell walls, have very small genomes, and previously were called mycoplasma-like organisms (MLOs). There are two groups of mollicutes that are pathogenic on plants. Spiroplasmas are mollicutes that are single-celled and form helical structures. Phytoplasmas are the more commonly encountered mollicute on plants and take on various shapes (called pleomorphism). Plant pathogenic phytoplasmas are usually transmitted by insects such as leafhoppers for the aster yellows phytoplasma, and plant hoppers for clover phyllody phytoplasma.

Diseases caused by phytoplasmas often cause virus-like symptoms to develop on plants; therefore, for many years researchers attributed these problems to viruses. In 1967, MLOs were found in plants for the first time, thereby documenting that such diseases were caused by these bacteria and not by viruses. MLOs that are found in plants are now called phytoplasmas or spiroplasmas. Phytoplasmas thus far have not been cultured *in vitro*. For the most part, confirmation of phytoplasmas requires molecular or serological tests and examination of plant tissues using electron microscopy.

Bacteria find their way to plant hosts in a number of ways. Most of these organisms survive on plant surfaces as epiphytes until sufficient population numbers are achieved. Bacteria reproduce by cell division, or fusion. With the exception of *Streptomyces*, these plant pathogenic bacteria do not make spores or other differentiated reproductive cells or structures. When high epiphytic populations are attained, the bacteria are able to enter the plant via natural openings (stomata, hydathodes) or wounds, reproduce inside plant tissues, and cause disease. Bacteria present on diseased plants are dispersed to other plants by splashing rain or irrigation water. Winds generally do not spread bacteria, so this pathogen does not travel far like fungal spores. Bacteria are readily spread by contact. Transplant trays, greenhouse benches, shears used to prune plants, workers who handle diseased plants, and even equipment that passes through fields can all become contaminated, harbor bacteria, and spread these pathogens to clean plants. A few bacterial pathogens are vectored by insects that become contaminated by feeding, then move and spread the pathogen to other plants. A significant number of bacteria are seedborne in vegetables because they are inside the seed or present on the seed as external contaminants. Germination of the infested seed can result in a plant that has the bacterial disease. Plant pathogenic bacteria rarely survive free in the soil, but can persist on crop residues that are not completely decomposed. There are a few exceptions, as *Agrobacterium tumefaciens*, *Ralstonia solanacearum*, *Rhizomonas suberifaciens*, and *Streptomyces scabies* are true soilborne organisms.

TABLE 3 Common prokaryotes that cause diseases of vegetables

Bacteria	Acidovorax
	Clavibacter
	Erwinia
	Pseudomonas
	Ralstonia
	Rhizomonas
	Serratia
	Streptomyces
	Xanthomonas
Phytoplasma	Aster yellows phytoplasma

VIRUSES

Viruses are sub-cellular entities and are made up of some form of nucleic acid that is usually encased in a coat of protein or other biochemical substance. Virus genomes can be single- or double-stranded DNA or RNA, and can consist of single (monopartite) or multiple (multipartite) components. Viroids are similar to viruses except that they consist of only an RNA genome; viroids do not have any protective outer coat. The outer virus coats take on diverse shapes such as flexuous threads, spherical polyhedrals, cylindrical rods, and others. Some viruses consist of multiple, attached particles, with each sub-unit containing part of the genome.

Being sub-microscopic entities, virus particles cannot be observed with the light microscope. Therefore, to confirm the presence and identification of a particular virus, plant tissue must be observed with an electron microscope and tested using serological and molecular techniques. Extensive development of virus-specific antisera and the enzyme-linked immunosorbent assay (ELISA) detection methods have facilitated confirmation of virus agents. The now commonly used molecular method called the polymerase chain reaction (PCR) is a powerful tool used for identifying viruses and other organisms. PCR matches pieces of known genetic sequences (primers) with the unknown virus sample. If the match is positive, then the identity of the virus agent can be confirmed. Labeling viruses with fluorescing substances has proven useful in fluorescent antibody microscopy. In addition, some plant viruses induce the host to form various crystals and other structures within the cell; these objects are called virus inclusions and can aid in the identification of the virus. The use of indicator plants, in which a test plant such as tobacco is rubbed with a plant extract that may contain viruses, may show that the plant sap was infected with a virus-like agent. However, the resulting symptoms on these indicator plants are generally not specific enough to pinpoint precisely which virus was present in the sap.

Plant virus taxonomy divides viruses into different groups depending on the natures of the genome and external coat. Therefore there are DNA and RNA viruses. Within each of these two groups there are single- or double-stranded nucleic acid viruses. Further dividing criteria are the shape of the virus external coat and whether there are one or more strands of nucleic acid per virus. Using these and other criteria, viruses are now grouped into families of closely related virus pathogens.

In general, viruses do not exist and survive in nature without another organism. The plant host is the primary refuge of these pathogens. Viruses typically infect plants in a systemic way, and thereby make their way into plant leaves, stems, seeds, and even flowers, pollen, and roots. Cuttings, crown divisions, and other vegetative means of dividing plants will therefore result in progeny that are already infected with the virus agent. Only certain viruses have the capability of entering plant seeds and becoming seedborne pathogens. A limited number of plant viruses are carried in pollen.

With the exception of pollen-borne viruses, in the field viruses mostly move between plant hosts via a mobile organism, called a vector, that carries the virus. Vectors usually feed on the infected plant, become contaminated or infested with the virus, then move to healthy plants and infect them by feeding. Arthropods are the most common plant virus vectors and include aphids, leafhoppers, whiteflies, thrips, beetles, and spider mites. Soilborne nematodes in the Longidoridae family (genera are *Trichodorus, Paratrichodorus, Longidorus, Paralongidorus, Xiphinema*) and protozoan soil microorganisms such as *Olpidium* and *Polymyxa* species are also virus vectors. Two viruses in the tombusvirus group, *Tomato bushy stunt virus* and *Lettuce necrotic stunt virus*, present an interesting exception to the dependence on vectors. These two pathogens have no known vector, but apparently exist freely in soil and water. The viruses are distributed when infested soil is moved and infested water floods fields or is used in irrigation. Some viruses are readily spread mechanically in plant sap and fluid. The rubbing and abrasion that takes place between plants in the field can therefore theoretically spread virus agents; however, this means of virus spread is not significant in the field. More significant is the spread of virus that takes place when people handle infected plants, become contaminated with plant sap having viruses, and then touch uninfected plants.

Viruses do not reproduce independently. Once a virus has been introduced into a host cell, the viral genome is released into that cell and induces the cell to manufacture the nucleic acid and protein components needed to assemble more virus particles. Newly assembled viruses then spread from cell-to-cell via

TABLE 4 Common viruses that cause diseases of vegetables

Virus family	Virus
Alfamovirus	*Alfalfa mosaic virus*
Begomovirus	*Sinaloa tomato leaf curl virus*
Carlavirus	*Pea streak virus* *Shallot latent virus*
Caulimovirus	*Cauliflower mosaic virus*
Closterovirus	*Beet pseudo-yellows virus* *Beet yellows virus* *Cilantro yellow blotch virus*
Comovirus	*Broad bean stain virus* *Broad bean true mosaic virus* *Squash mosaic virus*
Cucumovirus	*Cucumber mosaic virus*
Enamovirus	*Pea enation mosaic virus*
Furovirus	*Beet necrotic yellow vein virus*
Geminivirus	*Beet curly top virus* *Squash leaf curl virus* *Tomato yellow leaf curl virus*
Ilavirus	*Asparagus virus 2* *Tobacco streak virus*
Luteovirus	*Bean leaf roll virus*
Nepovirus	*Artichoke Italian latent virus* *Artichoke yellow ringspot virus* *Tobacco ringspot virus*
Ophiovirus	*Mirafiori lettuce virus*
Polerovirus	*Beet chlorosis virus* *Beet mild yellowing virus* *Beet western yellows virus* *Carrot red leaf virus*
Potexvirus	*Artichoke curly dwarf virus* *Asparagus virus 3*
Potyvirus	*Asparagus virus 1* *Bean common mosaic virus* *Bean yellow mosaic virus* *Beet mosaic virus* *Celery mosaic virus* *Garlic yellow streak virus* *Garlic yellow stripe virus* *Leek yellow stripe virus* *Lettuce mosaic virus* *Onion yellow dwarf virus* *Papaya ringspot virus type W* *Pea seedborne mosaic virus* *Potato virus Y* *Tobacco etch virus* *Turnip mosaic virus* *Watermelon mosaic virus* *Zucchini yellow mosaic virus*
Rhabdovirus	*Beet leaf curl virus*
Rymovirus	*Onion mite-borne latent virus*
Sequivirus	*Parsnip yellow fleck virus*
Tobamovirus	*Pepper mild mottle virus* *Tobacco mosaic virus* *Tomato mosaic virus*
Tobravirus	*Pea early browning virus*
Tombusvirus	*Lettuce necrotic stunt virus* *Watercress yellow spot virus*
Tospovirus	*Tomato spotted wilt virus*
Umbravirus	*Carrot mottle virus*

	Viroid
	Watercress chlorotic leaf spot viroid

cytoplasmic connections between adjacent cells (plasmodesmata). When viruses reach phloem vascular tissue, the viruses are rapidly dispersed to other parts of the plant, in particular the actively growing shoots and new leaves.

Pathogen life cycle and disease development can be complex and must take into account multiple factors. The primary source of the virus can be any number of plants in the field (crops, weeds, volunteer hosts), vegetative plant propagation units (transplants, cuttings, crown divisions), vectors that have obtained the virus, pollen, virus-infected seed, and in some cases, infested soil and water. Disease development and rate of spread will depend on the biology, distribution, and movement of the vector. Virus diseases initiated by seedborne viruses will depend on the number of infested seeds in any one seed lot. Transmission of the virus depends either on how the vector feeds on plants and how the virus is able to survive on or in the vector, or on the degree that infested plant sap is infectious. Therefore, a clear understanding of virus disease epidemiology requires extensive research and information.

FUNGI

This group of microorganisms encompasses the great majority of plant pathogens. Fungi are predominantly multicellular organisms, lack chlorophyll, have cell walls, and have DNA in a membrane-bound nucleus. These organisms usually produce a filamentous, branching structure (hyphae) that collectively is called the mycelium. A great variety of reproductive structures are found in this kingdom of organisms. The taxonomy of fungi is primarily based on the shape and nature of sexual spores, asexual conidia and sporangia, fruiting structures, structure of the hyphae, and other physical features (morphology). Molecular and biochemical profiles are now being incorporated into taxonomic schemes as well.

Kingdom Fungi contains the phyla Chytridiomycota, Zygomycota, Ascomycota, and Basidiomycota. Fungi are placed in these phyla based on morphological structures and genetic characteristics. Chytrids produce zoospores and gametes that have flagella and can therefore swim in water, fungal structures without cell cross walls (coenocytic), and are often single-celled. Few chytrids cause plant diseases, and no chytrid directly causes important vegetable diseases. However, the chytrid *Olpidium* vectors a virus pathogen of lettuce. Zygomycetes have asexual sporangia that form on sporangiophores and dark, thick-walled zygospores. A few zygomycetes are plant pathogens: *Choanephora, Mucor, Rhizopus.* Ascomycetes form sexual spores in a sac-like structure called the ascus; asci are usually borne within a distinct fungal structure such as a perithecium, apothecium, or cleistothecium. Basidiomycetes form the sexual basidiospores on a microscopic, club-shaped structure called the basidium. A great many of the vegetable diseases are caused by Ascomycte and Basidiomycete fungi.

Plant pathogenic fungi either penetrate host tissues directly or enter via wounds or plant openings like stomata. The pathogens then colonize host tissues and usually produce reproductive structures and infective propagules that can disperse and infect other plants. Propagules generally fall into one of two categories. Sexual propagules result from the fusion of two gametes or two compatible mycelia, which is followed by meiosis. Meiotic division results in sexually produced spores, with the fungi in Zygomycota having zygospores, Ascomycota producing ascospores, and Basidiomycota forming basidiospores.

Many of the fungi in these three phyla also produce a different, asexual spore type. These spores are not the result of sexual recombination. For fungal pathogens of vegetables, the asexual spore is often the most prominent inoculum type and is classified as either a sporangiospore or conidium. Sporangiospores are formed within enclosed structures (sporangia) and do not share a common cell wall with the sporangia. In some cases the mature sporangiospore has flagella and can swim in water; these spores are called zoospores. Conidia are asexual spores that form as a result of cell division on the tips of conidiophores. For this spore type the conidium and conidiophore at some point shared a common cell wall. Conidiophores can be formed singly, in clusters (fascicles), or on a variety of fungal structures, including cushion-shaped pads of mycelium (sporodochia), open cup-shaped structures (acervuli), or enclosed spherical bodies (pycnidia). Another asexual structure, the chlamydospore, is a vegetative cell that has enlarged, formed thick walls, and functions as a survival structure for the fungus.

Ascomycete and Basidiomycete taxonomy is complicated by the fact that many species produce both sexual and asexual phases. The sexual form of a species is called the teleomorph (or perfect stage), and includes the ascospore or basidiospore stage. The teleomorph name of a fungus generally has priority and is the primary name used in taxonomy. The same fungus, however, can produce an asexual phase called the anamorph (or imperfect stage). The anamorph includes those forms of fungi that make conidia. All asexual forms are placed in the Fungi Imperfecti, or Deuteromycete, category. Further, not all Fungi Imperfecti species produce a known teleomorph stage, making fungal taxonomy even more complex. When referring to a fungus in scientific publications, the current convention is to use the teleomorph name. However, in the field, the teleomorph form may have a limited or unknown role in disease development. Therefore, in many cases it is practical to refer to the anamorph name because this is the fungal form that will be more readily present on the crop and play the more significant role in disease development.

TABLE 5 Common fungi and fungus-like organisms that cause diseases of vegetables*

FUNGI

Group	Genus
Zygomycota	Choanephora Mucor Rhizopus
Ascomycota	Didymella Erysiphe Golovinomyces Leptosphaeria Monosporascus Mycosphaerella Pleospora Podosphaera Sclerotinia Sphaerotheca Zopfia
Basidiomycota	Helicobasidium Itersonilia Puccinia Urocystis Uromyces Ustilago
Fungi imperfecti	Acremonium Alternaria Ascochyta Aspergillus Botrytis Cercospora Cercosporidium Chalaropsis Cladosporium Colletotrichum Cylindrosporium Fusarium
Fungi imperfecti (cont.)	Macrophomina Microdochium Mycocentrospora Oidiopsis Oidium Penicillium Phloeospora Phoma Pseudocercosporella Pyrenochaeta Ramularia Rhizoctonia Sclerotium Septoria Stemphylium Thielaviopsis Verticillium

FUNGUS-LIKE ORGANISMS

Group	Genus
Oomycetes	Albugo Aphanomyces Bremia Peronospora Phytophthora Plasmopara Pseudoperonospora Pythium
Plasmodiophoromycetes	Plasmodiophora Spongospora

* For fungi having both teleomorph and anamorph stages, if one of the stages is not commonly found on vegetables or is not involved in pathogenesis, that name is not included in this table.

Plant pathogenic fungi occupy diverse and interesting ecological niches. Some fungi, like the powdery mildews, are obligate pathogens and can only survive and remain active on a living plant host. Other fungi are pathogens on plant foliage and flowers, and can also live as saprobes on dead and decaying organic matter. Soilborne plant pathogenic fungi survive as pathogens in host roots and other tissues, soil inhabitants that persist on soil organic matter, or free living in the soil. Some fungi, such as rusts, have evolved in specialized ways and can only infect one or a few plant hosts. Other fungi, like *Sclerotinia*, retain the ability to infect many different types of plants. Fungi are dispersed in a variety of ways. Most species produce spores that are spread in the air and via winds. Airborne and other spore types can also be splash-dispersed by rain and sprinkler irrigation. Soilborne fungi are spread by spores and mycelia that are carried in water or moved in soil and mud. Some species form specialized, hardened mycelial aggregates called sclerotia that are readily spread in soil and infested plant material. A number of important fungal pathogens are seedborne in vegetables.

FUNGUS-LIKE PATHOGENS

Two groups of plant pathogenic microorganisms were traditionally classified as fungi but are now considered to be taxonomically distinct based on morphological, biochemical, and molecular criteria. The first group is the phylum Plasmodiophoromycota, which includes the clubroot pathogen of crucifers (*Plasmodiophora brassicae*) and crook root pathogen of watercress (*Spongospora*). These organisms are now placed in kingdom Protozoa. Plasmodiophoromycetes are soilborne organisms, obligate pathogens, and produce swimming zoospores.

The second group, phylum Oomycota, contains extremely important vegetable pathogens such as the downy mildews (*Bremia, Peronospora, Plasmopara, Pseudoperonospora*), water molds (*Pythium, Phytophthora, Aphanomyces*), and the white rust organism (*Albugo*). These organisms are mostly soilborne, with the exception of most downy mildews and some species of *Phytophthora.* Swimming zoospores are produced by most species, vegetative hyphae are nonseptate (coenocytic), and sexual oospores are produced by the fusion of male (antheridia) and female (oogonia) gametangia.

Recent phylogenetic investigations indicate that reclassification of oomycetes was appropriate because these organisms lack chitin in their cell walls, usually have zoospores with heterokont flagella (one whiplash, one tinsel type), and possess other features that separate them from the true fungi and more closely align them with certain types of primitive golden algae, diatoms, and giant kelp. Researchers are divided on the name that should be used for the taxonomic kingdom that contains the oomycete group; kingdom Chromista and kingdom Straminipila have both been suggested.

For many of the soilborne oomycete species, sporangia will only form in water, so water cultures will be necessary for identification. The biology and ecology of oomycetes remains very similar to those of the true fungi, even though the two groups are now considered phylogenetically distinct from each other. Oomycete dispersal is dependent on airborne sporangia (such as for the downy mildews and the late blight pathogen, *Phytophthora infestans*) or water- and soil-borne zoospores.

TABLE 6 Selected physiological and abiotic problems of vegetable crops

Crop	Disorder	Cause
Celery	Black heart	Calcium deficiency
Lettuce	Ammonium toxicity	Ammonium buildup
Lettuce	Tipburn	Calcium deficiency
Tomato	Blossom end rot	Calcium deficiency
Artichoke	Tipburn	Calcium deficiency

Disorders caused by abiotic factors and physiological conditions

Apart from the living organisms that cause plant diseases, there are a number of non-living factors that contribute to problems and disorders of vegetable crops. Such abiotic factors include environmental extremes, nutrient deficiencies and toxicities, damage from crop production steps, soil and water conditions, and others. In addition, problems with the plant's physiology and genetics can result in disorders and mutations. This book covers only a few of the abiotic and physiological problems that can affect vegetable crops; a summary can be found in ***Table 6.***

References on general pathology

Agarwal, V. K. and Sinclair, J. B. 1997. *Principles of Seed Pathology*. Second Edition. CRC Press, Boca Raton, Florida.

Agrios, G. N. 1997. *Plant Pathology.* Fourth edition. Academic Press.

Alexopoulos, C. J., Mims, C. W., and Blackwell, M. 1996. *Introductory Mycology*. Fourth Edition. John Wiley & Sons.

Barnett, H. L., and Hunter, B. B. 1998. *Illustrated Genera of Imperfect Fungi*. Fourth Edition. American Phytopathological Society Press.

Bradbury, J. F. 1986. *Guide to Plant Pathogenic Bacteria.* CAB International Mycological Institute.

Bridge, P. D., Arora, D. K., Reddy, C. A., and Elander, R. P. 1998. *Applications of PCR in Mycology*. CAB International.

Bruehl, G. W. 1987. *Soilborne Plant Pathogens*. Macmillan Publishing Company.

Brunt, A. A., Crabtree, K., Dallwitz, M. J., Gibbs, A. J., and Watson, L. 1996. *Viruses of Plants: Descriptions and Lists from the VIDE Database*. CAB International.

Campbell, C. L. and Madden, L. V. 1990. *Introduction to Plant Disease Epidemiology.* John Wiley & Sons.

Christie, R. G. and Edwardson, J. R. 1994. *Light and electron microscopy of plant virus inclusions*. Monograph 9. Revised. University of Florida.

Deacon, J. W. 1997. *Modern Mycology.* Third edition. Blackwell Science, Oxford. 303 pp.

Dhingra, O. D. and Sinclair, J. B. 1995. *Basic Plant Pathology Methods*. Second edition. CRC Press.

Dick, M. W. 1995. *The straminipilous fungi: a new classification for the biflagellate fungi and their uniflagellate relatives with particular reference to Lagenidiaceous fungi*. C. A. B. Int. Mycol. Paper 168.

Dijkstra, J. and de Jager, C. P. 1998. *Practical Plant Virology: Protocols and Exercises*. Springer.

Domsch, K. H., Gams, W., and Anderson, T. H. 1993. *Compendium of Soil Fungi*. Vol. 1 and 2. IHW-Verlag.

Eriksson, O. E., Baral, H-O., Currah, R. S., Hansen, K., Kurtzman, C. P., Rambold, G., and Laessøe, T. (eds) 2003. *Outline of Ascomycota* – 2003 Myconet. http://www.umu.se/myconet/curr/outline.03.html

Fahy, P. C. and Persley, G. J. 1983. *Plant Bacterial Diseases: A Diagnostic Guide*. Academic Press.

Farr, D. F., et al. 1989. *Fungi on Plants and Plant Products in the United States*. American Phytopathological Society Press.

Goto, M. 1992. *Fundamentals of Bacterial Plant Pathology*. Academic Press.

Hawksworth, D. L. (editor). 1994. *The Identification and Characterization of Pest Organisms*. CAB International.

Hawksworth, D. L., Kirk, P. M., Sutton, B. C., and Pegler, D. N. 1995 *Ainsworth's and Bisby's Dictionary of the Fungi*. Eighth Edition. CAB International, Wallingford, UK.

Horst, R. K. 2001. *Westcott's Plant Disease Handbook*. Sixth edition. Kluwer Academic Publishers.

Kiffer, E. and Morelet, M. 2000. *The Deuteromycetes: Mitosporic Fungi Classification and Generic Keys*. Science Publishers, Inc.

Lelliott, R. A. and Stead, D. E. 1987. *Methods for the Diagnosis of Bacterial Diseases of Plants*. Blackwell Scientific Publications.

Leonard, K. J. and Fry, W. E. 1986. *Plant Disease Epidemiology, Volume 1: Population Dynamics and Management*. Macmillan Publishing Company.

Leonard, K. J. and Fry, W. E. 1989. *Plant Disease Epidemiology, Volume 2: Genetics, Resistance, and Management*. McGraw-Hill Publishing Company.

Maclean, D. J., Braithwaite, K. S., Manners, J. M., and Irwin, J. A. G. 1993. How do we identify and classify fungal plant pathogens in the era of DNA analyses? *Advances in Plant Pathology* 10: 207–244.

Matthews, R. E. F. 1991. *Plant Virology*. Third edition. Academic Press.

Matthews, R. E. F. 1992. *Fundamentals of Plant Virology*. Academic Press.

Money, N. P. 1998. Why oomycetes have not stopped being fungi. *Mycological Research* 102: 767–768.

Saettler, A. W., et al. 1989. *Detection of Bacteria in Seed*. American Phytopathological Society Press.

Schaad, N. W., Jones, J. B., and Chun, W. 2001. *Laboratory Guide for Identification of Plant Pathogenic Bacteria*. Third edition. American Phytopathological Society Press.

Sutic, D. D., Ford, R. E., and Tosic, M. T. 1999. *Handbook of Plant Virus Diseases*. CRC Press.

Sutton, B. C. 1980. *The Coelomycetes*. Commonwealth Mycological Institute.

Walkey, D. 1991. *Applied Plant Virology*. Second edition. Chapman and Hall.

Waller, J. M., Lenné, J. M., and Waller, S. J. (eds) 2001. *Plant Pathologist's Pocketbook*. Third Edition. CABI Publishing, Wallingford, UK, 516 pp.

References on vegetable diseases

Blancard, D. 1994. *A Colour Atlas of Tomato Diseases: Observation, Identification, and Control*. Manson Publishing (Halsted Press: John Wiley & Sons).

Blancard, D., Lecoq, H., and Pitrat, M. 1994. *A Colour Atlas of Cucurbit Diseases: Observation, Identification, and Control*. Manson Publishing (Halsted Press: John Wiley & Sons).

Blancard, D., Lot, H., and Maisonneuve, B. 2005. *A Colour Atlas of Diseases of Lettuce and Related Salad Crops: Observation, Identification, and Control*. INRA.

Cook, A. A. 1978. *Diseases of Tropical and Subtropical Vegetables and Other Plants*. Hafner Press.

Davis, R. M., Subbarao, K. V., Raid, R. N., and Kurtz, E. A. 1997. *Compendium of Lettuce Diseases*. American Phytopathological Society Press.

Dixon, G. R. 1981. *Vegetable Crop Diseases*. AVI Publishing Company.

Hall, R. 1991. *Compendium of Bean Diseases*. American Phytopathological Society Press.

Howard, R. J., Garland, J. A., and Seaman, W. L. 1994. *Diseases and Pests of Vegetable Crops in Canada*. Entomological Society of Canada.

Jones, J. B., Jones, J. P., Stall, R. E., and Zitter, T. A. 1991. *Compendium of Tomato Diseases*. American Phytopathological Society Press.

Kraft, J. M. and Pfleger, F. L. 2001. *Compendium of Pea Diseases*, Second edition. American Phytopathological Society Press.

Pernezny, K., Roberts, P. D., Murphy, J. F., and Goldberg, N. P. 2003. *Compendium of Pepper Diseases*. American Phytopathological Society Press.

Schwartz, H. F. and Mohan, S. K. 1995. *Compendium of Onion and Garlic Diseases*. American Phytopathological Society Press.

Sherf, A. F. and MacNab, A. A. 1986. *Vegetable Diseases and Their Control*. Revised edition. John Wiley & Sons.

Snowdon, A. L. 1990. *A Color Atlas of Post-Harvest Diseases and Disorders of Fruits and Vegetables, Volume 1: General Introduction and Fruits*. CRC Press.

Snowdon, A. L. 1992. *A Color Atlas of Post-Harvest Diseases and Disorders of Fruits and Vegetables, Volume 2: Vegetables*. CRC Press.

Stevenson, W. R., Loria, R., Franc, G. D., and Weingartner, D. P. 2001. *Compendium of Potato Diseases*. Second Edition. American Phytopathological Society Press.

University of California. 1985. *Integrated Pest Management for Cole Crops and Lettuce*. DANR Publication 3307.

University of California. 1986. *Integrated Pest Management for Potatoes*. DANR Publication 3316.

University of California. 1998. *Integrated Pest Management for Tomatoes*. Fourth edition. DANR Publication 3274.

University of California. *UC IPM Pest Management Guidelines*. UC IPM Online. http://www.ipm.ucdavis.edu/PMG/crops-agriculture.html

Walker, J. C. 1952. *Diseases of Vegetable Crops*. McGraw-Hill Book Company.

TABLE 7 Vegetable diseases discussed and illustrated in this book

CROP	BACTERIAL DISEASES	FUNGAL DISEASES	VIRAL DISEASES
Alliums	Bacterial blight	Black mold Botrytis leaf blight Cladosporium leaf blotch Downy mildew Fusarium basal plate rot Neck rot Penicillium blue mold Pink root Purple blotch Rust Smut White rot White tip	Garlic mosaic Onion yellow dwarf Shallot latent
Artichoke	Bacterial crown rot	Ascochyta rot Gray mold Powdery mildew Pythium root rot Ramularia leafspot Verticillium wilt	Artichoke curly dwarf Artichoke Italian latent Artichoke yellow ringspot
Arugula	Bacterial blight	Downy mildew White rust	
Asparagus		Cercospora blight Fusarium crown, root rot Phytophthora spear rot Purple spot Root rots Rust	Asparagus virus 1 Asparagus virus 2 Asparagus virus 3 Tobacco streak
Basil	Bacterial leaf spot	Fusarium wilt Gray mold White mold	Tomato spotted wilt
Bean	Common bacterial blight Halo blight	Anthracnose Fusarium wilt Gray mold Powdery mildew Root and foot rot diseases Rust White mold	Bean common mosaic Bean yellow mosaic
Beet	Bacterial leaf spot	Black leg Cercospora leaf spot Damping-off Downy mildew Powdery mildew Ramularia leaf spot Rust Scab	Beet chlorosis Beet curly top Beet leaf curl Beet mild yellowing Beet mosaic Beet necrotic yellow vein Beet pseudo-yellows Beet western yellows Beet yellows Cucumber mosaic

TABLE 7 Vegetable diseases discussed and illustrated in this book (continued)

CROP	BACTERIAL DISEASES	FUNGAL DISEASES	VIRAL DISEASES
Brassicas	Bacterial blight	Alternaria head rot	Beet western yellows
	Bacterial head rot	Alternaria leaf spot	Cauliflower mosaic
	Bacterial leaf spot	Black leg	Turnip mosaic
	Black rot	Clubroot	
		Downy mildew	
		Fusarium yellows	
		Gray mold	
		Light leaf spot	
		Phytophthora root rot	
		Phytophthora storage rot	
		Powdery mildew	
		Rhizoctonia diseases	
		Ring spot	
		Verticillium wilt	
		White leaf spot	
		White mold	
		White rust	
Broad bean		Chocolate spot	Bean leaf roll
		Downy mildew	Bean yellow mosaic
		Fusarium root rot	Broad bean stain
		Leaf and pod spot	Broad bean true mosaic
		Rust	
		White mold	
Broccoli raab	Bacterial blight	Alternaria leaf spot	
		Powdery mildew	
		White rust	
Carrot	Bacterial leaf blight	Alternaria leaf blight	Carrot motley dwarf
		Black mold	Parsnip yellow fleck
		Black rot	
		Cavity spot	
		Cercospora leaf blight	
		Crater rot	
		Crown rot	
		Licorice rot	
		Powdery mildew	
		Rust	
		Scab	
		Southern blight	
		Violet root rot	
		White mold	
Catnip	Bacterial leaf spot		
Celery	Aster yellows	Crater spot	Celery mosaic
	Bacterial leaf spot	Damping-off	Cucumber mosaic
		Early blight	
		Fusarium yellows	
		Late blight	
		Phoma crown, root rot	
		Pink rot	

TABLE 7 Vegetable diseases discussed and illustrated in this book (continued)

CROP	BACTERIAL DISEASES	FUNGAL DISEASES	VIRAL DISEASES
Celery *continued...*		Powdery mildew Pythium root rot	
Chervil		Powdery mildew White mold	
Cilantro	Bacterial leaf spot	Fusarium wilt	Cilantro yellow blotch
Corn salad		Powdery mildew White mold	
Cucurbit	Angular leaf spot Bacterial fruit blotch Bacterial wilt Cucurbit yellow vine disease	Anthracnose Charcoal rot Damping-off Downy mildew Fusarium crown, foot rot Fusarium wilt Gummy stem blight, black rot Monosporascus root rot Phytophthora crown & root rot Powdery mildew Scab Verticillium wilt	Cucumber mosaic Papaya ringspot Squash leaf curl Squash mosaic Tobacco ringspot Watermelon mosaic Zucchini yellow mosaic
Dill		Itersonilia canker Powdery mildew	
Endive/escarole	Bacterial soft rot	Alternaria leaf spot Powdery mildew Rhizoctonia blight Rust White mold	Beet western yellows
Fennel	Bacterial leaf spot	Cercosporidium blight White mold	
Horseradish	Black rot	Cercospora leaf spot Downy mildew Ramularia leaf spot Verticillium wilt White rust	Turnip mosaic
Jerusalem artichoke		Powdery mildew Sclerotinia rot Southern blight	
Lemongrass		Rust	
Lettuce	Aster yellows Bacterial leaf spot Corky root Varnish spot	Anthracnose Bottom rot Downy mildew Fusarium wilt Gray mold Lettuce drop Phoma basal rot Powdery mildew Verticillium wilt	Beet western yellows Lettuce mosaic Lettuce dieback Lettuce big vein Tomato spotted wilt Turnip mosaic

TABLE 7 Vegetable diseases discussed and illustrated in this book (continued)

CROP	BACTERIAL DISEASES	FUNGAL DISEASES	VIRAL DISEASES
Marjoram		Rust	
Mint		Powdery mildew Rust Verticillium wilt	
Mustards		Alternaria leaf spot White leaf spot White rust	
Oregano		Rust	
Parsley		Powdery mildew Root rot Septoria blight White mold	
Parsnip		Downy mildew Itersonilia canker Phloeospora leaf spot Phoma canker Powdery mildew Ramularia leaf spot	Celery mosaic Parsnip yellow fleck
Pea	Bacterial blight	Aphanomyces root rot Ascochyta blight Downy mildew Foot rot complex Fusarium wilt Gray mold Leaf and pod spot Powdery mildew Pythium root rot, damping-off White mold	Bean leaf roll Pea early browning Pea enation mosaic Pea seedborne mosaic Pea streak
Pepper	Bacterial spot	Anthracnose Damping-off Gray leaf spot Gray mold Phytophthora blight, root rot Powdery mildew Southern blight Verticillium wilt White mold	Beet curly top Cucumber mosaic Pepper mild mottle Potato virus Y Sinaloa tomato leaf curl Tobacco etch Tobacco mosaic Tomato mosaic Tomato spotted wilt
Radicchio	Bacterial leaf spot	Alternaria leaf spot Powdery mildew White mold	Tomato spotted wilt
Radish		Alternaria leaf spot Downy mildew Fusarium wilt White rust	

TABLE 7 Vegetable diseases discussed and illustrated in this book (continued)

CROP	BACTERIAL DISEASES	FUNGAL DISEASES	VIRAL DISEASES
Rhubarb	Bacterial soft rot Crown rot	Anthracnose Ascochyta leaf spot Downy mildew Gray mold Ramularia rot Root rots Rust	Turnip mosaic
Sage		Phytophthora root rot Powdery mildew	
Salsify		Rust White rust	
Spinach	Bacterial leaf spot	Anthracnose Cladosporium leaf spot Downy mildew Fusarium wilt Root rot complex Stemphylium leaf spot Verticillium wilt White rust	Beet curly top Beet western yellows Cucumber mosaic
Sweetcorn		Maize smut Stalk rot	
Swiss chard	bacterial leaf spot	Cercospora leaf spot Damping-off Downy mildew Powdery mildew Rust	Beet curly top
Tomatillo		Fusarium wilt	Turnip mosaic
Tomato	Bacterial canker Bacterial speck Bacterial spot Tomato pith necrosis	Alternaria stem canker Anthracnose Black mold Corky root rot Damping-off Early blight Fusarium crown and root rot Fusarium wilt Gray mold Late blight Phytophthora root rot Powdery mildew Southern blight Verticillium wilt White mold	Alfalfa mosaic Beet curly top Cucumber mosaic Potato virus Y Tobacco mosaic Tomato mosaic Tomato spotted wilt Tomato yellow leaf curl
Watercress		Crook root Downy mildew Septoria leaf spot	Turnip mosaic Watercress chlorotic leaf spot Watercress yellow spot

Diagnosis of disease

The first step in managing vegetable diseases is to identify them correctly. Plant disease diagnosis is the science and art of identifying the causal agents behind these problems. Prompt and accurate diagnosis is of the utmost importance in vegetable production systems.

The importance of timely, accurate diagnoses

Diseases, disorders, and other problems of vegetable crops are critical concerns for agricultural and horticultural production worldwide. The perishable nature of vegetable commodities and the market demand for high quality, virtually defect-free produce places great pressure on growers, field managers, pest control advisors, and other personnel to minimize damage caused by these problems. Growers and other field personnel who have the ability to accurately and rapidly diagnose vegetable disease problems will have a competitive advantage over those who have not developed such skills.

There are two general categories of plant problems. Biotic problems are caused by living organisms such as pathogens, nematodes, and insects and other pests. Abiotic problems are caused by nonliving factors such as temperature and moisture extremes, mechanical damage, chemicals, nutrient deficiencies or excesses, salt damage, and other environmental factors. Disease control is most efficient if the causes of biotic and abiotic problems are both identified in a timely and accurate way.

Overall strategy

Diagnosis is the science and art of identifying the agent or cause of the problem under investigation. When one renders a diagnosis, one has collected all available information, clues, and observations and then arrives at an informed conclusion as to the causal factor(s). Hence, plant problem diagnosis is an investigative, problem-solving process. This process relies on current observations as well as historical records. Growers and field personnel should therefore maintain records of what problems occurred, when problems developed, and any other information pertinent to the case.

It is essential to conduct the diagnostic process in a systematic, organized way. Such a system includes the following elements:

- Ask and answer the appropriate questions so as to define the problem and obtain information that is relevant to the case under investigation.
- Conduct a detailed, thorough examination of the plants and production areas.
- Use appropriate laboratory tests to obtain clinical information on possible causal agents and factors.
- Compile all the collected information and consult additional resources and references. Keep an open mind as the information is analyzed and do not make unwarranted assumptions; in particular, do not assume that only one causal factor is involved.
- Finally, make an informed diagnosis.

Asking the right questions

As a problem-solving process, the diagnostic strategy requires that information be obtained in three major areas of investigation:

- Comparison of the plant in its healthy state with its diseased condition.
- The presence or nature of the possible agents responsible for the problem.
- The surrounding environment, growing conditions, and production practices that form the context of the case. Such information is best obtained by asking and then answering, as completely as possible, a series of questions. Examples of such questions follow.

Questions about the healthy plant

- What are the genus, species, and cultivar names of the plant in question?
- Is this particular plant suited to the production area? Is the cultivar resistant or especially susceptible to diseases and other problems?
- Is the plant sensitive to certain environmental factors (salinity, excess or deficient soil moisture, etc.)?
- What are the characteristics, appearance, and growth habits of a healthy plant?
- How does the plant normally appear when grown under various conditions (greenhouse vs. outdoors, coastal vs. inland production locations, winter vs. summer) or at different stages of growth and development (seedling vs. transplant vs. mature plant)?
- What is the normal growth rate?

Questions about the unhealthy plant

SYMPTOMS

- What are the symptoms of the affected plant?
- Which plant parts are affected?
- Are symptoms restricted to external plant surfaces (spots and lesions) or are there also internal symptoms (vascular streaking, discolored pith or crown tissue)?
- Are symptoms present only on exposed plant surfaces or also on protected, covered tissues such as unexpanded inner leaves or unopened flowers?
- What is the distribution of the symptoms on any one particular plant (do symptoms occur on one side of the plant, only on older or newer leaves, on secondary roots but not on primary roots, etc.)?
- What were the initial symptoms?
- How do early symptoms differ from more advanced symptoms?
- How rapidly do early symptoms change into advanced ones?
- How long have the symptoms been present?

GROWTH STAGE

- What is the growth stage of the affected plant (seedling, new transplant, mature plant, flowering/fruiting plant, senescent plant)?
- Is a particular growth stage associated with the problem?
- What was the condition of the plant when first placed in the production area?
- How does the growth rate of the affected plant compare with that of a healthy plant?

SYMPTOMS ON OTHER PLANTS

- Are symptoms restricted to one species or one cultivar of a vegetable plant or are multiple cultivars and different species involved?
- Do the same symptoms occur in only one field or one greenhouse, while adjacent or nearby plantings of the same crop remain symptomless, or are many plantings of that particular vegetable affected?
- Do adjacent plantings, weeds, or nearby crops exhibit similar symptoms?
- If other plants are affected, do they belong to a common group or family of plants?
- Are symptomatic plants associated with a particular set of transplants or lot of seed?

PATTERNS

- How are the symptoms distributed within the specific production area of concern?
- Are there patterns (repeating numbers of plants or plant rows) to the symptoms or are they completely random throughout the planting?
- Are symptomatic plants found in clustered groups?
- Do the symptomatic plants occur in lines, streaks, circles, or other discernable pattern?
- Are symptomatic plants found mostly along the edges of the planting?
- Are affected plants next to buildings, roads, ditches, weedy areas, other crops, or other production areas?
- Are symptoms associated with sub-sets of plants within the planting, indicating an association with plants from certain transplant trays, different sources of plant material, or other production factors?

- Are symptoms associated with physical features in the field such as low or high spots of the field, places where water does not drain well, presence or absence of underlying gravel or clay, changes in soil types?
- Do such areas become flooded after rains or receive irrigation runoff? Take note of the irrigation system and possible patterns associated with each type. For example, in the field a plant growth pattern can occur in which linear stretches of plants grow poorly depending upon how the sprinkler lines were arranged.

TIMING

- What is the timing of symptom occurrence in relation to other factors?
- When did the symptoms first occur?
- Are there various stages of symptoms indicating new infections vs. older ones?
- Have symptom features or severity changed over time?
- Do symptoms appear to have developed gradually over a period of time or rapidly and all at once?
- Have the same or similar symptoms occurred before?

BIOTIC OR ABIOTIC

- Do symptoms resemble those caused by biotic agents such as pathogens, nematodes, arthropods, or vertebrate pests, or are symptoms more suggestive of physiological or abiotic factors such as nutritional problems, physiological disorders, genetic mutations (chimeras), chemical damage, or environmental extremes?
- Do the symptoms provide evidence that more than one factor or pathogen is involved?

Questions about possible agents and factors

BIOTIC

- Which pathogens, nematodes, and arthropod pests are known to occur on the host?
- Which biotic agents occur in the geographic area of concern? Compile a list of common biotic agents that occur in the area.

ABIOTIC AND PHYSIOLOGICAL

- Compile a list of physiological and abiotic factors known to cause problems for the plant in question (examples: tipburn in lettuce, blossom end rot of tomato).

SIGNS

- A 'sign' is the visible presence of a biotic causal agent. Are such signs present? Examples of signs are fungal growth and spores, bacterial ooze, and insect bodies or frass (insect droppings).
- Are there multiple signs that indicate more than one factor may be involved?
- What is the distribution of signs on the affected plant (present on all or only on certain plant parts)?
- Are signs present on all symptomatic plants?
- Are signs present on adjacent plantings or non-symptomatic plants?
- Are there signs on surrounding soil, irrigation pipe, or other inanimate objects in the area?

Questions about the surrounding environment, growing conditions, production practices

TIMING

- What time of year did the problem occur?
- From year to year, does the problem recur during the same month or time of year?

ENVIRONMENTAL CONDITIONS AND CONTEXT

- What are the current and past weather conditions?
- Have there been any unusual weather patterns, changes, or developments recently or in the past few weeks or months?
- Have there been any conditions that would hinder plant growth or favor pathogen and pest development?
- Is there evidence of abiotic stress factors (temperature extremes, water stress or excess, salt buildup, mineral deficiencies and toxicities, pollution, wind or other mechanical damage, etc.)?
- What is the general location of the field or greenhouse (coastal vs. inland, next to other crops or production areas, next to roads, etc.)?

Rooting Materials and Field Soils

- For greenhouse-produced transplants, what type of rooting medium is used?
- What is the condition of the mix (pH, porosity, salinity, nutrient level, etc.)?
- Has the rooting medium been subject to any environmental extremes (excess water or salts, water deficit conditions, high temperatures in the container, etc.)?
- What supplements have been added to or omitted from the rooting medium?
- Has the rooting medium been used previously, recycled, or exposed to contaminating factors?
- What is the condition and cleanliness of the pots, trays, or other containers used to hold the rooting medium?
- For field or greenhouse crops grown in the ground, what is the soil type (heavy clay, porous sand) and soil condition (compacted layers, change in soil type from affected to unaffected areas, pH, salinity levels, alkalinity, etc.)?
- Is there a history of this transplant crop grown at this site?

Water

- What is the water source and quality?
- How does the water drain off the site?
- Are there low spots where water collects or drains poorly?
- What type of irrigation system is used?
- Does the system deliver water uniformly?
- What is the frequency and duration of irrigations?
- How does the irrigation schedule correspond to other production practices such as pesticide sprays, fertilizer applications, transplanting, and other procedures?

General Production Practices

- What are the normal, typical production practices for the crop (propagation steps, planting arrangement, irrigating, fertilizing, pruning, pest management)?
- What is the condition of the facility, including structures, fertilizer injection systems, irrigation systems, spray equipment, etc.?
- What experience do you have with this crop or particular cultivar?
- What production steps were completed or omitted?
- What was previously planted or placed in the area of concern?
- What equipment, particularly spray equipment, was recently used in the area?
- Were any new production practices recently implemented?
- Did the onset of symptoms correspond with any cultural practice?
- Compile a complete record of production practices used on and around the symptomatic plants.
- Which fertilizers, pesticides, or other chemicals were applied to the plants, adjacent crops, greenhouse benches for transplants, and non-production areas in the vicinity? Compile information on all materials, formulations, rates, additives, and spray volumes used on symptomatic, non-symptomatic, and surrounding plants. Special attention should be placed on herbicide (selective, non-selective, pre-emergent, post-emergent) and pesticide tank mix applications.
- What weather patterns occurred before, during, and after the spray applications?
- What chemicals and additives were used previously in the spray equipment and how was the equipment cleaned afterwards? Chemical damage may be the subject of future litigation and careful records and photographs should be taken when such damage is suspected. Careful analysis will be required to determine if there is evidence that damage is associated with spray patterns or if untreated plants lack such symptoms.

HISTORY

- What is the sequence of crops that was planted at this site in the past few years?
- Has the same crop been placed here frequently in the past few seasons?
- Have cover crops, composts, or other amendments and inputs been used here?
- What is the history of plant diseases that occur at this site?
- Are there field notes, laboratory reports, or other information that document the presence and problems due to soilborne or foliar diseases?

Conducting the examination

Gather appropriate equipment for use in examining the production site. Equipment includes the following: notebook, hand lens, knife, pruning shears, shovel, plastic bags for plant and soil samples, bottles for water samples, labels and marking pens, soil sampling tube, flags for tagging plants for future observations, camera, ice chest/cooler, disinfectant for tools.

Examine all relevant propagation and production areas in an attempt to answer the pertinent information-gathering questions. Tour off-site areas such as adjoining roads, fields, and landscaped areas.

Thoroughly examine symptomatic plants. Carefully examine all roots, above-ground parts of the plants, and internal tissues of stems. Examine and dissect multiple examples of affected plants. Conduct a similar examination of healthy or asymptomatic plants to make a comparison.

Record all observations. Draw maps to indicate patterns of symptoms and locations of plantings having the problem. Take photographs to document symptoms and distribution of patterns. Photograph asymptomatic plants for comparison.

Collect and label representative plant samples for possible laboratory analysis. Samples should include various stages of affected plants (initial and more advanced symptoms) taken from multiple sites. In many cases and if feasible, the entire plant should be collected and not just a few leaves and stems; foliar symptoms may be caused by root and crown problems, so laboratory personnel need to see the entire specimen. Include a healthy plant for comparison. Keep samples cool and deliver them in a timely manner to the lab. If the plant being examined is a sample brought in from the field or greenhouse, what is the condition of the sample? Samples that are in poor condition, incomplete, or of limited size and number may not yield useful information. Is the sample representative of the problem? Have unaffected plants also been included in the sample for comparison? If warranted, collect appropriate soil and water samples for various analyses.

Incorporating laboratory tests

Because diagnosis based solely on symptoms is risky and may lead to inaccurate conclusions, laboratory analysis is usually highly recommended. If pathogens are possibly involved, send good quality, representative samples to a plant pathology laboratory. Pathology labs use various tests and techniques to identify biotic causes of plant problems. Direct examination of plant tissues is used to search for signs such as fungal structures and bacterial ooze. Culturing techniques using microbiological media are designed to isolate and recover fungal or bacterial pathogens from symptomatic tissues.

Various extraction methods can recover nematodes from plant tissues and soils. Serological methods, such as the enzyme-linked immunosorbent assay (ELISA), employ antisera that will identify specific pathogens from either plant tissues or from cultures. Molecular identification techniques for plant pathogens are constantly being developed and improved. In some cases these molecular tests can be used directly on symptomatic plant tissues to confirm the presence of the target pathogen. Note, however, that the detection or recovery of a pathogen does not necessarily mean that this agent caused the symptoms of the disease. Interpretation of lab results using any method often requires the expertise of diagnosticians and plant pathologists.

For many fungi, fungal-like organisms such as oomycetes, and bacteria, isolating suspect pathogens from symptomatic plant tissue is a critical step in the diagnosing of plant pathogens. Different strategies are adopted for this purpose. The plant sample is first usually washed to remove debris and soil. Typically the sample is then surface sterilized to reduce interference from secondary decay organisms and non-pathogenic microflora that are present on the plant surfaces. Small pieces of diseased tissue, taken from the edges of the infection, are placed in or on solid agar media. For isolating fungi and oomycetes, general purpose microbiological media are usually recommended: potato dextrose agar, corn meal agar, water agar. These media can be acidified if secondary decay bacteria are problematic and might interfere with recovery of the pathogen. In some cases the use of semi-selective media can help recover suspect pathogens. Semi-selective media are used to inhibit the growth of non-target organisms while facilitating growth of the target pathogen. Examples of semi-selective media are PARP for *Phytophthora*, Sorensen's NP-10 for *Verticillium*, and Komada's medium for *Fusarium*.

Isolating plant pathogenic bacteria follows a similar strategy. Samples are prepared in similar ways as for fungal diseases. Small pieces of symptomatic tissue are taken from the edges of the infection and are macerated in a drop of water. A few microliters of this water are streaked onto general purpose media such as nutrient agar, yeast extract dextrose calcium carbonate agar, or sucrose peptone agar. At times a semi-selective medium is useful if a particular pathogen is suspected. Examples of semi-selective media for bacteria are SX agar for *Xanthomonas campestris* pv. *campestris* and KBC medium for several *Pseudomonas syringae* pathovars.

Many book and journal references provide specific details on the appropriate media to consider for isolating and growing plant pathogenic fungi and bacteria. Several books are particularly useful in this regard and are listed in the references below.

Diagnosing virus diseases

For any pathogen group (fungi, bacteria, viruses), the symptoms may vary due to any number of factors. However, for viruses, the incidence and expression of disease symptoms can fluctuate and vary a great deal depending on the strain and virulence of the agent, the particular crop and cultivar host, age of host when infected, mode of infection (mechanical abrasion, seedborne inoculum, or arthropod vector), factors involving the biology of the vector (type, strain, and population), and environmental conditions. With some exceptions, symptoms caused by different viruses often resemble each other, thereby making field diagnosis difficult and ill advised. Virus disease diagnosis is further complicated when more than one viral agent infects the host plant. Clinical tests are required to positively identify viral agents in plants.

Digitally assisted diagnosis

Mention should be made of new technology involving digitally assisted diagnosis (DAD). DAD is the process of acquiring digital images of plant problems and sending these images to researchers, diagnosticians, and other experts for viewing and diagnosing. This diagnostic approach therefore relies almost exclusively on digital photographs. In some cases, such images may be very helpful in reducing the time required to identify the cause of the problem. Digital images can also be appropriate when distinctive symptoms and signs characterize the problem (such as powdery mildews, downy mildews, clubroot of crucifers, and others). However, care should be taken not to rely on such images alone. DAD can be a useful supplement to the diagnostic process, but will not be able to completely replace field visits and hands-on examination of plant samples.

Compiling information and drawing a conclusion

Compile all notes, observations, maps, laboratory results, photographs, and other information. Consult with and record information from printed references and books, on-line website resources, and university and other experienced professionals. This compilation will be the information base for the present diagnosis and can also be a useful resource for future diagnostic cases. After considering all the information, render an informed diagnosis.

References on disease diagnosis

Barnes, L. W. 1994. The role of plant clinics in disease diagnosis and education: A North American perspective. *Annual Review of Phytopathology* 32:601–609.

Fox, R. T. V. 1993. *Principles of Diagnostic Techniques in Plant Pathology.* CAB International.

Green, J. L., Maloy, O., and Capizzi, J. 1990. A systematic approach to diagnosing plant damage. *Plant Diagnostics Quarterly* 11(3):139–165.

Grogan, R. G. 1981. The science and art of plant disease diagnosis. *Annual Review of Phytopathology* 19:333–351.

Hansen, M. A. and Wick, R. L. 1993. Plant disease diagnosis: present status and future prospects. *Advances in Plant Pathology* 10:65–126.

Henson, J. M. and French, R. 1993. The polymerase chain reaction and plant disease diagnosis. *Annual Review of Phytopathology* 31:31–109.

Holmes, G. J., Brown, E. A., and Ruhl, G. 2000. What's a picture worth? The use of modern telecommunications in diagnosing plant diseases. *Plant Disease* 84:1256–1265.

Horne, C. W. 1989. Groundwork for decision: Developing recommendations for plant disease control. *Plant Disease* 73:943–948.

Kabashima, J. N., MacDonald, J. D., Dreistadt, S. H., and Ullman, D. E. 1997. *Easy on-site tests for fungi and viruses in nurseries and greenhouses.* UC DANR Publication no. 8002.

Kim, S. H. 1988. Technological advances in plant disease diagnosis. *Plant Disease* 72:802.

Marshall, G. 1996. *Diagnostics in Crop Protection.* Symposium Proceedings No. 65. British Crop Protection Council. Major Print Ltd.

Newenhouse, A. C. 1991. How to recognize wind damage on leaves of fruit crops. *HortTechnology* 1:88–90.

Putnam, M. L. 1995. Evaluation of selected methods of plant disease diagnosis. *Crop Protection* 14:517–525.

Schubert, T. S. and Breman, L. L. 1988. *Basic concepts of plant disease and how to collect a sample for disease diagnosis.* Plant Pathology Circular No. 307. Florida Dept. Agric. & Consumer Services.

Shurtleff, M. C. and Averre, C. W. 1997. *The Plant Disease Clinic and Field Diagnosis of Abiotic Diseases.* American Phytopathological Society Press.

Stowell, L. 1999. Digital disaster and the ethics of virtual plant pathology. *Phytopathology News* 33:62.

Thomas, M. B., Crane, J. H., Ferguson, J. J., Beck, H. W., and Noling, J. W. 1997. Two computer-based diagnostic systems for diseases, insect pests, and physiological disorders of citrus and selected tropical fruit crops. *HortTechnology* 7:293–298.

Walker, S. E. and Schubert, T. S. 1997. *Assessing plant problems in cropping systems: A systematic approach.* Plant Pathology Circular No. 381. Florida Dept. Agric. & Consumer Services.

Wallace, H. R. 1978. Diagnosis of plant diseases of complex etiology. *Annual Review of Phytopathology* 16:379–402.

Waller, J.M., Lenné, J.M., and Waller, S.J. 2001. *Plant Pathologist's Pocketbook.* Third edition. CABI Publishing, Wallingford, UK.

Waller, J.M., Ritchie, B.J., and Holderness, M. 1998. *Plant Clinic Handbook.* (IMI Technical Handbooks No.3). CABI Publishing, Wallingford, UK.

References on culture media

Atlas, R. M. 1997. *Handbook of Microbiological Media.* Second edition. CRC Press.

Dhingra, O. D. and Sinclair, J. B. 1995. *Basic Plant Pathology Methods.* Second edition. CRC Press.

Kirsop, B. E. and Doyle, A. 1991. *Maintenance of Microorganisms and Cultured Cells: A Manual of Laboratory Methods.* 2nd Edition. Academic Press.

Lelliott, R. A. and Stead, D. E. 1987. *Methods for the Diagnosis of Bacterial Diseases of Plants.* Volume 2. Blackwell Scientific Publications.

Schaad, N. W., Jones, J. B., and Chun, W. 2001. *Laboratory Guide for Identification of Plant Pathogenic Bacteria.* Third edition. American Phytopathological Society Press.

Shurtleff, M. C. and Averre, C. W. 1997. *The Plant Disease Clinic and Field Diagnosis of Abiotic Diseases.* American Phytopathological Society Press.

Singleton, L. L., Mihail, J. D., and Rush, C. M. 1992. *Methods for Research on Soilborne Phytopathogenic Fungi.* American Phytopathological Society Press.

Controlling disease

AS WITH DIAGNOSIS, management of diseases depends on a thorough knowledge of the three major components of a disease: susceptible host plant, virulent pathogen, and favorable environment. For disease to develop, all three factors must be present. Therefore, understanding these host-pathogen-environment dynamics is essential in devising disease management strategies. Such strategies target these three areas and manipulate them so that disease is not possible or is hampered in its development.

Criteria for commercial control of vegetable diseases

For commercial agriculture, the practical disease management options that ultimately are deployed must meet four main criteria. Management steps must first of all be effective at achieving commercially acceptable levels of control. Some disease control options may reduce disease severity and incidence, or result in the death of pathogens. However, if such effects do not result in vegetable crops that have commercially acceptable levels of yield and quality, that disease control measure will not be practical for the farmer.

Secondly, the disease control option must be economical to use. The cost of disease control must be significantly lower than the final value of the marketed crop. For example, the practice of using effective but expensive soil fumigation with tarps can significantly reduce soilborne diseases of most vegetables. However, such treatments are usually prohibitively expensive in relation to the values and profit margins of many of these crops. Therefore, such fumigation is usually not used on a regular basis for most vegetables.

Thirdly, a disease control option must be available for use at a commercial level. A number of research efforts have identified possible biocontrol agents in which a microorganism can be used to kill or manage plant pathogens. However, most of these agents have not been formulated or commercialized into viable products. Hence, such agents are of limited use to commercial agriculture. Finally, disease control measures must be amenable to being set within the context of an integrated disease management system. Few diseases are controlled by only one measure. Integrating host resistance, cultural practices, application of disease control materials, and other steps is the modern approach to disease control in vegetables.

Disease control options

SITE SELECTION

Select sites that do not have a history of problematic diseases. Pick fields that contain features that discourage pathogen survival and disease development. Considerable time and planning are required for strategic selection of such locations. Growers and field advisors must monitor and record the distribution and occurrence of persistent, soilborne plant pathogens such as *Fusarium, Plasmodiophora, Sclerotium, Verticillium*, and others. Susceptible crops should not be placed in locations having known, high populations of these pathogens. For some soilborne fungi such as *Phytophthora, Pythium*, and *Rhizoctonia*, distribution is often so widespread that site selection and avoidance is usually not possible. In some cases, soil tests can give estimated population levels for some soilborne pathogens. These tests can be used to give some indications of possible disease risk, though precise pathogen populations cannot be known for certain.

Other site-specific situations create risks that should be avoided. Pastures, foothills, riverbanks, and grasslands support weeds and natural vegetation that can be reservoirs for viruses and other pathogens. The aster yellows phytoplasma and its leafhopper vector can be found in weedy grasslands in coastal California. Once this vegetation dries up in the summer, the leafhoppers migrate from the grasslands and move into nearby lettuce or celery fields, resulting in aster yellows disease in these fields. Lettuce fields should not be placed adjacent to plantings of perennial hosts, such as *Gazania* species, of *Lettuce mosaic virus*. Crucifer vegetables may experience increased pressure from several diseases if planted near oilseed rape fields.

Consider other pertinent environmental factors that are related to sites. Crops planted close to an ocean may be more at risk from downy mildew diseases due to the consistently high humidity and cool temperatures. However, moving a few miles inland from the ocean can change these conditions and reduce downy mildew severity. Choosing a site that has lighter textured soils that drain well reduces the risk of damping-off and root rot for sensitive crops such as spinach.

EXCLUSION

Exclusion is preventing any contaminated, infested, or infected materials from entering the propagation, production, and harvest systems. Because seedborne pathogens are a primary means of pathogen introduction for a number of vegetable diseases, do not allow infested or infected seed to be used in the propagation system or production field. Growers should purchase seed that has been tested and certified to be below a certain infestation threshold level, or seed that has been treated to reduce pathogen infestation levels. Some seedborne diseases have well defined seed infestation levels, such as black rot of crucifers (caused by *Xanthomonas campestris* pv. *campestris*) and *Lettuce mosaic virus* of lettuce. For many others, however, seedborne thresholds have not been established. Note that the designations 'pathogen-free seed' and 'disease-free seed' are convenient marketing terms only, as it is not possible to scientifically prove that a seed lot is actually void of all pathogens; pathogen-free seed usually means that the pathogen incidence in a seed lot is below that which can be detected with standard methods.

The field of seed pathology and seed treatments is a highly developed one. The nature of seedborne pathogens and how to manage them has been extensively researched and studied. Many refinements have been made in producing seed so that pathogens on seed plants are minimized, pathogen detection is improved, and seed treatments are more effective. Some key steps in growing seed having minimal pathogen populations are the following: selecting and placing seed production sites in areas where the pathogen is not present and conditions do not favor disease development (dry, arid conditions with no rain during the crop cycle); careful, regular monitoring of seed fields for disease symptoms; roguing (removing) symptomatic and off-type seed plants; applying preventative spray treatments; employing appropriate harvest and processing methods so as to avoid contaminating seed; using seed health testing to evaluate seed for pathogens and viability. These advances in seed pathology are important to the vegetable industry because a number of damaging diseases are seedborne in these crops (see ***Table 8***, page 40).

Seed treatments are an important means of excluding pathogens from the seed used to initiate transplants and crops. Effective seed treatments that do not significantly reduce germination of the seed depend on the following factors: the species of vegetable seed being treated; the target pathogen(s); the treatment itself (hot water, steam, chlorine, other chemicals); dose of the treatment substance; length of treatment time; volume of seed being treated at any one time; post treatment handling of the seed (cooling, rinsing, drying, coating, storing, etc.). Examples of seed treatments commonly used to deal with seedborne pathogens include the following: hot water soaks (carrot: 50° C for 30 minutes; celery 48° C for 30 minutes; crucifers: 50° C for 20–30 minutes); sodium hypochlorite (tomato: 1.05% for 40 minutes; pea: 10% for 1–5 minutes); antibiotics such as agrimycin; fermentation and acid treatments (various cucurbits); other chemical treatments such as trisodium phosphate (tomato: 10% for 15 minutes for treating for *Tomato mosaic virus*); dry heat treatments (tomato: 70° C for 2–4 days for treating for *Tomato mosaic virus*).

Other seed treatments are intended to deposit protecting fungicides onto the seed coat. Such deposits will protect the seed during the first few days after planting and will protect seed and newly emerged seedling from soilborne damping-off pathogens. These seed treatments usually consist of fungicides applied to seeds such as beans, sweetcorn, and cole crops.

Because of the intense international marketing and transporting of vegetable seed, the importance of seed health needs to be continually examined. International standards for seedborne pathogen detection, testing methodology, seed treatment, seed viability levels, and other pertinent parameters will require continual research and subsequent discussion and acceptance by nations producing and selling vegetable seed.

TABLE 8 Important seedborne pathogens of vegetables*

CROP	PATHOGEN
Alliums	*Aspergillus niger*
	Botrytis allii
	Pseudomonas syringae pv. *porri*
Asparagus	*Asparagus virus 2*
	Fusarium oxysporum f. sp. *asparagi*
Basil	*Fusarium oxysporum* f. sp. *basilicum*
Bean	*Bean common mosaic virus*
	Colletotrichum lindemuthianum
	Pseudomonas syringae pv. *phaseolicola*
	Xanthomonas campestris pv. *phaseoli*
Beet	*Beet mild yellowing virus*
	Beet western yellows virus
	Cercospora beticola
	Peronospora farinosa f. sp. *betae*
	Phoma betae
Brassicas	*Alternaria brassicae*
	Alternaria brassicicola
	Phoma lingam
	Mycosphaerella brassicicola
	Pseudomonas syringae pv. *alisalensis*
	Pseudomonas syringae pv. *maculicola*
	Xanthomonas campestris pv. *campestris*
Broad bean	*Ascochyta fabae*
	Broad bean stain virus
	Broad bean true mosaic virus
Broccoli raab	*Alternaria brassicae*
	Pseudomonas syringae pv. *alisalensis*
Carrot	*Alternaria dauci*
	Alternaria radicina
	Cercospora carotae
	Xanthomonas campestris pv. *carotae*
Celery	*Cercospora apii*
	Phoma apiicola
	Pseudomonas syringae pv. *apii*
	Septoria apiicola
Cilantro	*Cilantro yellow blotch virus*
	Pseudomonas syringae pv. *coriandricola*
Cucurbit	*Acidovorax avenae* subsp. *citrulli*
	Cladosporium cucumerinum
	Colletotrichum orbiculare
	Cucumber mosaic virus
	Didymella bryoniae
	Fusarium oxysporum
	Fusarium solani f. sp. *cucurbitae*
Cucurbit continued...	*Pseudomonas lachrymans*
	Squash mosaic virus
	Tobacco ringspot virus
	Zucchini yellow mosaic virus
Lettuce	*Lettuce mosaic virus*
	Septoria lactucae
	Verticillium dahliae
	Xanthomonas campestris pv. *vitians*
Parsley	*Septoria petroselini*
Parsnip	*Phoma complanata*
Pea	*Ascochyta pinodes*
	Ascochyta pisi
	Fusarium oxysporum f. sp. *pisi*
	Pea early browning virus
	Pea seedborne mosaic virus
	Pseudomonas syringae pv. *pisi*
Pepper	*Colletotrichum* species
	Cucumber mosaic virus
	Xanthomonas campestris pv. *vesicatoria*
	Pepper mild mottle virus
	Tobacco mosaic virus
	Tomato mosaic virus
Spinach	*Cladosporum variabile*
	Cucumber mosaic virus
	Stemphylium botryosum
	Verticillium dahliae
Sweetcorn	*Ustilago maydis*
Swiss chard	*Cercospora beticola*
Tomato	*Alfalfa mosaic virus*
	Alternaria solani
	Clavibacter michiganensis subsp. *michiganensis*
	Colletotrichum coccodes
	Cucumber mosaic virus
	Fusarium oxysporum f. sp. *lycopersici*
	Pseudomonas syringae pv. *tomato*
	Tomato mosaic virus
	Xanthomonas campestris pv. *vesicatoria*
Watercress	*Septoria sisymbrii*

* This table lists those pathogens that have been documented to be seedborne, and in which the seedborne aspect plays a role in vegetable disease epidemiology.

Diseased or contaminated transplants likewise should not be purchased or used to plant production fields. The growing of high-quality, healthy transplants entails the use of many disease management steps, many of which are discussed in the sanitation section of this chapter. For a few vegetable crops such as asparagus and artichoke, vegetative crown tissue is divided and used as propagation material to start new fields. The disease control principle of exclusion demands that only healthy, uninfected crown divisions be used to propagate such crops.

Exclusion also means preventing contaminated equipment, water, soil, and other objects from entering the vegetable production area. While not always practical or easy to achieve, tractors and vehicles should be washed or cleaned of contaminated soil prior to entering a clean field. For at least two diseases, varnish spot (caused by *Pseudomonas cichorii*) and lettuce dieback (*Lettuce necrotic stunt virus*) of lettuce, the pathogen is found in infested water; such water should be excluded and not be used to irrigate the crop. If livestock are fed crucifer residues containing the clubroot organism (*Plasmodiophora brassicae*), the pathogen's resting spores survive passage through animal digestive systems and can infest manure. Such manure should be excluded from the field. Aspects of contaminated objects are also discussed under sanitation and cultural practices.

Resistant Plants and Cultivars

Resistant plants are an obvious and effective control measure and are one the most important components in an integrated disease control program. Cultivars should be selected that are resistant to the main pathogens of concern. The most valuable cultivars will also have resistance to other pathogens, desirable horticultural characteristics, and be suitable for the particular region and climate where it is placed. Likewise, growers can select cultivars that can tolerate the pathogen even if such plants are not technically resistant. For Verticillium wilt of cauliflower in California, some vigorous hybrids showed excellent tolerance to the pathogen in the field and produced high yields, even though under experimental conditions the cultivar was susceptible to the pathogen (*V. dahliae*).

Resistance in a plant can be expressed through the action of a single gene that confers immunity (resistance to certain races of the Fusarium wilt pathogen, for example) or through multiple genes that result in a broad resistance to many pathogens. Single gene resistance is called vertical resistance, and it limits the initial level of infection and subsequent production of inoculum. However, single gene resistance can be overcome by new strains of the pathogen. The breakdown of resistance due to such changes in the pathogen poses a constant concern for growers. For example, during the past 50 years in California, a new race of spinach downy mildew (caused by *Peronospora farinosa* f. sp. *spinaciae*) would periodically occur and cause significant damage to the previously resistant cultivars. Plant breeders would counter with new cultivars having resistance genes to the new race. Growers would then enjoy several years of mildew-free spinach until the development of yet another race. This back-and-forth dynamic has taken place for each of the races that so far has been found in California. Similar dynamics exist for lettuce downy mildew (caused by *Bremia lactucae*) in both Europe and the USA. In contrast, multiple gene resistance is called horizontal resistance, and it limits the rate of disease development, meaning that some disease may develop but at a low, generally tolerable level.

Perhaps the greatest limitation of resistant plants as a disease control option is that resistance is not available for all crops. For several of the most damaging plant diseases, such as late blight of tomato (caused by *Phytophthora infestans*) and white rot of onion and garlic (caused by *Sclerotium cepivorum*), growers do not yet have cultivars with high degrees of acceptable resistance. There are no known disease resistant cultivars for most of the smaller acreage, specialty vegetables such as the following: arugula, broccoli raab, cilantro, fennel, jicama (*Pachyrhizus erosus*), leafy mustards, radicchio, Swiss chard, tomatillo, and many Asian vegetables and herbs. Another major limitation is that resistance may be present in cultivars that lack adequate horticultural characteristics. There are celery cultivars with acceptable resistance to the Fusarium yellows pathogen (*Fusarium oxysporum* f. sp. *apii*); however, some of these selections lack the color, yield, and appearance qualities that the celery market currently requires. Finding a cultivar with multiple resistances can also be difficult for growers. Lettuce that is resistant to *Lettuce mosaic virus* may be quite susceptible to corky root disease (caused by *Rhizomonas suberifaciens*); a lettuce selection that resists corky root may be very susceptible to downy mildew.

Modern molecular technology and the production of genetically modified plants will provide novel sources of disease resistance and other benefits. One example is the development of transgenic summer squash (*Cucurbita pepo*) cultivars that are resistant to several important virus pathogens. This resistance is an example of pathogen-derived resistance in which genes from the virus pathogen itself (in this case genes that code for the virus coat protein) are introduced and integrated into the genome of the squash host. In 1994 a yellow crookneck summer squash was the first virus-resistant transgenic plant to be marketed in the USA. However, this technology has not yet gained general public support, so most vegetable breeding efforts apparently will rely, in the short term, on conventional breeding methods.

CULTURAL PRACTICES

The disease control category of cultural practices is a broad and diverse collection of production practices and choices used to reduce the effects of diseases. Such practices are designed to help plants avoid contact with pathogens, reduce inoculum in the environment of the host plant, and create environmental conditions unfavorable for disease development.

Crop rotations

A significant factor in disease problems is the growing of the same crop or closely related crops in consecutive plantings, or growing the same crop too frequently over a period of a few seasons. Growers need to rotate crops so that the pathogens of one crop do not continue to increase and survive on that particular crop. Implementing crop rotations that have diverse species will also encourage diversity in the soil microbe population. Crop rotations are useful, advisable strategies for all crops and for all diseases; however, rotating crops is particularly important for combating soilborne pathogens. Crop rotations should also include, when possible, the use of cover crops that encourage soil microbe diversity and add organic matter to soil. Accurate records must be maintained so that crop rotation schemes are documented for future planning.

The subject of crop rotations also encompasses the strategy of host-free periods. Researchers have found that some virus diseases are controlled much more readily if the crop rotation strategy includes a period of time in which no fields of the host plant are present in the region. For example, in coastal California's Salinas Valley, government regulations enforce a host-free period in which no celery can be planted in the field for the month of January and no lettuce can be present in fields from December 6 to 20. These mandatory host-free periods greatly assist in the management of *Celery mosaic virus* and *Lettuce mosaic virus*, respectively, because of the elimination of the primary virus hosts during a time in the winter when the aphid vectors are inactive and reduced in populations. This step prevents the virus pathogens from bridging from the fall production season into the following spring plantings. In another example, researchers found that overwintered carrot fields in the Salinas Valley were the primary source of the virus complex that causes carrot motley dwarf of carrot. When carrot growers stopped the practice of keeping carrot fields through the winter, the virus disease almost disappeared from the region.

Research indicates that certain plants, in addition to being revenue-generating crops, also have partial suppressive effects on various pathogens. For example, after broccoli crops are harvested and the plant residue is plowed into the soil, the decomposition of the broccoli stems and leaves releases chemicals that either directly inhibit soilborne pathogens or perhaps alter soil microflora populations that subsequently compete with pathogens. Broccoli as a rotation crop and even as a cover crop is now being used by California growers to take advantage of this suppressive effect. Cabbage crop residues and mustard cover crops show similar effects on soilborne pathogens.

When devising crop rotation strategies, growers must consider which crops and cover crops might increase disease problems. Vetch cover crops, if planted in fields having populations of *Sclerotinia minor*, can greatly increase the number of infective sclerotia of this pathogen. Oilseed radish cover crops can be used as trap crops to reduce cyst nematode (*Heterodera* species) populations in the soil; however, oilseed radish could cause increases in clubroot disease.

Fertilizers, soil amendments, and composts

Adding amendments and composts to the soil is beneficial for a number of fertility and soil conditioning reasons. However, with few exceptions, there is a lack of empirical data that clearly document a commercial level disease control benefit from such additions to soil. One exception is the application of lime that successfully reduces clubroot disease of crucifers. Amendments and composts should continue to be used, however, for plant nutrition and growth considerations. Implement balanced, appropriate fertilizer programs to encourage vigorous growth. Do not over apply fertilizers such as nitrogen, as too much nitrogen can result in excessive, succulent foliage that can be more susceptible to foliar pathogens.

Planting

Time of planting can offer an opportunity for minimizing diseases. In California, susceptible cauliflower that is planted in *Verticillium*-infested fields in the spring or summer will likely experience significant disease; however, cauliflower planted in the same fields in the late fall or winter will exhibit no Verticillium wilt symptoms. This difference in disease severity is attributed to soil temperatures; winter soil temperatures are too cool for the fungus to develop and cause significant problems. In the UK, delayed planting can reduce the impact of *Aphanomyces* on beet because warmer soils encourage rapid germination of seedlings. Early planting can reduce rhizomania on beet and clubroot on crucifers because of reduced pathogen activity in cold soils. Therefore, choose planting dates that might reduce disease pressure for the particular crop under consideration.

Proper soil preparation prior to planting can reduce seed decay and seedling damping-off diseases by tilling to reduce plant residues left from previous crops and by making raised beds with good soil tilth and drainage. Proper bed preparation will also assist in the establishment of transplants. At planting, place seed and transplants at proper depths. Placing plants too deeply can delay plant emergence or establishment, and thus increase disease problems.

Irrigation

For most foliar diseases, overhead sprinkler irrigation enhances pathogen survival and dispersal, and subsequent disease development. Bacterial diseases are especially dependent on rain and sprinkler irrigation. Therefore, eliminate or reduce the use of sprinkler irrigation if possible. The use of surface or buried drip tape for vegetable production has increased greatly in California and other areas in recent years and helps reduce the severity of many diseases. Drip irrigation usually allows for a more precise delivery of water, resulting in better water management, reduced soil saturation, and a lowered risk of soilborne diseases such as root rots. Where overhead irrigation is required, application should be made early in the day so that foliage can dry during the remainder of the day. For all irrigation schedules, carefully monitor irrigations so that excess water is not applied to the crop. For a few diseases, the pathogen can be present in irrigation water. Examples of such pathogens are *Pseudomonas cichorii* and *Lettuce necrotic stunt virus* in lettuce and *Phytophthora capsici* in several vegetable crops. In such cases, exercise caution when using such infested water.

Sanitation

Sanitation is the general practice of cleaning up or removing diseased or contaminated materials. During the process of producing vegetable transplants, for example, sanitation involves the use of clean or sanitized transplant trays, bench tops, and mowing equipment (used to mow the tops of transplants and encourage thicker stem development). Workers should wash or sanitize their hands or gloves before moving to and working with different transplant lots.

Sanitation can include removal of diseased material from fields. Roguing, or the removal of diseased plants from a crop, is not often done in production fields but can help prevent spread of diseases and inoculum increase. Roguing is a common practice used in vegetable seed crops. For asparagus diseases, the removal of diseased asparagus fern foliage can significantly reduce inoculum of foliar pathogens. Once vegetable crops are harvested, the plowing under and destroying of the remaining plants and crop residues aids in the destruction of inoculum. Destroying old plants also helps reduce virus reservoirs if such plants were infected with viruses or infested with virus vectors. Plowing and disking the soil assists in the breakdown of old crop residues, which will help reduce some seed and seedling diseases for the next crop and will decrease inoculum levels for foliar diseases.

As mentioned under exclusion, as much as possible prevent the movement of contaminated field equipment from infested to clean fields. While not always practical, attempt to wash or otherwise sanitize tractors and farm implements after using these in fields having important soilborne pathogens.

A form of sanitation is used for combating lettuce drop caused by *Sclerotinia minor*. This disease occurs when sclerotia from infected lettuce residues remain in the top few inches of soil and are positioned for the next lettuce planting. A sanitation step is deep plowing in which a mold-board plow inverts the soil and buries the sclerotia. Note that this procedure is effective only if sclerotia are low to moderate in number.

ENVIRONMENTAL MANIPULATION

Growers should create conditions that are unfavorable for disease development. Some cultural methods that achieve this include using optimum plant spacing to reduce relative humidity around plants, providing good soil drainage through proper soil preparation and irrigation practices, and using mulches to physically isolate above-ground plant parts from contact with the soil. Solarization can be included here. This practice works best in regions having high summer temperatures and solar radiation. Fallow fields are covered with plastic tarps, and the resulting heat generated below the tarp reduces soilborne pathogen populations.

Environmental manipulation is much more attainable in greenhouse settings. For greenhouse-grown transplants, the use of heaters and coolers allows growers to control temperatures. Venting the greenhouse reduces humidity and helps manage gray mold (caused by *Botrytis*). While most transplants are still watered with overhead sprinklers, other systems can be used. For example, absorbent irrigation mats are placed on benches; transplants are placed on top of the mats and water moves up into the tray cells by capillary action. Greenhouses also employ fans to increase air movement and enhance drying of foliage.

CONTROLLING OTHER BIOTIC FACTORS

Weeds, volunteer plants, and other hosts can harbor pathogens that spread to the vegetable production field. Therefore, these inoculum sources should be eliminated. Another biotic factor is the vector of pathogens. Vectors, primarily insects (aphids, whiteflies, leafhoppers, thrips) and nematodes that transmit viruses, should be monitored and managed. However, managing these organisms will not prevent virus transmission because in many cases only a brief feeding time is needed for insects to inject the virus pathogen into plants. Despite this difficulty in preventing virus transmission, control insect and other vectors as much as possible by applying appropriate insecticides, planting crops on reflective mulches to repel vectors, and planting crops under netting, fabric, or plastic tunnels to exclude vectors.

USING FUNGICIDES, BACTERICIDES, AND OTHER DISEASE CONTROL CHEMICALS

Applying fungicides, bactericides, and other disease control materials is an important option that must be used judiciously and integrated into other disease control choices. There is a current worldwide trend to reduce the use of synthetic chemicals on food commodities for both human health and environmental concerns. Therefore, use such chemicals only when needed. New products often are safer, reduced risk materials that are effective at much lower volumes than older fungicides.

Pre-plant fumigants are often highly successful in reducing soilborne inoculum, though their use is expensive and strictly regulated. The following fumigants are most commonly used to control soilborne pathogens: methyl bromide alone, methyl bromide plus chloropicrin combinations, chloropicrin alone, chloropicrin plus telone, metham sodium. The application of fumigants is often combined with the placing of plastic tarps over the treated soil. These tarps help prolong exposure periods at effective fumigant concentrations. Fumigants, however, do not eradicate soilborne pathogens because of limitations inherent in the chemical, failure to penetrate all parts of the soil profile, survival of pathogen propagules that are protected by plant residues, and other factors. In addition, even if the fumigant treatment is successful, soilborne pathogens can recolonize treated soils. For example, researchers found that the Fusarium wilt pathogen of melon, *F. oxysporum* f. sp. *melonis*, readily recolonized soils that had been treated with various fumigants.

Fungicide-treated seed is an important tool in combating certain seed and seedling diseases. In some situations fungicides applied immediately post-planting can be effective for managing diseases. For example, the application of fungicides at planting to spinach seed lines can effectively prevent damping-off caused by *Pythium*. A way of controlling lettuce drop caused by *Sclerotinia minor* is to apply fungicides to plants at the thinning stage. However, for the majority of soilborne pathogens, field-applied fungicides are usually not very effective.

TABLE 9 Historical list of some disease control materials used on vegetables*

Chemical group	Name	Date**
1, 2 Thiadazole	Etridiazole	1969
2, 6 Dinitroaniline	Fluazinam	1987
Acetamide	Cymoxanil	1977
Anilinopyrimidine	Cyprodinil	1994
	Pyrimethanil	1990
Antibiotic	Kasugamycin	1965
	Mildiomycin	1979
	Oxytetracycline	1974
	Polyoxin	1965
	Streptomycin	1944
	Validamycin	1970
Aromatic hydrocarbon	Dicloran	1959
	Pentachloronitrobenzene	1930
	Tolclofos methyl	1973
Benzamide	Zoxamide	1994
Benzene-sulfonamide	Flusulfamide	1986
Benzimidazole	Benomyl	1968
	Carbendazim	1972
	Thiabendazole	1962
Benzothiadiazole	Acibenzolar-S-methyl	1995
C-14 DMI: Imidazole	Imazalil	1973
	Prochloraz	1974
C-14 DMI: Piperazine	Triforine	1967
C-14 DMI: Pyrimidine	Fenarimol	1975
C-14 DMI: Triazole	Cyproconazole	1986
	Difenoconazole	1988
	Fenbuconazole	1988
	Myclobutanil	1984
	Propiconazole	1979
	Tebuconazole	1986
	Triadimefon	1975
	Triadimenol	1977
Carbamate	Propamocarb	1978
Carboxamide	Boscalid	2003
	Carboxin	1966
	Oxycarboxin	1966
Chloronitrile	Chlorothalonil	1964
Cinnamic acid	Dimethomorph	1988
Dicarboxamide	Iprodione	1970
	Procymidone	1976
	Vinclozolin	1975
Dinitrophenyl crotonate	Dinocap	1946
Dithiocarbamate	Ferbam	1931
	Flutolanil	1981
	Mancozeb	1961
	Maneb	1950
	Metiram	1958
	Thiram	1931
	Zineb	1943
	Ziram	1930
Guanidine	Dodine	1954
	Guazatine	1968
Hydroxy (2-amino) pyrimidine	Bupirimate	1975
	Dimethirimol	1968
	Ethirimiol	1968
Hydroxyanilide	Fenhexamid	1989
Inorganic	Calcium polysulfides	1852
	Copper sulfate	1880s
	Potassium bicarbonate	1990s
	Sulfur	1800s
	Zinc	1932
Morpholine	Fenpropimorph	1979
	Tridemorph	1968
Organo tin	Triphenyl tin hydroxide	1954
Phenylamide	Mefenoxam	1996
	Metalaxyl	1977
	Oxadixyl	1983
Phenylpyrrole	Fludioxonil	1990
Phosphonate	Fosetyl aluminium	1977
	Phosphorous acid	1980s
Phthalimide	Captafol	1961
	Captan	1949
$Q^{o}I$ (Quinone-outside inhibitor)	Azoxystrobin	1990
	Kresoxim-methyl	1990
	Pyraclostrobin	2002
	Trioxystrobin	1997
Quinoline	Quinoxyfen	1992
Thiophanate	Thiophanate methyl	1969

* This is not a complete list of materials used to control diseases of vegetables. Many of these products are no longer approved for use.

** Approximate discovery or release date

The effectiveness of fungicides applied to foliage and above ground parts of plants depends on several factors. Timing is the first consideration. Most available fungicides must be applied to the plant surface prior to infection and are classified as protectants. If applied after infection has already occurred, such materials will be less effective. Timing also deals with the interval between multiple sprays. Most materials need to be used several times on the vegetable plant. Environmental conditions, susceptibility of the plant, plant growth rate, inoculum level and pressure, and nature of the fungicide product will dictate how many days can elapse between applications. Because of the importance of timing, it is essential to have thorough and regular monitoring of fields for early symptoms and signs of disease.

Product rates are important factors as well and are set by the manufacturer. Maximum label rates are established by the fungicide producer and cannot be exceeded. However, in some cases lower rates are allowable. Such reduced rates should be used based on manufacturer recommendations, disease pressure and situation, and prior experience with the product. The final factor for successful deployment of fungicides is the selection of the product and formulation. Choose products based on label information on how the product controls a particular pathogen, recommendations by university, extension, and other professionals, prior experience of industry members and other growers, and published information and research. It is important to use multiple products, if available, to delay or prevent the development of fungicide resistance.

Most vegetable crops are considered minor crops when compared with the extensive plantings of field crops, grains, soybean, and cotton. Therefore, the choices of registered fungicides are often limited for vegetable crops. For specialty crops and herbs, few fungicides will be labeled for use. Before purchasing or using any fungicide product, check the label for use restrictions and safety guidelines. Consult with local extension, regulatory, and manufacturer agents for information and regulations on product application.

There are several categories of fungicides. Systemic fungicides move within the plant after application, usually in the vascular xylem tissue. Benzimidazoles were the first widely used systemic products, and phenylamide fungicides are probably the most important systemic group in current use. Protectant fungicides are usually non-systemic and remain only on the plant surfaces that received the spray. Protectants usually prevent pathogen infection before it takes place, and provide little benefit if the pathogen already entered the host. Therefore, protectants must be applied multiple times to protect new growth. Eradicant fungicides are usually systemic materials capable of localized (e.g. translaminar) or more general movement within the plant. Eradicants stop or reduce fungal growth after host penetration has occurred. Only eradicant fungicides can control some latent infections; infections are latent if the pathogen has succeeded in penetrating and initially colonizing the host, but is not actively growing and progressing in the host. See ***Table 9*** (previous page)for a list of some fungicides and other materials used to control diseases of vegetables and other plants.

The chemical mode of action of fungicides targets metabolic sites in the pathogen. Some chemicals act on a single site while others act on multiple sites. Researchers still have not identified all the mechanisms involved. Fungicides may affect processes such as energy production, major biosynthetic pathways, or membrane function. The triazoles or demethylation inhibitors (DMIs) interfere with sterol biosynthesis and impair membrane synthesis and function. Strobilurins are $Q^{o}I$ inhibitors that inhibit mitochondrial respiration by binding to the Q^{o} site of cytochrome b. Phenylamides affect RNA polymerase, and the aminopyrimidines inhibit the enzyme adenosine deaminase. Protectant, broad-spectrum fungicides such as the dithiocarbamates, copper, and tin have multi-site activity and interfere with energy production.

Mutations in the organism that result in changes at the site of action can render the once effective pesticide ineffective, and pathogens carrying such mutations are considered resistant or insensitive to that material. Pesticide resistant individuals exist naturally within pathogen populations. When a fungicide or bactericide is applied to plants, individuals that are resistant to that chemical survive and increase in number. With repeated use, these resistant individuals increase in proportion within the population and become sufficiently numerous so as to cause damaging levels of disease.

The development of resistance is most rapid against pesticides having only a single site of action. A single mutation enables the fungus or bacterium to block or overcome the toxic effects of the material. Resistance in fungal pathogens occurred rapidly, within a few years of product introduction, with benzimidazoles, pyrimidines, and phenylamides. Pathogens with short, repeating life cycles and abundant spore production develop fungicide resistance problems faster than slower cycling pathogens. Powdery mildews and *Botrytis cinerea* are examples of pathogens prone to developing resistance to new fungicides. Fungicides with multiple sites of action generally develop resistance less frequently or remain effective for longer periods. For example, pathogens have not developed resistance to the dithiocarbamates probably because these products are multi-site inhibitors. A number of bacterial pathogens readily develop resistance to copper and antibiotic materials used for disease control.

To combat and delay onset of pathogen resistance (or insensitivity), several strategies must be employed. Only apply fungicides when needed; the overuse of chemicals contributes greatly to the development of resistance. In general, fungicides should be applied before pathogens become established on the host plant. Different fungicide products having different modes of action should be applied either in tank mix combinations in one spray or in alternating schedules. Careful monitoring of unusually severe disease problems that occur despite fungicide applications and close collaboration with extension researchers and other professionals can help the industry detect outbreaks of resistant pathogens.

In addition to fungicides and bactericides, other types of chemical treatment may be used to control diseases. The majority of these alternative substances, however, are not yet commercially effective or viable products that consistently provide adequate control. Plant stimulants or conditioners such as plant or seaweed extracts and compost teas may show disease suppressive effects under certain experimental or small scale conditions. Some chemicals are elicitors, which induce resistance to pathogens in the host plant. This resistance is distinct from that resulting from gene-for-gene interactions and is termed systemic acquired resistance (SAR). Salicyclic acid, jasmonates, some fatty acids, and yeast cell wall components can induce such resistance. At least one synthetic plant activator, acibenzolar-S-methyl, is used commercially. Some chemicals stimulate fungal propagules to germinate and die in the absence of the host plant. Sclerotia of the allium white rot pathogen, *Sclerotium cepivorum*, normally only germinate in the presence of allium crops and are stimulated to do so by sulfur compounds released by roots. Diallyl disulfide, derived from allium plants, can be applied to fallow soils to stimulate sclerotia germination. Germinated sclerotia will die because no allium crop is present. Oils, plant extracts, microorganism byproducts and metabolites, and other natural plant products are being investigated for use for disease control.

BIOLOGICAL CONTROL

Classical biological controls, in which a product containing a viable antagonistic organism is applied to combat the target pathogen, are still being developed but are not yet available or effective for most diseases of vegetables. A *Coniothyrium minitans* product has shown some field efficacy for controlling *Sclerotinia sclerotiorum*. Other products containing *Trichoderma, Gliocladium, Streptomyces, Bacillus*, and other microorganisms have not demonstrated consistent, commercially acceptable levels of control.

DISEASE FORECASTING

A number of forecasting systems have been developed for diseases in vegetable crops. These weather data and computer supported programs attempt to predict either when pathogens will threaten crops or the degree of disease severity that might be expected given current conditions and inoculum load. Research on forecasting systems has been on-going for many years. Successful systems depend on a satisfactory understanding of the biology of the pathogen, development of the plant host, and environmental parameters that favor pathogen and disease development. Some proposed forecasting systems may function in a research context, but not provide sufficient information that results in commercially acceptable levels of disease control. The biological systems involving plant host, pathogen organism, and surrounding environment are extremely complex; predictive models generated by researchers often fail to capture this complexity. Validation is therefore required to test forecasting systems over several seasons, and reliance on new systems should be carefully considered.

TABLE 10 Checklist of vegetable disease management options following the general production sequence

Options before planting

- Examine the history and records of the proposed site. If possible, choose a different site if significant problems occurred on previous plantings of the same or related vegetable.
- Avoid this site if it has features that will be problematic for some diseases, such as heavy, poorly draining soils that will enhance damping-off problems or consistently high humidity that favors foliar diseases.
- If available, collect soil samples and have them tested for important soilborne pathogens and problem factors such as high salt content.
- The vegetable being considered might be a high risk crop due to factors such as the following: the previous crop was the same or related vegetable and subject to the same pathogens; poor crop rotation has been practiced in the past; this vegetable has been frequently cropped at this location. If these factors apply, consider another site or plant a different, unrelated crop.
- Consider if the target planting date might cause disease severity to be worse. If this might be the case, alter the planting date.
- Though usually prohibitively expensive for vegetable crops, treat the soil with fumigants to reduce inoculum of soilborne pathogens.
- Select a resistant cultivar. Select cultivars suitable to the region and to the production system being used. If resistant cultivars are not used, seek tolerant cultivars.
- For seedborne pathogens, seek to use seed that does not have significant levels of the pathogen. If appropriate, treat seed with hot water, effective fungicides, or other treatments. Have seeds treated with fungicides if directly seeded in the field.
- If transplants are to be used, arrange for vigorous, healthy, symptom-free transplants.
- Remove weeds, volunteer plants, and any other possible host plants from the field and surrounding areas.
- Work the soil and time the planting so that previous crop residues are absent or minimized.
- Add appropriate materials (lime, gypsum, fertilizers) that will enhance plant growth or create conditions unfavorable to the pathogen.
- Ensure that tractors and equipment used for soil preparation are not contaminated with infested soil from other fields.
- Properly prepare the soil so that it drains well, does not have low spots, and is in suitable condition to encourage rapid seed germination, transplant establishment, and subsequent growth of the crop.
- Devise and install an appropriate irrigation system.

Options at planting

- Ensure that the bed tops are suitably prepared for good seed germination and transplant establishment.
- Ensure that clean seed and disease-free, high-quality transplants are brought on site. Exclude unacceptable plant materials.
- Place seed and transplants at appropriate depths. Plant materials that are placed too deeply are subject to additional disease problems.
- In some cases, apply post-plant fungicides to the seed lines or transplants after they are placed in the ground.
- Schedule appropriate irrigations so that plant material is not over or under watered.

Options after planting

- After plants are established, attempt to eliminate the use of overhead sprinkler irrigation systems.
- Practice good growing methods to enhance plant development and reduce stress. Especially monitor irrigations so that excess water is not applied.
- Regularly and thoroughly monitor the field for signs of disease and for the presence of insects and weeds.
- Manage insect and weed pests in and around the field.
- If contagious diseases occur in other fields, avoid bringing possibly contaminated equipment into a healthy field.
- Apply fungicides as available, needed, and appropriate. As disease prediction systems become available, use these to time the applications. Use diverse fungicide products so as to reduce the onset of resistance.

Options at and after harvest

- Harvest crops when commodities are still healthy, in good condition, and not over mature. If diseases begin to occur in a field, harvest dates should be moved up and plants removed early.
- Harvested commodities should be kept clean, free of soil, and undamaged. Commodities should be properly packed and rapidly placed in cold storage or another appropriate storage environment.
- Plow under and otherwise destroy unharvested plants and weeds in and around the field.
- Schedule disking and other soil preparation steps so that crop residues break down in a timely manner. Allow sufficient time for residues to dissipate prior to the next crop.
- Consider planting a cover crop or broccoli that can reduce soilborne pathogen populations.
- Consider adding compost and other beneficial amendments to the soil prior to planting the next crop.
- In warmer regions, try soil solarization treatments for reducing inoculum from soilborne pathogens.

Other aspects of disease management

POSTHARVEST HANDLING

When vegetable commodities reach harvestable stage, the crop can be healthy but exposed to pathogen inoculum, infected by pathogens but not yet showing symptoms, or diseased and showing symptoms in various stages. In all cases, these commodities may be harvested, packed, and placed in storage. Depending on length of storage, temperature, humidity, and other environmental factors, the stored commodities can develop postharvest diseases that originated from field pathogens. To minimize postharvest disease development, harvest crops when commodities are undamaged, still healthy, in good condition, and not over mature. If diseases begin to occur in a field, harvest dates should be moved up and plants collected early. Harvested commodities should be kept clean, free of soil, and undamaged. Commodities should be properly packed and rapidly placed in cold storage or other appropriate storage environment. Time in storage is a critical factor, and the longer commodities are stored, the greater is the chance that postharvest diseases can occur.

CONTROLLING DISEASES IN ORGANIC SYSTEMS

The challenge of disease control is accentuated for organic vegetable growers because they do not use synthetic fungicides and fumigants. These growers have fewer insecticides to use against pathogen vectors and fewer herbicides to manage weeds. Yet the world market will continue to be extremely competitive and require organic growers to supply high-quality, disease-free produce having acceptable shelf-life. This makes disease control a challenging task for organic vegetable growers. In principle and with the exception of synthetic chemicals, the integrated disease management strategy for organic vegetable producers should be similar to that for conventional growers. All options that do not involve synthetic chemicals are applicable and appropriate for this segment of the industry.

In addition, organic growers generally try to emphasize strategies that have an ecological basis. For example, the organic system encourages as much as possible the growth and diversity of soil-inhabiting and leaf-epiphytic microorganisms that might have beneficial and pathogen-antagonistic influences. Increasing the genetic diversity of the crop rotation is another ecological management step. Integrating disease management decisions with insect and weed control and with general production practices is another aspect consistent with this approach.

For organic growers, the application of disease control materials is limited. Mineral-based control materials, primarily copper and sulfur fungicides, are generally inexpensive and widely available. However, disease control efficacy varies. Copper fungicides have some activity against a wide range of fungal and bacterial pathogens but generally are not extremely effective. Sulfurs also exhibit some activity against many pathogens, but usually only provide excellent control against certain pathogens such as powdery mildew fungi. Bicarbonate based fungicides have recently become available for control of plant diseases and have shown activity primarily against powdery mildews. Of note is recent research indicating that copper and sulfur materials, previously thought to have minimal impact on the environment, may actually have deleterious effects on the ecosystem.

INTEGRATED DISEASE MANAGEMENT

The optimum way to control diseases of vegetables is to use all available disease management tools in an integrated, strategic system. Reliance on only chemicals or some other one-dimensional approach will not provide the best means of controlling damaging problems. To be most successful, growers must integrate and coordinate all of the aspects discussed in this chapter: site selection, exclusion, resistant and tolerant cultivars, a wide range of cultural practices, sanitation, manipulation of the environment, control of reservoir hosts and vectors, fungicides and other disease control materials, disease forecasting systems, and proper postharvest handling. See ***Table 10*** for a chronological checklist of these options that is arranged in a sequence that fits general vegetable production.

A good case study that illustrates an integrated disease management system is the control strategy for *Lettuce mosaic virus* (LMV) on lettuce grown in California's Salinas Valley. A productive partnership between University of California campus-based researchers, Cooperative Extension researchers, lettuce growers, and regulatory agencies resulted in a management program that successfully keeps lettuce mosaic at minimal levels in this valley, which is the world's largest lettuce growing region. County government ordinances enforce the first four aspects of the integrated program outlined overleaf.

- *Lettuce seed assays for LMV*
  All lettuce seed to be planted in the Salinas Valley is tested for seedborne-LMV. Researchers found that a zero in 30,000 seed infection threshold is the key for LMV control here; if no infected seed are found in a 30,000 lettuce seed sample serologically tested by the enzyme-linked immunosorbent assay (ELISA), the seed lot is approved for planting.
- *Weed control*
  Because weeds can be a significant reservoir of the virus and the source from which aphids obtain the virus, weeds must be regularly controlled and removed in the lettuce production areas.
- *Plow-down of old lettuce plantings*
  Old, infected lettuce plants, like weeds, are a source of virus and vectors. Aphids can obtain LMV from these old plants and transport the virus to younger, nearby lettuce plantings. Old plantings, once harvested, must be plowed down in a timely manner.
- *Lettuce host-free period*
  To prevent continuous, year-to-year buildup of LMV, an annual host-free period is enforced for two weeks in December. This step is effective because LMV is an obligate pathogen and cannot survive in nature without living plant hosts. The ban on lettuce production in this winter period helps reduce the amount of virus that might 'bridge' over from one season to the next. Vector activity during this winter period is limited, which also contributes to the effectiveness of the lettuce free period.
- *Site selection*
  Over time, growers find that certain fields are prone to developing lettuce mosaic disease due to virus reservoirs in the area. Such sites are not planted to lettuce, if possible. In specific examples, lettuce mosaic tended to occur in fields adjacent to a business storage area where weeds were not controlled, and in another area where LMV-infected *Gazania* species ground covers were a part of the landscape.
- *Resistant lettuce cultivars*
  While not used extensively in the Salinas Valley, cultivars resistant to LMV are available and contribute to this integrated program.
- *Aphid control*
  Spraying for the aphids that vector LMV does not prevent the transmission of virus because aphids can transfer this pathogen to lettuce before the insecticides act to kill the insects. However, aphid control clearly helps to slow LMV spread and therefore should be included.
- *Research on LMV*
  University and extension researchers continue to examine the LMV system, attempting to increase understanding of disease dynamics, improve upon antiserum used for seed assays, delve into the genetic nature of LMV strains, and watch for changes in the local and worldwide situation regarding this important virus.

References on disease control

Agarwal, V.K. and Sinclair, J. B. 1997. *Principles of Seed Pathology*. Second Edition. CRC Press.

Brent, K. J. 1995. *FRAC Monograph No. 1 Fungicide resistance in crop pathogens: How can it be managed?* GIFAP, Brussels, 48pp.

Brown, R. G. 1999. *Plant Diseases and Their Control*. Sarup & Sons.

Brown, S., Koike, S. T., Ochoa, O. E., Laemmlen, F., and Michelmore, R. W. 2004. Insensitivity to the fungicide fosetyl-Aluminium in California isolates of the lettuce downy mildew pathogen, *Bremia lactucae*. *Plant Disease* 88:502–508.

Clarkson, J.P., Kennedy, R., and Bowtell, J. 1998. A methodology for evaluation of the efficacy of fungicide dosage and plant resistance in the control of fungal diseases of vegetable crops. *Proceedings of the 1998 Brighton Conference. Pests & Diseases* 3: 869–874.

Crute, I. R. 1998. British Society for Plant Pathology Presidential Address 1995 The elucidation and exploitation of gene-for-gene recognition. *Plant Pathology* 47: 107–113.

Ebbels, D. L. 2003. Principles of Plant Health and Quarantine. CABI Publishing, Wallingford, UK, 320 pp.

Fry, W. E. 1982. *Principles of Plant Disease Management*. Academic Press.

Fuchs, M., Tricoli, D. M., Carney, K. J., Schesser, M., McFerson, J. R., and Gonsalves, D. 1998. Comparative virus resistance and fruit yield of transgenic squash with single and multiple coat protein genes. *Plant Disease* 82:1350–1356.

Gaskell, M., Fouche, B., Koike, S., Lanini, T., Mitchell, J., and Smith, R. 2000. Organic vegetable production in California—science and practice. *HortTechnology* 10:699–713.

Grogan, R. G. 1980. Control of lettuce mosaic with virus-free seed. *Plant Disease* 64:446–449.

Hadidi, A., Khetarpal, R. K., and Koganezawa, H. 1998. *Plant Virus Disease Control*. American Phytopathological Society Press.

Hall, R. 1996. *Principles and Practice of Managing Soilborne Plant Pathogens*. American Phytopathological Society Press.

Hao, J. J., Subbarao, K. V., and Koike, S. T. 2003. Effects of broccoli rotation on lettuce drop caused by *Sclerotinia minor* and on the population density of sclerotia in soil. *Plant Disease* 87:159–166.

Hickey, K. D. 1986. *Methods for Evaluating Pesticides for Control of Plant Pathogens*. American Phytopathological Society Press.

Janse, J. D. and Wenneker, M. 2002. Review. Possibilities of avoidance and control of bacterial plant diseases when using pathogen-tested (certified) or –treated planting material. *Plant Pathology* 51: 523–536.

Koike, S. T., Smith, R. F., Jackson, L. E., Wyland, L. J., Inman, J. I., and Chaney, W. E. 1996. Phacelia, lana woollypod vetch, and Austrian winter pea: three new cover crop hosts of *Sclerotinia minor* in California. *Plant Disease* 80:1409–1412.

Lyr, H. 1995. *Modern Selective Fungicides: Properties, Applications, Mechanisms of Action*. Second edition. Gustav Fischer Verlag.

Lyr, H., *et al.* 1999. *Modern Fungicides and Antifungal Compounds II*. 12th International Reinhardsbrunn Symposium, Thuringia Germany. Intercept.

Maloy, O. C. 1993. *Plant Disease Control: Principles and Practice*. John Wiley & Sons.

Maude, R. B. 1996. *Seedborne diseases and their control: Principles and Practice*. CAB International, Wallingford, UK, 280 pp.

Ristaino, J. R. and Thomas, W. 1997 Agriculture, methyl bromide, and the ozone hole: can we fill the gap? *Plant Disease* 81: 964–977.

Skylakakis, G. 1981. Effects of alternating and mixing pesticides on the buildup of fungal resistance. *Phytopathology* 71: 1119–1121.

Smith, I. M., Dunez, J., Lelliot, R. A., Phillips, D. H., and Archer S. A. (eds). 1988. *European Handbook of Plant Diseases*. Blackwell Scientific Publications, Oxford, UK, 583 pp.

Soper, D. 1995. *A Guide to Seed Treatments in the UK*. British Crop Protection Council. Major Print Ltd.

Staub, T. 1991. Fungicide resistance: practical experience with anti-resistance strategies and the role of integrated use. *Annual Review of Phytopathology* 29:421–442.

Subbarao, K. V., Hubbard, J. C., and Koike, S. T. 1999. Evaluation of broccoli residue incorporation into field soil for Verticillium wilt control in cauliflower. *Plant Disease* 83:124–129.

Tomalin, C. D. S. (ed.). 2000. *The Pesticide Manual*. 12th Edition. British Crop Protection Council, Farnham, UK.

Thompson, D. C., Baron, J. J., and Kunkel, D. L. 2000. The IR-4 Project – a minor-use program for pest management solutions in the United States. *Proceedings of the BCPC Conference. Pests & Diseases* 3: 1253–1260.

Thurston, H. D. 2004. Ten thousand years of experience with sustainable plant disease control. *Plant Disease* 88:550–551.

University of California. *UC IPM Pest Management Guidelines*. UC IPM Online. http://www.ipm.ucdavis.edu/PMG/crops-agriculture.html

Zerbini, F. M., Koike, S. T., and Gilbertson, R. L. 1995. Biological and molecular characterization of lettuce mosaic potyvirus isolates from the Salinas Valley of California. *Phytopathology* 85:746–752.

PART 2

Diseases of Vegetable Crops

- **ALLIACEAE** (onion family)
- **APIACEAE** (parsley family)
  Apium graveolens (celery)
  Daucus carota (carrot)
  Pastinaca sativa (parsnip)
  Petroselinum crispum (parsley)
- ASPARAGUS OFFICINALIS (asparagus)
- BETA VULGARIS (beet)
- **BRASSICACEAE**
  (cabbage and mustard family)
- CAPSICUM (pepper)
- **CUCURBITACEAE** (gourd family)
- **FABACEAE** (pea family)
  Phaseolus species (beans)
  Pisum sativum (pea)
  Vicia faba (broad bean)
- LACTUCA SATIVA (lettuce)
- SOLANUM LYCOPERSICUM
  (tomato)
- SPINACIA OLERACEA (spinach)
- SPECIALTY AND HERB CROPS
  (horseradish, dill, chervil, Swiss chard, broccoli raab, mustards, endive/escarole, radicchio, coriander, lemongrass, artichoke, arugula, fennel, Jerusalem artichoke, mint, catnip, basil, marjoram, oregano, tomatillo, rhubarb, watercress, sage, salsify, scorzonera, corn salad, sweetcorn)

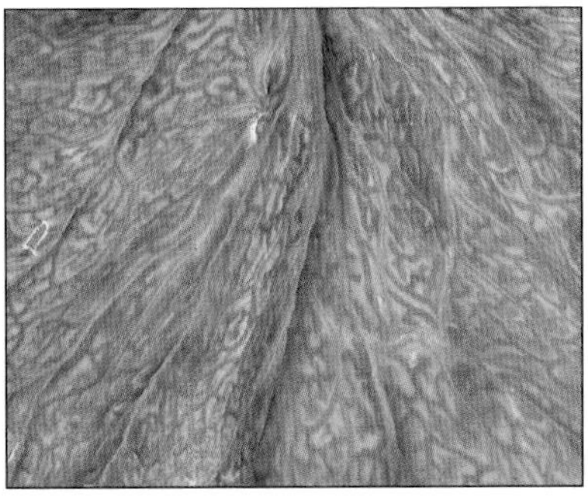
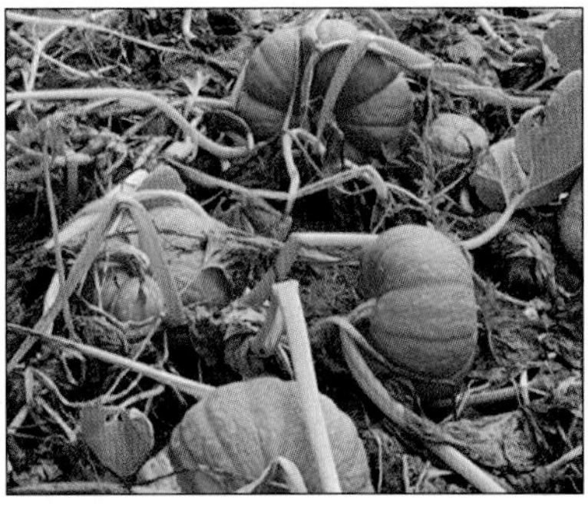

Alliaceae Onion family

ALLIUMS, formerly classified with the Liliaceae (lilies and relatives), now have their own family, the Alliaceae. The plants in this crop group are widely used for seasoning and cooking. The most familiar allium crops include bulb onion (*Allium cepa*), spring or salad onion (*A. cepa*), bunching onion (*A. fistulosum*), garlic (*A. sativum*), leek (*A. porrum = A. ampeloprasum* var. *porrum*), elephant garlic (*A. ampeloprasum*), shallot (*A. cepa* var. *ascalonicum*), chives (*A. schoenoprasum*), Chinese chives (*A. tuberosum*), and Egyptian tree onion (*A. cepa* var. *aggregatum*). Allium plants are monocots and may have originated in Asia.

Pseudomonas syringae pv. *porri*

BACTERIAL BLIGHT

Introduction and significance

In commercial settings this bacterium affects primarily leek and has been reported from the USA, England, France, and New Zealand. Onion, chives, and garlic have developed bacterial blight when inoculated with this pathogen in experimental situations.

Symptoms and diagnostic features

Young leaves show water-soaked then yellow longitudinal lesions or stripes that later split and rot. The leaves can become curled and twisted as growth continues (**1**, **2**). On older leaves, the pathogen causes yellow spots around wounds. Flowering stalks are very susceptible and develop deep, water-soaked lesions that ooze a bacterial exudate. Older stalk lesions are sunken, first yellow, then finally brown in color. Leek transplants can develop the disease while growing in greenhouses. Leaves of transplants develop yellow, then brown, elongated lesions (**3**). Lesions usually involve the tips of the leaves .

Causal agent

Bacterial blight is caused by *Pseudomonas syringae* pv. *porri*. The designation pathovar (pv.) indicates this pathogen is host specific to the allium group and does not infect other crops, such as tomato, celery, and bean, which are susceptible to other *P. syringae* pathogens. The pathogen is an aerobic, Gram-negative bacterium and can be isolated on standard microbiological media. It produces the cream-colored colonies typical of most pseudomonads. When cultured on Kings medium B, this organism produces a diffusible pigment that fluoresces blue under ultraviolet light.

A number of other bacterial pathogens affect alliums. Slippery skin disease is caused by *Pseudomonas gladioli* pv. *alliicola* and results in a water-soaked bulb rot that initially only affects a few fleshy scales. At room temperature, however, these infected scales rapidly lead to a extensive rot of the bulb. Sour skin is caused by *Pseudomonas cepacia*, which produces a pale yellow or light brown rot of the inner fleshy scales. Bacterial streak and bulb rot of sweet onion occurs in the USA and is incited by *Pseudomonas*

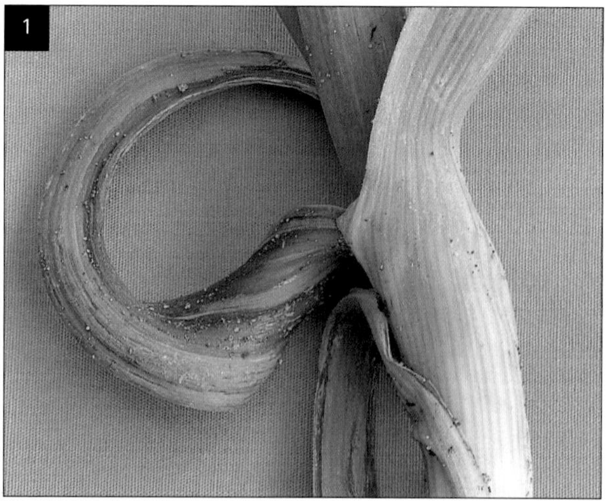

1 Water-soaked lesion and leaf deformity of early symptoms of bacterial blight of leek.

viridiflava. Soft rot is caused by *Erwinia carotovora* subsp. *carotovora* and is associated with very wet conditions in the field or poor handling at harvest. The bulb develops a watery rot with a pungent odor. Another bacterial disease is onion bacterial blight caused by a seedborne *Xanthomonas* species. This problem occurs mostly in tropical and subtropical areas and causes elongated yellow lesions that may have water-soaked edges; lesions later turn brown and necrotic.

Disease cycle

Pseudomonas syringae pv. *porri* is seedborne, so disease can be initiated in the field with direct seeded leek, or begin at the transplant stage in greenhouses. Spread of the pathogen at the seedling, transplant, or field stage is dependent on splashing water from rain or overhead sprinkler irrigation. The bacterium can survive in leek crop residues but will not persist in soil once residues are completely decomposed.

Control

Use seed that does not have significant levels of the pathogen. Appropriate seed treatments can also contribute to the management of seedborne inoculum. Rotate away from allium crops to reduce inoculum from crop residues. Irrigate using drip or furrow systems. Copper sprays may reduce spread. Avoid over-fertilizing with high nitrogen materials.

References

Alvarez, A. M., Buddenhagen, I. W., Buddenhagen, E. S., and Domen, H. Y. 1978. Bacterial blight of onion, a new disease caused by *Xanthomonas* sp. *Phytopathology* 68:1132–1136.

Gitaitis, R., MacDonald, G., Torrance, R., Hartley, R., Sumner, D. R., Gay, J. D., and Johnson, W. C. 1998. Bacterial streak and bulb rot of sweet onion: II. Epiphytic survival of *Pseudomonas viridiflava* in association with multiple weed hosts. *Plant Disease* 82:935–938.

Koike, S. T., Barak, J. D., Henderson, D. M., and Gilbertson, R. L. 1999. Bacterial blight of leek: a new disease in California caused by *Pseudomonas syringae*. *Plant Disease* 83:165–170.

Nunez, J. J., Gilbertson, R. L., Meng, X., and Davis, R. M. 2002. First report of *Xanthomonas* leaf blight of onion in California. *Plant Disease* 86:330.

Paulraj, L. and O'Garro, L. W. 1993. Leaf blight of onions in Barbados caused by *Xanthomonas campestris*. *Plant Disease* 77:198–201.

Roberts, P. 1973. A soft rot of imported onions caused by *Pseudomonas alliicola* (Burkh.) Starr and Burkh. *Plant Pathology* 22:98.

Roumagnac, P., Gagnevin, L., and Pruvost, O. 2000. Detection of *Xanthomonas* sp., the causal agent of onion bacterial blight, in onion seeds using a newly developed semi-selective isolation medium. *European Journal of Plant Pathology* 106:867–877.

Samson, R., Poutier, F., and Rat, B. 1981. Une nouvelle maladie du poireau: la graisse à *Pseudomonas syringae*. *Revue Horticole* 219:20–23.

Samson, R., Shafik, H., Benjama, A., and Gardan, L. 1998. Description of the bacterium causing blight of leek as *Pseudomonas syringae* pv. *porri*. *Phytopathology* 88:844–850.

Teviotdale, B. L., Davis, R. M., Guerard, J, P., and Harper, D. H. 1989. Effect of irrigation management on sour skin of onion. *Plant Disease* 73:819–822.

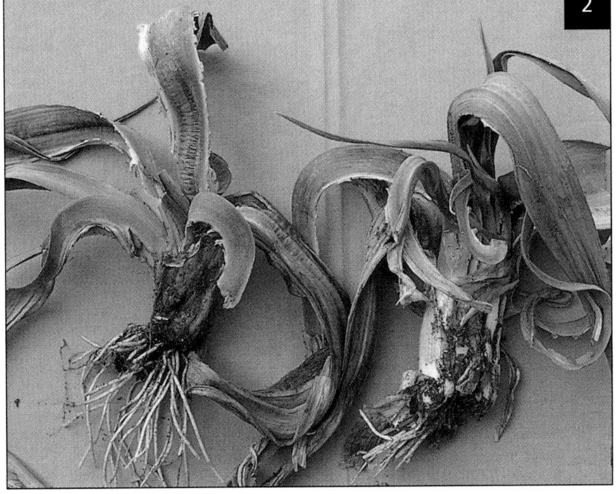

2 Deformed leeks severely affected by bacterial blight.

3 Brown lesions of advanced symptoms of bacterial blight of leek.

Alternaria porri, Stemphylium vesicarium

PURPLE BLOTCH

Introduction and significance

This disease occurs throughout the world but is most damaging in areas with warm, humid climates. It is an important concern on onion, garlic, and leek. Purple blotch is caused by two fungal pathogens which may co-infect the plant or act independently in causing disease.

Symptoms and diagnostic features

The first symptoms are small, water-soaked leaf spots, which gradually enlarge, turn yellow, and develop a pale tan center. Older lesions turn brown or purple, are oblong in shape, and may be several centimeters long (**4, 5, 6**). Under humid conditions dark fungal sporulation occurs in the center of the lesion. Large lesions are produced on flower stalks and these can be dark purple. When numerous lesions occur on leaves, dieback of the foliage can take place. Microscopic examination is required to confirm diagnosis of purple blotch, as symptoms may be confused with those of Cladosporium leaf blotch, white tip, and other leaf diseases.

Causal agents

Purple blotch is caused by two pathogens in the fungi imperfecti group. Purple blotch lesions can be caused by either fungus alone or by both fungi together. *Alternaria porri* forms conidia that are brown to gold-brown, with the main spore body being ellipsoidal with 8–12 transverse septa and usually a few longitudinal septa. Conidia have a long, tapering beak and are borne singly. Overall conidial dimensions are 100–300 x 15–20 µm.

The other purple blotch pathogen is *Stemphylium vesicarium*. This fungus produces olive-brown to gold-brown, oblong to oval conidia that have up to four longitudinal septa and varying numbers of transverse septa. Conidia measure 25–48 x 12–22 µm and have a length-to-width ratio of 1.5–3.0. Conidia are borne singly on conidiophores that have a distinctly swollen tip. In some regions the pseudothecia and ascospores of the sexual stage, *Pleospora allii*, can be found associated with crop debris.

Disease cycle

Infection requirements are similar for both pathogens, and disease severity is worse with warm and humid or wet conditions. Spores are airborne and infect the leaves via stomata or directly through the epidermis. Inoculum comes from infested crop residue in the field.

The optimum temperature for *A.porri* is 18–25° C for germination and 15–25° C for infection. The pathogen can still be active at lower temperatures (4–13° C), however, enabling disease to develop in autumn and winter. Conidia are produced at night and released early in the day as the humidity decreases. More than 8 hours of leaf wetness at 15–25 °C are required for infection. When dew duration lasts for 16 hours or more, *A. porri* conidia infect leaves and result in typical lesions; if there is less than 12 hours of dew, infections result only in small leaf flecks. The older leaves are more susceptible to infection than the young leaves, although damage by thrips may increase the susceptibility of young leaves. As symptoms appear in 1–4 days and new spores are produced after only 5 days, epidemics develop quickly under favorable conditions.

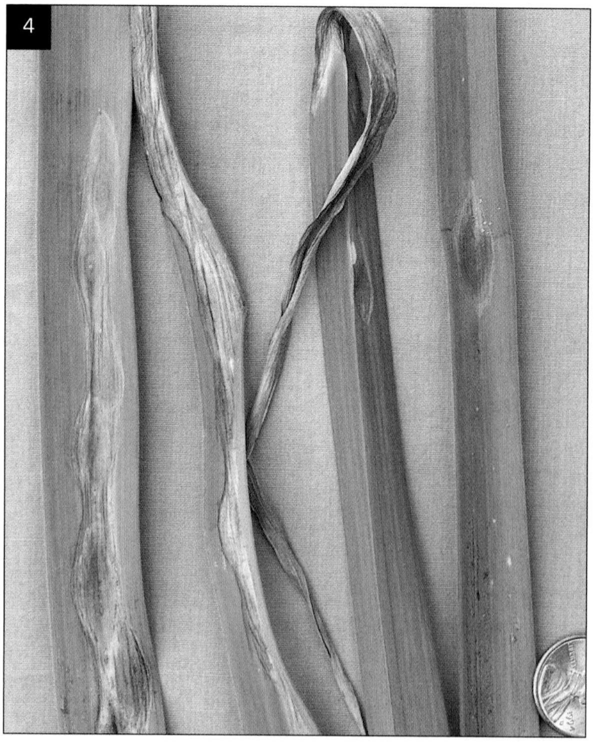

4 Leaf spots of purple blotch, caused by *Alternaria porri*, on leek.

Control

Plant resistant or tolerant varieties. For example, avoid using sweet Spanish onions because these are reportedly very sensitive to *A. porri*. Plow under crop residues after harvest to reduce spread and survival of inoculum. Rotate crops with non-hosts. Select sites and practice irrigation so that foliage drying is enhanced. Apply fungicides, especially as the crop canopy ages and becomes more dense and if leaf-wetness periods favor infection.

References

Basallotte-Ureba, M. J., Prados-Ligero, A. M., and Melero-Vara, J. M. 1999. Aetiology of leaf spot of garlic and onion caused by *Stemphylium vesicarium* in Spain. *Plant Pathology* 48:139–145.

Bisht, I. and Agrawal, R. C. 1993. Susceptibility to purple leaf blotch (*Alternaria porri*) in garlic (*Allium sativum*). *Annals of Applied Biology* 122:31–38.

Boiteux, L. S., Lima, M. F., de Menezes Sobrinho, J. A., and Lopes, C. A. 1994. A garlic (*Allium sativum*) leaf blight caused by *Stemphylium vesicarium* in Brazil. *Plant Pathology* 43:412–414.

Everts, K. L. and Lacy, M. L. 1990. The influence of dew duration, relative humidity, and leaf senescence on conidial formation and infection of onion by *Alternaria porri*. *Phytopathology* 80:1203–1207.

Everts, K. L. and Lacy, M. L. 1996. Factors influencing infection of onion leaves by *Alternaria porri* and subsequent lesion expansion. *Plant Disease* 80:276–280.

Gladders, P. 1980. New or unusual records of plant diseases and pests – Purple blotch of leeks caused by *Alternaria porri*. *Plant Pathology* 30:61.

Gupta, R. B. L. and Pathak, V. 1988. Yield losses in onions due to purple blotch disease caused by *Alternaria porri*. *Phytophylactica* 20:21–23.

Koike, S. T. and Henderson, D. M. 1998. Purple blotch, caused by *Alternaria porri*, on leek transplants in California. *Plant Disease* 82:710.

Meredith, D. S. 1966. Spore dispersal in *Alternaria porri* (Ellis) Neerg. on onions in Nebraska. *Annals of Applied Biology* 57:67–73.

Prados-Ligero, A. M., Melero-Vara, J. M., Corpas-Hervías, C., and Basallotte-Ureba, M. J. 2003. Relationships between weather variables, airborne spore concentrations and severity of leaf blight of garlic caused by *Stemphylium vesicarium* in Spain. *European Journal of Plant Pathology* 109:301–310.

Suheri, H. and Price, T. V. 2000. Infection of onion leaves by *Alternaria porri* and *Stemphylium vesicarium* and disease development in controlled environments. *Plant Pathology* 49:375–382.

Suheri, H. and Price, T. V. 2001. The epidemiology of purple leaf blotch on leeks in Victoria, Australia. *European Journal of Plant Pathology* 107:503–510.

5 Leaf spots of purple blotch, caused by *Stemphylium vesicarium*, on leek.

6 Close-up of leaf spots of purple blotch, caused by *Stemphylium vesicarium*, on leek.

Aspergillus niger

BLACK MOLD

Introduction and significance

This disease is most prevalent in subtropical and tropical production areas where high temperatures favor its development. While black mold can cause some problems in the field, most losses occur in storage. Black mold concerns in temperate areas, such as the UK in the 1980s, have been associated with poor storage conditions and elevated storage temperatures.

Symptoms and diagnostic features

This disease causes unsightly, black, dusty fungal growth on and between the bulb scales (7, 8). The entire surface of the bulb may be turned black in severe cases, and a bulb rot may follow. Secondary bacterial organisms usually contribute to the soft rot symptoms.

Causal agent

Black mold is caused by the fungus *Aspergillus niger*. The pathogen is dark brown to black in culture. The fungus produces a readily recognized conidiophore that consists of an erect unbranched stalk, a single swollen spherical vesicle, prophialides and phialides that line the surface of the vesicle, and densely packed chains of spherical black, spiny conidia (2–5 μm in diameter). Conidia are dry and readily dispersed in the air.

Disease cycle

The pathogen is seedborne in onion. In addition, the fungus occurs widely in field soils and is a saprophyte that colonizes dead plant foliage. *Aspergillus niger* first grows on senescing onion leaves, then progresses to the neck of the bulb and finally into the bulb itself. Bulb infection in the field is usually associated with damage to the fleshy scales, such as when growth splits occur. During harvesting operations, spores on infected foliage are dispersed and provide another source of the pathogen within the crop and to nearby fields. Optimum temperatures for growth are 28–34° C and minimum temperatures for spore germination are 17° C. Humidity greater than 80% is required for spore germination. The presence of free water for 6–12 hours on the onion surface is required for infection.

Control

Use seed that does not have significant levels of the pathogen. If clean seed is not available, broad-spectrum treatments such as thiram have shown good efficacy as seed treatments. Foliar fungicides can reduce black mold when applied towards the end of the crop. Harvest the crop under dry conditions and minimize damage to the neck tissue of the bulbs. Maintain temperature and relative humidity at appropriate levels during storage. Monitor storage conditions when using high temperatures to dry out the onions after harvest;

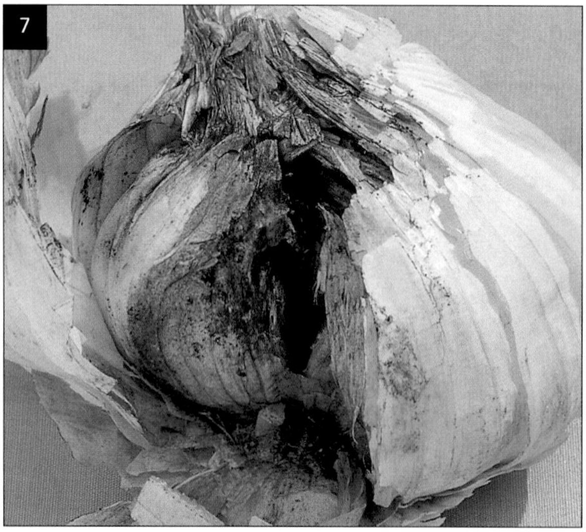

7 Sporulation of *Aspergillus niger* on garlic bulb.

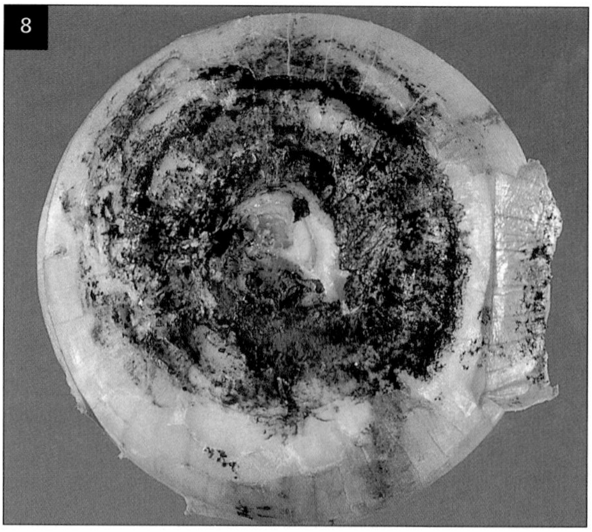

8 Severe symptoms caused by *Aspergillus niger* on a bulb onion.

temperatures should not exceed 30–32° C and the relative humidity should be kept below 80% as long as possible. Long-term storage at low temperatures stops growth of *A. niger*, but the fungus becomes active at temperatures above 15° C.

References

Hayden, N. J. and Maude, R. B. 1992. The role of seed-borne *Aspergillus niger* in transmission of black mould of onion. *Plant Pathology* 41:573–581.

Hayden, N. J., Maude, R. B., and Proctor, F. J. 1994. Studies on the biology of black mould (*Aspergillus niger*) on temperate and tropical onions. 1. A comparison of sources of the disease in temperate and tropical field crops. *Plant Pathology* 43:562–569.

Hayden, N. J., Maude, R. B., El Hassan, H. S., and Al Magid, A. A. 1994. Studies on the biology of black mould (*Aspergillus niger*) on temperate and tropical onions. 2. The effect of treatments on the control of seedborne *A. niger*. *Plant Pathology* 43:570–578.

Botryotinia squamosa
(anamorph = *Botrytis squamosa*)

BOTRYTIS LEAF BLIGHT, BOTRYTIS BLAST

Introduction and significance

This common foliar disease is an important cause of quality loss in green onion, where the condition of the foliage is important for market. It can cause serious loss of foliage in bulb onion, which can impact on yields.

Symptoms and diagnostic features

The first symptoms are leaf spots that are small, white, and have pale green halos. Spots are usually elliptical in shape and can increase to 5–10 mm in diameter (**9, 10**). Multiple spots coalesce and result in bleached foliage and general collapse and death of leaves (**11**).

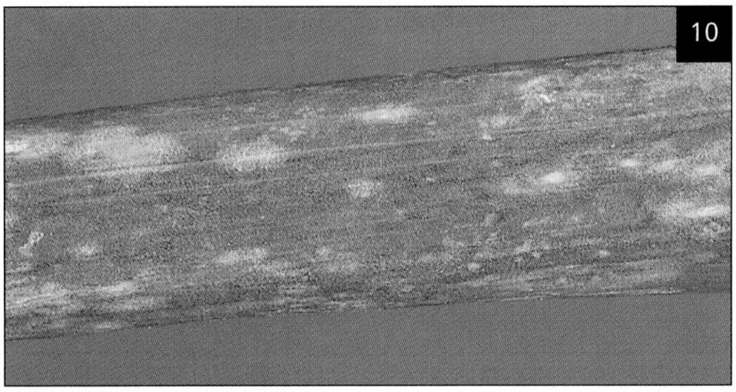

10 White spots with halos on salad onions, caused by caused by *Botrytis squamosa*.

9 Leaf symptoms of Botrytis leaf blight on bulb onions.

11 Severe leaf dieback of onion caused by *Botrytis squamosa*.

12 Onion seed stalk infected with gray mold (*B. cinerea*).

Leaf dieback can be found 5 to 12 days after the first spotting symptoms. Older leaves are more susceptible than younger foliage. The white spot symptoms may be confused with hail damage or physical abrasion from wind-blown soil. However, these abiotic problems are usually confined to the exposed side of the leaf, are more irregular in shape, and lack the diagnostic halo.

Causal agent

Botrytis leaf blight or Botrytis blast disease is caused by *Botrytis squamosa*. *Botrytis squamosa* has larger conidia (14–24 x 9–18 µm) than the other common *Botrytis* pathogens on onion; *B. cinerea* (**12**) has conidia measuring mostly 9–14 x 7–10 µm, and *B. allii* conidia are 7–11 x 5–6 µm. *Botyrtis squamosa* is difficult to isolate from young lesions, so examine senescent leaves and older lesions for *B. squamosa* spores. This pathogen produces black sclerotia, measuring 3–10 mm, on leaf and bulb tissues. *Botrytis squamosa* appears to be host specific to *Allium* species. In some cases the fungus produces a perfect stage, *Botryotinia squamosa*. It is not clear what role this teleomorph has in disease development.

Disease cycle

Initial inocula are airborne spores that originate from nearby crops, debris of previously infected crops, and sclerotia in the soil. Inoculum production within the crop usually increases after the oldest leaves have declined. At least 6 hours of leaf wetness is required for infection by conidia, and the number of lesions that develop increases as leaf wetness duration is prolonged. Optimum conditions for development are temperatures of 12–24° C. Disease development is favored by high crop density; hence spring onion plantings are more prone to problems than the wider spaced bulb onion.

Control

Do not plant successive crops, especially of spring onion, where Botrytis leaf blight occurs regularly. Early season onion crops should not be placed close to overwintered onion. Use cultivars that are less susceptible. Because this pathogen is most severe on *A. cepa* varieties, if possible plant other types of allium crops. For example, chives are reported to be immune. Increase row spacing to improve airflow through the crop. Avoid excessive nitrogen fertilizations, as this causes thick leaf canopies and increased susceptibility.

Irrigate early in the day so that leaves dry quickly. If available, use forecasting systems to identify periods of disease risk and hence guide timing of fungicides. BOTCAST, the first prediction system for Botrytis leaf blight, was developed in North America and has been modified to suit conditions in various other countries. Multiple fungicide applications may be required for spring onion, while bulb onion may not need any treatment because of low susceptibility. Incorporate crop residues after harvest so that inoculum levels are reduced.

References

Clarkson, J. P., Kennedy, R., and Phelps, K. 2000. The effect of temperature and water potential on the production of conidia by sclerotia of *Botrytis squamosa*. *Plant Pathology* 49:119–128.

De Visser, C. L. M. 1996. Field evaluation of a supervised control system for Botrytis leaf blight in spring sown onions in the Netherlands. *European Journal of Plant Pathology* 102:795–805.

Ellerbrock, L. A. and Lorbeer, J. W. 1977. Survival of sclerotia and conidia of *Botrytis squamosa*. *Phytopathology* 67:219–225.

Ellerbrock, L. A. and Lorbeer, J. W. 1977. Sources of primary inoculum of *Botrytis squamosa*. *Phytopathology* 67:363–372.

Lorbeer, J. W. and Vincelli, P. C. 1990. Efficacy of dicarboximide fungicides and fungicide combinations for control of Botrytis leaf blight of onion in New York. *Plant Disease* 74:235–237.

Sutton, J. C., James, T. D. W., and Rowell, P. M. 1986. BOTCAST: a forecasting system to time the initial fungicide spray for managing Botrytis leaf blight of onions. *Agriculture Ecosystems and Environment* 18:123–143.

Walters, T. W., Ellerbrock, L. A., van der Heide, J. J., Lorbeer, J. J., and LoParco, D. P. 1996. Field and greenhouse procedures to evaluate onions for Botrytis leaf blight resistance. *HortScience* 31:436–438.

Botrytis allii (= *B. aclada*)

NECK ROT

Introduction and significance

Neck rot is mainly a storage disease of allium crops. The disease caused significant losses for the UK bulb onion industry for many years until improvements in harvesting and storage techniques were developed. Producers who lack modern drying facilities still incur significant losses. Neck rot affects garlic, shallot, multiplier onion, and leek, but it is most important on bulb onion where losses can sometimes be greater than 50%.

Symptoms and diagnostic features

Typical symptoms are a water-soaked or light brown decay at the neck of stored bulb onion. Dense gray mycelium is produced around the neck and between the bulb scales. Numerous black sclerotia, measuring approximately 5 mm in diameter, also develop on the shoulders of the bulb (**13, 14**). This fungal growth only becomes apparent after several weeks in storage. The pathogen continues to spread from the neck into the bulb and eventually most of the bulb becomes soft and rotted. A brown rot may occur on the fleshy scales of maturing bulb onion in the field, and this is usually associated with splitting or damage of the bulb. The pathogen can affect any part of the bulb and is not limited to the bulb necks. It is possible to find *B. allii* sporulating on senescent foliage prior to crop harvest.

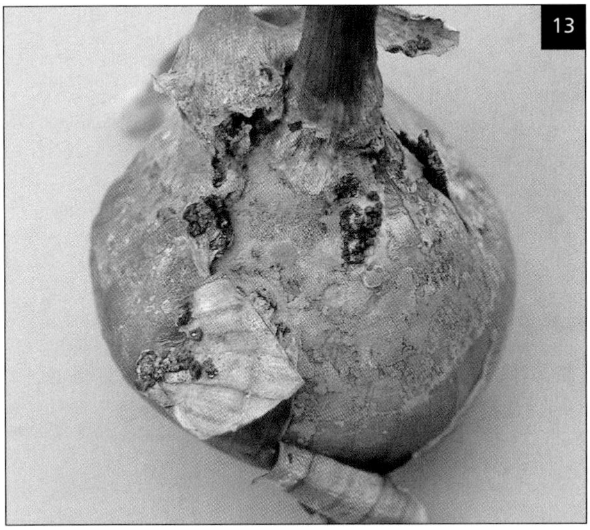

13

13 Neck rot on bulb onion showing black sclerotia and dense gray sporulation.

14

14 Onion bulb with neck rot showing sclerotia and gray sporulation.

Causal agent

Neck rot is caused by *Botrytis allii*, which is synonymous with *Botrytis aclada*. This fungus produces conidia that measure 7–11 x 5–6 μm; these spores are smaller than those produced by most other *Botrytis* species found on alliums. The conidia of *B. allii* are readily produced in culture. In contrast, the Botrytis leaf blight pathogen *B. squamosa* does not readily sporulate in culture and has significantly larger conidia (14–24 x 9–18 μm). Another *Botrytis* pathogen, *B. porri*, is important on garlic and leek. *Botrytis byssoidea* is another species that causes a neck rot disease on onion. This pathogen is distinct from *B. allii*, and causes a disease known as gray mold neck rot.

Disease cycle

The pathogen is transmitted on seed and sets of onion and shallot. If seed is stored at sufficiently low temperatures, the fungus may survive on seed for more than three years. The pathogen spreads from the seed coat to seedling leaves while the seed coat remains attached to the cotyledon. *B. allii* infects these leaves and then sporulates when tissue senesces. During the growing season, symptoms may be limited on foliage. However, just prior to harvest the bulb onion foliage is topped or bent, creating wounds that are open to inoculum. While the neck tissue retains moisture, the pathogen grows down to the bulb where it causes scale rot. Soilborne sclerotia are another source of inoculum as these germinate to produce conidia. Sclerotia can survive for two years in buried debris.

Control

Use seed that does not have significant levels of the pathogen. Use disease-free sets that are grown from such seed. A 1% seed infection is regarded as an economic threshold. Treatment for seedborne *B. allii* has been very effective. Use heat treatments to control the pathogen in sets. Plant resistant types or cultivars of alliums. When topping onions at harvest, leave a relatively long neck (10 cm) that can be thoroughly dried before the fungus spreads down the neck and into the bulb. Bulb necks can be dried by using forced air drying or by harvesting and field drying the onions under warm, dry conditions. After drying, maintain stored commodities below 75% relative humidity and at 0–1° C. Fungicides applied during the growing season may provide some suppression of neck rot. Rotate with non-hosts so that soilborne sclerotia of *B. allii* are reduced in number.

References

Bertolini, P. and Tian, S. P. 1997. Effect of temperature of production of *Botrytis allii* conidia on their pathogenicity to harvested white onion bulbs. *Plant Pathology* 46:432–438.

Gladders, P., Carter, M. A., and Owen, W. F. 1994. Resistance to benomyl in *Botrytis allii* from shallots. *Plant Pathology* 43:410–411.

Maude, R. B. 1983. The correlation between seed-borne infection by *Botrytis allii* and neck rot development in store. *Seed Science and Technology* 11:829–834.

Maude, R. B. and Presly, A. H. 1977. Neck rot (*Botrytis allii*) of bulb onions. I. Seed-borne infection and its relationship to the disease in the onion crop. *Annals of Applied Biology* 86:163–180.

Maude, R. B. and Presly, A. H. 1977. Neck rot (*Botrytis allii*) of bulb onions. II. Seed-borne infection and its relationship to the disease in store and the effect of seed treatment. *Annals of Applied Biology* 86:181–188.

Maude, R. B., Shipway, M. R., Presly, A. H., and O'Connor, D. 1984. The effects of direct harvesting and drying systems on the incidence and control of neck rot (*Botrytis allii*) in onions. *Plant Pathology* 33:263–268.

Netzer, D. and Dishon, I. 1967. Selective media to distinguish between two *Botrytis* species on onion. *Phytopathology* 57:795–796.

Nielsen, K., Justensen, A. F., Jensen, D. F., and Yohalem, D. S. 2001. Universally primed polymerase chain reaction alleles and internal transcribed spacer restriction fragment length polymorphisms distinguish two serotypes in *Botrytis aclada* distinct from *Botrytis byssoidea*. *Phytopathology* 91:527–533.

Yohalem, D. S., Nielsen, K., and Nicolaisen, M. 2003. Taxonomic and nomenclatural clarification on the onion neck rotting *Botrytis* species. *Mycotaxon* 85:175–182.

Fusarium culmorum, F. oxysporum f. sp. *cepae, F. proliferatum*

FUSARIUM BASAL PLATE ROT

Introduction and significance

This disease is important worldwide in bulb onion, chives, garlic, and shallot. It is becoming increasingly important in the UK and on leek in the USA.

Symptoms and diagnostic features

Foliar symptoms develop at any stage during the growing season and include a general yellowing, necrosis of the leaves from the leaf tip downward, and wilting. There may be a tan to pink root rot on onion and red purple discoloration of the stems and bulbs of garlic. For all affected *Allium* species there is a rot of the basal plate tissue where roots are attached to the crown (**15**). Such a rot is initially water-soaked, light tan to darker brown, with tissues remaining firm. With further development, the basal infection turns into a soft rot, extends up into the fleshy scales, and causes the plant to collapse. Under humid conditions, a fluffy, white mycelium is produced on the affected tissues. Bulbs showing no obvious symptoms may still rot in storage. On leek, affected roots are initially gray and water-soaked in appearance and later become pink, soft, and rotted. In addition to the basal plate discoloration, leek can develop a tan to pink lesion on outer leaf sheaths in contact with soil (**16**).

Causal agents

Fusarium basal plate rot is caused by several species of the fungus *Fusarium*. *Fusarium oxysporum* f.sp. *cepae* causes basal plate rot on onion and *F. culmorum* is the causal agent of the same disease on garlic and leek. *Fusarium proliferatum* is reported to affect the bulbs of onion and garlic. *Fusarium oxysporum* f. sp. *cepae* has morphology and colony characteristics that are similar to other *F. oxysporum* fungi. The fungus forms one- or two-celled, oval to kidney shaped microconidia on monophialides, and four- to six-celled, fusiform, curved macroconidia. Macroconidia are usually produced in cushion-shaped structures called sporodochia. Chlamydospores are formed in culture. This pathogen is apparently host specific to *Allium* species.

Fusarium culmorum produces abundant macroconidia that are stout, thick walled, and have prominent septa. Macroconidia foot cells are indistinctly to slightly notched. Conidiophores are monophialides. Microconidia are absent and chlamydospores are present in culture. Colonies on potato dextrose agar produce abundant, dense, white aerial mycelia, and the undersurfaces of these cultures are distinctly carmine red.

Fusarium proliferatum produces mostly microconidia in culture. These are one- or two-celled, borne in long chains from polyphialides, and have slightly flattened bases because of their formation in the chains. Chlamydospores are not formed. Semi-selective media like Komada's medium can help isolate all *Fusarium* pathogens if secondary rot organisms are present.

15 Discolored basal plate of leek infected with *Fusarium culmorum*.

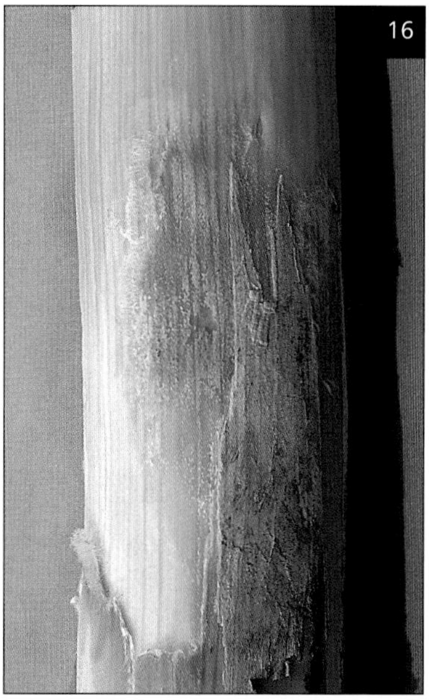

16 Stem lesion of leek infected with *Fusarium culmorum*.

Disease cycle

Fusarium basal plate rot is a soilborne disease; the pathogens persist in soil by means of chlamydospores. Pathogens may be carried on onion sets and garlic cloves. Seedborne inoculum may also exist: in California, leek transplants in a soilless medium in trays can develop extensive basal plate rot – circumstantial evidence of seedborne pathogen introduction. The optimum temperature for disease development is 25–28° C and few problems occur when soil temperatures are less than 15° C. These *Fusarium* species invade the roots and spread to the basal plate, but can also infect the bulb directly. Damage from insect feeding can increase problems. Likewise, *Fusarium* infection can attract pests and allow for extensive secondary decay from soil microorganisms. Wet weather close to harvest favors disease development, particularly in garlic.

Control

Use resistant cultivars, as this is the most important component in the control strategy. Rotate crops to non-hosts for at least 4 years to help reduce soilborne inoculum and lessen potential disease pressure. It would be desirable to use seed that is tested and found to not have detectable levels of the pathogen; however, seed assays for these pathogens are not yet fully developed. Plant only transplants which appear to be disease free. The storage of bulbs at temperatures below 4° C will reduce further development in storage. In some regions fungicides have been successfully used on seed or sets and as a dip treatment on seedlings.

References

Abawi, G. S. and Lorbeer, J. W. 1972. Several aspects of the ecology and pathology of *Fusarium oxysporum* f. sp. *cepae*. *Phytopathology* 62:870–876.

Armengol, J., Vicent, A., Sales, R., Garcia-Jimenez, J., and Rodriguez, J. M. 2001. First report of basal rot of leek caused by *Fusarium culmorum* in Spain. *Plant Disease* 85:679.

Dugan, F. M., Hellier, B. C., and Lupien, S. L. 2003. New Disease report. First report of *Fusarium proliferatum* causing rot of garlic bulbs in North America. *Plant Pathology* 52:426.

Everts, K. L., Schwartz, H. F., Epsky, N. D., and Capinera, J. L. 1985. Effects of maggots and wounding on occurrence of Fusarium basal rot of onions in Colorado. *Plant Disease* 69:878–882.

Koike, S. T., Gordon, T. R., and Aegerter, B. J. 2003. Root and basal rot of leek caused by *Fusarium culmorum* in California. *Plant Disease* 87:601.

Tamietti, G. and Garibaldi, A. 1977. Observations on basal rot of leek caused by *Fusarium culmorum*. *Rivista di Patologia Vegetale* 13:69–75.

Mycosphaerella allii (anamorph = *Cladosporium allii*),
M. allii-cepae (anamorph = *C. allii-cepae*)

CLADOSPORIUM LEAF BLOTCH

Introduction and significance

Cladosporium leaf blotch is a common, but generally minor, foliar disease of onion and leek. This disease emerged as a serious problem of onion in Ireland and southern England in the late 1970s and early 1980s when it caused significant loss of yield, but has been of limited importance since then. Onion and leek are infected by different *Cladosporium* species.

Symptoms and diagnostic features

The symptoms on onion foliage are conspicuous, large white spots or blotches that measure as large as 1.5 x 0.5 cm. These spots can resemble burn from herbicides or nitrogen fertilizers. These lesions take on a brown or dark brown color as the fungus sporulates on the affected tissue. The foliage collapses as lesions merge (**17**); foliage collapse is especially rapid when crops start to senesce. On leek, leaf lesions are large (up to 2.5 x 1.5 cm), elliptical white blotches that turn brown as the fungus sporulates (**18**) . For both leek and onion, lesions are difficult to distinguish from white tip and purple blotch, which can all occur on the same plant.

Causal agents

The cause of Cladosporium leaf blotch of **onion** is *Cladosporium allii-cepae*. Conidia are echinulate, usually one septate, cylindrical with rounded ends, slightly constricted in the middle, and measure 65–95 x 13–17 µm. The perfect stage, *Mycosphaerella allii-cepae*, has been produced in culture. This pathogen was formerly known as *Heterosporium allii-cepae*. *Cladosporium allii-cepae* also infects chives, shallot, garlic, and the weed crow garlic (*A. vineale*).

The cause of Cladosporium leaf blotch of **leek** is *Cladosporium allii*. Conidia are echinulate, usually one to two septate, cylindrical with rounded ends, slightly constricted in the middle, and measure 28–42 x 12–16 µm. The perfect stage is *Mycosphaerella allii*. Pseudothecia and bicellular ascospores (measuring 25–40 x 10–20 µm) are produced on host tissue. This pathogen was formerly known as *Heterosporium allii*. It primarily infects leek.

Disease cycle

Inoculum is mostly found on infested crop residues, though spores can move from overwintered to spring planted crops when they are in close proximity. For both species, optimum temperatures for spore germination and infection are 15–20° C, and germination requires 18 to 20 hours of 100% relative humidity. Germination can be slightly inhibited if free water is present. Leaf symptoms can be detected after two days under optimum conditions. Infection occurs more readily on damaged or senescing leaves. Interestingly, the production of conidia is sensitive to light and even moonlight can inhibit spore production. A minimum of 8 hours darkness, greater than 95% relative humidity, and temperatures of 9–12° C must occur for large numbers of conidia to develop. The conidia are released in the day, mostly during late morning to early afternoon. In England, most spore production, and hence most disease development, happens during October to April when relative humidity tends to be high. There are indications that these pathogens may be seedborne, though this aspect requires further investigation.

Control

Incorporate infected crop residues promptly after harvest, as survival in buried debris may be as short as 2 months. Rotate crops to non-hosts. Apply fungicides when necessary.

17 General dieback from *Cladosporium allii-cepae* on bulb onions.

18 Dieback symptoms caused by *Cladosporium allii* on leek.

References

Jordan, M. M., Burchill, R. T., and Maude, R. B. 1990. Epidemiology of *Cladosporium allii* and *Cladosporium allii-cepae*, leaf blotch pathogens of leek and onion. I. Production and release of conidia. *Annals of Applied Biology* 117:299–312.

Jordan, M. M., Burchill, R. T., and Maude, R. B. 1990. Epidemiology of *Cladosporium allii* and *Cladosporium allii-cepae*, leaf blotch pathogens of leek and onion. II. Infection of host plants. *Annals of Applied Biology* 117:313–336.

Jordan, M. M., Maude, R. B., and Burchill, R. T. 1990. Sources, survival and transmission of *Cladosporium allii* and *C. allii-cepae*, leaf blotch pathogens of leek and onion. *Plant Pathology* 39:237–242.

Kirk, P. M. and Crompton, J. G. 1984. Pathology and taxonomy of Cladosporium leaf blotch of onion (*Allium cepa*) and leek (*Allium porrum*). *Plant Pathology* 33:317–324.

Ryan, E. W. 1978. Leaf spot of onions caused by *Cladosporium allii-cepae*. *Plant Pathology* 27:200.

Penicillium hirsutum

PENICILLIUM MOLD, BLUE MOLD

Introduction and significance

This is mainly a storage disease, but field symptoms can occur on maturing bulbs. Blue mold can be an important problem on garlic.

Symptoms and diagnostic features

On garlic, Penicillium mold can cause the death of planted cloves prior to emergence. Infected cloves first show a water-soaked, irregularly shaped lesion that later becomes tan to light brown. The fungus can produce a blue-green, clustered mycelial growth in the lesions. In advanced stages the cloves are extensively rotted, and the blue-green sporulation is widespread on the affected tissues and around the base of young garlic plants. Sporulating *Penicillium* can be present between the scales of onion extending from the neck into the shoulder of the bulb (**19, 20, 21**) where it can cause a rot. If the pathogen infects commodities near harvest, the presence of sporulating blue-green mold is unsightly and can lead to crop rejection.

Causal agent

Penicillium mold is caused by *Penicillium hirsutum*, previously known as *P. corymbiferum*. Several other *Penicillium* species are regularly found on allium bulbs. On diseased tissue and in culture, *P. hirsutum* produces conidiophores that are shaped like a brush (penicillus) and bear long chains of dry, windborne conidia. Conidiophores are conspicuously roughened, form from surface hyphae or in fascicles, and have complex branching patterns (terverticillate or quaterverticilliate). In culture the colonies appear gray-green to dull green.

Disease cycle

For garlic, the fungus is carried on the cloves that are planted in the field. In addition, inoculum can be present on crop residue in the field; infection of bulbs takes place if there has been damage or growth splitting. The pathogen is spread during harvesting, particularly if there is damage to onion bulbs or garlic cloves. High temperatures and dry soil conditions favor disease development. The optimum conditions for pathogen development are temperatures of 21–25° C and high relative humidity.

19 Penicillium blue mold under the outer papery scale of bulb onion.

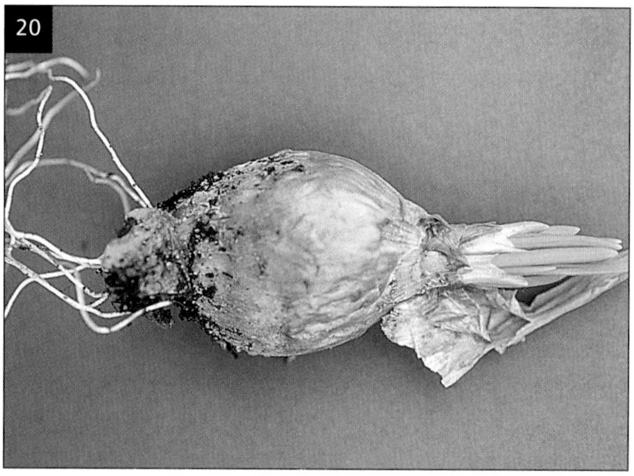

20 Penicillium blue mold on onion plant developing shortly after planting.

21 Penicillium blue mold on onion sets.

Control

Carefully handle bulbs and cloves during harvest to minimize damage. Avoid high humidity levels in storage. During storage or shipment, a combination of low temperatures (less than 5° C) and low relative humidity are required to restrict disease development.

References

Brammall, R. A. 1989. Resistance to benomyl in isolates of *Penicillium* sp. causing clove decay of garlic. *Canadian Journal of Plant Pathology* 11:409–414.

Peronospora destructor

DOWNY MILDEW

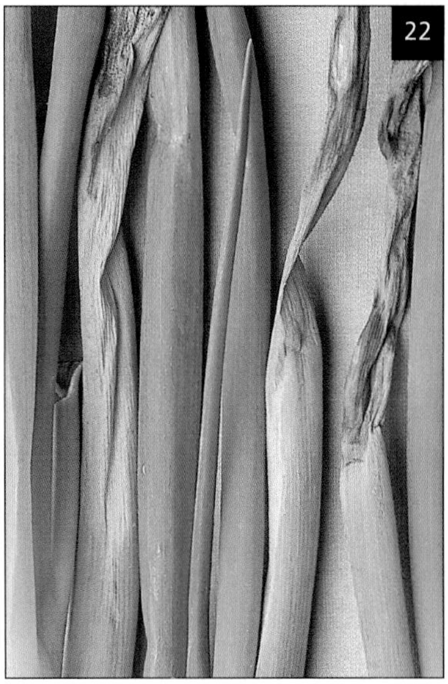

22 Onion leaves infected with downy mildew.

Introduction and significance

Downy mildew is a destructive disease of onion and reduces yield and quality of foliage and bulbs. Severe attacks occur sporadically because long periods of cool, wet weather are required for epidemics to develop. Problems with downy mildew have increased in recent years in the UK, notably since the increase in production of overwintered bulb onions. It is a difficult disease to control with fungicides, and improved disease management strategies are a priority for research.

23 Sporulation of downy mildew of onion.

Symptoms and diagnostic features.

Early symptoms of downy mildew are bleaching of leaf tips and small, irregularly shaped, chlorotic blotches on leaves. As disease develops, chlorotic blotches enlarge and coalesce into extensive lesions that can reach up to 10–15 cm long. These diseased areas then turn brown as they age (**22**). Characteristic purple, fuzzy sporulation grows on the affected lesions (**23**) and such growth often has concentric zones. Lesions can girdle the leaf, causing them to bend over and collapse (**24**). Lesions also occur on the flowering stalks, which likewise can become girdled and collapse. In major epidemics the entire crop can be defoliated in only four disease cycles. If downy mildew is severe during bulb formation, final bulb yields can be significantly reduced. Lesions can be colonized by *Stemphylium* fungi that produce conspicuous dark brown to black sporulation and sometimes purple pigmentation inside the downy mildew lesion. Plants may be systemically infected by downy mildew and affected bulbs become soft and discolored in storage; some sprout prematurely.

24 Bulb onions showing severe downy mildew.

Causal agent

Downy mildew is caused by the oomycete *Peronospora destructor*. This obligate pathogen produces thin walled sporangia that measure 40–72 x 18–29 µm. Sporangiophores emerge through stomata over a temperature range of 4–25° C. Globose oospores, measuring 30–44 µm in diameter, occur in onion bulbs, crop residues, and volunteer plants. *Peronospora destructor* is restricted to allium hosts and affects many wild and cultivated alliums, including onion, garlic, leek, chives, and shallot. *Allium cepa*, however, is the most seriously affected host.

Disease cycle

Infection requires cool temperatures (less than 22° C) and the presence of free moisture on the leaf for at least 3 hours. The optimum temperature for germination of sporangia is 10–12° C. Germ tubes form in 2 to 4 hours, produce an appressorium, and penetrate the leaf through stomata. Spores are produced at night and released by wind currents during the day, surviving on leaf surfaces for up to 3 days. Periods of dry weather will halt disease development. The pathogen overwinters via oospores and mycelium inside bulbs; oospores are also able to survive in soil. Colder winters appear to reduce the severity of downy mildew epidemics in the following season, presumably by reducing survival of the pathogen. The pathogen can be seedborne, though this factor is not considered important.

Control

Plant onion sets and transplants that are disease free. Heat treatment of sets may reduce viability of the pathogen inside infected sets. Some differences exist in cultivar susceptibility, so plant cultivars and *Allium* species that are less susceptible. Rotating onion crops 3 to 4 years may be helpful in reducing soil inoculum, though this strategy does not eliminate airborne sporangia. Incorporate crop residues soon after harvest. Control volunteer plants. Select fields that receive good air movement; sheltered locations should be avoided because these create conditions conducive to downy mildew development. Crop rows that are orientated parallel with the prevailing winds will also enhance drying of the crop environment. Apply fungicides at an early stage and prior to significant disease development. Protectants such as dithiocarbamates and copper have no curative activity to check established lesions, hence the systemic product mefenoxam is often used in combination with the protectants. Disease forecasting or decision support systems such as DOWNCAST and ZWIPERO are undergoing field evaluation.

References

Develash, R. K. and Sughai, S. K. 1997. Management of downy mildew (*Peronospora destructor*) of onion (*Allium cepa*). *Crop Protection* 16:63–67.

De Visser, C. L. M. 1998. Development of a downy mildew advisory model based on Downcast. *European Journal of Plant Pathology* 104:993–943.

Friedrich, S., Leinhos, G. M. E., and Löpmeier, F.-J. 2003. Development of ZWIPERO, a model forecasting sporulation and infection periods of onion downy mildew based on meteorological data. *European Journal of Plant Pathology* 109:35–45.

Gilles, T., Phelps, K., Clarkson, J. P., and Kennedy, R. 2004. Development of MILIONCAST, an improved model for predicting downy mildew sporulation on onions. *Plant Disease* 88:695–702.

Jesperson, G. D. and Sutton, J. C. 1987. Evaluation of a forecaster for downy mildew of onion (*Allium cepa* L.). *Crop Protection* 6:95–103.

Wright, P. J., Chynoweth, R. W., Beresford, R. M., and Henshall. W. R. 2002. Comparison of strategies for timing protective and curative fungicides for control of onion downy mildew (*Peronspora destructor*) in New Zealand. *Proceedings of the BCPC Conference – Pests & Diseases 2002* 1:207–212.

Phoma terrestris

PINK ROOT

Introduction and significance

Pink root is a severe disease in tropical and sub-tropical regions. This pathogen affects many crops, including eggplant, pepper, tomato, legumes, spinach, carrot, crucifers, and cucurbits. Even grain crops, such as maize, millet, and sorghum, are susceptible.

Symptoms and diagnostic features

Infected roots are pink in color; as disease develops, the pink coloration becomes more intense and can even turn dark purple (**25**). Roots are water-soaked and rotted. If allowed to dry out, such roots are fragile and papery in texture. New roots are affected as they emerge. Plants appear stunted and foliage may show symptoms of nutrient deficiencies and stress (**26**). Severely affected seedlings may die. The outer bulb scales may become infected and develop pink or red blemishes or water-soaked rots within 1 to 3 weeks of planting. Symptoms may resemble those of Fusarium basal plate rot.

25 Diseased onion roots infected with pink root.

Causal agent

Pink root is caused by *Phoma terrestris*. This soilborne fungus produces globose, dark brown to black pycnidia on infected roots or fleshy tissue. Setae grow on these pycnidia, especially around the ostiole, and measure 8–120 µm long. Conidia are ovoid, single-celled, hyaline, biguttulate, and measure 3.7–5.8 x 1.8–2.4 µm. The fungus produces chlamydospores. In older literature the pathogen is called *Pyrenochaeta terrestris*.

26 Onion plants infected with pink root.

Disease cycle

Pink root is a soilborne disease that is enhanced by warm soil temperatures (24–28° C). There is little infection when soil temperatures are below 16° C. The pathogen survives in crop residues or as chlamydospores in soil for only a few years. Invasion of the root occurs just behind the root tip, then spreads upward into the main body of the root.

Control

Rotate onion with non-host crops to reduce soil populations of the pathogen. An optimum crop rotation scheme allows at least 3 years between susceptible crops. Use resistant cultivars if such are available. Because of cooler soil temperatures present during autumn, plant at this time of year to reduce disease severity. Fumigation of soil may be justified on badly infested fields. Soil solarization can be effective in some regions, especially if combined with soil fumigation. Fungicide-treated seed and soil fungicide applications are not effective. There is some evidence that reducing the interval between irrigation applications has lessened the impact of pink root on yield.

References

Gorenz, A. M., Walker, J. C., and Larson, R. H. 1948. Morphology and taxonomy of the onion pink-root fungus. *Phytopathology* 38:831–840.

Levy, D. and Gornik, A. 1981. Tolerance of onions to the pink root disease caused by *Pyrenochaeta terrestris*. *Phytoparasitica* 9:51–57.

Porter, I. J., Merriman, P. R., and Keane, P. J. 1989. Integrated control of pink root (*Pyrenochaeta terrestris*) of onions by dazomet and soil solarization. *Australian Journal of Agricultural Research* 40:861–869.

Phytophthora porri

WHITE TIP

Introduction and significance

White tip is one of the most important foliar diseases on leek in western Europe. It has become more important in the UK with increasing leek production and loss of effective fungicides. The disease has been mainly reported from Europe, Canada, and Japan. *Phytophthora porri* affects various alliums, including bulb and green onion, and garlic. In Japan, losses of 70% or more can occur in onion crops.

Symptoms and diagnostic features

Early symptoms on leek leaves consist of irregularly shaped, water-soaked lesions. Lesions enlarge to several

27

27 Severe symptoms of white tip on leeks.

centimeters long, appear as elliptical blotches, and are usually most prevalent towards the tip of the leaf (**27**, **28**). Older lesions develop a bleached white center with a water-soaked margin when the disease is active (**29**). In dry conditions, the water-soaked margin may not be evident. Lesions are readily colonized by secondary fungi and sooty molds, making diagnosis difficult in the field because of symptom similarities with other foliar problems such as *Cladosporium* leaf blotch, purple blotch, or frost damage. Badly affected leaves rot and plants may be stunted or even killed. The disease spreads rapidly in cool, wet weather. On onion and garlic, this pathogen causes water-soaked leaf blight and root rot symptoms.

Causal agent

The cause of white tip is the oomycete organism *Phytophthora porri*. The pathogen produces sporangia that are usually non-papillate and measure 37–75 x 31–48 µm. Sporangia germinate directly to produce a germ tube or release zoospores that measure 10–12 µm. This pathogen can be observed in the lab by floating small pieces of infected leaf tissue, taken from the leading edge of lesions, in water for a few days. Microscopic examination of the leaf pieces will reveal the sporangia. Note that sporangia form sparsely or not at all when cultured on solid agar media. Oospores are also produced and measure 19–36 µm.

Earlier reports indicated that *P. porri* caused white tip of alliums and was also able to infect brassicas. However, researchers now know that *P. porri* is host specific to allium hosts and does not infect brassicas. *P. porri* tends to have paragynous (rather than amphigynous) antheridia, produces oogonia more consistently, and has fluffy aerial mycelium on V8 agar. The brassica pathogen has been given the name *P. brassicae*, produces large numbers of sporangia on solid agar medium, has a non-fluffy (appressed) appearance on agar media, and infects only brassicas.

Disease cycle

The disease is soilborne and infection can occur from sporangia or oospores when infested soil is splashed onto the leaves, or when leaves are in direct contact with the soil. In the autumn, severe infection of onion seedlings has occurred when heavy rains flatten the seedlings onto the ground. Mycelial growth can occur between 0–25 °C, with an optimum of 15–20 °C.

Control

Rotate onion and leek crops with at least a 3-year break with non-host crops. Use fungicides as appropriate, and apply products having different modes of action to discourage development of resistant strains. Avoid planting into wet or poorly draining sites. Prepare soils so that crops are planted into well-draining beds.

References

Griffin, M. J. and Jones, O. W. 1977. *Phytophthora porri* on autumn-sown salad onions. *Plant Pathology* 26:149–150.

Man in t'Veld, W. A., de Cock, A. W. A. M., Ilieva, E., and Lévesque, C. A. 2002. Gene flow analysis of *Phytophthora porri* reveals a new species: *Phytophthora brassicae* sp. nov. *European Journal of Plant Pathology* 108:51–62.

Smilde, W. D., van Nees, M., and Frinking, H. D. 1996. Effects of temperature on *Phytophthora porri* in vitro, in planta and in soil. *European Journal of Plant Pathology* 102:687–695.

28 Close-up of white tip disease of leek.

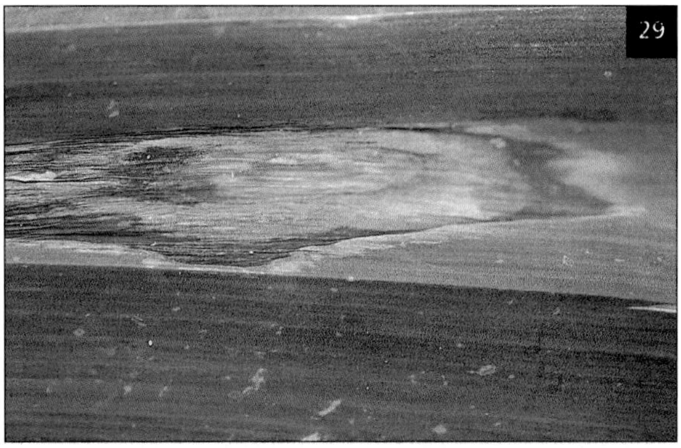

29 Leek leaf lesions of white tip disease showing water-soaked margins.

Puccinia allii

RUST

Introduction and significance

Rust diseases of allium crops occur wherever these crops are grown and are important production factors that cause significant crop damage to onion, leek, garlic, and chives. However, the rust pathogen is comprised of genetically distinct sub-groups; therefore, rust in one part of the world may or may not be the same pathogen as rust elsewhere. For example, rust is the most important and widespread disease of leek (**30**) in western Europe. Leek crops in California, though, are free from rust concerns while neighboring garlic plantings have been devastated by rust. Examination of the California garlic rust isolates demonstrated that these fungi are different morphologically and genetically from garlic rust isolates in the Middle East region. Hence *Puccinia allii* is a complex pathogen.

31 Leaf pustules of rust on onion.

Symptoms and diagnostic features

Initial symptoms consist of small, 1–2 mm in diameter, leaf flecks and spots that are irregularly shaped and white or light tan in color. These lesions increase to 3–5 mm spots that develop the typical, bright orange pustules (**31**, **32**) of rust diseases. Pustules erupt through the leaf surface between the veins, develop on both upper and lower leaf surfaces, and release copious amounts of dusty orange spores. There may be chlorotic halos around the pustules. Rust usually first occurs on the older foliage and subsequently spreads to newer leaves. Severely infected leaves can be almost covered with pustules, turn chlorotic, then finally become tan, dry, and dead (**33**). Reduced plant size and yield loss accompany severe rust. In garlic, rust causes severe stunting of plants and bulbs (**34**). At harvest the outer garlic bulb sheaths can be split and weak, resulting in shattering of the bulb and overall poor quality. Late in the infection cycle the teliospore phase can also be seen in darker brown pustules on diseased foliage (**35**).

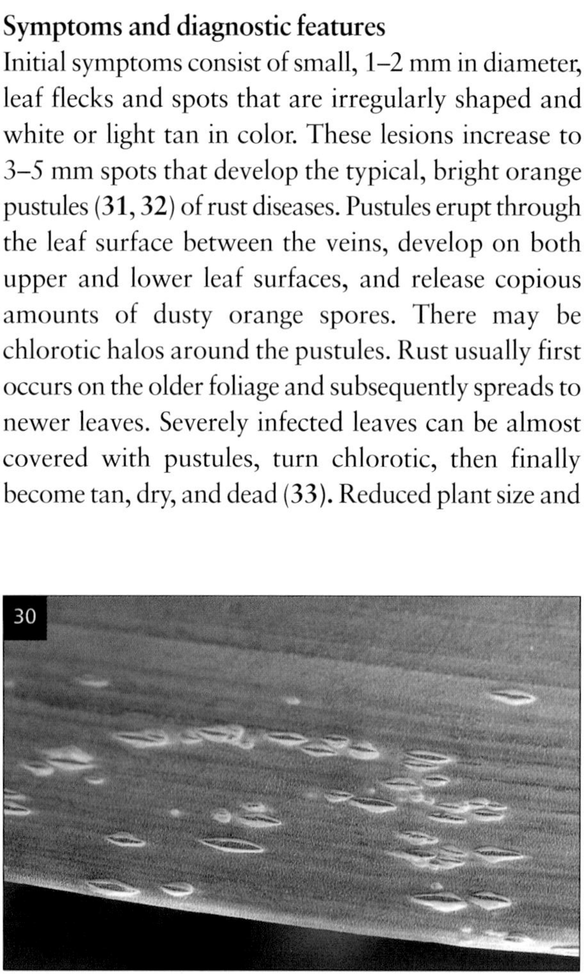

30 Rust uredinia on leek leaf.

Causal agent

Rust is caused by the fungus *P. allii*, which is an autoecious macrocyclic (full-cycled) rust. However, isolates with shortened life cycles, which lack pycnia and aecia stages, are common and referred to as having hemiform cycles. The prominent orange pustules are the uredinia stage that produces roughened, one-celled urediniospores measuring 23–29 x 20–24 µm. The dark brown, two-celled teliospore is produced later and measures 28–45 x 20–26 µm. In the UK, teliospores are uncommon on leek but readily occur on chives. In California, teliospores are very common on garlic and onion. Pycnia and aecidia stages occur on some hosts such as chives and *A. fistulosum*.

Because of variation in rust isolate host ranges and morphologies, over the years the allium rust pathogen has been classified into various genera and species, including the following: *P. allii*, *P. porri*, *P. mixta*, *P. blasdalei*, *Uromyces ambiguous*, or *U. duris*. Currently, it appears justified to group all garlic, chives, and leek isolates under *P. allii*, and to consider *P. allii* as a species

32 **Leaf pustules of garlic rust.**

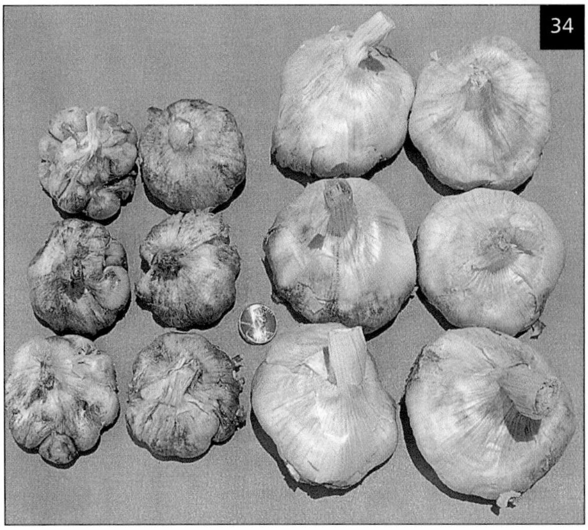

34 **Stunted garlic bulbs from severe rust. Healthy bulbs are on the right.**

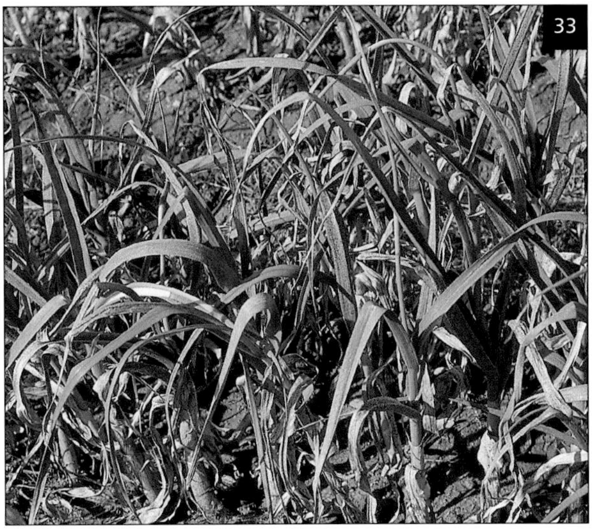

33 **Field symptoms of rust on garlic.**

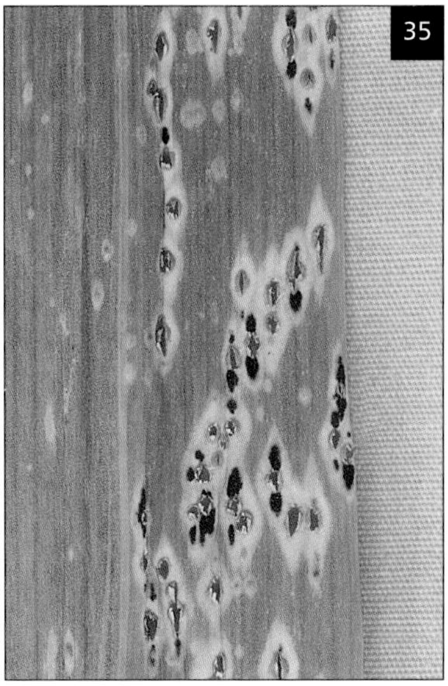

35 **Close-up of garlic leaf with uredinia and telia of rust.**

complex. Within this *P. allii* complex, physiological specialization does occur as leek isolates from Europe do not infect onion or chives, and the California garlic isolates do not infect leek.

Disease cycle

In Europe, overwintered crops are a significant source of rust inoculum for spring planted alliums. In the UK, rust is normally active from July onwards and is only inhibited by cold periods in winter. Weedy alliums can also be hosts and provide inoculum for commercial crops. Urediniospores are the primary inoculum and are spread long distances via winds. Optimum conditions for infection are 15° C and 100% relative humidity for 4 hours. The pathogen is active between temperatures of 10–24° C. Rust infection is favored by high nitrogen applications and low potash levels. The fungus does not survive in the soil. For the devastating garlic rust outbreaks in California in the late 1990s, a source of primary inoculum was not identified.

Control

Crop rotation to nonhosts may help reduce, but not eliminate, rust pressure. Control allium weeds and plow under infected crop residues to help reduce inoculum. Avoid planting consecutive host crops; for example, in some regions in Europe a single leek crop planted in the spring can often be grown with no or little rust. Apply fungicides, such as morpholine or triazole products, where rust occurs. Frequent applications are required to maintain good control in long season crops. For leek, plant resistant cultivars, which are widely used in Europe.

References

Anikster, Y., Szabo, L. J., Eilam, T., Manisterski, J., Koike, S. T., and Bushnell, W. R. 2004. Morphology, life cycle biology, and DNA sequence analysis of rust fungi on garlic and chives from California. *Phytopathology* 94:569–577.

Clarkson, J. and Kennedy, R. 1997. Quantifying the effect of reduced doses of propiconazole (Tilt) and initial disease incidence on leek rust development. *Plant Pathology* 46:952–963.

Harrison, J. M. 1987. Observations on the occurrence of telia of *Puccinia porri* on leeks in the UK. *Plant Pathology* 36:114–115.

Jennings, D. M., Ford-Lloyd, B. V., and Butler, G.M. 1990. Effect of leaf age, leaf position and leaf segment on infection of leek by leek rust. *Plant Pathology* 39:591–597.

Jennings, D. M., Ford-Lloyd, B. V., and Butler, G.M. 1990. Rust infections of some *Allium* species: an assessment of germplasm for utilizable rust resistance. *Euphytica* 49:99–109.

Koike, S. T. and Smith, R. F. 2001. First report of rust caused by *Puccinia allii* on wild garlic in California. *Plant Disease* 85:1290.

Koike, S. T., Smith, R. F., Davis, R. M., Nunez, J. J., and Voss, R. E. 2001. Characterization and control of garlic rust in California. *Plant Disease* 85:585–591.

Niks, R. E. and Butler, G. M. 1993. Evaluation of morphology of infection structures in distinguishing between different *Allium* rust fungi. *Netherlands Journal of Plant Pathology* 99, Supplement 3:139–149.

Smith, B. M., Crowther, T. C., Clarkson, J. P., and Trueman, L. 2000. Partial resistance to rust (*Puccinia allii*) in cultivated leek (*Allium ameloprasum* ssp. *porrum*): estimation and improvement. *Annals of Applied Biology* 137:43–51.

Uma, N. U. and Taylor, G. S. 1986. Occurrence and morphology of teliospores of *Puccinia allii* on leek in England. *Transactions of the British Mycological Society* 87:320–323.

Sclerotium cepivorum

WHITE ROT

Introduction and significance

This is a major root rot disease of alliums, with onion and garlic being particularly affected. White rot can also be a common problem in garden settings. In some regions white rot is an increasing problem due to intense crop rotations and shortage of uninfested land. In the USA and other regions, another *Sclerotium* species, *S. rolfsii*, causes southern blight disease on alliums (**36**).

Symptoms and diagnostic features

White rot affects roots and crowns, and overall symptoms are usually first noticed only after root infection is well established. Foliage of infected plants turns yellow, wilts, collapses, and eventually dies and becomes brown and dry (**36, 37**). In badly affected areas, foliage growth is poor and patches of plants die rapidly. White, persistent mycelium develops on diseased roots and the base of bulbs in contact with soil. Numerous tiny, black, spherical sclerotia, measuring less than 1 mm in diameter, form on mycelium and diseased tissue (**38**). In advanced stages of disease, the roots and bulbs become soft and rotted due to activity from secondary decay organisms. Symptoms on leek are usually less severe than on onion or garlic.

36 Onion infected with southern blight.

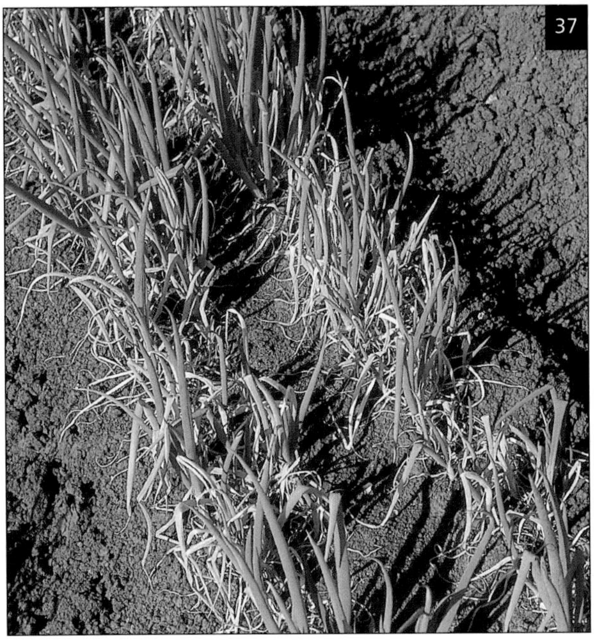

37 Onion field affected by white rot.

38 Garlic infected with white rot.

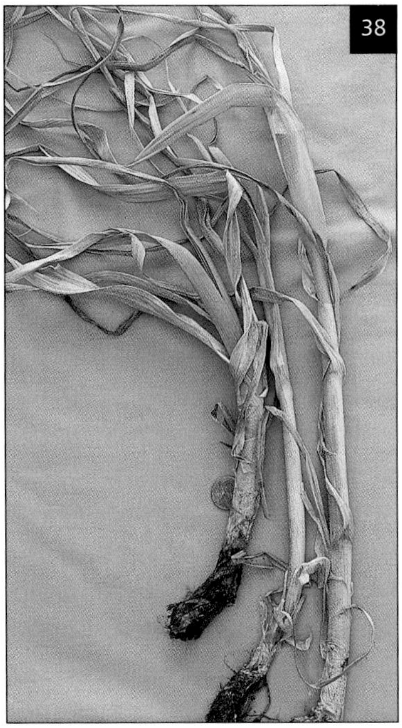

Causal agent

The cause of white rot is *Sclerotium cepivorum*, which has no known perfect stage. Conidia are likewise not produced, though small spermatia occur on germinating hyphae but appear to have no role in disease development. *Sclerotium cepivorum* reproduces, survives, and infects by sclerotia (**39**) that consist of a smooth black rind that is two to five cell layers deep and an inner medulla of closely packed hyphae. The pathogen can be readily isolated on solid media such as potato dextrose agar. The white rot fungus is host specific to allium crops. The other species, *S. rolfsii*, causes a watery rot of alliums and produces brown, spherical sclerotia that are usually significantly larger (1–2 mm in diameter) than those of *S. cepivorum*. *Sclerotium rolfsii* has a very broad host range.

Disease cycle

White rot is associated with soilborne inoculum, so affected areas and disease incidence increase as alliums are cropped in infested fields. Sclerotia can persist in soil without a plant host for over 20 years. These propagules remain dormant until allium root exudates, such as propyl and alyl cysteine amino acids that are specific to this group of plants, are present in soil. Soil microorganisms metabolize the amino acids into the stimulatory compounds alkyl and alkenyl thiols and sulphides. Sclerotia germinate and the resulting mycelium grows 1–2 cm through soil and invades the host root and basal plate. Secondary spread can occur by mycelial growth from plant to plant if their roots are in close proximity. Temperature is a key factor in disease development as sclerotia show little activity below 9° C or above 24° C; the optimum range is 14–18° C. Crops planted in the autumn may therefore

39 Sclerotia and mycelium of white rot on onion.

experience reduced disease severity. Sclerotia germinate under moist conditions (not below –100 kPa [= –1bar]), but germination is inhibited in very wet soils. Bulb onion crops established from transplants are more severely affected than direct seeded crops because of more vigorous root systems that develop earlier in the season and hence encounter more soilborne inoculum.

Control

Crop rotation is of limited value because sclerotia persist in soil for many years. Rather, select fields and plant crops where there is no history of the disease. Implement sanitation measures to prevent introduction of infested soil and contaminated equipment into clean fields. Use only disease-free transplants. Be aware that sclerotia survive passage through the digestive tracts of grazing animals and may therefore be present in manures. Truly resistant cultivars are not yet available. Soil fumigation with dazomet, metam sodium, methyl bromide, and chloropicrin on heavily infested soils can be useful treatments, though the fungus is not eradicated. Fungicides used as onion seed or garlic clove treatments or transplant drenches can be partially effective. Drenches with dicarboximide fungicides were initially useful, but enhanced microbial degradation of these compounds in some soils led to a decline in their performance. Triazole fungicides are now used for white rot control in some countries.

Soil inoculum can be reduced if sclerotia can be stimulated to germinate in the absence of host plants. Such germination is triggered by diallyl disulphide (DADS), which mimics the stimulatory activity of allium roots and is now commercially available in some countries. Effectiveness of this treatment relies on thorough application to soil and appropriate temperature and moisture conditions. Note that not all sclerotia respond to this treatment. Composted onion waste is also showing promise for white rot control in the UK when incorporated back into the field. Composting temperatures must reach 50° C to kill any sclerotia present in onion waste.

Though not completely effective, flooding of infested fields is used in some regions to manage white rot. This is most effective when temperatures are above 20° C. Solarization reduced sclerotial populations in Australia and Egypt and might form part of an integrated control strategy where climatic conditions are suitable.

References

Brix, H. D. and Zinkernagel, V. 1992. Screening for resistance of *Allium* species to *Sclerotium cepivorum* with special reference to non-stimulatory responses. *Plant Pathology* 41:308–316.

Coley Smith, J .R. 1986. Interactions between *Sclerotium cepivorum* and cultivars of onion, leek, garlic and *Allium fistulosum*. *Plant Pathology* 35:362–369.

Coley Smith, J. R. 1990. British Society for Plant Pathology Presidential Address 1989. White rot disease of *Allium*: problems of soil-borne disease in microcosm. *Plant Pathology* 39:214–222

Coley Smith, J. R. and Entwistle, A. R. 1988. Susceptibility of garlic to *Sclerotium cepivorum*. *Plant Pathology* 37:261–264.

Couch, B. C. and Kohn, L. M. 2000. Clonal spread of *Sclerotium cepivorum* in onion production with evidence of past recombination events. *Phytopathology* 90:514–521.

Crowe, F. J .and Hall, D. H. 1980. Vertical distribution of sclerotia of *Sclerotium cepivorum* and host root systems relative to white rot of onion and garlic. *Phytopathology* 70:70–73.

Entwistle, A. R. 1986. Controlling allium white rot (*Sclerotium cepivorum*) without chemicals. *Phytoparasitica* 20:121–125.

Entwistle, A. R. 1992. Loss of control of Allium white rot by fungicides and its implications. *Aspects of Applied Biology* 12:201–209.

Esler, G. and Coley Smith, J. R. 1983. Flavour and odour characteristics of species of *Allium* in relation to their capacity to stimulate germination of sclerotia of *Sclerotium cepivorum*. *Plant Pathology* 32 :13–22.

Gladders, P., Wafford, J. D., and Davies, J. M. L. 1987. Control of allium white rot in module raised bulb onions. In: T. J. Martin (ed.). Application to Seeds and Soil. *BCPC Monograph* No. 39:371–378.

Koike, S. T., Gonzales, T. G., and Oakes, E. D. 1994. Crown and root rot of chives in California caused by *Sclerotium rolfsii*. *Plant Disease* 78:208.

Leggett, M. E. and Rahe, J. E. 1985. Factors affecting the survival of sclerotia of *Sclerotium cepivorum* in the Fraser Valley of British Columbia. *Annals of Applied Biology* 106:255–263.

Melero-Vara, J. M., Prados-Ligero, A. M., and Basallotte-Ureba, M. J. 2000. Comparison of physical, chemical and biological methods of controlling garlic white rot. *European Journal of Plant Pathology* 106:581–588.

Smolinska, U. 2000. Survival of *Sclerotium cepivorum* sclerotia and *Fusarium oxysporum* chlamydospores in soil amended with cruciferous residues. *Journal of Phytopathology* 148:343–349.

Smolinska, U. and Horbowicz, M. 1999. Fungicidal activity of volatiles from selected cruciferous plants against resting propagules of soilborne fungal pathogens. *Journal of Phytopathology* 147:119–124.

Urocystis cepulae (= *U. colchici* var. *cepulae*)

SMUT

Introduction and significance

Smut occurs in many onion growing areas and is occasionally important. It affects bulb and salad onion, leek, shallot, and chives. Garlic appears to be immune and resistance is present in *Allium fistulosum*.

Symptoms and diagnostic features

Symptoms can appear as early as seedling emergence, when cotyledons show infections. Cotyledons, true leaves, and leaf sheaths develop oblong to elongated, dark, raised blisters on outer surfaces (**40**). These growths can cause downward curling of the leaves. Mature blisters break open to expose black, powdery fungal growth (**41**). There is progressive spread of the disease inwards, which can kill seedlings within 3 to 4 weeks. Infected bulb tissue remains firm, though secondary decay organisms can penetrate damaged areas and cause rot. Another disease, smudge (caused by *Colletotrichum circinans*), also produces black fungal growth on onion, leek, and shallot, so differentiating smut from smudge in the field may be difficult.

Causal agent

Smut is caused by *Urocystis cepulae* (synonymous with *Urocystis colchici* var. *cepulae*). It is a basidiomycete fungus belonging to the Ustilaginales. This pathogen produces distinctive black spore masses that consist of chlamydospores. These are spherical, single-celled, brown to black, 12–15 μm in diameter, with an outer layer of small sterile cells, 46 μm in diameter.

Disease cycle

The pathogen is soilborne and can survive for up to 20 years in soil. The pathogen may be introduced on sets or transplants. Seedborne infection is not considered important. Chlamydospores can be spread by winds and water, and optimum temperatures for germination are 13–22° C. Most plant infection occurs at 10–12° C and disease activity is greatly reduced above 25° C.

Control

Use fungicide-treated seed and resistant cultivars if available. Use healthy transplants because these are able to resist infection from soilborne inoculum. Plant when soil temperatures are higher.

40 Smut infection on green onion.

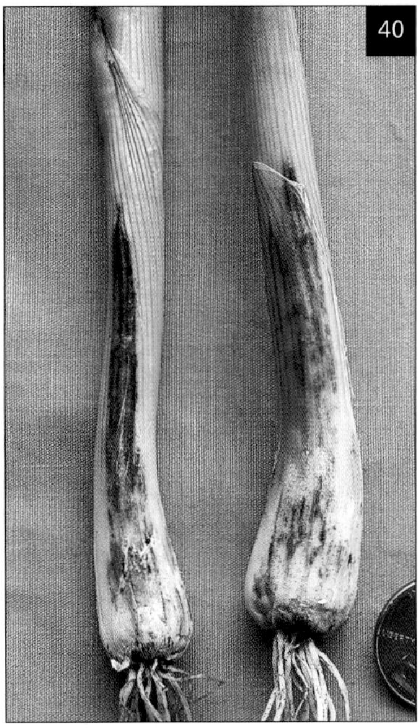

41 Black sporulation of smut on green onion.

References

Utkhede, R. S. and Rahe, J. E. 1980. Screening world onion germplasm collection and commercial cultivars for resistance to smut. *Canadian Journal of Plant Science* 60:157–161.

Garlic yellow stripe virus, Garlic yellow streak virus, Leek yellow stripe virus

GARLIC MOSAIC

Introduction and significance

Garlic mosaic is the common name of a virus disease that affects garlic. There are several viruses that infect garlic and cause this disease, and many of these pathogens are not yet fully identified or characterized.

Symptoms and diagnostic features

Foliar symptoms can vary greatly, but mostly consist of mild to severe mosaics, streaking, striping, and chlorotic mottling. Symptoms are often most evident in the youngest leaves. The overall effect is generally smaller bulb size and yield reductions of up to 50%.

Causal agent

Some of the viruses in the garlic mosaic complex are aphid-borne potyviruses that include *Garlic yellow streak virus, Leek yellow stripe virus*, and *Garlic yellow stripe virus*. The latter two viruses are both reported from California. Onion mite-borne latent virus is a rymovirus transmitted by the mite *Aceria tulipae* and is difficult to eliminate from stocks by meristem tip culture.

Disease cycle

Very little information is available on disease development of this problem.

Control

The fact that garlic is propagated by vegetative cloves makes virus management a challenge, and a high percentage of garlic cloves can be infected with one or more viruses. Control of all garlic viruses relies on production of virus-free stocks in areas well away from commercial production areas. Planting larger cloves may help maintain yield even when viruses are present in the cloves.

References

Conci, V. C., Canavelli, A., Lunello, P., Di Rienzo, J., Nome, S. F., Zumelzu, G., and Italia, R. 2003. Yield losses associated with virus-infected garlic plants during five successive years. *Plant Disease* 87:1411–1415.

Dovas, C. I., Hatziloukas, E., Salomon, R., Barg, E., Shiboleth, Y., and Katis, N. I. 2001. Incidence of viruses infecting *Allium* spp. in Greece. *European Journal of Plant Pathology* 107: 677–684.

Mohamed, N. A. and Young, B. R. 1981. Garlic yellow streak virus, a potyvirus infecting garlic in New Zealand. *Annals of Applied Biology* 97:65–74.

van Dijk, P., Verbeek, M., and Bos, L. 1991. Mite-borne virus isolates from cultivated *Allium* species, and their classification into two new rymoviruses in the family Potyviridae. *Netherlands Journal of Plant Pathology* 97:381–399.

Onion yellow dwarf virus

ONION YELLOW DWARF

Introduction and significance

Onion yellow dwarf virus (OYDV) is thought to be present in all production areas.

Symptoms and diagnostic features

Initial symptoms are yellow streaks at the base of the youngest leaves. All the new leaves are affected and result in a general yellowing. The leaves may be crinkled, flattened, and tend to fall over. Bulbs are small but remain firm. Flower stalks are yellow and twisted. In garlic, OYDV probably contributes to mosaic symptoms in combination with other viruses.

Causal agent

OYDV is a potyvirus with filamentous particles that measure 722–820 x 16 nm. Another allium virus pathogen is *Shallot latent virus*, a carlavirus first reported in the Netherlands in 1978. This virus has long filamentous particles (650–652 nm) and is transmitted by the shallot aphid (*Myzus ascalonicus*). Vegetatively propagated shallot is frequently affected. *Shallot latent virus* also occurs in onion, garlic, and leek.

Disease cycle

OYDV is spread by the green peach aphid (*Myzus persicae*) and other aphid vectors in a nonpersistent manner. The virus has a narrow host range within *Allium* species.

Control

Control is achieved by using healthy planting material. Clean stocks of garlic may be obtained by meristem tip culture and maintained by virus indexing of stocks. Healthy onion crops may be raised from seed because OYDV is not seed transmitted. Rotation away from allium crops and control of volunteers are required to break the cycle of virus spread between crops.

References

Bos, L. 1982. Viruses and virus diseases of *Allium* species. *Acta Horticulturae* 127:11–29.

Bos, L., Huttinga, H., and Maat, D. Z. 1978. Shallot latent virus, a new carlavirus. *Netherlands Journal of Plant Pathology* 84:227–237.

Conci, V. C. and Nome, S. F. 1991. Virus free garlic (*Allium sativum* L.) plants obtained by thermotherapy and meristem tip culture. *Journal of Phytopathology* 132:186–192.

Lot, H., Chovelon, V., Souche, S., and Delecole, B. 1998. Effects of onion yellow dwarf and leek yellow stripe viruses on symptomatology and yield from three French garlic cultivars. *Plant Disease* 82:1381–1385.

Shiboleth, Y. M., Gal-On, A., Koch, M., Rabinowitch, H. D., and Salomon, R. 2001. Molecular characterization of *Onion yellow dwarf virus* (OYDV) infecting garlic (*Allium sativum* L.) in Israel: Thermotherapy inhibits virus elimination by meristem tip culture. *Annals of Applied Biology* 138:187–195.

Walkey, D. G. A., Webb, M. J. W., Bolland, C. J., and Miller, A. 1987. Production of virus-free garlic (*Allium sativum* L.) and shallot (*A. ascalonicum* L.) by meristem-tip culture. *Journal of Horticultural Science* 62: 211–220.

Apiaceae Parsley family

PLANTS IN THE APIACEAE (parsley family), formerly known as the Umbelliferae, are characterized by distinctive umbrella-shaped flowering structures; they are a major group of vegetables with many uses. Carrot (*Daucus carota*) and parsnip (*Pastinaca sativa*) are primarily grown for their edible roots. Celery (*Apium graveolens* var. *dulce*) and fennel (*Foeniculum vulgare*) produce thick edible petioles, while celeriac (*Apium graveolens* var. *rapaceum*) is grown for its large storage root and hypocotyl. The foliage of other species is used as herbs, seasonings, garnish, and other culinary purposes: parsley (*Petroselinum crispum*), chervil (*Anthriscus cerefolium*), dill (*Anethum graveolens*), cilantro (also known as coriander or Chinese parsley) (*Coriandrum sativum*). The seeds of celery, fennel, coriander, anise (*Pimpinella anisum*), caraway (*Carum carvi*), and cumin (*Cuminum cyminum*)are also widely used. The different apiaciae plants suffer from generally distinct sets of diseases, as described in the sections that follow, though they also have some in common.

APIUM GRAVEOLENS (CELERY)

42 Stunted, chlorotic symptoms of aster yellows of celery.

Aster yellows phytoplasma

ASTER YELLOWS

Introduction and significance

Aster yellows occurs sporadically in Apiaceae crops, notably on celery and carrot. While symptoms can be unusual and striking, economic losses are rarely experienced.

Symptoms and diagnostic features

In celery, the foliar symptoms are yellowing, stunting, and overall poor growth (**42**). The heart of the plant sometimes becomes necrotic. The petioles are very brittle and there is peeling back of the epidermis and cracking on older plants. A characteristic symptom is the extreme twisting and curling of celery petioles (**43**, **44**). On carrot, parsnip, and parsley, aster yellows causes leaf chlorosis, proliferation of shoots from the crown, and abnormal greening (virescence) and development of leaf structures (phyllody) in flowers of seed crops.

Causal agent

Aster yellows disease is caused by the aster yellows phytoplasma. Phytoplasmas, like typical bacteria, are prokaryotes but are placed in a distinct category called mollicutes. Mollicutes are single-celled organisms that lack a cell wall, appear in various shapes (called pleomorphism), and have very small genomes. Phytoplasmas inhabit the phloem tissue of their host plants. This pathogen affects a very wide host range of cultivated and wild plants.

Disease cycle

The aster yellows phytoplasma is vectored primarily by the aster leafhopper (*Macrosteles fascifrons*), but many other species of leafhopper can also be vectors. These insects acquire the aster yellows phytoplasma by feeding on infected crops or weeds such as dandelion, plantain, sowthistle, and wild lettuce. After an incubation period, the insects can transmit the phytoplasma in a persistent manner. The phytoplasma reproduces by division or budding in the phloem sieve cells of host plants and within their leafhopper vectors. This pathogen is not seedborne.

Control

Control strategies are rarely needed. Remove inoculum sources by controlling weeds. Some areas such as pastures and river or stream banks naturally harbor the phytoplasma in resident vegetation; avoid planting celery and other susceptible crops in these areas. Using insecticides to control leafhoppers rarely affects aster yellows incidence.

References

Chiykowski, L. N. 1977. Transmission of a celery-infecting strain of aster yellows by the leafhopper *Aphrodes bicinctus*. *Phytopathology* 67:522–524.

Khadhair, A. H. and Evans, I. R. 2000. Molecular and microscopical detection of aster yellows phytoplasma associated with infected parsnip. *Microbiological Research* 155:53–57.

Zhang, J., Hogenhout, S. A., Nault, L. R., Hoy, C. W., and Miller, S. A. 2004. Molecular and symptom analyses of phytoplasma strains from lettuce reveal a diverse population. *Phytopathology* 94:842–849.

Zhou, X., Hoy, C. W., Miller, S. A., and Nault, L. R. 2002. Spacially explicit simulation of aster yellows epidemics and control on lettuce. *Ecological Modeling* 151:293–307.

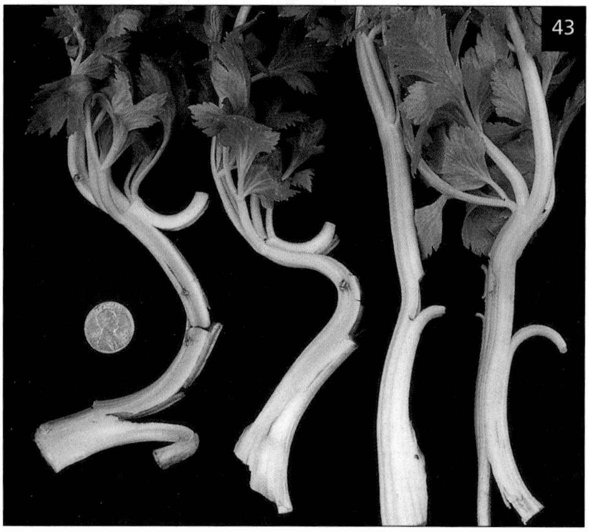

43 **Brittle, cracked petioles of celery infected with aster yellows.**

44 **Twisted petioles of aster yellows of celery.**

Pseudomonas syringae pv. *apii*

BACTERIAL LEAF SPOT

Introduction and significance

Although bacterial leaf spot has been recognized since 1921 in New York State, it remained a relatively minor disease until 1989 when it was first observed in California on greenhouse-grown transplants and field-planted celery. The disease was found in all celery-growing regions of California by 1991. Further research is required to clarify the relationship between the celery bacterial leaf spot pathogen and the *Pseudomonas* pathogens causing leaf spots on fennel and parsley. A similar disease caused by *Pseudomonas syringae* pv. *coriandricola* occurs on cilantro (coriander) in California, Florida, and parts of Europe.

Symptoms and diagnostic features

Initial symptoms are small (2–5 mm in diameter), angular shaped, water-soaked leaf spots that have a greasy appearance and appear on both sides of an infected leaf. As spots develop, they enlarge and can turn darker brown (**45**, **46**). Under dry conditions the lesions are light brown and papery. As disease progresses, lesions increase in number, coalesce, and can result in extensive leaf death. Plant vigor can be reduced, but plants are not killed. Bacterial leaf spot is strictly a leaf-spot disease. Symptoms can be confused with those of bacterial blight and brown stem caused by *P. cichorii*, but bacterial leaf spot does not cause symptoms on celery petioles.

45 Leaf spots of bacterial leaf spot of celery.

46 Close-up of leaf spots of bacterial leaf spot of celery.

Causal agent

Bacterial leaf spot is caused by *P. syringae* pv. *apii*, which is an aerobic, Gram-negative bacterium. The pathogen can be isolated on standard microbiological media and produces cream-colored colonies typical of most pseudomonads. When cultured on Kings medium B, this organism produces a diffusible pigment that fluoresces blue under ultraviolet light. Strains of this pathogen are host specific to celery. This pathogen is seedborne, which results in infection of celery transplants. Distinct races have not been documented.

Disease cycle

P. syringae pv. *apii* can survive on celery seed for at least 2 to 3 years; hence seed is the main source of primary inoculum. The pathogen survives epiphytically on leaves until environmental conditions allow populations to increase and infection to occur through stomata or wounds. Symptoms are most frequently found in transplants (**47**) having soft lush growth due to warm growing conditions, high humidity, and high nitrogen fertilization. Mowing of transplants to promote vigorous and uniform growth and irrigating with high-pressure systems readily spread the pathogen. Symptoms appear 7 to 10 days after long periods of leaf wetness (greater than 7 hours/day) over a 2- to 3-day period. In the field, disease activity is limited unless there is rain or overhead irrigation.

47 Celery transplants infected with bacterial leaf spot of celery.

Control

Use seed that does not have significant levels of the pathogen. Treat infested seed with hot water (50° C for 25 minutes). Practice careful sanitation at transplant nurseries. Disinfect benches, seed trays, mowers, and other production equipment. Prevent transfer of the pathogen on worker hands, clothing, and footwear by using sanitizing agents. Irrigate so that plants dry quickly after watering, and avoid using high-pressure irrigation systems that lead to water-soaking of leaves. Chemical control in the form of copper sprays is of limited benefit.

References

Koike, S. T., Gilbertson, R. L., and Little, E. L. 1993. A new bacterial disease of fennel in California. *Plant Disease* 77:319.

Lacy, M. L., Berger, R. D., Gilbertson, R. L., and Little, E. L. 1996. Current challenges in controlling diseases of celery. *Plant Disease* 80:1084–1091.

Little, E .L., Koike, S. T., and Gilbertson, R. L. 1997. Bacterial spot of celery in California: etiology, epidemiology, and role of contaminated seed. *Plant Disease* 81:892–896.

Pernezny, K., Datnoff, L., and Sommerfeld, M. L. 1994. Brown stem of celery caused by *Pseudomonas cichorii*. *Plant Disease* 78:917–919.

Thayer, P. L. 1965. Temperature effect on growth and pathogenicity to celery of *Pseudomonas apii* and *P. cichorii*. *Phytopathology* 55:1365.

Thayer, P. L. and Wehlburg, C. 1965. *Pseudomonas cichorii*, the cause of bacterial blight of celery in the Everglades. *Phytopathology* 55:554–557.

Cercospora apii

EARLY BLIGHT

Introduction and significance

This disease can cause considerable damage, and severely affected crops have been abandoned, particularly in the southeastern USA and Mexico. Early blight also affects celeriac.

Symptoms and diagnostic features

Early symptoms are small, yellow flecks that enlarge into round to oval, gray-brown spots that can measure from 1–2 cm in diameter (**48, 49**). Under humid conditions, lesions become light gray and fuzzy when the

48 Early blight symptoms on celery leaf.

49 Close-up of early blight of celery.

50 Severe blighting due to early blight of celery.

pathogen sporulates. The lesions have indistinct margins and do not contain discrete, dark fruiting structures as seen with late blight. Disease usually progresses from the oldest leaves to the newer foliage. Celery petioles also become infected and have elongated lesions. Severe infection results in necrosis and death of foliage and typical blight-like symptoms (**50**). Seedlings can exhibit symptoms. Field symptoms of early blight usually appear earlier in the season than those of late blight.

Causal agent

Early blight is caused by the fungus *Cercospora apii*. The pathogen produces hyaline, multicelled, filiform spores that measure 22–290 x 3.5–4.5 µm and are borne on clusters of brown conidiophores. *Cercospora apii* is seedborne in celery. This pathogen appears to be primarily a pathogen of celery and celeriac, though *C. apii* has been reported to infect lettuce and the ornamental plant bells-of-Ireland (*Moluccella laevis*). The relationship between these different isolates is unknown.

Disease cycle

Both seed and infected celery debris are important sources of inoculum. *Cercospora apii* can survive as mycelium on seed for more than 2 years. It is favored by warm (15–30° C) temperatures and long periods of leaf wetness or high relative humidity. Sporulation is greatest when there have been at least 10 hours of leaf wetness. Spores are released during the morning as relative humidity decreases and are subsequently dispersed by wind. Spores can also be spread by splashing water and via field operations. Spores germinate and penetrate through stomata after only 5 hours of leaf wetness. The disease cycle takes 5 to 14 days.

Control

Use seed that does not have significant levels of the pathogen. If transplants are still produced in the ground, rotate seedbeds to avoid soilborne inoculum. Irrigate early in the day to allow foliage to dry quickly. Plant resistant or tolerant cultivars if such are available. Use wider plant spacing and raised beds to improve air circulation. Apply fungicides if celery is produced in high-risk areas. A disease forecasting system has been developed to help schedule fungicide sprays.

References

Berger, R. D. 1969. A celery early blight spray program based on disease forecasting. *Florida State Horticultural Society Proceedings* 81:107–111.

Berger, R. D. 1973. Early blight of celery: analysis of disease spread in Florida. *Phytopathology* 63:1161–1165.

Koike, S. T., Tjosvold, S. A., Groenewald, J. Z., and Crous, P. W. 2003. First report of a leaf spot disease of bells-of-Ireland (*Moluccella laevis*) caused by *Cercospora apii* in California. *Plant Disease* 87:203.

Strandberg, J. O. and White, J. M. 1978. *Cercospora apii* damage of celery – Effects of plant spacing and growth on raised beds. *Phytopathology* 68:223–226.

Fusarium oxysporum f. sp. *apii*

FUSARIUM YELLOWS

Introduction and significance

Fusarium yellows is the most important soilborne disease of celery in North America and parts of Europe. Susceptible cultivars can be severely damaged and unharvestable. From the 1930s through 1959, Fusarium yellows was controlled by using resistant cultivars. However, race 2 of the pathogen subsequently occurred, overcame this resistance, and caused substantial damage. Race 2 is now established throughout the USA. Though race 2 tolerant cultivars are now available, the disease continues to be an important concern.

Symptoms and diagnostic features

Affected plants are severely stunted and foliage turns bright yellow(**51**). The internal vascular and pith tissue of crowns and roots takes on a dark red to brown discoloration (**52**). This vascular discoloration rarely extends far up into the attached petioles. As disease progresses, the central pith of the crown and root will break down, become colonized by secondary decay organisms, and turn soft and rotted. The pathogen infects large roots and causes them to become gray, water-soaked, and soft (**53**). Severely affected plants can collapse and die.

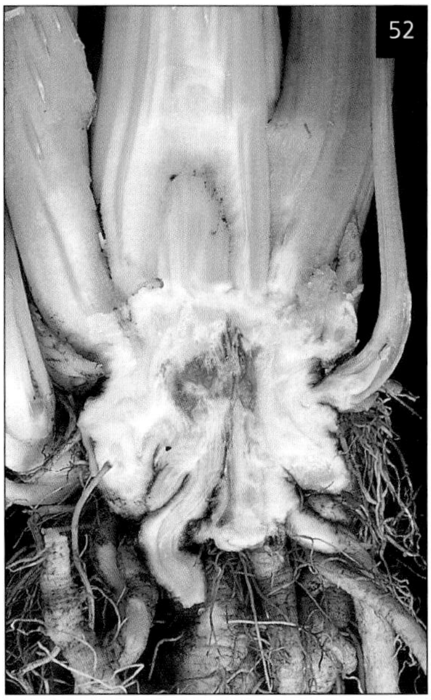

52 Discolored vascular tissue of Fusarium yellows of celery.

Causal agent

Fusarium yellows is caused by the soilborne fungus *Fusarium oxysporum* f.sp. *apii*. Pathogen morphology and colony characteristics are similar to other *F. oxysporum* fungi. The fungus forms one- or two-celled, oval to kidney-shaped microconidia on monophialides, and four- to six-celled, fusiform, curved macroconidia. Microconidia measure 5–12 x 2–4 µm,

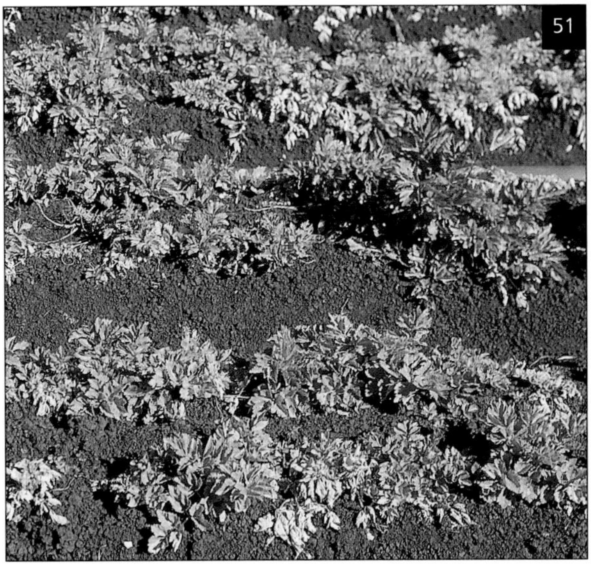

51 Decline and stunting of celery plants infected with Fusarium yellows.

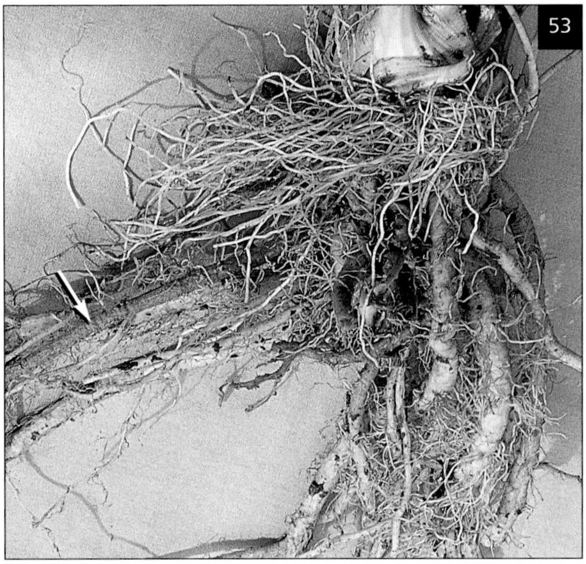

53 Dark, decayed celery root infected with Fusarium yellows.

while macroconidia range from 25–45 x 3–5 µm (four-celled) to 35-60 x 3–5 µm (six-celled). Macroconidia are usually produced in cushion shaped structures called sporodochia. Chlamydospores are also formed. The pathogen is usually readily isolated from symptomatic vascular tissue. Semi-selective media like Komada's medium can help isolate the pathogen if secondary rot organisms are present. This pathogen is apparently host specific to celery. Two distinct races have been documented. Race 1 affects only self-blanching cultivars; race 2 was first recorded in California in 1978 and affects both self-blanching and green cultivars. There is some diversity between race 2 isolates. For example, isolates from Ohio appear to be significantly more virulent than California isolates.

Disease cycle

The pathogen is soilborne and can survive for many years in soil as resilient chlamydospores. Chlamydospores germinate in response to celery root exudates and infect the roots. The pathogen then invades the vascular system and spreads through the plant. Fusarium yellows is favored by warmer temperatures (20–32° C) and dry soils. Under field conditions, symptoms can appear 30 to 40 days after planting. This disease is caused only by primary inoculum that is soilborne and is therefore considered a monocyclic disease.

Control

Do not plant susceptible celery cultivars into fields having a history of Fusarium yellows. Therefore, maintain good records and monitor where the disease occurs. Prevent introduction of the pathogen into uninfested fields by completing production procedures in clean fields before entering known infested areas. If equipment and vehicles have been in infested fields, clean such equipment prior to entering other areas. The most important management step is to use resistant or tolerant cultivars. Few celery cultivars have complete resistance to this pathogen, so tolerant cultivars may still develop some symptoms.

References

Awuah, R. T., Lorbeer, J. W., and Ellerbrock, L. A. 1986. Occurrence of Fusarium yellows of celery caused by *Fusarium oxysporum* f.sp. *apii* Race 2 in New York and its control. *Plant Disease* 70:1154–1158.

Cerkauskas, R. F. and Chiba, M. 1991. Soil densities of *Fusarium oxysporum* f.sp. *apii* race 2 in Ontario, and association between celery cultivar resistance and photocarcinogenic furocoumarins. *Canadian Journal of Plant Pathology* 13:305–314.

Correll, J. C., Puhalla, J. E., and Schneider, R. W. 1986. Identification of *Fusarium oxysporum* f.sp. *apii* on the basis of colony size, virulence, and vegetative compatibility. *Phytopathology* 76:396–400.

Elmer, W. H. and Lacy, M. L. 1987. Effects of inoculum densities of *Fusarium oxysporum* f.sp. *apii* in organic soil on disease expression in celery. *Plant Disease* 71:1086–1089.

Elmer, W. H., Lacy, M. L., and Honma, S. 1986. Evaluations of celery germ plasm for resistance to *Fusarium oxysporum* f. sp. *apii* Race 2 in Michigan. *Plant Disease* 70:416–419.

Elmer, W. H. and Lacy, M. L. 1987. Effects of crop residues and colonization of plant tissues on propagule survival and soil populations of *Fusarium oxysporum* f. sp. *apii* Race 2. *Phytopathology* 77:381–387.

Hart, L. P. and Endo, R. M. 1978. Reappearance of Fusarium yellows of celery in California. *Plant Disease Reporter* 62:138–142.

Lacy, M. L., Berger, R. D., Gilbertson, R. L., and Little, E. L. 1996. Current challenges in controlling diseases of celery. *Plant Disease* 80:1084–1091.

Opgenorth, D. C. and Endo, R. M. 1985. Abiotic factors and chlamydospore formation in *Fusarium oxysporum* f. sp. *apii*. *Transactions of the British Mycological Society* 84:740–742.

Orton, T. J., Durgan, M. E., and Hulbert, S. D. 1984. Studies on inheritance of resistance to *Fusarium oxysporum* f. sp. *apii* in celery. *Plant Disease* 68:547–578.

Puhalla, J .E.1984. Races of *Fusarium oxysporum* f. sp. *apii* in California and their genetic interrelationships. *Canadian Journal of Botany* 62:546–550.

Toth, K. F. and Lacy, M. L. 1991. Increasing resistance in celery to *Fusarium oxysporum* f. sp. *apii* with somaclonal variation. *Plant Disease* 75:1034–1037.

Phoma apiicola

PHOMA CROWN AND ROOT ROT

Introduction and significance

This disease, caused by *Phoma apiicola*, occurs in North America and Europe and affects celery, celeriac, and fennel. Carrot, caraway, parsley, and parsnip are reported to be susceptible to this pathogen, but are also are infected by *P. complanata*. Yet another species, *P. rostrupii*, has been reported on carrot from Russia and Ukraine, where it is an important seedborne disease and causes a dry brown root rot. Overall, Phoma crown and root rot is considered a minor disease, though losses occur both in the field and in storage. Phoma problems on celeriac are known as scab or rust, and recently have become more damaging in the UK.

Symptoms and diagnostic features

The roots and crowns of celery show light brown lesions that turn black and spread to the petiole bases. With decline of the petioles, yellowing and death of the outer leaves follows. Leaf spot symptoms are rarely seen. Early root infection results in stunted growth and possible plant death. On celeriac, the roots develop a dry brown rot. The presence of black pycnidia in rotted tissue is an important diagnostic feature, as symptoms can resemble those caused by licorice rot or black rot.

Causal agent

Phoma crown and root rot is caused by the fungus *P. apiicola*. On rotted tissues the pathogen produces black, spherical pycnidia (measuring up to 250 µm in diameter) that have cylindrical necks and produce rod-shaped, hyaline conidia that measure 3–14 x 1–2 µm.

Disease cycle

The pathogen is seedborne, so the disease can be initiated on transplants prior to being field planted. *Phoma apiicola* can survive in infested debris for several years. Conidia are splashed by rain or overhead sprinkler irrigation and infect plants directly or through wounds. The optimum temperature for activity is 16–18° C, but under wet conditions disease development can take place at much lower temperatures.

Control

Rotate crops for at least 2 years away from Apiaceae crops. Use seed that does not have significant levels of the pathogen. Fungicide sprays may give some control.

References

Hewett, P. D. 1970. *Phoma apiicola* on celery seed. *Plant Pathology* 20:123–126.

Williams, P. H. and Wade, E. K. 1967. Scab of celeriac. *Plant Disease Reporter* 51:427–429.

Pythium spp.

DAMPING-OFF, PYTHIUM ROOT ROT

Introduction and significance

Apiaceae are affected by various *Pythium* species throughout most production areas. Poor sanitation at nurseries and adverse growing conditions are usually responsible for the most severe problems. Disease severity differs depending on the particular region. In the UK, celery seedlings are very susceptible to damping-off, though in California this disease is rare.

Symptoms and diagnostic features

Poor emergence and collapse of seedlings soon after emergence are characteristic of damping-off. Infection can spread rapidly within transplant trays, resulting in patches of dead plants. Roots are initially water-soaked and later develop reddish-brown lesions (**54**). In the field, transplants can become infected from soilborne inoculum. In such situations, poor or stunted growth is the most obvious symptom. Affected plants have chlorotic foliage, brown roots, and reduced feeder roots

Causal agent

Damping-off and Pythium root rot are caused by several *Pythium* species, including the following: *P. artotrogus*, *P. debaryanum*, *P irregulare*, *P. mastophorum*, *P. paroecandrum*, *P. ultimum*. *Pythium paroecandrum* also causes a root rot of parsley. These oomycete organisms are common soil inhabitants that survive by means of resilient oospores. Most species produce sporangia, which mature and release infective swimming zoospores.

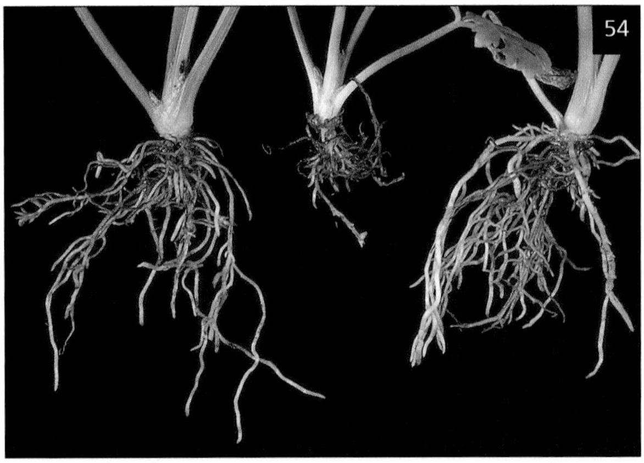

54 Dark, rotted celery roots affected by *Pythium*.

Disease cycle

Oospores germinate and grow in response to seed and root exudates. Infection and disease development require wet soil conditions that allow for zoospore movement and contact with host roots. Severe infection occurs in low lying or poorly drained areas. Disease symptoms can appear rapidly on small seedlings, but for larger transplants the foliar symptoms may not appear for several weeks. Established celery plants are generally not affected by these pathogens. The pathogen can contaminate trays, equipment, bench tops, and other surfaces and become problematic in transplant production systems.

Control

Practice thorough sanitation in greenhouse transplant environments. Regularly clean and disinfect transplant trays, planters, bench tops, and other equipment and containers that come in contact with transplants. Use pathogen-free rooting medium and irrigation water. Use soilless, peat-based rooting media. Set the growing conditions so that rapid germination and growth of celery transplants is enhanced. Avoid applying excess water and nutrients. In the field, rotate with non-host crops, though such strategies will not eradicate soil-borne pathogens such as *Pythium*. Place transplants in fields that are properly prepared and leveled and that drain well. In some production regions, such as the UK, fungicides may be required at planting.

Refrences

Hershman, D. E., Varney, E. H., and Johnston, S. A. 1986. Etiology of parsley damping-off and influence of temperature on disease development. *Plant Disease* 70:927–930.

McCracken, A. R. 1984. *Pythium paroecandrum* associated with a root rot of parsley. *Plant Pathology* 33:603–604.

Starr, J. L. and Aist, J. R. 1977. Early development of *Pythium polymorphon* on celery roots infected by *Meloidogyne hapla*. *Phytopathology* 67:497–501.

Vazquez, M .R., Davis, R. M., and Greathead, A. S. 1996. First report of *Pythium mastophorum* on celery in California. *Plant Disease* 80:709.

Sclerotinia sclerotiorum

PINK ROT

Introduction and significance

Pink rot is the name given to *Sclerotinia* disease on celery. *Sclerotinia* occurs wherever celery is grown and losses can be high. *Sclerotinia scerlotiorum* has a wide host range and can be difficult to manage. *Sclerotinia minor* also occurs on Apiaceae crops such as celery and fennel, but is less prevalent than *S. sclerotiorum*.

Symptoms and diagnostic features

Symptoms on celery can develop at any stage of growth and in storage. If celery transplants are grown in seedbeds in the ground, a watery rot can develop where plants are in contact with the soil. Infected seedlings soon collapse. The pathogen can grow from diseased seedlings to neighboring plants. Airborne ascospores can land on celery transplants grown in trays inside greenhouses and cause brown, foliar blighting and collapse of plants (**55**). Fluffy white mycelium and black sclerotia usually form on diseased plants (**56**).

For field planted celery, *Sclerotinia* species infect the base of the petioles and cause a soft, brown spreading lesion that usually has a pink color or border (**57**). Pink rot develops rapidly and plants soon rot and collapse. Another feature of this disease is a foliar blight that is initiated by airborne ascospores and results in brown, soft decay of foliage. Aerial infections often follow damage to young foliage caused by the calcium deficiency disorder known as blackheart. In most cases, *Sclerotinia* damage occurs as crops near maturity and the plant canopy becomes extensive. Pink rot may resemble bacterial soft rot (**58**); however, bacterial soft rot causes extensive water soaking of tissues, an extremely mushy decay of petioles, and a very strong disagreeable odor.

Causal agent

On celery, pink rot is primarily caused by *Sclerotinia sclerotiorum*, though the small sclerotial species, *S. minor*, can also infect celery. *Sclerotinia sclerotiorum* can infect the flowering stalks of parsley seed crops. Fennel is susceptible to both pathogens. For detailed descriptions of these pathogens, see the bean white mold section in the chapter on legume diseases.

Disease cycle

For detailed descriptions of the disease cycles, see the bean white mold section in the chapter on legume diseases. Ascospores of *S. sclerotiorum* require exogenous nutrients prior to infecting healthy tissues; this phenomenon is demonstrated on celery when the physiological condition blackheart predisposes foliage to infection.

Control

For an integrated approach to *Sclerotinia* control, see the bean white mold section in the chapter on legume diseases.

References

Budge, S. P. and Whipps, J. M. 1991. Glasshouse trials of *Coniothyrium minitans* and *Trichoderma* species for the biological control of *Sclerotinia sclerotiorum* in celery and lettuce. *Plant Pathology* 40:59–66.

Clarkson, J. P., Staveley, J., Phelps, K., Young, C. S., and Whipps, J .M. 2003. Ascospore release and survival in *Sclerotinia sclerotiorum*. *Mycological Research* 107:213–222.

Griffin, M. J. 1985. Evaluation of dicarboximide fungicides for control of pink rot (*Sclerotinia sclerotiorum*) of glasshouse celery. *Tests of Agrochemicals and Cultivars No. 6 (Annals of Applied Biology* 106, *Supplement)*:48–49.

Koike, S. T. 1999. Stem and crown rot of chervil, caused by *Sclerotinia sclerotiorum*, in California. *Plant Disease* 83:1177.

Koike, S. T., Daugovish, O., and Downer, J. A. 2006. Sclerotinia petiole and crown rot of celery, caused by *Sclerotinia minor*, in California. *Plant Disease* 90: In press.

Reyes, A. A. 1988. Suppression of *Sclerotinia sclerotiorum* and watery soft rot of celery by controlled atmosphere storage. *Plant Disease* 72:790–792.

56 Fungal growth on celery petioles caused by *Sclerotinia sclerotiorum*.

57 Celery petioles infected with *Sclerotinia sclerotiorum*.

55 Celery transplants infected by ascospores of *Sclerotinia sclerotiorum*.

58 Bacterial soft rot: Decayed petioles of soft rot of celery (compare with pink rot, above).

Septoria apiicola

LATE BLIGHT

Introduction and significance

Late blight is probably the most important foliar disease of celery worldwide and is capable of causing complete crop loss in extreme cases. Problems are usually related to seedborne infection and disease severity therefore varies greatly from season to season. This celery pathogen also affects celeriac. Septoria blight of parsley is caused by a different species, *S. petroselini*. Late blight of Apiaceae plants is unrelated to late blight disease, caused by *Phytophthora infestans*, of tomato and potato.

Symptoms and diagnostic features

Initial symptoms are small, chlorotic leaf spots that measure less than 5 mm in diameter. These lesions soon become brown and necrotic and contain small, discrete, black fruiting bodies (pycnidia) of the fungus (**59**). The leaf spots sometimes have chlorotic margins. Spots usually are angular in shape. In the field, older, lower foliage is infected first and the symptoms move up onto younger foliage with time. As the disease develops, leaf spots increase in number, the spots coalesce, and the entire leaf can turn brown and take on a blighted appearance. The petioles are also susceptible and develop grayish-brown, irregularly shaped lesions that harbor numerous pycnidia (**60**). Diseased petioles are not marketable and must be trimmed off the harvested celery. If crops are subject to rain or sprinkler irrigation, disease spread can be rapid and result in extensive damage to the crop. Because the pathogen is seedborne, symptoms can be found on transplants. Under humid conditions, spores ooze out of the pycnidia and form white, curled filaments on leaf surfaces. Symptoms on celeriac are similar to those on celery.

Causal agent

Late blight is caused by the fungus *Septoria apiicola*. On plant tissue the pathogen is readily identified by its black pycnidia that are immersed in plant tissues. Pycnidia are ostiolate, measure 75–195 µm in diameter, and contain hyaline, multicelled, filamentous conidia (22–56 x 2–2.5 µm). No perfect stage has been reported. The pathogen is seedborne. Celery and celeriac are not susceptible to the Septoria blight pathogen of parsley, *S. petroselini*.

Disease cycle

Infested seed is usually the most important source of the pathogen. The pathogen can survive for 2 to 3 years on seed when storage conditions are dry and cool (5 –15° C). Under ambient conditions, survival on seed appears to be about one year. Storage conditions suitable for celery seed also favor survival of *S. apiicola*. Spores penetrate through leaf stomata and epidermis, producing leaf spots in 7 to 8 days under optimum conditions. Disease severity increases with rising temperatures over the range of 5–25° C. Increasing leaf wetness duration, from 72 to 96 hours, also increases disease severity. Leaf wetness duration for at least 24 hours at 20° C provides optimal conditions for disease development. Severe epidemics are associated with prolonged periods of rain and season-long overhead sprinkler irrigation. Late blight development, as the name implies, occurs mostly after canopy closure and as crops near maturity.

59 Leaf infections of late blight of celery.

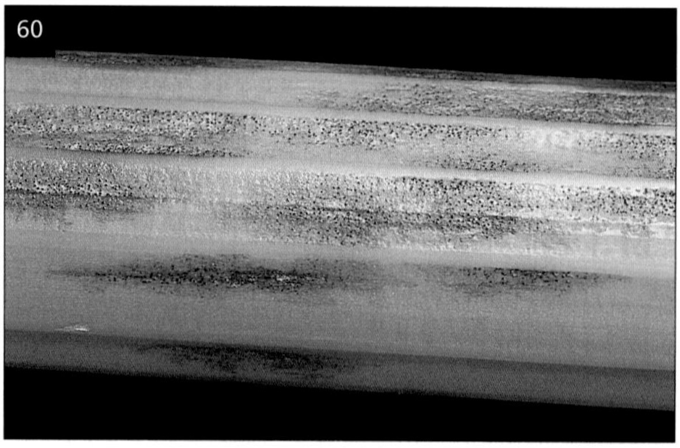

60 Petiole infections of late blight of celery.

Control

Use seed that does not have significant levels of the pathogen. Before planting seed for transplants, treat with hot water (30 minutes at 48–49° C), aerated steam, or fungicides. An effective chemical treatment is soaking seed in a 0.2% aqueous suspension of thiram for 24 hours at 30° C. Store infested seed for at least 2 years to significantly reduce viable seedborne inoculum. Inspect transplants and do not use any that show symptoms. Avoid planting consecutive celery crops so that diseased crop residue can break down. Destroy volunteer celery or celeriac plants. After transplants are established in the field, irrigate crops with furrow or drip irrigation. Do not walk through or drive equipment through diseased fields prior to entering healthy fields, as inoculum can adhere to clothing and equipment. Apply fungicides as preventative measures prior to plant infection. There is some progress in using predictive models to schedule sprays, and leaf wetness durations of 12 hours or more have been used as a threshold for treatment in North America. Resistant cultivars in commercial lines are not yet available.

References

Ataga, A. E., Epton, H. A. S., and Frost, R. R. 1999. Interaction of virus-infected celery and *Septoria apiicola*. *Plant Pathology* 48:620–626.

Edwards, S. J., Collin, H. A., and Isaac, S. 1997. The response of different celery genotypes to infection by *Septoria apiicola*. *Plant Pathology* 46:264–270.

Gabrielson, R. L. and Grogan, R. G. 1964. The celery late blight organism, *Septoria apiicola*. *Phytopathology* 54:1251–1257.

Green, K. R., O'Neill ,T. M., and Wilson, D. 2002. Effect of leaf wetness duration and temperature on the development of leaf spot (*Septoria apiicola*) on celery. *Proceedings of the BCPC Conference – Pests & Diseases 2002* 1:225–230.

Kavanagh, T. and Ryan, E. W. 1971. Methods of assessment of celery leaf spot (*Septoria apiicola*) in relation to fungicide evaluation. *Annals of Applied Biology* 68:263–270.

Lacy, M. L. 1994. Influence of wetness periods on infection of celery by *Septoria apiicola* and use in timing sprays for control. *Plant Disease* 78:975–979.

Mathieu, D. and Kushalappa, A. C. 1993. Effect of temperature and leaf wetness duration on the infection of celery by *Septoria apiicola*. *Phytopathology* 83:1036–1040.

Maude, R. B. and Shuring, C. G. 1970. The persistence of *Septoria apiicola* on diseased celery debris in soil. *Plant Pathology* 19:177–179.

Mudita, I. W. and Kushalappa, A. C. 1993. Ineffectiveness of the first fungicide application at different incidence levels to manage Septoria blight in celery. *Plant Disease* 77: 1081–1084.

Sheridan, J. E. 1966. Celery leaf spot: sources of inoculum. *Annals of Applied Biology* 57:75–81.

Thanatephorus cucumeris
(anamorph = *Rhizoctonia solani*)

CRATER SPOT

Introduction and significance

This disease affects the petioles of celery and parsley and causes disfiguring blemishes, which necessitates extra trimming at harvest. The disease occurs sporadically, but when it appears significant damage can result. This disease should not be confused with crater rot of carrot that is caused by *Rhizoctonia carotae*.

Symptoms and diagnostic features

Symptoms are confined to the lower portions of petioles near the soil and range from small (3–8 mm in diameter), pale brown water-soaked spots to large red-brown sunken lesions up to 5 cm in diameter (**61**). Lesions occur on both the outer and inner surfaces of the petiole; lesions on the outer petioles are more rectangular because of the prominent veins in the tissue that delimit the lesion (**62**). The disease produces a dry,

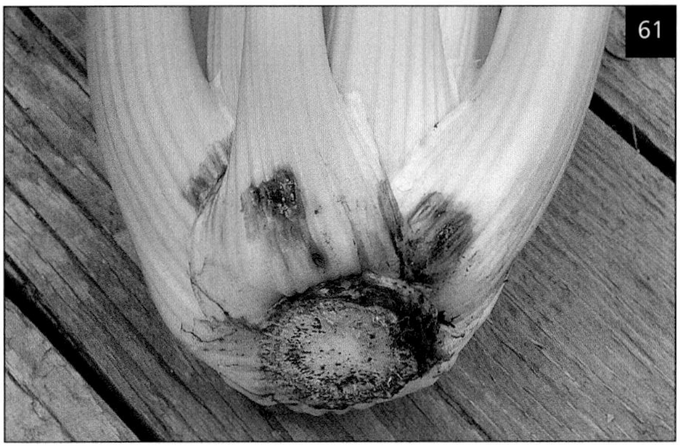

61 Crater spot lesions at base of celery plant.

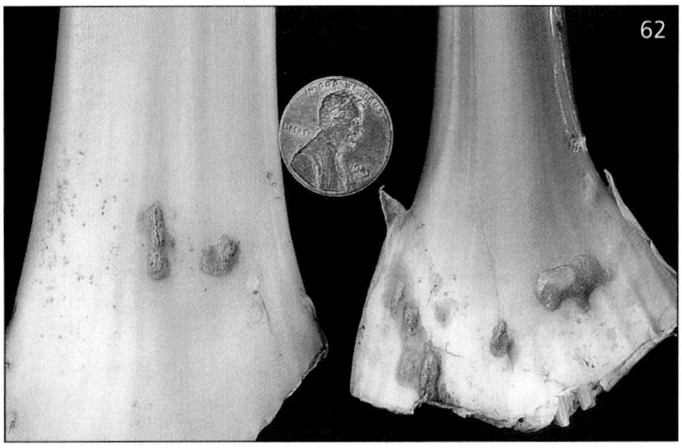

62 Close-up of crater spot lesions of celery.

firm rot, though secondary organisms sometimes colonize lesions and cause soft rots. Symptoms are usually detected close to harvest.

Causal agent

Crater spot is caused by the basidiomycete fungus *Thanatephorus cucumeris*. The anamorph stage, *R. solani*, is more commonly observed. This pathogen is a true soil inhabitant, survives in soil as a saprophyte, and has a very broad host range. For a description of *R. solani*, see the carrot section on crown rot in this chapter. It appears that some *R. solani* isolates from celery can infect diverse crops such as bean, cotton, and tomato; likewise, isolates from sugar beet and potato can cause crater spot. The white or gray white mat of mycelium of the teleomorph *T. cucumeris* can sometimes be seen on the lower parts of celery petioles.

Disease cycle

Rhizoctonia solani persists in organic matter and as sclerotia in soil. Infection is particularly favored by warm, moist soil conditions. However, in the UK problems can occur on late season crops when temperatures are cool and when cultivation moves soil into close contact with celery petioles. The *T. cucumeris* stage may not be involved in disease development.

Control

Do not plant celery in fields where there are fresh or partially decomposed crop residues that provide substrates for *R. solani*. Avoid planting transplants too deeply into the soil, as these are more likely to develop severe crater spot. Fields should be well drained and not kept overly wet. Avoid moving soil into and around the plant crowns during cultivation and weeding operations. Where problems occur regularly, apply fungicides. Rotate celery with non-host crops such as cereals to reduce soil populations of *R. solani*. However, short crop rotations will not eradicate populations of soilborne *R. solani*.

References

Houston, B. R. and Kendrick, J. B. 1949. A crater spot of celery petioles caused by *Rhizoctonia solani*. *Phytopathology* 39:470–474.

Pieczarka, D. J. 1981. Shallow planting and fungicide application to control Rhizoctonia stalk rot of celery. *Plant Disease* 65: 879–880.

Celery mosaic virus

CELERY MOSAIC

Introduction and significance

Celery mosaic virus (CeMV) occurs in production areas all over the world and can cause significant damage in extreme cases. Carrot, celeriac, celery, cilantro, dill, parsley, and parsnip are all affected.

Symptoms and diagnostic features

Younger leaves show vein clearing, chlorotic spotting, and interveinal mottling (**63, 64**). Older leaves exhibit narrowing, twisting, and cupping of the leaflets. The petioles have a green mottle or streaking, but do not have the necrotic spotting characteristic of *Cucumber mosaic virus* infection. Infection of young plants can cause stunting and horizontal, prostrate growth of the outer petioles. Crinkle strains cause raised green blisters and crinkling of leaves.

63 Celery leaves infected with celery mosaic virus.

64 Celery leaves infected with celery mosaic virus.

Causal agent

CeMV is a potyvirus and is made up of filamentous rods measuring approximately 750 x 15 nm and single-stranded RNA genomes. CeMV is transmitted in a non-persistent manner by over 20 different aphid species. Two distinct strains, common and crinkle leaf mosaic, are recognized.

Disease cycle

There are many wild Apiaceae hosts, including cow parsley (*Anthriscus sylvestris*), poison hemlock (*Conium maculata*), wild celery (*Apium graveolens*), and wild parsnip (*Pastinaca sativa*). The virus can be acquired by aphids after only 5–30 seconds of feeding, but infectivity is lost within 24 hours. Symptoms appear within days of inoculation. There is no seed transmission.

Control

Avoid planting celery in areas were the virus is known to cause significant disease. Control weed hosts and spray for the aphid vectors. Control in high-risk areas of the USA relies on an annual celery-free period of at least one month. Follow general suggestions for managing virus diseases (see Part 1).

References

D'Antonio, V., Falk, B., and Quiros, C. F. 2001. Inheritance of resistance to celery mosaic virus in celery. *Plant Disease* 85:1276–1277.

Purciful, D. E. and Shepard, J. F. 1967. Western celery mosaic virus in Florida celery. *Plant Disease Reporter* 51:502–505.

Sutabutra, T. and Campbell, R. N. 1971. Strains of celery mosaic virus from parsley and poison hemlock in California. *Plant Disease Reporter* 55:328–332.

65 Celery petioles infected with *Cucumber mosaic virus*.

66 Celery petioles infected with *Cucumber mosaic virus*.

Cucumber mosaic virus

CUCUMBER MOSAIC

Introduction and significance

Cucumber mosaic virus (CMV) occurs worldwide in many crops and weeds. Celery and other Apiaceae crops are infected only occasionally.

Symptoms and diagnostic features

Affected plants show outward and downward curling of young petioles that gives the center of plants a conspicuous flattened appearance. Leaves have vein clearing and dark green, thickened interveinal tissue that produces a crinkled effect. When young plants are infected, they may show leaf yellowing, together with vein necrosis and distinctive elongated, necrotic, or translucent sunken lesions on the petioles (**65, 66**).

Causal agent

CMV has one of the widest host ranges of any pathogen, with over 800 reported plant (crop and weed) hosts. CMV is a cucumovirus with virions that are isometric (29 nm in diameter) and contain three single stranded RNAs. Various strains exist, with the type strain and the calico strain occurring in celery.

Disease cycle

The virus is transmitted by many different aphid species in a non-persistent, stylet-borne manner. Aphids can acquire the virus in less than a minute and transmit it immediately, but they remain infective for only a few hours. CMV is not reported to be seedborne in celery.

Control

Control is difficult to achieve because of the numerous sources of CMV in other crops and weeds, such as chickweed (*Stellaria media*). Follow general suggestions for managing virus diseases (see Part 1).

References

Bruckart, W. L. and Lorbeer, J. W. 1976. Cucumber mosaic virus in weed hosts near commercial fields of lettuce and celery. *Phytopathology* 66:253–259.

Douine, L. and Devergne, J. C. 1979. Recensement des espèces végétables sensibles au virus de la mosaïque du concombre (CMV). Etudes bibliographiques. *Annales de Phytopathologie* 11:439–475.

Lord, K .M., Epton, H. A. S., and Frost, R. R. 1988. Virus infection and furcoumarins in celery. *Plant Pathology* 37:385–389.

Pemberton, A. W. and Frost, R. R. 1986. Virus diseases of celery in England. *Annals of Applied Biology* 108:319–331.

Tomlinson, J. A. and Carter, A. L. 1970. Studies on the seed transmission of cucumber mosaic virus in chickweed (*Stellaria media*) in relation to the ecology of the virus. *Annals of Applied Biology* 66:381–386.

67 Blackheart symptoms on young celery foliage.

BLACKHEART

Introduction and significance

Blackheart is a physiological disorder of celery that is similar to tipburn of lettuce, spinach, and other leafy vegetables. If environmental conditions favor development of this abiotic disorder, significant quality loss can take place. In addition, some pathogens such as *Sclerotinia sclerotiorum* readily colonize blackheart-affected tissue, causing extensive decay and further plant damage.

Symptoms and diagnostic features

Symptoms initially develop exclusively on the margins of developing leaf tips deep within the celery head. Such symptoms consist of light to dark brown speckling, lesions, and necrosis. As the plant grows, the affected foliage grows up and out of the inner plant whorl and becomes visible. Leaf tips continue to die back and the brown tissue expands into the rest of the leaflets (**67**).

Causal factor

Blackheart develops when the margins of young leaves contain low calcium levels. Conditions that favor rapid plant growth may trigger this disorder. Such predisposing conditions include warm temperatures and high fertilization rates. Ascospores of the pink rot pathogen (*Sclerotinia sclerotiorum*) can land on blackheart-affected foliage and cause a soft, brown rot having characteristic white mycelium and black sclerotia.

Control

Plant celery cultivars that are not prone to blackheart. Avoid applying excessive amounts of fertilizer, and irrigate with drip irrigation systems. Drip irrigation systems provide more even irrigation rates and help reduce blackheart risk. Foliar calcium supplements have limited benefit.

References

Gubbels, G. H. and Carolus, R. L. 1971. Influence of atmospheric moisture, ion balance, and ion concentration on growth, transpiration, and blackheart of celery. *Journal of the American Society for Horticultural Science* 96:201–204.

DAUCUS CAROTA (CARROT)

Xanthomonas campestris pv. *carotae*

BACTERIAL LEAF BLIGHT

Introduction and significance

This common bacterial disease occurs in most carrot producing areas. Damaging outbreaks are associated with high rainfall or use of overhead irrigation.

Symptoms and diagnostic features

The first symptoms of this disease are angular yellow leaf spots that later develop into irregularly shaped, brown, water-soaked spots with yellow halos (**68**). These lesions dry out and become brittle. Older lesions sometimes appear black. Lesions develop particularly at the leaf margins. Formation of a gummy exudate and browning of the petioles also occurs. The pathogen can cause lesions and a blight of the flower umbels and flower stalks; a yellow exudate can ooze from such lesions. Bacterial leaf blight symptoms may resemble those of other foliar diseases of carrot such as Alternaria leaf blight and Cercospora leaf blight. Examination by microscope and lab analysis are required to differentiate these various foliar blights.

68 Leaf spots of bacterial leaf blight of carrot.

Causal agent

Bacterial leaf blight is caused by the bacterium *Xanthomonas campestris* pv. *carotae*. The pathogen can be isolated on standard microbiological media and produces yellow, mucoid, slow growing colonies typical of most xanthomonads. The pathogen is host specific to carrot. Other bacterial problems of carrots are postharvest soft rots of roots caused by *Pseudomonas marginalis*, *P. viridiflava*, *Erwinia carotovora* ssp. *carotovora*, *E. chrysanthemi*, and other species.

Disease cycle

The pathogen is mostly dispersed between plants by splashing rain and overhead irrigation water, though some spread occurs by insects and passing animals and machinery. Considerable numbers of bacteria must be present on the leaf surface before symptoms occur. *Xanthomonas campestris* pv. *carotae* is seedborne but also survives in soil in association with infested carrot debris. The severity of epidemics is related to the amount of infested seed. Under Californian conditions, 1 x 10^7 colony forming units/g seed are required for severe epidemics to occur. Lower numbers of contaminated seed may still produce significant disease but would require especially favorable temperatures and rainfall. Optimum temperatures for disease development are 25–30° C.

Control

Use seed that does not have significant levels of the pathogen. Grow seed crops in dry regions without using overhead irrigation to reduce the risk of seedborne inoculum. Treat infected seed with hot water. Copper sprays may partially reduce disease severity if treatments are applied from an early stage of plant development. Rotate carrots with non-susceptible crops on a 2- or 3-year cycle to eliminate infested crop debris.

References

Cubeta, M. A. and Kuan, T. L. 1986. Comparison of MD5 and XCS media and development of MD5A medium for detecting *Xanthomonas campestris* pv. *carotae* in carrot seed. *Phytopathology* 76:1109.

Godfrey, S. A. C. and Marshall, J. W. 2002. Identification of cold-tolerant *Pseudomonas viridiflava* and *P. marginalis* causing severe carrot postharvest bacterial soft rot during refrigerated export from New Zealand. *Plant Pathology* 51:155–162.

Kuan, T.L., Minsavage, G. V., and Gabrielson, R. L. 1985. Detection of *Xanthomonas campestris* pv. *carotae* in carrot seed. *Plant Disease* 69:758–760.

Meng, X. Q., Umesh, K. C., Davis, R. M., and Gilbertson, R. L. 2004. Development of PCR-based assays for detecting *Xanthomonas campestris* pv. *carotae*, the carrot bacterial leaf blight pathogen, from different substrates. *Plant Disease* 88:1226–1234.

Umesh, K. C., Davis, R. M., and Gilbertson, R. L. 1998. Seed contamination thresholds for development of carrot bacterial blight caused by *Xanthomonas campestris* pv. *carotae*. *Plant Disease* 82:1271–1275.

Williford, R. E. and Schaad, N. W. 1984. Agar medium for selective isolation of *Xanthomonas campestris* pv. *carotae* from carrot seeds. *Phytopathology* 74:1142. Abstract.

Alternaria dauci

ALTERNARIA LEAF BLIGHT

Introduction and significance

Alternaria leaf blight is one of the most important foliar diseases of carrot and occurs worldwide. Severe epidemics reduce carrot root size and yields, and additional losses occur when weakened and diseased foliage cannot be pulled by top-lifting machines, leaving roots in the ground. In European regions with cool climates, serious epidemics only develop in seasons with above-average rainfall.

Symptoms and diagnostic features

Initial symptoms are greenish-brown, water-soaked, angular spots (**69**). These spots become dark brown to black and may have a yellow halo. Lesions often occur on or near the edge of older leaflets. Extensive spotting results in an overall general browning and yellowing of the entire leaf. As lesions enlarge and coalesce, the leaf may die. Severely affected crops exhibit large patches where the foliage has a scorched or blighted appearance. Dark, rectangular, elongated lesions are produced on the petioles. The pathogen also affects flowers, bracts, and developing seeds in carrot seed plants. Alternaria leaf blight symptoms may resemble those of other foliar diseases of carrot such as bacterial leaf blight and Cercospora leaf blight. Examination by microscope and lab analysis are required to differentiate these various foliar blights.

If carrot seedlings are infected shortly after emergence, as might be the case if inoculum is seedborne or soilborne, such seedlings can die. Affected seedlings have a gray or black rot of the upper root at soil level, similar to that caused by *Pythium* species. There are, however, few reports of infected roots in harvested, stored carrots. *Alternaria* pathogens of other Apiaceae crops include *A. petroselini* and *A. smyrnii* on celery and parsley, and *A. selini* on parsley.

Causal agent

Alternaria leaf blight is caused by *Alternaria dauci*. Conidia of *A. dauci* are brown and club-shaped. The main conidial body measures 50–100 x 12–24 μm, and has seven to eleven transverse septa and one or more longitudinal septa per segment. Each spore has one very long, septate, hyaline or pale brown, apical beak that measures 50–250 x 2–5μm.

69 Leaf spots of Alternaria leaf blight of carrot.

Disease cycle

Spores can germinate in 2 hours on wet leaves and infect via stomata. Sporulation and infection can occur over a broad temperature range (8–30° C), though the optimum is 25° C. Completion of the disease cycle from infection to sporulation requires only 8 to 10 days, which allows epidemics to develop rapidly. The pathogen is seedborne and is found in the seed mericarp and on the seed surface. It can survive on seed for several years and on plant debris for several months. Infested crop residues may be an important source of inoculum, especially if carrots are grown in successive years. Where adjacent crops are grown sequentially, there can be significant spread of wind-borne inocula from the older to the newer crop.

Control

Use seed that does not have significant levels of the pathogen. Where there is a low level of seedborne infection, treat carrot seed with hot water or fungicides. The development of fungicide-resistant strains has reduced the effectiveness of dicarboximide seed treatments. Because carrot cultivars differ in susceptibility, plant those having more tolerance or resistance. Reduce soilborne infection by rotating carrots with non-hosts. Avoid placing newer carrot fields adjacent to older, possibly diseased fields. Regularly monitor carrot crops for Alternaria leaf blight symptoms and apply fungicides in a timely manner. A predictive scheme for Alternaria leaf blight management has been developed in Canada. This system uses leaf-wetness duration and temperature to calculate infection indices. These indices indicate risk of slight, moderate, or severe infection. For example, severe infection is likely with greater than 72 hours leaf wetness at 7° C, or greater than 12 hours leaf wetness at 16–20° C. The use of infection indices will require validation for local conditions. Gibberellic acid has been used to stimulate carrot foliar growth and may be another means for managing Alternaria leaf blight.

References

Ben-Noon, E., Shtienberg, D., Shlevin, E., and Dinoor, A. 2003. Joint action of disease control measures: a case study of Alternaria leaf blight of carrot. *Phytopathology* 93:1320–1328.

Ben-Noon, E., Shtienberg, D., Shlevin, E., Vintal, H., and Dinoor, A. 2001. Optimization of chemical suppression of *Alternaria dauci*, the causal agent of Alternaria leaf blight of carrot. *Plant Disease* 85:1149–1156.

Chelkowski, J. and Visconti, A. (eds). 1992 .*Topics in Secondary Metabolism – Volume 3. Alternaria: Biology, Plant Diseases and Metabolites.* Elsevier, Amsterdam, 573pp.

Farrar, J. J., Pryor, B. A., and Davis, R. M. 2004. Alternaria diseases of carrot. *Plant Disease* 88:776–784.

Gillespie, T. J. and Sutton, J. C .1979. A predictive scheme for timing fungicide applications to control Alternaria leaf blight of carrots. *Canadian Journal of Plant Pathology* 1:95–99.

Maude, R. B. 1966. Studies on the etiology of black rot, *Stemphylium radicinum* (Meir, Drechsl & Eddy) Neerg., and leaf blight, *Alternaria dauci* (Kuhn) Groves & Skolko, on carrot crops; and on fungicide control of their seedborne infection phases. *Annals of Applied Biology* 57:83–93.

Maude, R. B., Drew, R. L. K., Gray, D., Petch, G. M., Bujalski, W., and Nienow, A. W. 1992. Strategies for control of seed-borne *Alternaria dauci* (leaf blight) of carrots in priming and process engineering systems. *Plant Pathology* 41:204–214.

Netzer, D. and Kenneth, R. G. 1969. Persistence and transmission of *Alternaria dauci* (Kuhn) Groves & Skolko in semi-arid conditions in Israel. *Annals of Applied Biology* 63:289–294.

Pryor, B. M., Strandberg, J. O., Davis, R. M., Nunez, J. J., and Gilbertson, R. L. 2002. Survival and persistence of *Alternaria dauci* in carrot cropping systems. *Plant Disease* 86:1115–1122.

Strandberg, J. O. 1977. Spore production and dispersal of *Alternaria dauci*. *Phytopathology* 67:1262–1266.

Strandberg, J. O. 1987. Isolation, storage and inoculum production for *Alternaria dauci*. *Phytopathology* 77:1008–1012.

Strandberg, J. O. 1988. Establishment of Alternaria leaf blight on carrots in controlled environments. *Plant Disease* 72: 522–526.

Strandberg, J. O. 1988. Detection of *Alternaria dauci* on carrot seeds. *Plant Disease* 72:531–534.

Wijnheijmer, E. H. M., Brandenburg,W. A., and Ter Borg, S. J. 1990. Interaction between wild and cultivated carrots (*Daucus carota* L.) in the Netherlands. *Euphytica* 40:147–154.

Zimmer, R.C. and McKeen, W. 1969. Interaction of light and temperature on sporulation of carrot foliage pathogen *Alternaria dauci*. *Phytopathology* 59:743–749.

Alternaria radicina

BLACK ROT

Introduction and significance

Although this disease is important as a storage disease of carrots, it also causes seedling damping-off, foliar and crown infection, and an umbel blight. Occasionally extensive crop loss occurs, though the importance of black rot has decreased where carrots are maintained and stored *in situ* in the field rather than harvested and kept in cold storage. Black rot is found in all the main carrot-production areas. The disease also affects other important Apiaceae crops such as parsnip.

Symptoms and diagnostic features

Initial symptoms are small, brown leaf spots having chlorotic halos that later enlarge to form black lesions. These lesions are similar to those caused by *Alternaria dauci*, though they tend to be more prominent and cause dark brown or black lesions at the base of the petiole. Emerging seedlings show weakening and blackening of the crowns and upper roots (**70**). In the field, crown symptoms usually consist of black lesions that extend below soil level (**71, 72, 73**). Secondary black lesions also develop on the taproot, particularly if there is any splitting or damage. In postharvest storage, *A. radicina* causes dry, black, sunken lesions that have sharply defined margins. Root-to-root spread can occur during prolonged storage.

70 Weakened carrot crowns caused by black rot.

72 Blackening of internal carrot crown and upper root tissue caused by black rot.

71 Diseased carrot crown affected by black rot.

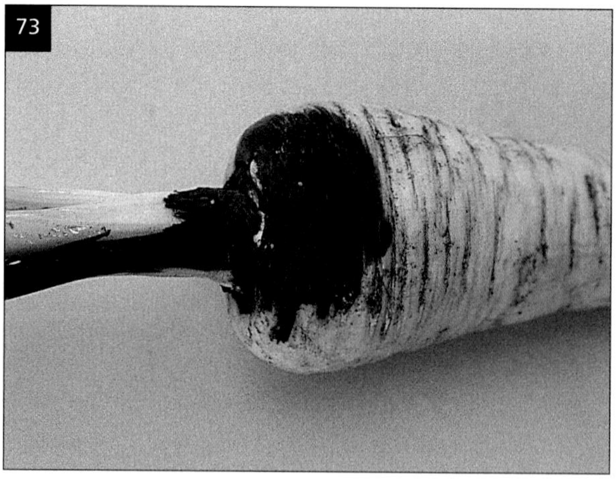

73 Black rot symptoms on parsnip.

Causal agent

Black rot is caused by the fungus *Alternaria radicina*. The spores of *A. radicina* are borne singly or in chains of two, are ellipsoid to oval, measure 20–50 x 10–25 µm, and have seven to eight lateral septa. Each spore may also have one or two longitudinal septa per segment. The conidia lack the long beaks of many other *Alternaria* species. *Alternaria radicina* also affects the crowns, petiole bases, and roots of celery and celeriac, and causes foliar symptoms in caraway, dill, fennel, parsley, and parsnip.

Disease cycle

Alternaria radicina is commonly seedborne and is found on the surface and in the pericarp and testa of seed. The pathogen survives in crop residues and also persists in the soil in the absence of debris for at least 8 years. *A. radicina* produces microsclerotia that function as survival structures. Disease development is favored by warm (greater than 20° C) temperatures and prolonged wet conditions. Growth of this pathogen occurs over the range -0.5–30 °C, but optimum temperature is 28 °C. Crops can be affected at any stage, but senescent leaves are most susceptible. Infections on petioles are followed by crown infections.

Control

Use seed that does not have significant levels of the pathogen. For low levels of seed infection, treat with hot water or fungicides. Plant resistant cultivars. Implement crop rotations to reduce soilborne inoculum. Foliar fungicides used for Alternaria leaf blight should have some effect on the spread of black rot, but specific recommendations for field use have not been developed. For stored crops, carefully handle carrots during harvesting and washing to avoid damage to roots. Maintain cool temperatures (0–1 °C) and high relative humidity during storage.

References

Benedict, W. G. 1977. Effect of soil temperature on the pathology of *Alternaria radicina* on carrots. *Canadian Journal of Botany* 55:1410–1418.

Chelkowski, J. and Visconti, A. (eds), 1992. *Topics in Secondary Metabolism – Volume 3*. Alternaria: Biology, Plant Diseases and Metabolites. Elsevier, Amsterdam, 573pp.

Farrar, J. J., Pryor, B. A., and Davis, R. M. 2004. Alternaria diseases of carrot. *Plant Disease* 88:776–784.

Maude, R. B. 1966. Studies on the etiology of black rot, *Stemphylium radicinum* (Meir, Drechsl & Eddy) Neerg., and leaf blight, *Alternaria dauci* (Kuhn) Groves & Skolko, on carrot crops; and on fungicide control of their seedborne infection phases. *Annals of Applied Biology* 57:83–93.

Pryor, B. M. and Gilbertson, R. L. 2001. A PCR-based assay for detection of *Alternaria radicina* on carrot seed. *Plant Disease* 85:18–23.

Pryor, B. M., Davis, R. M., and Gilbertson, R. L. 1994. Detection and eradication of *Alternaria radicina* on carrot seed. *Plant Disease* 78:452–456.

Pryor, B. M., Davis, R. M., and Gilbertson, R. L. 1998. Detection of soil-borne *Alternaria radicina* and its occurrence in California carrot fields. *Plant Disease* 82:891–895.

Pryor, B. M., Davis, R. M., and Gilbertson, R. L. 2000. A toothpick inoculation method for evaluating carrot cultivars for resistance to *Alternaria radicina*. *HortScience* 35:1099–1102.

Athelia arachnoidea
(anamorph = *Rhizoctonia carotae)*

CRATER ROT

Introduction and significance

This problem is a storage disease of carrots, but little is known about its activity in fields. Most records of this disease are from western Europe and North America. This disease should not be confused with crater spot of celery that is caused by *Rhizoctonia solani.*

Symptoms and diagnostic features

Typical symptoms develop after several weeks of low temperature storage and consist of small, 5–10 mm in diameter, sunken lesions or pits on the carrot surface. Over time the root develops patches of white, flattened mycelium (**74**). Brown sclerotia may also form on the stored carrot. Secondary bacterial rots and gray mold may develop on affected roots and lead to general breakdown and soft rot of roots.

74 Crater rot of stored carrots showing white fungal growth.

Causal agent

Crater rot is caused by the fungus *Rhizoctonia carotae*. The hyphae of *R. carotae* are hyaline and narrower (4–5 µm) than most other *Rhizoctonia* spp. and have clamp connections at most septa. Indeed, because of these morphological characters, this stage of the pathogen is not considered a true *Rhizoctonia* species. The fungus forms sclerotia that initially are white but later mature and are brown, measure 1–3 mm diameter, and consist of aggregates of mycelium. The sclerotia are produced in clusters on the surface of the carrot. The role of the basidiomycete perfect stage, *Athelia arachnoidea*, in disease development is not known.

Disease cycle

Mycelium spreads from root to root even in cold storage (1–3° C). High humidity and free moisture favors disease development. The pathogen is soilborne and some infection is thought to occur in the field prior to harvest. Mycelium on the crown or roots is thought to initiate infection during cold storage. The pathogen is also able to survive in wooden storage bins.

Control

Carefully handle and wash carrots prior to cold storage. Maintain sufficiently cold temperatures, but avoid overly wet storage conditions. Disinfest storage containers; line with polyethylene liners. Careful monitoring of carrots in storage will allow problems to be detected at an early stage and affected crops to be removed before serious problems develop.

References

Adams, G. C. and Kropp, B. R. 1996. *Athelia arachnoidea*, the sexual state of *Rhizoctonia carotae*, a pathogen of carrot in cold storage. *Mycologia* 88:459–472.

Derbyshire, D. M. and Crisp, A. F. 1978. Studies on treatments to prolong storage life of carrots. *Experimental Horticulture* 30:23–28.

Jensen, A. 1971. Storage diseases of carrots, especially Rhizoctonia crater rot. *Acta Horticulturae* 20:125–129.

Punja, Z. K. 1987. Mycelial growth and pathogenesis by *Rhizoctonia carotae* on carrot. *Canadian Journal of Plant Pathology* 9:24–31.

Ricker, M. D. and Punja, Z. K. 1991. Influence of fungicide and chemical salt dip treatments on crater rot caused by *Rhizoctonia carotae* in long term storage. *Plant Disease* 75:470–474.

Athelia rolfsii (anamorph = *Sclerotium rolfsii*)

SOUTHERN BLIGHT

Introduction and significance

Southern blight is important on a wide range of cultivated plants, including Apiaceae vegetables such as carrot, celery, parsley, and parsnip. It is most damaging in tropical and sub-tropical regions. Southern blight is of limited importance in southern Europe and is not established in the UK.

Symptoms and diagnostic features

This pathogen can cause water-soaked leaf lesions if foliage is in contact with the soil, leading to yellowing and wilting of foliage. Direct infection of the carrot root is more important and results in a pale brown, soft rot of the taproot. In advanced stages of the disease, the woody core of the root may be pulled out of the ground or the whole root may rot away, leaving a hole in the ground that is lined with mycelium and root tissue. White stringy mycelium forms on infected roots (**75**) and also on adjacent soil. Small (1–2 mm diameter), brown, spherical sclerotia appear on infected tissues and mycelium. Sclerotia are produced in 5 to 6 days when temperatures are optimal (27–30° C). Plant-to-plant spread occurs when the mycelium grows over the soil surface and contacts adjacent plants.

75 Mycelium and sclerotia on carrot root infected with southern blight.

Causal agent

Southern blight is caused by the soilborne pathogen *Athelia rolfsii*. This basidiomycete fungus is usually present in the field in its anamorph stage, *Sclerotium rolfsii*; the role of the basidiomycete stage in disease development is not clear. *Sclerotium rolfsii* produces abundant white mycelium, and hyphae have clamp connections. The diagnostic spherical, tan to light brown sclerotia form readily in culture. Basidiospores may be produced by some isolates, are hyaline and measure 6–12 x 1.0–1.7 µm.

Disease cycle

The pathogen is soilborne and can survive for many years in soil as sclerotia. *S. rolfsii* is a soil saprophyte and can colonize organic substrates. Disease occurs when temperatures are above 15° C. Wet conditions and acid soils are conducive to disease development.

Control

Implement crop rotation, though this strategy will be only partially effective because of the wide host range and soilborne nature of this pathogen. Growing vegetables during the cooler months may reduce damage. Fungicides and soil amendments are generally not very effective. Soil fumigation and heat treatments can be effective and may be affordable for high value crops.

References

Gurkin, R. S. and Jenkins, S. F. 1985. Influence of cultural practices, fungicides, and inoculum placement on southern blight and Rhizoctonia crown rot of carrot. *Plant Disease* 69:477–481.

Okabe, I. and Matsumoto, N. 2003. Phylogenetic relationship of *Sclerotium rolfsii* (teleomorph *Athelia rolfsii*) and *S. delphinii* based on ITS sequences. *Mycological Research* 107:164–168.

Punja, Z. K. 1985. The biology, ecology and control of *Sclerotium rolfsii*. *Annual Review of Phytopathology* 23:97–127.

Punja, Z. K., Carter, J. D., Campbell, G. M., and Rossell, E. L. 1986. Effects of calcium and nitrogen fertilizers, fungicides, and tillage practices on incidence of *Sclerotium rolfsii* on processing carrots. *Plant Disease* 70:819–824.

Cercospora carotae

CERCOSPORA LEAF BLIGHT

Introduction and significance

This is one of the more important foliar diseases of carrots, particularly in parts of North America. Cercospora leaf blight was first confirmed in the UK in 2005. Yield losses occur because of early damage to foliage and when weakened and diseased foliage cannot be pulled by top-lifting machines, leaving carrot roots in the ground.

Symptoms and diagnostic features

The first symptoms are small necrotic leaf flecks that measure 1–3 mm in diameter and tend to be angular in shape (**76**). These small flecks enlarge to form gray to tan spots (3–5 mm diameter) having chlorotic halos. As lesions increase in number and coalesce, leaves can wither and die. Petiole lesions are elliptical and brown with a paler center (**77**). Severely blighted foliage is weakened and snaps off during mechanical harvesting. In seed crops, early infection may prevent seed development, while later infection results in infested seed. The pathogen does not infect carrot roots. Cercospora leaf blight symptoms may resemble those of other foliar diseases of carrot such as bacterial leaf blight and Alternaria leaf blight. Examination by microscope and lab analysis are required to differentiate these various foliar blights.

Causal agent

Cercospora leaf blight is caused by the fungus *Cercospora carotae*. The pathogen produces hyaline, multicelled, filiform spores that measure 40–110 x 2.2–2.5 µm and are borne on clusters of brown conidiophores. Sporulation on the lesions causes these spots to take on a pale gray appearance. This pathogen appears to be host specific to carrot and other *Daucus* species.

76 Leaf spots of Cercospora leaf blight of carrot.

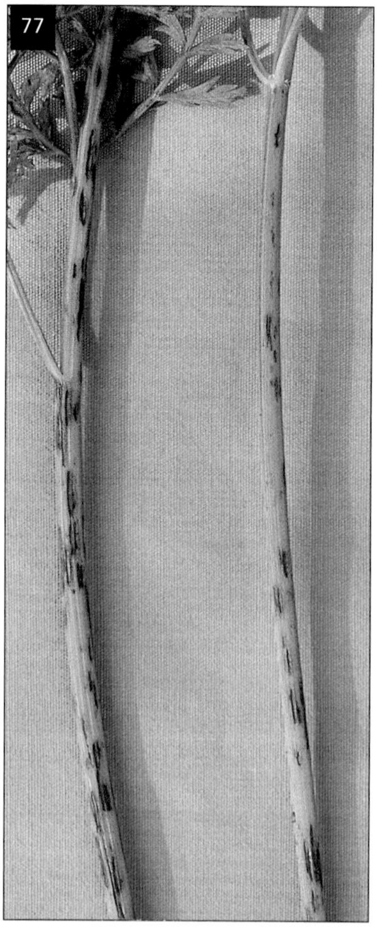

77 Petiole lesions of Cercospora leaf blight of carrot.

Disease cycle

The pathogen grows over the temperature range 7–37° C, with an optimum of 19–28° C. Spore germination requires a minimum 12 hours of leaf wetness at 20–30° C. The optimum temperature for infection is 28° C. Spores penetrate leaves via stomata and symptoms appear in as few as three days under ideal conditions, usually occuring first on the younger foliage. Young leaves of late planted crops are particularly susceptible to inoculum from older nearby crops. Seed, infested carrot debris, and wild *Daucus* plants are potential inoculum sources. Spores are airborne and also dispersed by rain or overhead irrigation.

Control

Use seed that does not have significant levels of the pathogen. Rotate carrots so that infected crop residues decay. Plant resistant cultivars. Use fungicides if available. Research indicates that fungicide treatments should be initiated when 50% of the middle leaves are affected; additional applications are needed at 7–10-day intervals if the temperature exceeds 16° C and leaf wetness exceeds 12 hours.

References

Carisse, O. and Kushalappa, A. C. 1992. Influence of interrupted wet periods, relative humidity, and temperature on infection of carrots by *Cercospora carotae*. *Phytopathology* 82: 602–606.

Carisse, O., Kushalappa A. C., and Cloutier, D. D. 1993. Influence of temperature, leaf wetness, and high relative humidity duration on sporulation of *Cercospora carotae* on carrot leaves. *Phytopathology* 83:338–343.

Carisse, O. and Kushalappa, A. C. 1990. Development of an infection model for *Cercospora carotae* on carrot based on temperature and leaf wetness duration. *Phytopathology* 80:1233–1238.

Kushalappa, A. C., Boivin, G. and Brodeur, L. 1989. Forecasting incidence thresholds of Cercospora blight in carrots to initiate fungicide application. *Plant Disease* 73:979–983.

Erysiphe heraclei, Leveillula spp.

POWDERY MILDEW

Introduction and significance

Powdery mildews are commonly found on many Apiaceae plants. *Erysiphe heraclei* infects carrot, celery, chervil, dill, parsley, and parsnip (see also pages 118 and 126). *Leveillula lanuginosa* or *L. taurica* has been reported from warmer regions in the Mediterranean, Asia, and Africa on carrot, caraway, celery, coriander, dill, fennel, and parsley. In western Europe, attacks show considerable seasonal variation and are often of limited importance. In Mediterranean countries, early attacks are more important and can cause significant yield losses.

Symptoms and diagnostic features

The first signs of powdery mildew caused by *E. heraclei* usually appear on the older leaves and consist of scattered white colonies having superficial, radiating fungal growth typical of most powdery mildews (**78**). Subsequently the younger leaves become affected and general colonization of the leaf surface ensues. Severe infection causes some leaf twisting and deformity and early senescence of the foliage. For powdery mildew caused by *Leveillula* species, symptoms are angular, chlorotic leaf spots. White sporulation usually occurs mainly on the corresponding undersides of these spots.

78 Powdery mildew of carrot infected by *Erysiphe heraclei*.

Causal agents

Erysiphe heraclei is an ectophytic powdery mildew with cylindrical conidia measuring 25–45 x 12–21 μm and borne singly. Cleistothecia are not always observed, but when present measure 80–120 μm, have long coralloid appendages, and enclose asci with three to five ascospores. Each ascospore is 18–30 x 10–16 μm. In contrast, *Leveillula* species are endophytic powdery mildews. Because the *Leveillula* perfect stage is sometimes not observed, the fungus is often called by its conidial anamorph name (*Oidiopsis*). *Leveillula lanuginosa* has cylindrical conidia measuring 40–80 x 13–20 μm with distinctive swellings near each end of the spore. Cleistothecia of this species measure 170–250 μm and have short hyaline or yellow appendages and numerous asci with two ascospores (30–35 x 15–20 μm). *Leveillula taurica* forms lanceolate primary conidia (30–80 x 12–22 μm) and more cylindrical secondary conidia; its cleistothecia are 140–250 μm in diameter with short hyaline or light brown appendages and contain asci with two ascospores (25–40 x 15–23 μm).

Disease cycle

The disease cycle can be completed in 7 days under optimum conditions. High humidity and moderate temperatures favor infection. Crops become more susceptible with age and carrots become more susceptible 7 weeks after planting. Disease severity is increased by drought stress and is reduced by rain or overhead irrigation. The *Leveillula* pathogens are mainly associated with dryer production areas. If cleistothecia are produced, these structures enable pathogen survival between crops. The cleistothecia have been recorded as contaminants in carrot seed, but it is not known if seed transmission occurs. There may be some host specialization, but isolates appear to be able to infect a number of host species. Wild and volunteer Apiaceae species and weeds are potential sources of conidial inoculum. Growing powdery mildew hosts throughout the year enables the pathogen to survive and spread from crop to crop.

Control

Maintain good crop growth through adequate fertilizer and irrigation programs. Mulching may reduce drought stress and likewise reduce powdery mildew severity. Use resistant cultivars when available. Fungicides such as sulfur or triazole products may be effective.

References

Citrulli, M. 1975. The powdery mildew of parsley caused by *Leveillula lanuginosa*. *Phytopathologia Mediterranea* 14:94–99.

Koike, S. T. and Saenz, G. S. 1994. Occurrence of powdery mildew on parsley in California. *Plant Disease* 78:1219.

Koike, S. T. and Saenz, G. S. 1997. First report of powdery mildew caused by *Erysiphe heraclei* on celery in North America. *Plant Disease* 81:231.

Netzer, D. and Katzir, R. 1966. Combined control of Alternaria blight and powdery mildew in carrot. *Plant Disease Reporter* 50:594–595.

Palti, J. 1975. Erysiphaceae affecting umbelliferous crops, with special reference to carrot, in Israel. *Phytopathologia Mediterranea* 14:87–93.

Soylu, E. M. and Soylu, S. 2003. First report of powder mildew caused by *Erysiphe heraclei* on dill (*Anethum graveolens*) in Turkey. *Plant Pathology* 52:423.

Helicobasidium brebissonii
(anamorph = *Rhizoctonia crocorum*)

VIOLET ROOT ROT

Introduction and significance

This disease occurs worldwide and is an important problem of carrot in Europe, New Zealand, Tasmania, and parts of North America. The pathogen has a wide host range and affects other major crops such as sugar beet, potato, and asparagus. Of vegetable root crops, carrot is perhaps the most susceptible to violet root rot. Other Apiaceae vegetables can be affected and include parsley, parsnip, and celeriac in the UK. Affected roots are unmarketable and even slight infection can induce an unpleasant flavor to the commodity.

Symptoms and diagnostic features

Patches of severely affected carrots may appear in early autumn when the foliage turns yellow and wilts. Dense, felted, purple to white fungal growth may be present on the crowns of affected plants and also on soil surfaces adjacent to such plants (**79**). Soil adheres to roots when they are harvested and an extensive, dark purple mycelial growth is present on the surface of affected roots (**80**, **81**). Severely affected roots usually exhibit some soft rot symptoms due to secondary decay organisms. More typically, violet root rot is detected on washed roots of plants having healthy foliage (**82**). The affected roots show patches of purplish black mycelium and small dark spots that are 150–200 μm in diameter. These spots are infection cushions that form after host

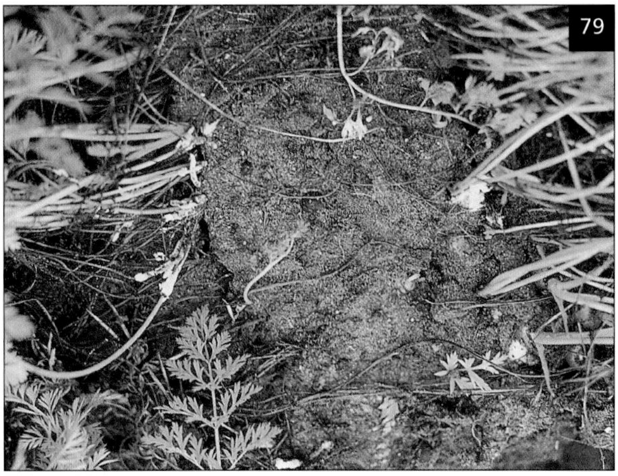

79 Growth of violet root rot pathogen around carrot crown.

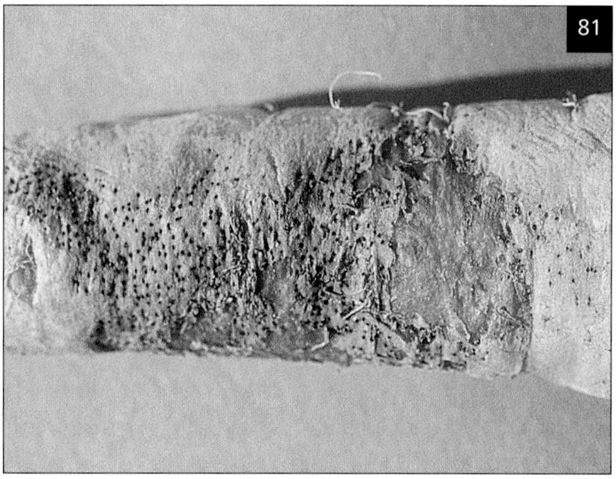

81 Violet root rot of parsnip.

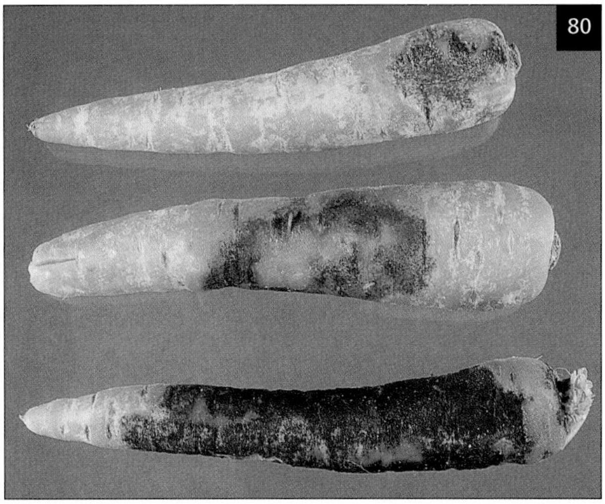

80 Violet root rot of carrot.

82 Severe violet root rot on parsley roots.

penetration. When large areas of the root have been colonized, the dense fungal growth on washed roots has a leathery appearance. Mycelial growth is slow, so fungal signs are rarely found on young seedlings.

Causal agent

Violet root rot is caused by the basidiomycete fungus *Helicobasidium brebissonii*. The anamorph is traditionally named *Rhizoctonia crocorum*, though recent taxonomic research may result in a name change. *Rhizoctonia crocorum* is the primary stage that causes the disease and produces coarse mycelial strands (up to 1 mm diameter) that enables it to spread through soil and infect new host plants. Sclerotia form mainly around the small lateral roots and are usually less than 5 mm in diameter. The presence of dark violet brown pigmentation and a hyphal diameter of 4–8 µm separates this pathogen from *R. solani*, which has brown hyphae that measure 5–10 µm. *H. brebissonii* is slow-growing and difficult to isolate. Attempt isolations by placing tissues from actively developing lesions onto malt extract agar. The *Helicobasidium* perfect stage usually produces basidiospores only in the spring.

There are taxonomic and etiological questions regarding *Helicobasidium* isolates from different parts of the world. Some research indicates that *H. mompa* (from east Asia and pathogenic on apple and other woody and herbaceous plants), *H. compactum* (from tropical and sub-tropical regions and synonymous with *H. mompa*), and *H. brebissonii* ([= *H. purpureum*] from Europe and North America) may all be one species. In England, this hypothesis is supported by observations that violet root rot could be detected on carrots grown in an apple orchard that was infected with violet root rot. Recent molecular studies indicate that *Helicobasidium* is involved in a complex life cycle with *Tuberculina* species, which are parasites of rust fungi. The *Helicobasidium/Rhizoctonia crocorum* organism appears to be a stage in the life cycle of *Tuberculina*.

Disease cycle

This pathogen is soilborne and occurs where root crops have been grown intensively. It is capable of long-term survival and disease can recur after a 10 to 20 year absence of root crops. The fungus survives by means of sclerotia and weed hosts. Perennial weeds such as sowthistle (*Sonchus* spp.), thistles (*Cirsium* spp.), and bindweed (*Convolvulus* spp.) are important alternate hosts. There are some reports that grass weeds, clover, and alfalfa are also hosts. The production of other vegetable crops may contribute to soil inoculum. These crops include asparagus, bean, beet, celery, crucifers, fennel, rhubarb, and sweet potato. Rapid disease development is associated with mild conditions in autumn. The pathogen is active over a wide range of temperatures (5–30° C) and has a temperature optimum of 20° C.

Control

Rotate crops using non-host plants. In the UK, a rotation of 3 to 4 years between root crops seems to provide satisfactory control for this disease on sugar beet. However, longer rotations may be required for carrots. Practice thorough weed control, particularly for perennial weeds. When violet root rot is detected in a crop, harvest the crop early to prevent further spread and losses. Crop residues and wash water from packing houses can harbor the pathogen; do not return these materials to land scheduled for carrot production. The pathogen may also survive animal digestive systems and be carried in manure from stock fed on diseased roots. Avoid moving infested soil between fields by thoroughly cleaning farm machinery after operations in infested fields. For acid soils, apply lime to raise soil pH to about 7.0. Resistant cultivars are not available.

References

Asher, M. 2001. Root rots: is there a problem? *British Sugar Beet Review* 69:18–21.

Dalton, I. P., Epton, A. S., and Bradshaw, N. J. 1981. The susceptibility of modern carrot cultivars to violet root rot caused by *Helicobasidium purpureum*. *Journal of Horticultural Science* 56:95–96.

Lutz, M., Bauer, R., Begerow, D., and Oberwinkler, F. 2004. *Tuberculina –Thanatophyton/Rhizoctonia crocorum – Helicobasidium*: a unique mycosparasitic-phytoparasitic life strategy. *Mycological Research* 108:227–238.

Moore, R. T. 1987. The genera of *Rhizoctonia*-like fungi: *Ascorhizoctonia, Ceratorhiza* gen. nov., *Epulorhiza* gen. nov., *Moniliopsis*, and *Rhizoctonia*. *Mycotaxon* 29:91–99.

Uetake, Y., Arakawa, M., Nakamura, H., Akahira, T., Sayama, A., Cheah, L.H., Okabe, I., and Matsumoto, N. 2002. Genetic relationships among violet root rot fungi as revealed by hyphal anastomosis and sequencing of the rDNA ITS regions. *Mycological Research* 106:156–163.

Valder, P.G. 1958. The biology of *Helicobasidium purpureum* Pat. *Transactions of the British Mycological Society* 41:283–308.

Mycocentrospora acerina

LICORICE ROT

Introduction and significance

Licorice rot is a storage disease of carrot and celery in temperate regions. The same pathogen also causes black canker on parsnip and anthracnose of caraway. In the field, the licorice-rot pathogen can occasionally be found growing in large root lesions caused by cavity spot. In the UK, its importance has declined as carrots are now frequently maintained and stored *in situ* under straw in the field rather than harvested and kept in cold storage.

Symptoms and diagnostic features

Symptoms usually develop around the crown or root tips after 5 to 6 weeks of storage, but may occur on other parts of the taproot. Infected areas consist of watery, dark brown, or black lesions that are sunken and penetrate deep into the root (83). Lesions do not have a discrete margin, unlike those caused by *Alternaria radicina*, and rotted tissues usually contain diagnostic chains of chlamydospores. On celery, the symptoms are sunken, elongated, black lesions at the base of the petioles and are referred to as black crown rot. On various hosts including carrot, this pathogen also causes dark brown or black leaf lesions that are similar to those caused by various other pathogens.

Causal agent

Licorice rot is caused by the fungus *Mycocentrospora acerina*. The pathogen produces distinctive conidia that measure 60–250 x 10–12 µm, are needle-shaped, strongly curved, and have whip-like terminal appendages and lateral appendages (resembling a germ tube) on the basal cell. The conidia have 4 to 24 transverse septa. Conidia usually form under humid conditions during storage. Their production under field conditions is stimulated by rainfall and occurs over a wide range of temperatures (2–25° C). Conidia are dispersed by splashing water over short distances, usually less than 9 meters. Hyphae are hyaline, but may appear pinkish when developing on lesions under damp conditions. The chlamydospores are ovate to spherical, dark brown, thick-walled single cells measuring 15–20 µm and produced within the mycelium.

Disease cycle

The fungus survives in soil for at least two years by means of chlamydospores. *Mycocentrospora acerina* may persist in soil longer when other susceptible hosts are present. It has a wide host range of over 90 species, including crops such as beet, pea, and many common weeds. It can infect plants at all stages of growth and is saprophytic on decaying plant residues. The fungus is favored by cool conditions, and solar radiation inhibits spore production. Chlamydospores on the surface of flooded soil can produce conidia that are then dispersed by water movement. There is little spread in storage and intact tissues are resistant to infection. Inoculum may spread during postharvest rinsing if wash water is not adequately chlorinated and changed.

Control

Rotate carrots and other Apiaceae crops with non-host plants on a 3- to 4-year cycle. Carefully handle carrots during harvesting and washing procedures. Store carrots in clean, cool, and humid conditions. For carrots with wounds, a short incubation period of high temperatures and high humidity prior to long-term storage should be beneficial.

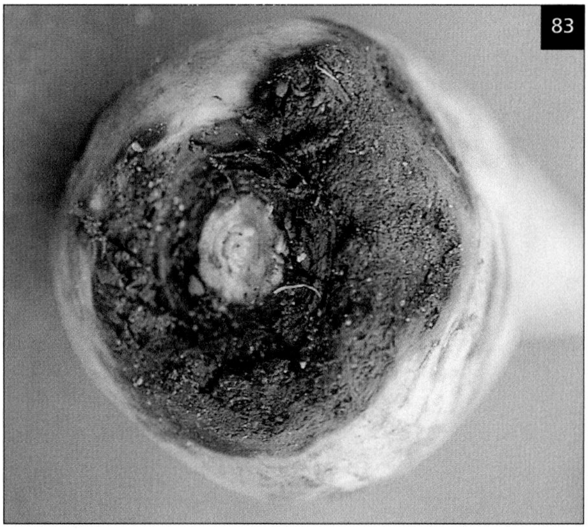

83 Licorice rot canker on parsnip root.

References

Davies, W. P. and Lewis, B. G. 1980. The inter-relationship between the age of carrot roots at harvest and infection by *Mycocentrospora acerina* in storage. *Annals of Applied Biology* 95:11–17.

Davies, W. P., Lewis, B. G., and Day, J. R. 1981. Observations on infection of stored carrot roots by *Mycocentrospora acerina*. *Transactions of the British Mycological Society* 77:139–151.

Day, J. R., Lewis, B. G., and Martin, S. 1972. Infection of stored celery plants by *Centrospora acerina*. *Annals of Applied Biology* 71:201–210.

Derbyshire, D. M. and Crisp, A. F. 1978. Studies on treatments to prolong storage life of carrots. *Experimental Horticulture* 30:23–28.

Evenhuis, A. 1997. Effect of root injury on lesion development of caraway roots infected by *Mycocentrospora acerina*. *European Journal of Plant Pathology* 103:537–544.

Evenhuis, A., Verdam, B. and Zadoks, J. C. 1997. Splash dispersal of conidia of *Mycocentrospora acerina* in the field. *Plant Pathology* 46:459–469.

Hermansen, A .1992. Weeds as hosts of *Mycocentrospora acerina*. *Annals of Applied Biology* 121:679–686.

Le Cam, B., Rouxel, F., and Villenueve, F. 1993. Post-harvest pathogens on cold stored carrots: *Mycocentrospora acerina*, the major spoilage fungus. *Agronomie* 13:125–133.

Lewis, B. G., Davies, W. P., and Garrod, B. 1981. Wound-healing in carrot roots in relation to infection by *Mycocentrospora acerina*. *Annals of Applied Biology* 99:35–42.

Wall, C. J. and Lewis, B. G. 1978. Quantitative studies on survival of *Mycocentrospora acerina* conidia in soil. *Transactions of the British Mycological Society* 71:143–146.

Wall, C. J. and Lewis, B. G. 1980. Infection of carrot leaves by *Mycocentrospora acerina*. *Transactions of the British Mycological Society* 75:163–165.

Wall, C. J. and Lewis, B. G. 1980. Survival of chlamydospores and subsequent development of *Mycocentrospora acerina* in soil. *Transactions of the British Mycological Society* 75:207–211.

Pythium violae, P. sulcatum

CAVITY SPOT

Introduction and significance

Cavity spot is an important disease of carrots in Europe, Australia, USA, and other temperate regions. The disease results in regular and significant economic losses. Modest infection levels of 10 to 20% affected roots can lead to rejection of the entire harvest because it is not economically feasible to remove diseased roots on the grading line. Cavity spot has been recognized since 1961 when it was described on carrot and parsnip in the USA. However, the cause of the problem remained unidentified until the late 1980s.

Symptoms and diagnostic features

Symptoms are usually first detected about 12 weeks after planting. The first symptoms are small, pale yellow or water-soaked elliptical spots on the root surface. A section through such lesions will reveal a small cavity just a millimeter or so below the periderm, or outer skin of the root. Cavity spot infections are particularly evident when there has been rotting and disintegration of the surface layer of the lesion, revealing an open sunken cavity that usually ranges from 4 – 13 mm wide. If roots are washed carefully before examination, the cavity spot lesion may still have remains of the periderm over or around the edge of the lesion (**84**).

84 Carrot roots showing periderm tissue extending over cavity spot lesions.

Cavity spot lesions often have a grayish or black margin (85, 86). Lesions increase in size as secondary organisms colonize damaged tissue and may become several centimeters in diameter. Diagnosis of cavity spot can be complicated by the presence of a number of soil fungi in the lesions, including *Pythium intermedium*, *P. sylvaticum*, *Fusarium* spp. and *Cylindrocarpon destructans*. As lesions enlarge and coalesce, large areas of the root may be covered with a shallow, soft rot.

On parsnip, lesions referred to as cavity spot are small, slightly sunken, and red-brown. When lesions are sectioned, the red-brown discoloration of the tissue penetrates several millimeters into the root. Lesions penetrate deeper in parsnip than carrot roots because parsnip roots do not produce a wound reaction that limits fungal invasion. Formal confirmation of a cavity spot pathogen for parsnip is still needed. However, because experiments demonstrated that metalaxyl treatments controlled cavity spot symptoms on parsnip, it was believed that an oomycete pathogen may be involved.

Causal agent

Cavity spot is caused by the soil oomycetes *Pythium violae* and *P. sulcatum*. *Pythium violae* is often the dominant pathogen, though *P. sulcatum* is important at some sites. Both pathogens are slow-growing in culture and difficult to isolate unless cavity spot lesions are young. Corn meal agar containing pimaricin (100 mg /liter) and rifampicin (= rifamycin; 30 mg/liter) is a useful medium for isolating these organisms. In culture, *P. violae* grows slowly at 5 to 7 mm per day at 25° C; in contrast, non-pathogenic, fast growing *Pythium* species can grow at rates of 25 to 30 mm per day at 25° C. Oogonia of *P. violae* average 29.5 µm diameter (range 16–34 µm), with the inner oospores measuring 11–28 µm. *Pythium sulcatum* produces aplerotic oospores averaging 14.5 µm diameter. *Pythium violae* has a wider host range than *P. sulcatum* and affects common weeds, ryegrass, and wheat. This wide host range appears to explain observations that cavity spot can occur even if there is no history of carrot production.

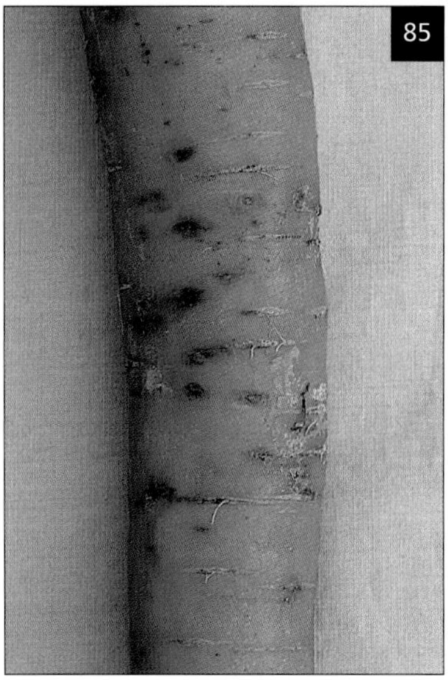

85 Root lesions of cavity spot of carrot.

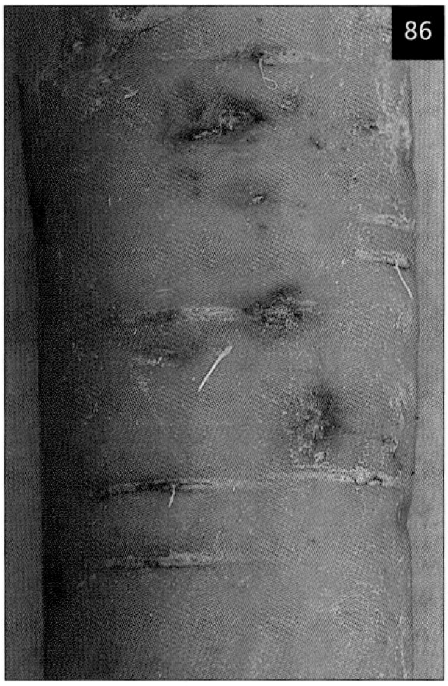

86 Close-up of root lesions of cavity spot of carrot.

Disease cycle

After *P. violae* initiates root infection, the pathogen grows within the carrot root for 3 to 4 days before a host resistance reaction is stimulated and wound periderm tissue, containing lignin and suberin, is formed beneath the lesion. This results in a dark discoloration around the edge of the cavity spot lesion.

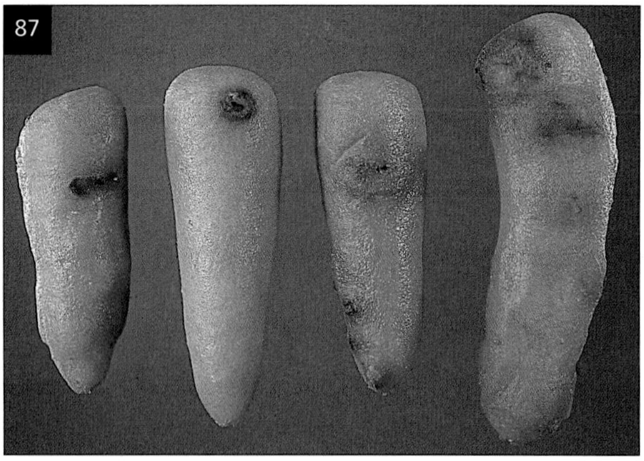

87 After peeling of canning carrots, cavity spot lesions appear as raised gray areas.

Wound periderm is resistant to the peeling process used for canning carrots, resulting in cavity lesions that remain as raised mounds on peeled roots (87). The cavity is formed when cells of the secondary phloem, periderm, and pericycle collapse following enzyme production by the pathogen. Other, fast growing *Pythium* species appear to be less effective at circumventing host resistance and inducing tissue collapse. However, these other species are capable of producing larger quantities of cell wall degrading enzymes than *P. violae*, enabling them to invade and extend lesions initiated by *P. violae*. *P. violae* is favored by soil temperatures below 20° C. Cavity spot is unlikely to spread between roots kept in cold storage.

Cavity spot is usually most prevalent in acid (pH 5 or lower) soils and where there is poor drainage or compacted soil. Development of cavity spot is favored by wet, anaerobic conditions that reduce host vigor and increase root exudates; such exudates stimulate germination of *Pythium* oospores in soil.

Control

Rotate carrots with at least 5 years between carrot or parsnip crops. Longer rotations may be required if cavity spot had been severe. Serological tests (ELISA) of soil for the cavity spot pathogen are available in the UK and may be used for risk assessment. Use cultivars with moderate to good levels of resistance to cavity spot. Fungicides, particularly the phenylamide chemical metalaxyl (and related isomer mefenoxam), were very effective as spray treatments for many years. Treatment timing is critical, so apply fungicides at planting or shortly after crop emergence. Applications made later than 6 weeks after planting are less effective. However, there are recent indications of rapid breakdown of metalaxyl in soil, resulting in lack of acceptable control. Avoid planting carrots in sites with poor drainage. Irrigate carrots carefully so that flooding and excess water do not occur. Avoid sites with acid soil pH, or amend such soils by applying calcium carbonate lime and adjusting soil pH to 7–7.5. Deep cultivation between carrot rows has reduced cavity spot development in some situations. Harvest crops early if cavity spot has started to appear.

References

Davison, E. M. and McKay, A. G. 1999. Reduced persistence of metalaxyl in soil associated with its failure to control cavity spot of carrots. *Plant Pathology* 48:830–835.

El-Tarabily, K. A., Hardy, G. E. St J., Sivasithamparam, K., Hussein, A. M., and Kurtboke, I. D. 1996. Microbiological differences between limed and unlimed soils and their relationship with cavity spot of carrot (*Daucus carota* L.) caused by *Pythium coloratum* in Western Australia. *Plant and Soil* 183:279–290.

Farrar, J. J., Nunez, J. J., and Davis, R. M. 2002. Repeated soil applications of fungicide reduce activity against cavity spot in carrots. *California Agriculture* 56(2):76–79.

Guba, E. F., Young, R. E., and Ui, T. 1961. Cavity spot disease of carrot and parsnip roots. *Plant Disease Reporter* 45:102–105.

Gladders, P. and McPherson, G. M. 1986. Control of cavity spot in carrots with fungicides. *Aspects of Applied Biology* 12:223–233

Hiltunen, L. H. and White, J. G. 2002. Review paper: Cavity spot of carrot (*Daucus carota*). *Annals of Applied Biology* 113:259–268.

Liddell, C. M., Davis, R. M., Nuñez, J. J., and Guerard, J. P. 1989. Association of *Pythium* species with carrot root dieback in the San Joaquin Valley of California. *Plant Disease* 73:246–249.

Lyshol, A. J., Semb, L., and Taksdal, G. 1984. Reduction of cavity spot and root dieback in carrots by fungicide applications. *Plant Pathology* 33:193–198.

Schrandt, J.K., Davis, R. M., and Nuñez, J. J. 1994. Host range and influence of nutrition, temperature, and pH on growth of *Pythium violae* from carrot. *Plant Disease* 78:335–338.

Vivoda, E., Davis, R. M., Nuñez, J. J., and Guerard, J. P. 1991. Factors affecting the development of cavity spot of carrot. *Plant Disease* 75:519–522.

White, J. G. 1988. Studies on the biology and control of cavity spot. *Annals of Applied Biology* 113:259–268.

White, J.G., Stanghellini, M. E., and Ayoubi, L. M. 1988. Variation in the sensitivity to metalaxyl of *Pythium* spp. isolated from carrots and other sources. *Annals of Applied Biology* 113:269–277.

Sclerotinia minor, S. sclerotiorum

WHITE MOLD, COTTONY ROT, WATERY SOFT ROT

Introduction and significance

White mold or cottony rot of carrots is perhaps most important as a storage rot. *Sclerotinia sclerotiorum* occurs commonly on harvested carrots in temperate regions, but it usually only becomes apparent when roots are stored under warm humid conditions for several days. The other white mold species, *S. minor*, has occasionally been reported on carrots in the USA and Japan. White mold infects most Apiaceae crops, especially causing economic damage in celery and fennel. *Sclerotinia sclerotiorum* can also cause blights and seed stalk dieback on Apiaceae crops such as celery and parsley.

Symptoms and diagnostic features

The first symptoms are small, 5–10 mm in diameter, water-soaked, brown lesions on leaves, crowns, or the upper exposed shoulders of taproots (**88**). Senescent foliage in contact with the soil is particularly susceptible to initial infections. White mycelial growth can be found on senescent leaves and other parts of the carrot when the soil is moist and humidity is high. The pathogen can spread extensively between plants via the mats of senescing lower foliage. As the pathogen develops, the white mycelium produces small mounds of hyphae that later become hard, black sclerotia (**89**). Infection of leaves can lead to root rots in the field. The roots develop a soft rot and sclerotia occur on and around the taproots. On stored roots, *S. sclerotiorum* produces extensive, fluffy white mycelium, and underlying tissues become soft and rotted (**90**). In contrast, bacterial soft rots are slimy and lack this white mycelial growth.

Causal agents

White mold is caused by two species of the ascomycete fungus *Sclerotinia*: *S. sclerotiorum* and *S. minor*. For detailed descriptions of these pathogens, see the bean white mold section in the chapter on legume diseases.

88 **Close up of *Sclerotinia* growing near carrot crown.**

89 **Sclerotia of *Sclerotinia sclerotiorum* forming on carrot roots.**

90 **Harvested carrots showing white mold infections.**

Disease cycle

The disease cycle is similar to that described for white mold on other vegetables such as bean and lettuce. For detailed descriptions of the disease cycles, see the bean white mold section in the chapter on legume diseases. Even though optium temperatures for this pathogen are 15–20° C, some activity continues even during cold storage of carrots at 0° C. Storage losses can be high because of extensive mycelial spread from root to root.

Control

For an integrated approach to white mold control, see the bean white mold section in the chapter on legume diseases. Handle carrots carefully at all stages of harvesting and washing. Clean and disinfect storage containers between batches of carrots. Fungicides are generally no longer used as pre-storage treatments. Maintain storage and shipment conditions at temperatures of 0° C and provide at least 95% relative humidity. Inspect stored carrots so that problems can be identified at an early stage and losses minimized by promptly disposing affected carrots.

References

Cheah, L. H., Page, B. B. C., and Shepherd, R.1997. Chitosan coating for inhibition of Sclerotinia rot of carrots. *New Zealand Journal of Crop and Horticultural Science* 54:63–70.

Clarkson, J. P., Staveley, J., Phelps, K., Young, C. S., and Whipps, J. M. 2003. Ascospore release and survival in *Sclerotinia sclerotiorum*. *Mycological Research* 107:213–222.

Finlayson, J. E., Rimmer, S. R., and Pritchard, M. K. 1989. Infection of carrots by *Sclerotinia sclerotiorum*. *Canadian Journal of Plant Pathology* 11:242–246.

Kora, C., McDonald, M. R., and Boland, G. J. 2003. Sclerotinia rot of carrot: an example of phenological adaptation and bicyclic development by *Sclerotinia sclerotiorum*. *Plant Disease* 87:456–470.

Streptomyces scabies

SCAB

Introduction and significance

This disease is occasionally a serious problem in carrot and can make fresh market crops unmarketable. Parsnip may also be affected. Scab is a familiar problem on potato, radish, beet, and turnip.

Symptoms and diagnostic features

Symptoms on carrot are raised, brown, corky lesions on the taproot. Lesions measure 0.5–2 cm and are often elongated horizontally across the carrot root. Sunken lesions are also produced (**91**).

Causal agent

The pathogen is considered to be the actinomycete *Streptomyces scabies*. Actinomycetes are Gram-positive bacteria that can form branching filaments and are common soil inhabitants. However, recent research on potato has identified several species of *Streptomyces* that cause common scab, so the etiology of scab disease may be more complex. *Streptomyces acidiscabies* has been reported from the USA and *S. europeiscabiei* from carrot in France and The Netherlands. Scab isolates from carrot are able to infect potato and radish.

Disease cycle

Streptomyces scabies is a soil organism that invades the root through root openings such as sites of lateral root emergence or wounds. The pathogen causes localized death of cells and the formation of corky tissue. This wound reaction results in the scab symptoms. Infection is favored by alkaline soils, dry conditions, and warm temperatures (optimum 20° C, with little activity below 11° C). Carrot is thought to be susceptible to infection for only 2 to 3 weeks during development, beginning when roots start to thicken (when root diameter is greater than 2 mm). This infection period is equivalent to 475 to 625 degree days after planting. At this stage, the periderm grows through the epidermis and the resulting small wounds are vulnerable to infection.

Control

Rotate carrots with crops that are not susceptible to scab; however, crop rotations are not entirely successful because *Streptomyces* spp. survive in soil for very long periods. Irrigate to maintain adequate soil moisture,

91 Carrot root showing scab symptoms.

especially during the early stages of crop growth. There are indications of differences in cultivar susceptibility, so plant more tolerant varieties if these can be identified. After harvest, corky tissue on raised scab lesions may be dislodged during washing operations, so low levels of infection may escape detection.

References

Bouchek-Mechiche, K., Pasco, C., Andrivon, D., and Jouan, B. 2000. Differences in host range, pathogenicity to potato cultivars and response to soil temperature among *Streptomyces* species causing common and netted scab in France. *Plant Pathology* 49:3–10.

Goyer, C. and Beaulieu, C. 1997. Host range of *Streptomyces* strains causing common scab. *Plant Disease* 81:901–904.

Hanson, L. E. and Lacy, M. L. 1990. Carrot scab caused by *Streptomyces* spp. in Michigan. *Plant Disease* 74:1037.

Janse, J. D. 1988. A *Streptomyces* species identified as the cause of carrot scab. *Netherlands Journal of Plant Pathology* 94:303–306.

Lambert, D. H. 1991. First report of additional hosts for the acid scab pathogen *Streptomyces acidiscabies*. *Plant Disease* 75:750.

Schoneveld, J. A. 1994. Effect of irrigation on the prevention of scab in carrots. *Acta Horticulturae* 354:135–144.

Wanner, L. A. 2004. Field isolates of *Streptomyces* differ in pathogenicity and virulence on radish. *Plant Disease* 88:785–796.

Thanatephorus cucumeris
(anamorph = *Rhizoctonia solani)*

CROWN AND ROOT ROT

Introduction and significance

This disease is occasionally a problem in North America. In the UK, the name crown rot has been used for both external symptoms associated with *Fusarium* spp. and for the internal symptoms caused by virus infections (e.g. *Parsnip yellow fleck virus*, see description and symptoms on page 124). Therefore, clarification will be necessary when this common term is used.

Symptoms and diagnostic features

This disease causes the crowns and lower portions of petioles to develop dry, dark brown, sunken lesions. Such infections can also occur further down the root. Symptoms on the taproot may resemble the lesions caused by the cavity spot pathogen (*Pythium* species). The disease results in premature senescence and death of foliage. Root infections may be colonized by secondary soil fungi and soft rot bacteria. The mycelium of *R. solani* occasionally may be present on the crown under certain conditions. The pathogen also causes damping-off and root rot of carrot seedlings (**92**).

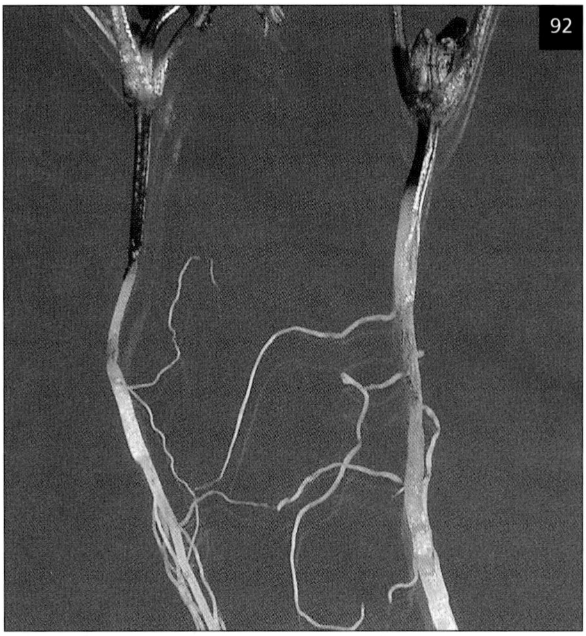

92 Brown lesions of crown and root rot on carrot seedlings.

Causal agent

Crown rot is caused by the basidiomycete fungus *Thanatephorus cucumeris*. On plants the anamorph stage, *Rhizoctonia solani*, is more often observed. This pathogen is a true soil inhabitant, survives in soil as a saprophyte, and has a broad host range. *Rhizoctonia solani* has no asexual fruiting structures or spores and produces characteristically coarse, brown, approximately right-angle branching hyphae. The hyphae are distinctly constricted at branch points, and a cross wall with dolipore septum is deposited just after each branch. Hyphal cells are multi-nucleate. Small, tan to brown, loosely aggregated clumps of mycelia function as sclerotia that enable the fungus to withstand unfavorable conditions. The teleomorph *T. cucumis* is not often observed, and its role in disease development is not known. *Rhizoctonia solani* isolates are divided into distinct groups, called anastomosis groups (AGs), that have differing virulence levels and in some cases preferred host ranges. The groups affecting carrots are mainly AG 2-2 and AG 4, and occasionally AG 1. In the US, AG 2-4 was confirmed on carrot in Georgia.

Disease cycle

Crown rot develops when temperatures are above 18° C and soils are moist. Spread from plant to plant can occur if there is a full leaf canopy. Lesion development on roots continues during storage. Problems tend to be most severe on organic soils and are favored by warm and wet conditions close to harvest. In the UK, during the 1980s, Rhizoctonia root rot of seedlings was particularly damaging in autumn-sown carrots that were overwintered under polythene covers, resulting in a high proportion of deformed roots. This growing practice is no longer used and planting in early winter has reduced root rot problems.

Control

Allow sufficient time for crop residues to break down in soil, thereby reducing soil inoculum prior to carrot planting. Crops such as alfalfa and legume cover crops appear to increase risks of *Rhizoctonia* problems, so avoid these rotation crops if this disease is a concern. Avoid cultivating or related procedures late in the crop cycle because such practices can deposit infested soil onto the crown or petioles. Fungicides may provide some control of the disease. In stored carrots, maintain temperatures close to 0° C.

References

Davis, R. M. and Nuñez, J. J. 1999. Influence of crop rotation on the incidence of *Pythium*- and *Rhizoctonia*-induced carrot root dieback. *Plant Disease* 83:146–148.

Grisham, M. P. and Anderson, N. A. 1983. Pathogenicity and host specificity of *Rhizoctonia solani* isolated from carrots. *Phytopathology* 73:1564–1569.

Mildenhall, J. P. and Williams, P. H. 1973. Effect of soil temperature and host maturity on infection of carrot by *Rhizoctonia solani*. *Phytopathology* 63:276–280.

Sumner, D. R., Phatak, S. C., and Carling, D. E. 2003. First report of *Rhizoctonia solani* AG–2–4 on carrot in Georgia. *Plant Disease* 87:1264.

Thielaviopsis basicola, Chalaropsis thielavioides

BLACK MOLD

Introduction and significance

These two fungi cause postharvest problems when washed carrot roots are not stored properly. *Thielaviopsis* may occasionally cause problems on seedlings and mature crops in the field. Many crops in the UK will have some black mold contamination. In North America these diseases are more prevalent on muck soils than on mineral soils.

Symptoms and diagnostic features

Superficial light to dark gray fungal growth develops on the surface of the affected carrot root (**93**). As disease progresses, the growth becomes black and irregular in outline. Root tissue below such mycelium remains relatively firm unless secondary bacterial organisms colonize the lesions.

Causal agents

Black mold is caused by two soilborne fungi. *Thielaviopsis basicola* (synonym *Chalara elegans*) is the more common of the two organisms and affects a range of vegetables including legumes and tomato. It survives in the soil by means of thick walled chlamydospores which are rectangular, dark brown, measure 7–12 x 10–17 µm, and form in chains of four to eight cells. These chains readily fragment into single cells. *T. basicola* also produces hyaline, rectangular endoconidia that measure 7–17 x 2.5–4.5 µm and are exuded in long chains. *Chalaropsis thielavioides* produces ovoid, dark brown chlamydospores which measure 14–19 µm and are borne singly or in short chains. This pathogen likewise produces long chains of hyaline, rectangular endoconidia measuring 8–15 x 2.5–4.5 µm.

93 **Black mold on carrot caused by *Thielaviopsis*.**

Disease cycle

Both these fungi are common soilborne organisms. Occasionally there may be symptoms of root infection in the field prior to harvest, but this phase of the disease is associated with damage to the root. Black mold is more often associated with roots damaged during harvesting, washing, and grading. Black mold is particularly a problem when roots are not cooled rapidly and stored at 5° C. Washed, pre-packed carrots are most vulnerable because of the high humidity in the bags.

Control

Avoid black mold by carefully handling roots to minimize damage during harvesting and postharvest handling. Cool carrots quickly to remove field heat. Chlorinate and change wash water regularly to reduce spread of inoculum. Fungicide dips are no longer used for controlling black mold, though other chemicals such as potassium sorbate and propionic acid have shown good activity against *T. basicola*.

References

Derbyshire, D. M. and Shipway, M. R. 1978. Control of post-harvest deterioration in vegetables in the UK. *Outlook on Agriculture* 9:246–252.

Paulin-Mahady, A. E., Harrington, T. C., and McNew, D. 2002. Phylogenetic and taxonomic evaluation of *Chalara, Charlaropsis*, and *Thielaviopsis* anamorphs associated with *Ceratocystis. Mycologia* 94:62–72.

Punja, Z. K., Chittaranjan, S., and Gaye, M.M. 1992. Development of black root rot caused by *Chalara elegans* on fresh market carrot. *Canadian Journal of Plant Pathology* 14:299–309.

Punja, Z. K. and Gaye, M.M. 1993. Influence of postharvest handling practices and dip treatments on development of black root rot on fresh market carrots. *Plant Disease* 77:989–995.

Uromyces graminis, U. lineolatus

RUST

Introduction and significance

Rust diseases are a minor problem in carrot and other Apiaceae vegetables, but they are common on wild Apiaceae. Little is known about this disease and monitoring is advisable if Apiaceae crops are introduced into new areas of production.

Symptoms and diagnostic features

Early symptoms of rust consist of chlorotic, irregularly shaped leaf lesions. Such lesions later develop typical raised pustules, often forming mostly on the undersides of infected leaves. The leaf epidermis covering the pustules will rupture and release aeciospores (**94**).

Causal agents

Two heteroecious rust fungi affect carrot; these rusts produce different spore types on two distinct host plants. *Uromyces graminis* produces aecia on Apiaceae plants such as carrot and fennel, and uredinia and telia on grass hosts (*Melica* spp.). This pathogen occurs in Asia, Mediterranean regions, Russia, and South America. *Uromyces lineolatus* (= *U. scirpi*) has aecidia on Apiaceae and uredinia and telia on club-rush (*Scirpus* spp.); it is found in wet and marshy areas in Europe and North America.

Autoecious rusts complete their life cycle on one host plant and include *Puccinia angelicae* on angelica, *P. nitida* on parsley, *P. pimpinellae* on fennel or anise, and *P. apii* on celery.

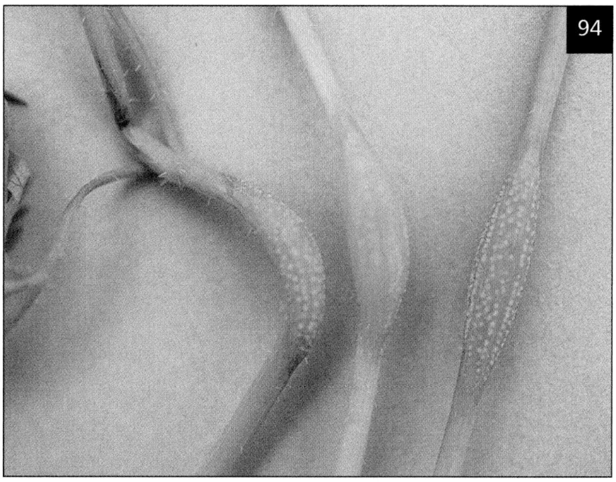

94 **Rust aecia on carrot petioles.**

Control

Investigate rust outbreaks to determine possible sources of inoculum; such information may allow for the elimination of alternate rust hosts near the crop. Apply fungicides if available.

References

Wilson, M. and Henderson, D. M. 1996. *British Rust Fungi.* Cambridge University Press, Cambridge, 384 pp.

Carrot red leaf virus, Carrot mottle virus

CARROT MOTLEY DWARF

Introduction and significance

Carrot motley dwarf (CMD) is a sometimes damaging disease of carrots grown in temperate regions. The disease occurs primarily in the USA (California) but is also common in the UK. CMD also affects parsley, cilantro, and dill.

Symptoms and diagnostic features

When young carrot plants are affected, growth is stunted and there is pronounced reddening, chlorotic mottling, and overall yellowing of the foliage (**95**). There may be twisting of the lower leaves. Roots are severely stunted (**96**) and some plants may die. Young parsley plants show reddening of the older leaves and a general yellowing of the foliage, and plants may become stunted (**97**). Infection of older carrot and parsley plants produces milder symptoms and plants may be symptomless when temperatures are above 24° C. In general, leaf discoloration of all Apiaceae plants may be caused by a variety of factors, so virus tests are required to confirm CMD disease.

Causal agents

CMD disease is caused by a co-infection of two viruses: *Carrot red leaf virus* (CRLV), which is a polerovirus with isometric particles (25 nm in diameter), and *Carrot mottle virus* (CMoV), an umbravirus having single

95 Carrot field affected with carrot motley dwarf.

stranded RNA but lacking a protein coat. Both viruses can infect plants individually, but CMD disease is only produced when both viruses are present. When present as single viruses, only CRLV can be transmitted by the aphid vector. CMoV cannot be vectored by the aphid unless this virus is accompanied by the helper virus CRLV; the CRLV proteins encapsidate the RNA of CMoV, enabling transmission by aphids. The mixture of viruses is transmitted by the willow carrot aphid (*Cavariella aegopodii*) in a persistent manner. CMD is restricted to cultivated and wild Apiaceae, and the viruses are not seedborne.

96 Stunted carrots infected with carrot motley dwarf. Healthy plant is on the right.

Disease cycle

For carrot, disease spreads when these viruses are vectored between successive or nearby carrot crops. Presence of the aphid vector and CMD disease in overwintered and volunteer carrots is an important source of the problem for newly planted carrots. Wild Apiaceae, such as wild carrot (*Daucus carota*) and cow parsley (*Anthriscus sylvestris*), are also potential sources of CMD viruses.

97 Carrot motley dwarf causing yellowing and reddening of parsley foliage.

Control

In some regions, control has been achieved by planting spring crops at least 1.5 km away from overwintered crops. In California, reducing the number of carrot fields that are kept intact through the winter has significantly reduced disease occurrence in new, spring planted carrots. Remove volunteer and weed hosts before new crops emerge. Carrot cultivars vary in their response to CMD, so use more tolerant cultivars if available.

References

Morton, A., Spence, N. J., Boonham, N., and Barbara, D. J. 2003. Carrot red leaf associated RNA in carrots in the United Kingdom. *Plant Pathology* 52:795.

Waterhouse, P. M. and Murant, A. F. 1983. Further evidence on the nature of the dependence of carrot mottle virus on carrot red leaf virus for transmission by aphids. *Annals of Applied Biology* 103:455–464.

Watson, M. T. and Falk, B. W. 1994. Ecological and epidemiological factors affecting carrot motley dwarf development in carrots grown in the Salinas Valley of California. *Plant Disease* 78:477–481.

Watson, M. T., Tian, T., Estabrook, E., and Falk, B. W. 1998. A small RNA resembling the beet western yellows luteovirus ST9-associated RNA is a component of the California carrot motley dwarf complex. *Phytopathology* 88:64–170.

PASTINACA SATIVA (PARSNIP)

Erysiphe heraclei

POWDERY MILDEW

Introduction and significance

Powdery mildew is probably the most important foliar disease of parsnip in the UK. Powdery mildew is more damaging to parsnip than to carrot or celery. Early attacks can cause significant yield losses.

Symptoms and diagnostic features

The symptoms usually appear first on the petioles and leaf blades of older foliage. The superficial white fungal growth is similar to that of other powdery mildews. In the UK, little infection is seen until crop canopies close over in July. Younger leaves become infected as disease progresses. When powdery mildew is severe, both upper and lower leaf surfaces are affected and develop yellowing and early loss of the foliage (**98, 99**).

98 Early symptoms of powdery mildew on underside of parsnip leaf.

99 Powdery mildew of parsnip.

Causal agent

Powdery mildew is caused by the ascomycete *Erysiphe heraclei*. For a description of this pathogen, see the carrot section on powdery mildew in this chapter (page 103). There may be some host specialization, but isolates appear to be able to infect a number of Apiaceae host species.

Disease cycle

Overlapping parsnip plantings, volunteer parsnip, and weed hosts are potential sources of infection. The disease cycle can be completed in 7 days under optimum conditions. High humidity and moderate temperatures favor infection. Crops become more susceptible with age. Disease severity is increased by drought stress and is reduced by rain or overhead irrigation.

Control

Apply fungicides such as sulfur or triazole products if severe disease is a possibility. Such materials are strictly protectants and must be applied prior to significant disease development. Avoid over-fertilizing because high nitrogen rates can result in succulent foliage that is very susceptible to powdery mildew.

References

Palti, J. 1975. Erysiphaceae affecting Apiaceae crops, with special reference to carrot, in Israel. *Phytopathologia Mediterranea* 14:87–93.

Itersonilia pastinacae, I. perplexans

ITERSONILIA CANKER, BLACK CANKER

Introduction and significance

Itersonilia or black canker is an important disease of parsnip in the UK, North America, and Australia. The pathogen also affects dill and wild Apiaceae plants. Affected parsnip roots are unmarketable.

Symptoms and diagnostic features

Infected parsnip develops dark brown, black, or purple-black lesions located around the crown, upper root, and especially around the bases of lateral roots (**100**). Lesions are usually superficial and extend only a few millimeters deep. Advanced lesions lose their smooth surface layer and have coarse exposed tissues. Secondary organisms are prevalent in advanced cankers. Foliar symptoms consist of small (1–2 mm diameter), brown spots with yellow halos. Spots coalesce into larger necrotic areas. Petioles and inflorescences are also affected. Root symptoms usually begin late in the summer and continue to develop as long as crops remain in the ground. Late harvested crops can therefore be severely affected. Other pathogens cause cankers on parsnip and laboratory tests are required to identify the pathogens, which include the following: *Mycocentrospora acerina*, *Alternaria radicina*, *Cylindrocarpon destructans*, *Sclerotinia sclerotiorum*.

Causal agent

Black canker is caused by the basidiomycete fungus *Itersonilia*. Two species, *I. perplexans* and *I. pastinacae*, are implicated in this disease. Some researchers believe the two species are synonymous. In culture, *Itersonilia* produces low growing, slimy colonies. Hyphae have clamp connections at septa and produce short sterigmata that bear spores. Spores are kidney shaped ballistospores that measure 11.4–20.0 x 7.2–11.4 µm and are forcibly discharged from the sterigmata. Thick walled chlamydospores, measuring 9–13 x 13–20 µm, are produced by some isolates. *Itersonilia* has a yeast phase that appears in culture under certain conditions. *Itersonilia* can be isolated from diseased tissue by suspending tissue samples directly over agar surfaces; ballistospores are discharged and will fall onto the agar. On Waksman albumen agar, *I. pastinacae* produces chlamydospores and exhibits slower, more irregular growth than *I. perplexans*, which does not produce chlamydospores.

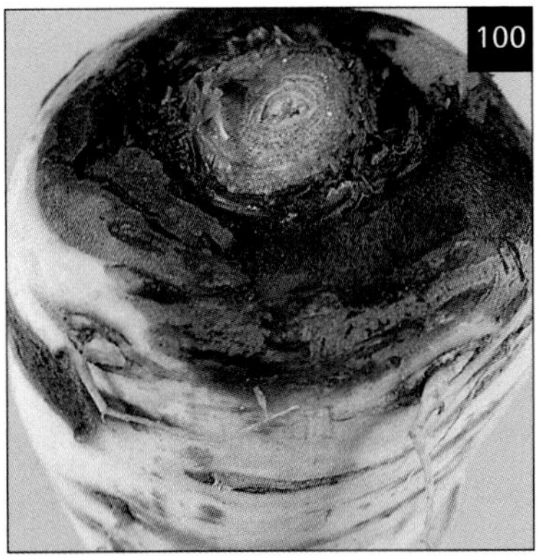

100 Parsnip root infected with *Itersonilia*.

Disease cycle

The pathogen survives in soil by means of chlamydospores and in crop residues. Airborne ballistospores are produced, released during the morning as humidity decreases, and land on susceptible tissues. Itersonilia canker increases significantly when leaf senescence starts, so infection in young crops is uncommon. Spores washed from leaves make contact with crown and upper root tissues, leading to canker lesions. Cankers are associated with extended periods of rainfall. Optimum temperatures are about 20° C. The pathogen may be seedborne.

Control

Rotate parsnip with non-host crops and control wild Apiaceae plant hosts. Use seed that does not have significant levels of the pathogen. Select and plant cultivars that are resistant to black canker. Protect roots by covering crowns and exposed upper-root tissues with soil. Regularly sample and examine parsnip roots as the crop nears maturity so that affected fields can be harvested early. Fungicide sprays for preventing root-canker symptoms are not effective.

References

Channon, A. G. 1963. Studies on parsnip canker. I. The causes of the disease. *Annals of Applied Biology* 51:1–15.

Channon, A. G. 1963. Studies on parsnip canker. II. Observations on the occurrence of *Itersonilia pastinacae* and related fungi on the leaves of parsnips and in the air within parsnip crops. *Annals of Applied Biology* 51:223–230.

Channon, A. G. 1964. Studies on parsnip canker. III. The effect of sowing date and spacing on canker development. *Annals of Applied Biology* 54:63–70.

Channon, A. G. 1965. Studies on parsnip canker. VI. *Cercospora acerina* (Hartig) Newhall – a further cause of black canker. *Annals of Applied Biology* 56:119–128.

Channon, A. G. 1969. Infection of the flowers and seed of parsnip by *Itersonilia pastinacae*. *Annals of Applied Biology* 64:281–288.

Channon, A. G. and Thomson, M. C. 1981. Parsnip canker caused by *Cylindrocarpon destructans*. *Plant Pathology* 30:181.

Koike, S. T. and Tjosvold, S. A. 2001. A blight disease of dill in California caused by *Itersonilia perplexans*. *Plant Disease* 85:802.

Phloeospora herclei

PHLOEOSPORA LEAF SPOT

Introduction and significance

Phloeospora leaf spot is generally a minor leaf disease of parsnip. In the UK severe infection usually occurs on late summer and autumn parsnip crops. A Phloeospora disease also has been reported from New Zealand.

Symptoms and diagnostic features

The first symptoms are small (1–2 mm) pale green or brown leaf spots. As disease develops, leaf spots increase in number, spots coalesce to form large necrotic blotches (2–5 cm), and infected leaf tissue becomes gray brown (**101**). The disease spreads rapidly through the crop, causing extensive leaf death and defoliation (**102**). No root symptoms have been reported. Severely affected leaf tissue has tiny black fruiting bodies called acervuli. Spore tendrils ooze from acervuli and form characteristic white patches on leaf surfaces (**103**).

Causal agent

Phloeospora leaf spot is caused by the fungus *Phloeospora herclei*. Fruiting bodies are minute, cup-shaped structures called acervuli. Conidia are formed in these structures and are curved, one to four septate (though most are single septate), and measure 50–80 x 3.5–5 µm. The early literature probably confused *P. herclei* with *Ramularia pastinaceae*. The Phloeospora leaf spot disease reported from New Zealand is attributed to *P. crescentium*. However, *P. herclei* and *P. crescentium* may be the same organism.

Disease cycle

Little is known about the epidemiology of this pathogen. Inoculum is spread by splashing water and direct contact between leaves. *Phloeospora herclei* probably survives on infected debris and on weed hosts such as hogweed (*Heracleum sphondylium*).

Control

Specific control measures have not been developed.

References

Laundon, G. F. 1970. Records of fungal plant diseases in New Zealand. *New Zealand Journal of Botany* 8:51–66.

Riley, E. A. 1952. Leaf spot of parsnip caused by *Phloeospora crescentium* (Barth.) n. comb. *Mycologia* 44:213–215.

101 Necrotic leaf blotches of Phloeospora leaf spot of parsnip.

102 General necrosis of parsnip foliage caused by Phloeospora leaf spot.

103 White spore masses of Phloeospora leaf spot of parsnip.

Phoma complanata

PHOMA CANKER

Introduction and significance

This disease has increased in importance since it was first reported in Canada in 1984. The pathogen is host specific to parsnip and related weeds, such as cow parsnip (*Heracleum lanatum*). Field diagnosis of Phoma canker is difficult because several pathogens produce similar symptoms and more than one causal agent may be present on diseased plants. In the UK, *P. complanata* may now be the most important cause of parsnip root cankers.

Symptoms and diagnostic features

If present, leaf spots are circular, brown, and often have yellow halos. As disease develops, leaf spots coalesce and result in a general yellowing, necrosis, and blighting of foliage. Pycnidia appear in leaf lesions about two weeks after infection. Dark brown to black lesions may develop on petioles, resulting in leaves that twist and bend over. Large, dark brown to black cankers form on parsnip crowns and on the sides of taproots (**104**). These cankers can enlarge and affect much or all of the crown tissue. Cankers result in splitting or cracking on the upper portion of the root. Careful examination of the surfaces of cankers will reveal pycnidia embedded in these tissues (**105**). Pycnidia ooze a white or creamy spore exudate. Affected roots are reported to have a sweet cinnamon odor.

Causal agent

Phoma canker is caused by *Phoma complanata*. Colonies are light to olivaceous gray, have even margins, and produce dense aerial mycelium. Pycnidia measure 35–49 µm in diameter and have thick walls. Conidia are hyaline, guttulate, and variable in shape. Conidia dimensions are 5–9 x 2–3.5 µm (mean 7.4 x 2.4 µm), though large (27 x 8 µm) one-septate spores occur in older cultures or lesions.

A second pathogen, *Phomopsis diachenii*, can be confused with the Phoma canker organism. Reddish brown leaf spots with deep purple margins measure 10–20 mm in diameter and contain black pycnidia. Microscopic examination shows that *P. diachenii*, like most *Phomopsis* species, produces both filiform and oval spores.

Disease cycle

Leaf spots and petiole lesions of Phoma canker appear 4 to 5 days after infection, and disease rapidly progresses if favorable conditions are present. Insects are reported to contribute to secondary spread. Root infection occurs when spores are washed down from the infected foliage. In the UK, foliar symptoms may be absent even though root cankers are present; in these situations, the source of infection is likely to be soilborne inoculum. *Phoma complanata* is seedborne and can reduce seedling establishment. Diseased roots contain elevated levels of furocoumarins. These substances are photocarcinogens, so take precautions to prevent skin contact when handling parsnip roots.

104 Root symptoms of Phoma canker of parsnip.

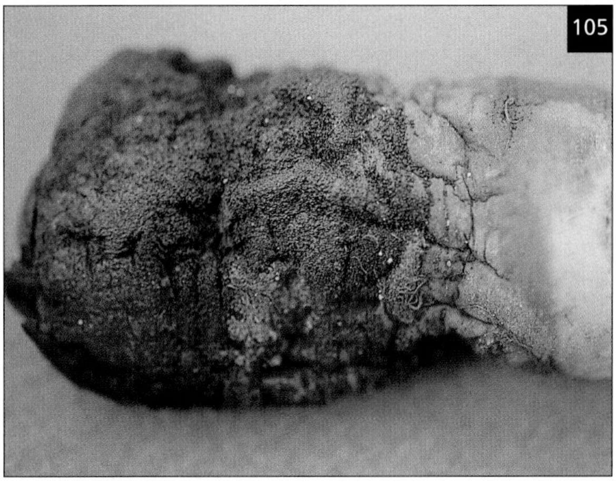

105 Decayed parsnip root with pycnidia of the Phoma canker pathogen.

Control

Rotate crops so that a 4-year break occurs between parsnip plantings. Use seed that does not have significant levels of the pathogen. Choose and plant cultivars that are known to have tolerance or reduced susceptibility to this pathogen. Monitor and sample crops regularly so that harvest can be scheduled early if Phoma canker starts to appear. Application of fungicides may help control foliar symptoms; however, chemicals are not effective at preventing disease caused by soilborne inoculum.

References

Cerkauskas, R. F. 1985. Canker of parsnip caused by *Phoma complanata*. *Canadian Journal of Plant Pathology* 7:135–138.

Cerkauskas, R. F. 1987. Phoma canker severity and yield loss in parsnip. *Canadian Journal of Plant Pathology* 9:311–318.

Cerkauskas, R. F. 1987. Phoma canker of parsnips. Ministry of Agriculture Fisheries and Food, Ontario, Canada. Leaflet 87-041, 3pp.

Cerkauskas, R. F. and Mc Garvey, B. D. 1988. Fungicidal control of Phoma canker of parsnips. *Canadian Journal of Plant Pathology* 10:252–258.

Plasmopara umbelliferarum,
(P. crustosa/P. nivea)

DOWNY MILDEW

Introduction and significance

Downy mildew occurs on parsnip, but in the UK severe crop damage is rare. It also occurs on carrot, celery, chervil, fennel, and parsley. Downy mildew isolates affecting carrot are reported only in Europe.

Symptoms and diagnostic features

Symptoms are yellow, angular shaped leaf blotches. The lesions soon turn brown and leaves develop a ragged appearance as the centers of lesions fall out. White sporulating growth develops on the undersides of the lesions (**106**). Young, succulent growth is most susceptible to infection and leaves with multiple lesions turn yellow and senesce (**107**).

Causal agent

Downy mildew is caused by an oomycete in the genus *Plasmopara*. This produces monopodial branched sporangiophores that emerge from leaf stomates. Smaller branches that bear sporangia are arranged at right angles to the supporting branches. Sporangia are ovoid, hyaline, have a single pore on the distal ends, and measure 23–27 x 17–19 µm. Oospores are produced within infected tissues and seeds. The nomenclature of the various downy mildews that infect Apiaceae is under review. Some researchers believe the names *P. crustosa* and *P. nivea* are not valid, and propose that *P. umbelliferarum* be used as the pathogen name.

106 Parsnip leaf infected with downy mildew.

107 Necrotic leaf tissue of parsnip caused by downy mildew.

Disease cycle

Infection is favored by cool, wet conditions. Free moisture is required for sporangia to germinate and release zoospores through the distal pore. These swimming spores then invade the leaf via stomata.

Control

Control measures are not usually required. Rotate to non-host crops to reduce the risk of infection from soilborne oospores. When symptoms appear early in the season, apply fungicides.

References

Constantinescu, O. 1990. The nomenclature of *Plasmopara* parasitic on Umbelliferae. *Mycotaxon* 43:471-477.

108 Leaf lesions caused by Ramularia leaf spot of parsnip.

Ramularia pastinacae

RAMULARIA LEAF SPOT

Introduction and significance

This leaf spot disease is commonly found on parsnip. However, economic impact is usually limited.

Symptoms and diagnostic features

This disease is characterized by small (3–7 mm in diameter), pale brown leaf spots with darker margins and chlorotic halos (**108**). The centers of the older lesions often fall out, resulting in shot-hole symptoms (**109**). The undersides of active lesions support white sporulation.

109 Shot hole symptoms of Ramularia leaf spot of parsnip.

Causal agent

Leaf spot is caused by the fungus *Ramularia pastinacae*. Conidiophores emerge from leaf tissues in fascicles. Conidia are hyaline, cylindrical, and zero to four septate, though most spores are bicellular. Conidia measure 14–42 x 3–5 µm.

Disease cycle

There has been little investigation of this disease, but warm and wet conditions favor infection. The disease occurs in small foci and then spreads to cause a more general infection of the crop.

Control

Broad-spectrum fungicides used for other diseases should provide some control of this problem. Rotate to non-hosts to avoid carry over of this pathogen to subsequent parsnip plantings. Plow in crop residues soon after harvesting, and control volunteer parsnip plants.

Parsnip yellow fleck virus

PARSNIP YELLOW FLECK

Introduction and significance

Parsnip yellow fleck virus (PYFV) is widespread, infects many Apiaceae plants, and occasionally causes severe losses in carrot crops in the Netherlands and in the UK. This is the most common virus disease of parsnip in Europe.

Symptoms and diagnostic features

On carrots, symptoms are browning and death of the youngest leaves of seedlings, followed by death of these plants (**110, 111**). Plants that continue to live can show stunting, early leaf senescence, and distortion and elongation of the top part of the root. Roots can show internal discoloration and external black spotting (**112**). In carrot seed crops, plants can decline prematurely. For parsnip, the first symptoms are bold yellow veins and vein netting. Foliage later shows yellow flecking and a yellow green mosaic (**113, 114**). It is possible that the problem known as 'crown rot,' which has resulted in serious losses in northwest England since the mid-1980s, was caused by PYFV (**115**).

Causal agent

PYFV is a sequivirus and has isometric particles that are 30 nm in diameter and separate into two components during sedimentation. It is transmitted in a semi-persistent manner for a period of 1 to 4 days by adults of several aphid species: willow carrot aphid (*Cavariella aegopodii*), *C. pastinaceae*, and possibly others. A helper virus, *Anthriscus yellows virus* (AYV), is required for aphid transmission of PYFV. There are two recognized serotypes of PYFV: the parsnip serotype infects celery, parsnip, and cow parsley (*Anthriscus sylvestris*); the anthriscus serotype infects carrot, chervil, cilantro, and dill.

Disease cycle

An interesting dynamic exists between these two viruses, the various Apiaceae hosts, and the aphid vectors. Only plants that are susceptible to both viruses can act as virus reservoirs. The major crops – carrot, celery, and parsnip – are all immune to AYV, the helper virus. Therefore, within plantings of these host plants there is no plant-to-plant spread of the problem. However, weed hosts such as cow parsley (*A. sylvestris*) and hogweed (*Heracleum sphondylium*) are hosts to both viruses and therefore can be sources of both pathogens.

Control

Avoid planting susceptible crops in areas were the virus is known to cause significant disease. Control weed hosts and spray for the aphid vectors.

110 Leaf necrosis, death of the growing point, and loss of carrots caused by *Parsnip yellow fleck virus*.

111 Death of carrot growing point caused by *Parsnip yellow fleck virus*.

References

Elnager, S. and Murant, A. F. 1976. Relations of the semi-persistent viruses parsnip yellow fleck and anthriscus yellows with their vector *Cavariella aegopodii*. *Annals of Applied Biology* 84:169-181.

Hemida, S. K. and Murant, A. F. 1989. Particle properties of parsnip yellow fleck virus. *Annals of Applied Biology* 114:87–100.

Runham, S. R., Town, S. J., and Gladders, P. 1995. Evaluation of fungicides against crown-rot disorder of carrots. Tests of Agrochemicals and Cultivars: *Annals of Applied Biology* 126 (Supplement) 16: 12–13.

Van Dijk, P. and Bos, L. 1985. Viral dieback of carrot and other Apiaceae caused by the *Anthriscus* strain of parsnip yellow fleck virus, and its distinction from carrot motley dwarf. *Netherlands Journal of Plant Pathology* 91:169–187.

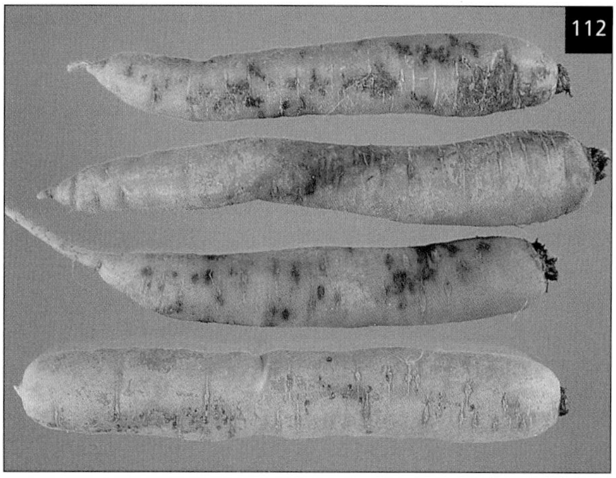

112 Black spotting on carrot roots associated with *Parsnip yellow fleck virus*.

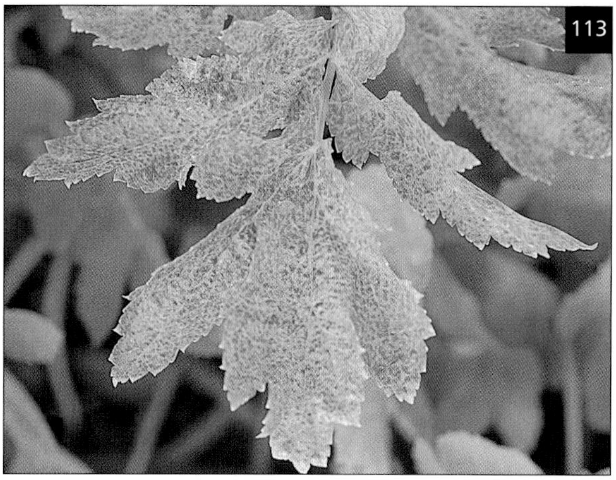

113 Mosaic symptoms on parsnip caused by *Parsnip yellow fleck virus*.

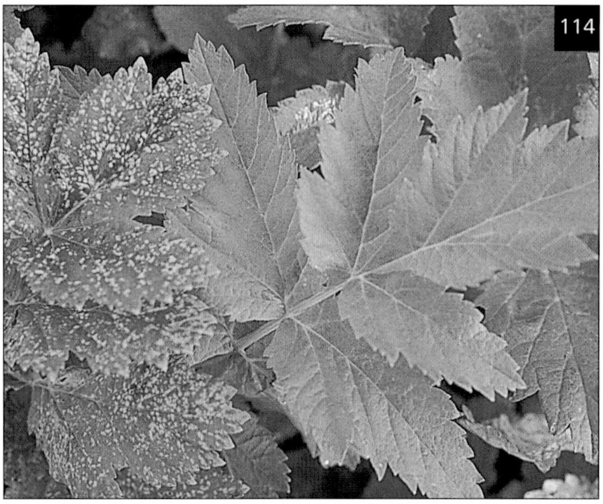

114 *Parsnip yellow fleck virus* causing yellow spotting on parsnip leaf.

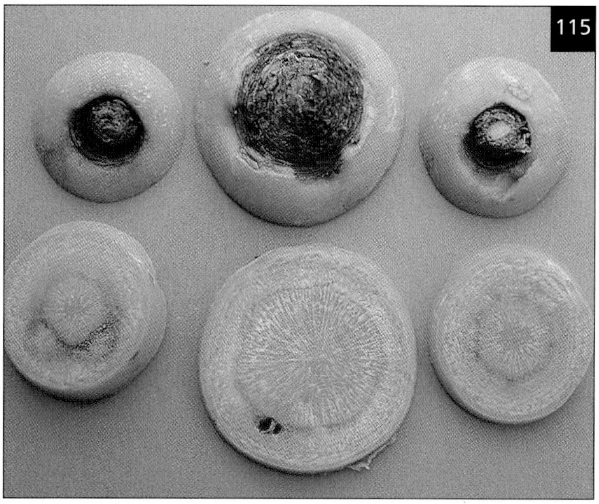

115 Crown rot and root core discoloration of carrot caused by *Parsnip yellow fleck virus*.

PETROSELINUM CRISPUM (PARSLEY)

Erysiphe heraclei, Leveillula lanuginosa

POWDERY MILDEW

116 Powdery mildew on parsley leaves.

On parsley, powdery mildew produces dense, white growth on leaves (**116**). Severe attacks cause yellowing and early senescence of the older leaves. The primary powdery mildew pathogen is *Erysiphe heraclei*. *Leveillula lanuginosa* is a second type of powdery mildew affecting parsley and causes angular yellow blotches on the upper leaf surface; white sporulation is found on both upper and lower leaf surfaces. For more details, see the carrot section on powdery mildew in this chapter (page 103).

References

Citrulli, M. 1975. The powdery mildew of parsley caused by *Leveillula lanuginosa*. *Phytopathologia Mediterranea* 14:94–99.

Koike, S. T. and Saenz, G. S. 1994. Occurrence of powdery mildew on parsley in California. *Plant Disease* 78:1219.

Phytophthora spp., *Pythium* spp.

ROOT ROT

Introduction and significance

Root rots caused by *Phytophthora* and *Pythium* pathogens are important field problems in parsley. These diseases have been reported from Europe, North America, and Australia. In parsley, losses of more than 50% can occur.

Symptoms and diagnostic features

The first symptoms of Phytophthora root rot are leaf yellowing and wilting of foliage several weeks after planting. Plants are stunted and finally collapse and die (**117, 118**). The main parsley roots have orange-brown lesions that enlarge and rot. Pythium root rots result in stunted plants and decayed feeder roots.

Causal agents

Several species of *Phytophthora* infect parsley, including *P. cryptogea* in California, *P. parasitica* in Hawaii, and *P. primulae* in the UK. Carrots are affected by other *Phytophthora* species, including *P. cactorum*, *P. cryptogea*, *P. megasperma*, *P. parasitica*, *P. porri*, and *P. primulae*. *Pythium paroecandrum* and perhaps other *Pythium* species affect parsley. *Phytophthora* and *Pythium* pathogens are oomycete organisms, and most species produce sporangia and swimming zoospores.

Disease cycle

These pathogens are soilborne and are mostly favored by wet, cool soil conditions. However, *P. parasitica* on carrots develops when temperatures are high (30–36° C). Some isolates show host specialization. For example, *P. parasitica* from parsley infects coriander but not carrot, celery, or tomato.

Control

Rotate with non-susceptible crops. Prepare the soil so that there is good drainage and reduced soil compaction. Planting on raised beds may be helpful. Poorly draining sites should be avoided. Fungicide treatments of metalaxyl have been used in the UK.

117 Collapsed parsley infected with *Phytophthora primulae*.

118 *Phytophthora primulae* root rot of parsley.

References

Davis, R. M., Winterbottom, C. Q., and Mirectich, S.M.1994. First report of Phytophthora root rot of parsley. *Plant Disease* 78:1122.

Hershman, D. E., Varney, E. H., and Johnston, S. A .1986. Etiology of parsley damping-off and influence of temperature. *Plant Disease* 70:927–930.

Hine, R. B. and Aragaki, M. 1963. Influence of soil temperature on a crown rot disease of parsley caused by *Phytophthora parasitica*. *Phytopathology* 53:1113–1114.

McCracken, A. R. 1984. *Pythium paroecandrum* associated with root-rot of parsley. *Plant Pathology* 33: 603–604.

Sclerotinia sclerotiorum

WHITE MOLD

White mold occurs on a number of Apiaceae crops, including parsley. This pathogen is perhaps more commonly found on parsley being grown for seed than on parsley grown for market.

Infected seed stalks of parsley become light tan to bleached in color. The leaves on affected stems initially wilt (**119**). Eventually the entire stem dries up. White mycelium may develop on the outside of such stems. However, cutting into the seed stalk will usually reveal profuse white mycelium and large black sclerotia in the central core of the stalk (**120**).

For more detail see the white mold section for carrot and the pink rot section for celery.

119 Parsley infected with *Sclerotinia sclerotiorum*.

120 Sclerotia of *Sclerotinia sclerotiorum* forming inside parsley stem.

Septoria petroselini

SEPTORIA BLIGHT

Introduction and significance

This is a common and important foliar disease of parsley in Europe, Asia, and North America. Severe disease can render the parsley crop unmarketable. Parsley is not susceptible to the late blight pathogen of celery, *Septoria apiicola*.

Symptoms and diagnostic features

Symptoms consist of irregularly shaped leaf spots that initially are small and measure 2–5 mm in diameter. As disease develops, the spots increase in number, coalesce, and result in extensive blighting of the overall foliage (**121, 122**). Leaf spots range in color from green-brown to bleached white and tend to be angular in shape. The presence of many tiny, dark fruiting bodies (pycnidia) in the spots is a key characteristic sign of *Septoria*.

Causal agent

Septoria blight is caused by the fungus *Septoria petroselini*. On plant tissue the pathogen is readily identified by its black pycnidia that are immersed in plant tissues. Pycnidia are ostiolate, measure approximately 100 μm in diameter, and contain hyaline, multicelled, filamentous conidia (30–40 x 1–2 μm) that are mainly one to three septate. No perfect stage has been reported. The pathogen is seedborne. *Septoria petroselini* does not infect celery or celeriac.

121 Parsley leaves infected with Septoria blight.

122 Parsley leaves infected with Septoria blight.

Disease cycle

Septoria blight of parsley is a seedborne disease. There is little research on the epidemiology of this pathogen. However, the biology will be very similar to that of *S. apiicola*; see the celery section on late blight in this chapter (page 90). Rain and overhead irrigation spread the pathogen in both transplant and field settings.

Control

See the celery section on late blight in this chapter (page 90). As with celery, it is critical to use seed that does not have significant levels of the pathogen. If rain or overhead sprinkler water does not occur, it is often possible to trim back symptomatic parsley plants and harvest the subsequent healthy regrowth. Parsley cultivars may vary in susceptibility; therefore, plant tolerant cultivars if available.

References

Miller, S. A., Colburn, G. C., and Evans, W. B. 1999. Management of Septoria leaf blight of parsley with fungicides and efficacy of a disease predictive model. *Phytopathology* 83:S53 (Abstract).

Asparagus officinalis Asparagus

ASPARAGUS (*Asparagus officinalis*) is a perennial monocot plant formerly included in the Liliaceae (lily family) but now separately classed within the Asparagaceae. It is grown for its immature, unexpanded shoots that are called spears. Spears originate from large underground crowns that consist of fleshy rhizomes. Asparagus spears are only harvested for a limited period during the spring season, after which the remaining spears are allowed to grow, expand, and develop into the foliage, or fern, of the plant. The fern stage replenishes crown energy and nutrients and allows the plant to grow and remain commercially productive. Asparagus is grown as a perennial crop with a productive life of 5 to 15 years. In addition to traditional green asparagus, there are purple spear varieties and white spears that are cut while they are still below ground and have not formed chlorophyll.

Cercospora asparagi

CERCOSPORA BLIGHT

Introduction and significance

This disease is most damaging in conjunction with high temperatures and humidity, and problems occur in parts of southern USA, South America, and Asia. Damage is much less severe in cooler climates.

Symptoms and diagnostic features

Pale brown to gray elliptical lesions that measure 1–4 mm in diameter, with a red-brown or purple margin, are produced on leaves and branches. Lesions appear before the canopy closes over the lower part of stems. As disease progresses, needles and smaller shoots turn yellow, then brown, and dry.

Causal agent

Cercospora blight is caused by the fungus *Cercospora asparagi*. The pathogen produces hyaline, multicelled, filiform spores that measure 35–130 x 2.5–5 µm and are borne on clusters of brown conidiophores. Conidiophores arise from black stromata. Sporulation on the lesions causes these spots to take on a pale gray appearance. This pathogen appears to be host specific to asparagus.

Disease cycle

Conidia are dispersed by wind or splashing water from rain and sprinkler irrigations. Disease develops rapidly if warm, humid weather persists. Infected crop residue is the main source of inoculum.

Control

Apply fungicide sprays. Avoid using overhead sprinkler irrigation. At the end of the growing season, reduce pathogen inoculum by managing the diseased fern. For smaller acreages, the fern can be removed from the field and destroyed. Otherwise, fern can be mowed or chopped, and then buried by cultivation or re-ridging (earthing up) during the dormant period.

References

Conway, K. E., Motes, J. E., and Foor, C. J. 1990. Comparison of chemical and cultural controls for Cercospora blight on asparagus and correlations between disease levels and yield. *Phytopathology* 80:1103–1108.

Cooperman, C. J., Jenkins, S. F., and Averre, C. W. 1986. Overwintering and aerobiology of *Cercospora asparagi* in North Carolina. *Plant Disease* 70:392–394.

Fusarium oxysporum f. sp. *asparagi*,
F. proliferatum (teleomorph = *Gibberella fujikuroi*)

FUSARIUM CROWN & ROOT ROT

Introduction and significance

Fusarium crown and root rot is implicated in the so-called 'decline' problem of asparagus. The decline syndrome results in a progressive yield loss over a period of years. Other factors, such as virus infection, are also thought to contribute to decline. Fusarium crown and root rot is recognized as an economically important problem worldwide.

Symptoms and diagnostic features

Fusarium crown and root rot causes the fern foliage to first turn chlorotic (**123**). Later the fern wilts, turns brown, and dies (**124**). Internal tissues of lower stems and crowns have a red-brown discoloration, but remain firm and unrotted (**125**). Red-brown, oval lesions also develop on asparagus roots and lower stems (**126**). Infected roots become dark, soft, and stringy. Plant growth is weakened and severely affected plants die. Postharvest decay of the spears may be caused by various *Fusarium* spp., which results in reduced quality and shelf-life. Seedling asparagus plants are stunted and grow poorly, and foliage can be bright yellow.

Causal agents

Fusarium crown and root rot is caused by two species of *Fusarium*: *F. oxysporum* f. sp. *asparagi* and *F. proliferatum*. For the first species, pathogen morphology and colony characteristics are similar to other *F. oxysporum* fungi. The fungus forms one- or two-celled, oval to kidney-shaped microconidia on monophialides, and four- to six-celled, fusiform, curved macroconidia. Microconidia measure 6–15 x 2.5–4.0 µm, while macroconidia range from 27–60 x 3.5–5.5 µm. Macroconidia are usually produced in cushion-shaped structures called sporodochia. Spherical chlamydospores are also formed and are 10–11 µm in diameter.

Fusarium proliferatum produces microconidia and macroconidia on polyphialides. An important identfication feature is that microconidia are borne in long chains and have distinctly truncated bases. Chlamydospores are not formed. The pathogens can be isolated from symptomatic vascular tissue. However, semi-selective media like Komada can help isolate them because saprophytic organisms are often present in asparagus crowns and surrounding tissues. *Fusarium oxysporum* f. sp. *asparagi* is apparently host specific to asparagus. Other *Fusarium* species can also be found in asparagus crown and root tissue or sporulating at the base of spears. *Fusarium culmorum* causes a foot and stem rot of asparagus. *Fusarium redolens* f. sp. *asparagi* has recently been separated from the *F. oxysporum* group and also causes a crown and root rot.

123 Yellowing of asparagus foliage caused by Fusarium wilt.

124 Dieback of asparagus foliage caused by Fusarium wilt.

Disease cycle

In addition to being a true soilborne pathogen able to survive long periods of time in soil, the Fusarium wilt pathogen is also seedborne. However, infected crowns used for propagation are often the most important sources of infection. The practice of propagating asparagus by cutting up existing plant crowns allows the pathogen to be distributed along with crown divisions used to establish new plantings. Diseased propagation crowns and infested seed result in new plants being already infected. In the case of soilborne inoculum, the fungus infects asparagus roots and then grows systemically in the plant. If asparagus plants are infected with viruses, root exudate release is increased and soilborne Fusarium can be stimulated. A similar interaction can occur when asparagus is replanted after a previous asparagus crop; autotoxins released by the previous asparagus crop's residues increase nutrient leakage from roots of the new crop.

Control

Use seed that does not have significant levels of the pathogen. Plant seed in propagation fields that do not have a history of the problem, and which have not been recently used for asparagus. If using crowns for propagation, use only healthy, symptomless plants that are taken from fields that do not have a history of this disease. In some countries soil tests are used to detect *Fusarium* prior to selecting uninfested propagation sites. Maintain records so that asparagus is not planted in fields having a history of this problem. Fungicide seed treatments may help prevent infection of seedlings. Manage the crop to reduce stress on the plants. Practice extended crop rotations, perhaps 5 or more years, between asparagus plantings. Some cultivars such as UC 157 have some tolerance to this pathogen. An older practice consisted of applying salt (NaCl) to the beds of infected plants. Though not widely practiced in the USA, the application of salts helped reduce disease severity and associated crop loss.

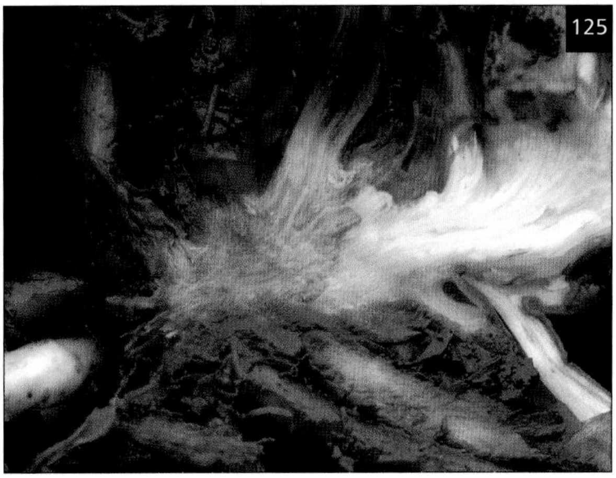

125 Vascular discoloration in asparagus crown infected with *Fusarium oxysporum*.

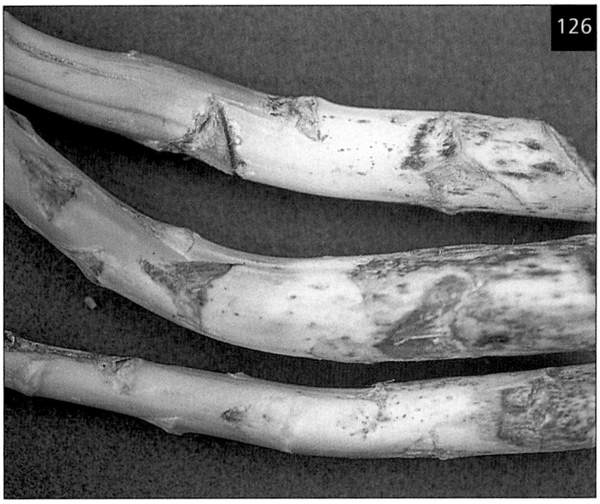

126 Reddish lesions at base of spears caused by Fusarium wilt.

References

Baayen, R. P., van den Boogert, P. H. J. F., Bonants, P. J. M., Poll, J. T. K., Blok, W. J., and Waalwijk, C. 2000. *Fusarium redolens* f.sp. *asparagi*, causal agent of asparagus root rot, crown rot and spear rot. *European Journal of Plant Pathology* 106:907–912.

Blok, W. J. and Bollen, G. J. 1975. Host specificity and vegetative compatibility of Dutch isolates of *Fusarium oxysporum* f. sp. *asparagi*. *Canadian Journal of Botany* 75:383–393.

Blok, W. J. and Bollen, G. J. 1993. The role of autotoxins from root residues of the previous crop in the replant disease of asparagus. *Netherlands Journal of Plant Pathology* 99, *Supplement* 3:29–40.

Didelot, D., Nourrisseau, J. G., and Bouhot, D. 1996. Fusarium root rot of asparagus: Development of a predictive test for root rot. *Proceedings of the VIII International Symposium on Asparagus. Acta Horticulturae* 415:373–375.

Elmer, W. H. 2004. Combining nonpathogenic strains of *Fusarium oxysporum* with sodium chloride to suppress Fusarium crown and root rot of asparagus in replanted fields. *Plant Pathology* 53:751–758.

Elmer, W. H. 2000. Incidence of infection of asparagus spears marketed in Connecticut by *Fusarium* spp. *Plant Disease* 84:831–834.

Elmer, W. H. 1992. Suppression of Fusarium crown and root rot of asparagus with sodium chloride. *Phytopathology* 82:97–104.

Elmer, W. H., Johnston, D. A., and Mink, G. I. 1996. Epidemiology and management of the diseases causal to asparagus decline. *Plant Disease* 80:117–125.

Elmer, W. H. and Stephens, C. T. 1989. Classification of *Fusarium oxysporum* f. sp. *asparagi* into vegetatively compatible groups. *Phytopathology* 79:88–93.

Elmer, W. H., Summerell, B. A., Burgess, L. W., and Nigh, E. L. 1999. Vegetative compatibility groups in *Fusarium proliferatum* from asparagus in Australia. *Mycologia* 91:650–654.

Evans, T. A. and Stephens, C. T. 1989. Increased susceptibility to Fusarium crown and root rot in virus-infected asparagus. *Phytopathology* 79:253–258.

Inglis, D. A. 1980. Contamination of asparagus seed by *Fusarium oxysporum* f. sp. *asparagi* and *Fusarium moniliforme*. *Plant Disease* 64:74–76.

Johnston, S. A., Springer, J. K., and Lewis, G. D. 1979. *Fusarium moniliforme* as a cause of stem and crown rot of asparagus and its association with asparagus decline. *Phytopathology* 69:778–780.

Knaflewski, M. and Sadowski, C. 1990. Effect of chemical treatment of seeds on the healthiness of asparagus seeds and crowns. *Acta Horticulturae* 271:383–387.

Nigh, E. L. 1990. Stress factors influencing Fusarium infection in asparagus. *Acta Horticulturae* 271:315–322.

Nigh, E .L., Guerrero, C., and Stanghellini, M. E. 1997. Evaluation of Fusarium infected asparagus spears for fumonisin mycotoxins. *Proceedings of IX International Asparagus Conference*, 100–106.

Schofield, P. 1991. Asparagus decline and replant problem in New Zealand. *New Zealand Journal of Crop and Horticultural Science* 19:213–220.

Schofield, P., Nichols, M. A., and Long, P. G. 1996. The involvement of *Fusarium* spp. and toxins in the asparagus replant problem. *Proceedings of the VIII International Symposium on Asparagus. Acta Horticulturae* 415:309–314.

Schreuder, W., Lamprecht, S. C., Marasas, W. F. O., and Calitz, F. J. 1995. Pathogenicity of three *Fusarium* species associated with asparagus decline in South Africa. *Plant Disease* 79:177–181.

Helicobasidium brebissonii* (anamorph = *Rhizoctonia crocorum*), *Zopfia rhizophila

ROOT ROTS

There are occasional problems with violet root rot, caused by *Helicobasidium brebissonii*, in asparagus. The anamorph, *Rhizoctonia crocorum*, is the primary stage that causes the disease and produces coarse mycelial strands (up to 1 mm diameter) that enable it to spread through soil and infect new host plants. Roots become covered with strands or dense mats of purple mycelium. This pathogen is soilborne, occurs where root crops have been grown intensively, and spreads from plant to plant. Because of the practice of growing asparagus as a perennial, even low levels of soil infestation can result in significant problems. Control relies on avoiding infested sites. Where disease has occurred, do not replant asparagus. See the violet root rot section under carrot in the Apiaceae disease chapter (page 105).

The fungus *Zopfia rhizophila* is common on the old fleshy or dead roots of mature asparagus plants. It produces numerous black perithecia on the root surface. Perithecia contain asci having two-celled, brown ascospores that measure 68–85 x 35–45 µm. This fungus is generally regarded as a weak pathogen. Plants growing in poorly drained areas are most likely to be affected.

References

Sadowski, C. 1990. The occurrence of *Zopfia rhizophila* Rabenh. on asparagus roots in Poland. *Acta Horticulturae* 271:377–381.

Phytophthora megasperma, Phytophthora spp.

PHYTOPHTHORA SPEAR AND CROWN ROT

Introduction and significance

Phytophthora species are important pathogens of asparagus in the Americas, Europe, Australia, and New Zealand. Losses can exceed 50%. The disease appears to be less important where white asparagus is produced. Problems in England have appeared only recently. Losses occur from reduced stands in new plantings, lowered yields in established crops, and postharvest rot of spears.

Symptoms and diagnostic features

This is an early season disease. Emerging spears initially develop small, light brown lesions. As disease progresses, lesions increase in size and can completely girdle the spear. Spears often bend or twist at the infection site, and eventually can collapse. Spear tissue becomes soft and supports the growth of white mycelium (**127**). If conditions dry after infection, diseased spears will shrivel. Infected roots are initially water-soaked, then later brown and rotted. The pathogen can also infect asparagus crowns and causes a browning and softening of internal tissues. The soft tissue contrasts with dry rot symptoms caused by *Fusarium*.

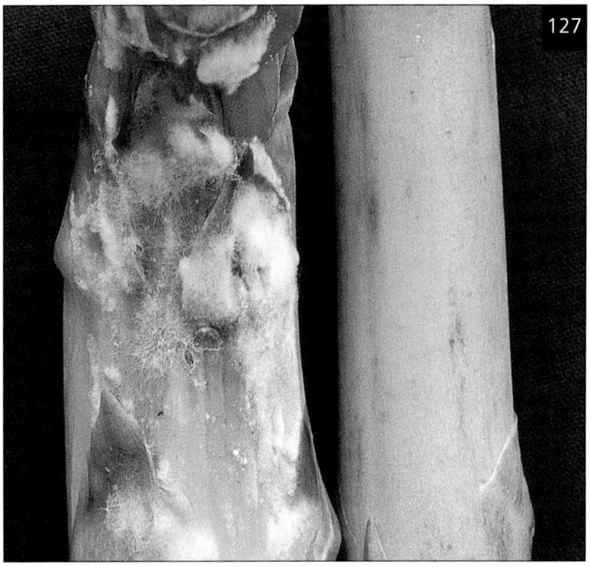

127 White mycelium and spear infection of asparagus infected with *Phytophthora*.

Causal agents

The main pathogen of Phytophthora spear and crown rot is the oomycete *Phytophthora megasperma*. Semi-selective media, such as corn meal agar amended with pimaricin, ampicillin, rifampicin, and PCNB, are helpful for recovering *Phytophthora*. *Phytophthora megasperma* is homothallic and produces oogonia with mainly paragynous antheridia. Sporangia are non-papillate, internally proliferating, and measure 35–60 x 25–45 µm. Other *Phytophthora* pathogens of asparagus include *P. cactorum* and *P. cryptogea* in California, and *P. richardiae* in Australia. All *Phytophthora* species affecting asparagus are soilborne organisms. Asparagus isolates of *P. megasperma* may be a distinct species and could be renamed in the future.

Disease cycle

Phytophthora species survive for long periods in soil by means of oospores. Zoospores are released from sporangia that are produced on infected host tissues or from germinating oospores. Zoospores swim in the soil water and infect shoots and roots. Young spears are the most susceptible. The optimum temperatures for infection are 10–12° C. Severe attacks are associated with wet conditions and poor drainage. Later in the season when temperatures are 20–25° C and fern growth is mature, there may be little disease. Infected crowns that are divided and used for propagation will result in new plantings that already have the disease.

Control

Resistant cultivars are not yet available, though new hybrids from New Zealand have partial resistance. Use seedlings and crown divisions that are disease-free. Avoid fields that have a history of the disease or poorly draining, heavy soils. Manage the field so that soils have good structure and drain well. Apply fungicides, particularly phenylamide and fosetyl aluminum products.

References

Falloon, P. G. 1990. Field screening of asparagus for tolerance to Phytophthora rot. *Proceedings of the 7th International Symposium on Asparagus. Acta Horticulturae* 271:69–76.

Falloon, P. G. and Grogan, R. G. 1988. Isolation, distribution, pathogenicity and identification of *Phytophthora* species on asparagus in California. *Plant Disease* 72:495–497.

Falloon, P. G., Mullen, R. J., Benson, B. L., and Grogan, R. G. 1985. Control of Phytophthora rot with metalaxyl in established asparagus. *Plant Disease* 69:921–923.

Pleospora herbarum
(anamorph = *Stemphylium vesicarium*)

PURPLE SPOT

128 Stem lesions of purple spot of asparagus.

129 Yellowing of asparagus leaves caused by *Stemphylium*.

Introduction and significance

This disease is an important cause of yield loss in some production areas, notably New Zealand, USA, and parts of Europe. In other areas, disease severity varies and is dependent on wet weather in the latter part of the growing season. Yield is reduced when purple spot causes premature defoliation and diseased, unmarketable spears.

Symptoms and diagnostic features

Two types of purple spot inoculum, ascospores and conidia, cause similar symptoms. Small, elliptical lesions that measure 2–6 mm in diameter develop on the emerging spears. Lesions are often sunken and brown with purple margins. These infection sites are shallow and do not penetrate deeply into the spear. However, the prominent lesions still reduce marketability of the product. Severe attacks result in large numbers of lesions. Damage to spears from windblown soil increases the degree of infection. On mature fern, the symptoms are small white lesions with red-brown margins that are on individual leaflets and stems (**128**). Disease causes yellowing and defoliation of the needles (**129**). Loss of the foliage can weaken the crown and reduce yields in subsequent seasons.

Causal agent

The cause of purple spot is the ascomycete fungus *Pleospora herbarum*. Spherical perithecia mature over several months and produce yellow-brown ascospores that have seven transverse septa and measure 38 x 18 µm. The anamorph of this pathogen is *Stemphylium vesicarium*. Conidia of *S. vesicarium* have verrucose walls, are yellow-brown to olive-brown, measure 25–42 x 12–22 µm, and have one to six (usually three) transeverse septa and one to three longitudinal septa. The spores have a conspicuous basal scar. Conidia are borne on mainly unbranched, yellow-brown to olive-brown conidiophores that have a distinctly swollen apical cell with a pore.

Disease cycle

The pathogen survives on crop residues and fern debris, and produces perithecia that release ascospores in the spring. Spear infection is associated with periods of wet weather during the harvest period. Symptoms may appear on spears in less than 24 hours after a rain. Infection occurs through stomata, and long periods (greater than 16 hours) of surface wetness are required for severe infection to develop. This pathogen does not persist in the soil.

Control

Remove and dispose of fern residues prior to emergence of spears in spring. Differences in cultivar susceptibility may exist, so select more tolerant cultivars. Short compact plants with low branches (for example, the culivar Cito) are generally more susceptible than tall erect types (cv. Jersey King). Apply protectant fungicides, such as chlorothalonil, to the fern.

References

Evans, T. A. and Stephens, C. T. 1984. First report in Michigan of the teleomorph of *Stemphylium vesicarium*, causal agent of purple spot of asparagus. *Plant Disease* 68:1099.

Falloon, P. G., Falloon, L. M., and Grogan, R. G. 1987. Etiology and epidemiology of Stemphylium leaf spot and purple spot of asparagus in California. *Phytopathology* 77:407–413.

Hausbeck, M. K., Hartwell, J., and Byrne, J. M. 1997. Epidemiology of Stemphylium leaf spot and purple spot in no-till asparagus. *Proceedings of IX International Asparagus Conference*, 48–53.

Johnson, D. A. 1990. Effect of crop debris management on severity of Stemphylium purple spot of asparagus. *Plant Disease* 74:413–415.

Johnson, D. A. and Lunden, J. D. 1986. Effects of wounding and wetting duration on infection of asparagus by *Stemphylium vesicarium*. *Plant Disease* 70:419–420.

Lacy, M. L. 1982. Purple spot: a new disease of young asparagus spears caused by *Stemphylium vesicarium*. *Plant Disease* 66:1198–1200.

Meyer, M. P., Hausbeck, M. K., and Podolsky, R. 2000. Optimal fungicide management of purple spot of asparagus and impact of yield. *Plant Disease* 84:525–530.

Suberi, H. and Price, T. V. 2000. Infection of leaves by *Alternaria porri* and *Stemphylium vesicarium* and disease development in controlled environments. *Plant Pathology* 49:375–382.

Sutherland, P. W., Hallet, I. C., Parkes, S. L., and Templeton, M. D. 1990. Structural studies of asparagus spear infection by *Stemphylium*. *Canadian Journal of Botany* 68:1311–1319.

Puccinia asparagi

RUST

Introduction and significance

Rust occurs worldwide wherever asparagus is grown. The disease is an important cause of early defoliation, and hence yield loss for subsequent years, in production areas in North America and Europe. In the UK, rust was of minor significance for many years, but it remains a threat as indicated by serious outbreaks in 1997.

Symptoms and diagnostic features

After asparagus harvest is completed, subsequent shoots are allowed to develop and grow into the foliage or fern. Rust disease infects this fern foliage and causes dark to brown-red pustules on the stems and needles (**130**). Rust can spread rapidly, causing extensive

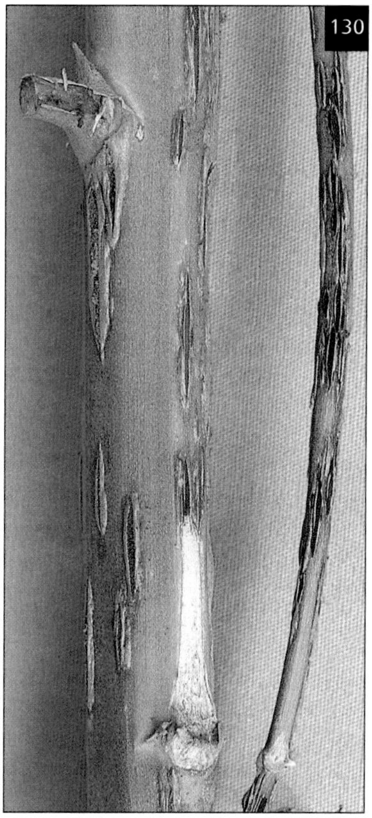

130 Stem pustules of rust of asparagus.

131 Dieback of asparagus foliage infected with rust.

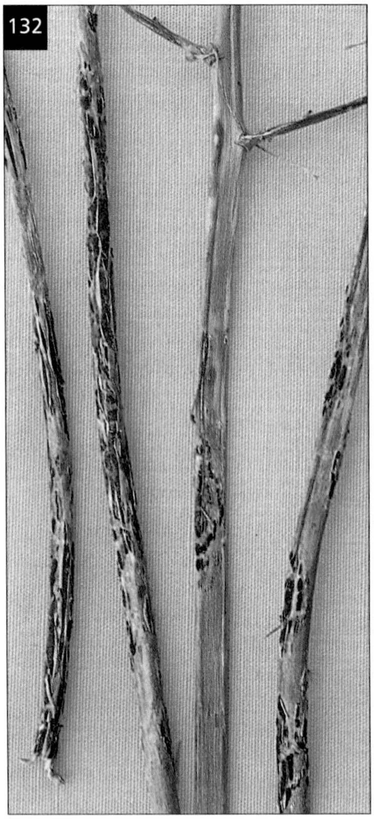

132 Black teliospores of rust of asparagus.

infection of the foliage (**131**). Later in the season, the pustules may become black as the teliospores are formed (**132**). Severe infection results in weakened plants and reduced yield.

Causal agent

Rust disease is caused by the basidiomycete fungus *Puccinia asparagi*, an autoecious, macrocyclic rust. The fungus forms uredinia pustules that release brown, one-celled urediniospores that measure 19–30 x 18–25 µm. Later in disease development, the same pustules will form black, two-celled teliospores that measure 30–50 x 19–26 µm. Two other spore types, spermagonia and aecidiospores, can also develop on yellowish lesions in the spring but are often overlooked.

Disease cycle

Like most rust fungi, the life cycle is complex. Disease is often initiated by teliospores that germinate to produce intermediate spore types that eventually form aecidia. Aecidiospores penetrate the asparagus host and result in uredinia and urediniospores during the summer. Teliospores are produced in increasing numbers towards the end of the season. Sporulation is optimal at 25–30° C. Spore germination requires surface moisture for only 3 hours (optimum 9 hours) at 15° C. The combination of warm days and cool nights with dew formation is particularly favorable for rust development.

Control

Remove and dispose of fern residues prior to emergence of spears in spring. Remove volunteer asparagus plants that might be infected. Apply fungicides such as triazole products. Resistant cultivars are not yet available.

References

Beraha, L., Linn, M. B., and Anderson, H. W. 1960. Development of asparagus rust pathogen in relation to temperature and moisture. *Plant Disease Reporter* 44:82–86.

Blanchette, B. L., Groth, J. V., and Waters, L. 1982. Evaluation of asparagus for resistance to *Puccinia asparagi*. *Plant Disease* 66:904–906.

Fantino, M. G., Granier, A., Solaini, J., and Di Carmine, D. 1990. Four years of trials (1985–1988) on asparagus rust (*Puccinia asparagi* D. C.) in Lazio. *Acta Horticulturae* 271:371–375.

Johnson, D. A. 1986. Two components of slow-rusting in asparagus infected with *Puccinia asparagi*. *Phytopathology* 76: 08–211.

Johnson, D. A. and Lunden, J. D. 1992. Effect of rust on yield of susceptible and resistant asparagus cultivars. *Plant Disease* 76:84–86.

Mullen, R. J. and Viss, T. C. 1996. Control of asparagus rust in the Sacramento-San Joaquin delta region of California. *Proceedings of the VIII International Symposium on Asparagus. Acta Horticulturae* 415:297–300

Asparagus virus 1, 2, 3, Tobacco streak virus

ASPARAGUS VIRUSES

Introduction and significance

Asparagus is affected by several viruses including *Arabis mosaic virus*, *Cucumber mosaic virus*, *Strawberry latent ringspot virus*, *Tobacco streak virus*, and three asparagus viruses: *Asparagus virus 1*, *Asparagus virus 2*, and *Asparagus virus 3*. The extent of virus problems on asparagus has not been thoroughly investigated. At present the asparagus viruses 1, 2, and 3 and *Tobacco streak virus* are considered the most important. *Asparagus virus 3* has only been reported from Japan.

Symptoms and diagnostic features

Asparagus plants do not exhibit typical virus-like symptoms when infected with these virus pathogens. Deformities, mosaic patterns, and other typical symptoms are not observed on diseased plants. However, infected asparagus plants will have reduced vigor and yield fewer spears, especially if plants were infected at an early stage. Virus-infected plants may be more susceptible to other pathogens.

Causal agents and disease cycle

Asparagus virus 1 (AV-1) is a potyvirus that has a 750 nm flexuous particle and is worldwide in distribution. In the USA (Washington state), older plantings can have 100% infection. It is transmitted in a nonpersistent manner by various aphid species, including *Myzus persicae*. AV-1 can be mechanically transmitted, but not by plant-to-plant contact in the field. Weed hosts include fat hen (*Chenopodium album*). AV-1 does not cause obvious symptoms by itself, but yield losses of up to 30% have been claimed. Asparagus plants often carry a mixture of viruses; AV-1 in combination with AV-2 has caused much more severe yield loss (up to 70%) than either virus alone.

Asparagus virus 2 (AV-2) has no known insect vector and is transmitted mechanically via sap on cutting knives, through seed, and by pollen. AV-2 is an ilarvirus with isometric particles and is widely distributed in North America, New Zealand, and Europe. In New Zealand, the average incidence of AV-2 in a survey of 67 fields was 44%. In the USA, seed infection in the late 1980s averaged 22%, but this has now been largely eliminated with virus-tested seed. AV-2 has a wider host range than AV-1 and includes other vegetables such as beet, cucumber, basil, and bean. Yield loss estimates for AV-2 alone are often 20–30%, but can be much higher. Single virus infections of AV-1 or AV-2 and virus combinations may increase nutrient exudate release from asparagus roots; this factor results in more severe damage from Fusarium crown and root rot disease.

Asparagus virus 3 (AV-3) is a potexvirus. It is transmitted mechanically and has filamentous, flexuous particles 580 nm long. *Tobacco streak virus* (TSV) is an ilavirus most prominent in North and South America and Australia. It is transmitted by thrips (*Frankliniella occidentalis, Thrips tabaci*) and mechanically. It may also be seedborne and transmitted to flowering plants by pollen. In the USA, TSV has been associated with plants that were infected by AV-2 from seed. It has a wider host range than AV-2 including beet, cucurbits, lettuce, tomato, pea, basil, bean, and spinach.

Control

AV-2 is the most important virus to control. Use seed, transplants, and divided crowns that do not harbor the virus. Disinfect cutting knives used in harvest, though this is not always feasible. Controlling the vectors will not provide good control of AV-1 or TSV as these are acquired and transmitted rapidly. However, insect management should be practiced. Maintain good weed control if weed hosts are present.

References

Betaccini A, Giunchini, L., and Poggi Pollini, C. 1990. Survey on asparagus virus diseases in Italy. *Acta Horticulturae* 271:279–284.

Elmer, W. H. 2001. *The economically important diseases of asparagus in the United States*. Online. Plant Health Progress doi:10.1094/PHP-2001-0521-01-RV.

Evans, T. A., DeVries, R. M., Wacker, T. L., and Stephens, C. T. 1990. Epidemiology of asparagus viruses in Michigan asparagus. *Acta Horticulturae* 271:285–290.

Falloon, P. G., Falloon, L. M. and Grogan, R. G. 1986. Survey of California asparagus for asparagus virus I, asparagus virus II and tobacco streak virus. *Plant Disease* 70:103–105.

Jaspers, M. V. and Falloon, P. G. 1996. Survey of asparagus crops in New Zealand for asparagus virus 2. *Acta Horticulturae* 415:301–307.

Jaspers, M. V., Falloon, P. G., and Pearson, M. N. 1999. Long-term effects of asparagus virus 2 infection on growth and reproductivity in asparagus. *Annals of Applied Biology* 135:379–384.

Uyeda, I. and Mink, G. I. 1981. Properties of asparagus virus II, a new member of the ilarvirus group. *Phytopathology* 40:832–846.

Beta vulgaris Beet

THE PRIMARY VEGETABLES in the former Chenopodiaceae (goosefoot family) – now a subfamily of the Amaranthaceae – include those grown as leafy vegetables and as root commodities. Leafy commodities are primarily spinach (*Spinacia oleracea*) and Swiss chard (*Beta vulgaris* subsp. *cicla*). The main root commodity is table beet or beetroot (*Beta vulgaris*). However, young beet foliage is also harvested for market and is called leaf beet or spinach beet. Cultivated beet and related commercial commodities are derived from *Beta maritima* (sea beet) that originated in the Mediterranean and North African areas. Table beet traditionally has been grown for its thick, edible red root. The red root color comes from a combination of very stable purple (betacyanin) and yellow (betaxanthin) pigments. Recently many more table beet choices are available as cultivars and can have roots that are white, yellow, and other colors. Table beets are also grown for the foliage and sold as beet greens. In production areas such as California, beet is planted in high density, wide bed (2 m across) configurations, grown for only a few weeks, then harvested for the young, tender foliage. This foliage is used as a fresh market commodity that is mixed with lettuce, mustards, and other leafy vegetables and bagged as a ready-to-eat salad product. Swiss chard is grown for its large leaves and prominent petioles. Chard cultivars are likewise diverse, and various cultivars have white, red, yellow, or orange-colored petioles. Like beets in California, Swiss chard is also grown for only a few weeks and then harvested for a small, 'baby leaf' vegetable commodity. As sugar beet (*Beta vulgaris*) is another closely related vegetable, information on sugar beet should be consulted when considering diseases of spinach, Swiss chard, and table beet. Because of the economic importance of spinach, spinach diseases are detailed in a separate chapter.

Aphanomyces cochlioides

DAMPING-OFF, BLACK ROOT ROT

Introduction and significance

Damping-off and black root rot diseases affect table beet, Swiss chard, sugar beet, spinach, and a number of weeds including *Chenopodium album*. The disease is known as 'caida' in South America. If conditions favor the pathogen, there can be widespread stand loss.

Symptoms and diagnostic features

Pre-emergence losses are usually limited, but seedlings that do emerge above ground become infected by soilborne inoculum or mycelium spreading from nearby infected seedlings. The pathogen invades the hypocotyl and spreads into the petioles and cotyledons, often causing plants to collapse and die. Infected tissues are initially brown and then turn black. As the fleshy tissues of the lower stem rot and dissipate, the remaining central core of the hypocotyl dries out, becoming dark, hard, and wire-like in appearance and texture (**133**). If temperatures are above 15° C and the soil is wet, extensive areas of the crop could be lost within a few days. On older plants the disease is called black root rot. These larger plants develop root rots, especially of the fine feeder roots, which result in stunted growth, yellowing, and wilting. Roots are dark in color. The shape of the swollen hypocotyl and upper taproot, the harvested beet, is distorted.

133 Beet seedlings affected by *Aphanomyces*.

134 Damping-off of red beet seedlings caused by *Pythium*.

Causal agent
Damping-off and black root rot diseases are caused by the oomycete *Aphanomyces cochlioides*. The life cycle of this pathogen is probably similar to that described for *A. euteiches*. The sexual stage for homothallic *A. cochlioides* develops when antheridia fuse with the oogonium and an oospore is produced. Oospores are hyaline to yellow, measure 20–23 µm in diameter, and have less sculpturing of the internal wall than *A. euteiches*. Oospores are formed in rotted tissues and are capable of surviving for many years in soil. The asexual stage consists of long, filamentous zoosporangia that may be up to 3–4 mm long. Swimming primary zoospores are released from the tips of the zoosporangia. These primary zoospores measure 6–15 µm and may encyst and later germinate and release secondary zoospores. A number of other soilborne pathogens cause damping-off of beet, including the following: *Pleospora bjoerlingii*, *Pythium* species (**134**), and *Rhizoctonia solani.*

Disease cycle
The pathogen is soilborne and persists for long periods of time in the field. Oospores germinate and produce zoosporangia, which subsequently release zoospores. These zoospores swim to, and make contact with, roots. Seedling infection becomes progressively more severe as temperatures increase from 13 to 25° C. Damage is limited if soil temperatures are below 15° C. The pathogen can also survive on weed hosts.

Control
Resistant cultivars are widely used for sugar beet, where susceptibility to *A. cochlioides* has been associated with susceptibility to *R. solani*. Table beet appears to be less sensitive to both pathogens. Rotate with non-host crops such as small grains, soybean, and maize. Bean, clover, and lucerne (alfalfa) may increase soil inoculum and thereby cause problems for table beet. Select sites with well-draining soils, prepare seedbeds so that water drains well, and provide adequate potassium to the crop. In Europe, plant early in the season so that seedlings establish while soil temperatures are unfavorable for the pathogen. Conversely, late planting increases the risk of attack. In the UK, seed treatment with hymexazol has been widely used on sugar beet; this treatment is registered for table beet seed in some countries.

References

Byford, W. J. 1985. A comparison of fungicide seed treatments to improve sugar beet seedling establishment. *Plant Pathology* 34:463-466.

Payne, P. A., Asher, M. .J C., and Kershaw, C. D. 1994. The incidence of *Pythium* spp. and *Aphanomyces cochlioides* associated with sugar beet growing soils of Britain. *Plant Pathology* 43:300–308.

Osburn, R. M. and Schroth, M. N. 1989. Effect of osmopriming sugar beet seed on germination rate and incidence of *Pythium ultimum* damping off. *Plant Disease* 73:21–24.

O'Sullivan, E. and Kavanagh, J. A. 1991. Characteristics and pathogenicity of isolates of *Rhizoctonia* spp. associated with damping-off of sugar beet. *Plant Pathology* 40:128–135.

O'Sullivan, E. and Kavanagh, J. A. 1992. Characteristics and pathogenicity of *Pythium* spp. associated with damping-off of sugar beet. *Plant Pathology* 41:582–590.

Cercospora beticola

CERCOSPORA LEAF SPOT

Introduction and significance

This disease occurs wherever table beets and Swiss chard are grown and is one of the most important diseases affecting plants in the chenopodium group. The disease results in significant damage in warm temperate areas, including southern Europe, the Mediterranean region, Japan, Russia, and the USA. Crop losses can exceed 40% on a root weight basis and, in extreme cases, almost complete crop loss can occur. The disease is particularly problematic when the crop is grown for its foliage.

135 Cercospora leaf spot on beet.

Symptoms and diagnostic features

Circular leaf spots are small initially, up to 2 mm in diameter, and can be very numerous. Spots have a pale brown or off-white center with a reddish margin (**135**); the latter is useful for distinguishing it from Ramularia leaf spot. Spots expand in size, remain circular or oblong, and can result in extensive loss of foliage. The centers of the spots become gray as the fungus produces dark mycelium and other structures within and on the leaf tissue.

Causal agent

Cercospora leaf spot is caused by the fungus *Cercospora beticola*. The conidia are borne on unbranched brown conidiophores growing in clusters from stomata. Conidia are hyaline, with three to 14 septa, and are long and thread-like in shape. Conidia range in length from 78–228 µm, and are slightly tapered from base (4.4–6.3 µm width) to tip (1.6–3.2 µm width). Spore length and septation are influenced by environmental conditions; spores may be up to 400 µm long and have up to 27 septa. In leaf tissue the fungus makes dark mycelial clumps that are called sclerotia. A perfect stage has not been reported. Hosts include sugar beet, table beet, Swiss chard, wild sea beet, spinach, and various *Atriplex* and *Chenopodium* weeds. The pathogen is seedborne. On sugar beet, various physiological races have been identified.

Disease cycle

Spores are splash and wind dispersed and may be carried in irrigation water. Sporulation, germination, and infection occur most rapidly at 25–35° C when night temperatures are above 16° C and relative humidity is above 90%. There is limited disease activity below 15° C. Leaf penetration occurs only via stomata. Dry periods, of up to 6 hours during the day, enhance infection compared with continuous wetness. Spots appear in 7 to 10 days after infection under optimal conditions. *Cercospora beticola* overwinters as sclerotia in infected leaves and can survive in soil for up to 2 years. Initial inoculum comes from infested seed, weed hosts, and spores produced on infected crop residues.

Control

Use resistant cultivars if available. For sugar beet, specific and non-specific resistance has been identified. Bury infected crop residues and destroy volunteer plants. Do not plant vegetable production crops close to seed crops. Use seed that does not have significant levels of the pathogen. For infested seed, apply seed treatments. Apply fungicides prior to infection and symptom development. Isolates resistant to sterol demethylation-inhibiting (DMI) fungicides have been documented, so use products with different modes of action to reduce the risks of further resistance problems. In the UK, strobilurin plus triazole fungicide combinations have recently been approved for use against *Cercospora* on sugar beet. Disease prediction systems are used in some countries to better time fungicide applications.

References

Calpouzos, L. and Stallknecht, G. F. 1967. Symptoms of Cercospora leaf spot of sugar beets influenced by light intensity. *Phytopathology* 57:799–800.

Karaoglanidis, G. S. and Thanassoulopoulos, C. C. 2003. Cross-resistance patterns among sterol biosynthesis inhibiting fungicides (SBIs) in *Cercospora beticola*. *European Journal of Plant Pathology* 109:929–934.

Karaoglanidis, G. S., Ioannidis, P. M., and Thanassoulopoulos, C. C. 2002. Changes in sensitivity of *Cercospora beticola* populations to sterol-demethylation-inhibiting fungicides during a 4-year period in northern Greece. *Plant Pathology* 51:55–62.

Moretti, M., Arnold, A., D'Agostini, A., and Farina, G. 2003. Characterization of field-isolates and derived DMI-resistant strains of *Cercospora beticola*. Mycological Research 107:1178–1188.

Pundhir, V. S. and Mukhopadhyay, A. N. 1987. Epidemiological studies on Cercospora leaf spot of sugar beet. *Plant Pathology* 36:185–191.

Wallin, J. R. and Loonan, D. V. 1971. Effect of leaf wetness duration and air temperature on *Cercospora beticola* infection of sugarbeet. *Phytopathology* 61:546–549.

Erysiphe betae

POWDERY MILDEW

Introduction and significance

This common disease occurs in Europe, the Middle East, and parts of the USA. It affects sugar beet, fodder beet, table beet, and Swiss chard, as well as wild sea beet (*Beta maritima*). Severe attacks can reduce yield by up to 25% and table beet crops with badly diseased foliage may be unmarketable.

Symptoms and diagnostic features

The first signs of powdery mildew are scattered, small, radiating colonies consisting of superficial white fungal growth. Such growth tends to start on the older leaves and occurs on both upper and lower leaf surfaces. As the disease develops, the entire leaf can become heavily colonized with extensive patches of the white powdery fungus (**136, 137**). Older leaves with severe powdery mildew will turn yellow and can later senesce and die. Younger leaves can also be affected if disease in the field is severe.

Causal agent

Powdery mildew is caused by the fungus *Erysiphe betae*. This pathogen is a typical powdery mildew fungus having both sexual and asexual stages on beet and related hosts. Asexual conidia are produced basipetally in short chains arising from mycelium on the leaf surface. Conidia are barrel-shaped, hyaline, and variable in size depending upon environmental conditions. Conidia are smaller (measuring 40 x 14 μm) when temperatures and relative humidity are low and larger (measuring 49 x 18 μm) when both factors are high. The sexual stage consists of a spherical cleistothecium, that is initially yellow, later turns dark brown to black, and measures 0.1 mm in diameter. Cleistothecia have numerous simple or branched hyphal appendages and contain four to six asci, each ascus bearing three to five ascospores. Ascospores measure 20–30 x 14–16 μm.

136 Beet leaf showing powdery mildew.

137 Beet severely affected by powdery mildew.

Disease cycle

Powdery mildew conidia have a high water content (60%) that enables these spores to germinate at low relative humidity. Germination occurs over a wide range of environmental conditions, with an optimum of 25° C and 70–100% relative humidity. Spores penetrate the leaf within a few hours and then produce hyphal growth and chains of conidia. As conidia are dislodged from the leaf and dispersed by winds, a new spore will mature at the basipetal end of the conidial chain. Spores are short-lived but may be transported in high altitude winds over long distances. Epidemics are associated with dry weather alternating with periods of high relative humidity and temperatures above 20° C. Plant susceptibility increases with age, and damage is more severe on drought-stressed plants because of leaf loss. Frequent rains slow the progress of the disease. Powdery mildew overwinters in buried roots (in Europe these are called groundkeepers) and on volunteer beet and weed hosts. Cleistothecia can be found on crop debris, which is another means of overwintering. Cleistothecia release ascospores during rain.

Control

Powdery mildew control relies on fungicides. Apply treatments at the onset of symptoms. Sulfur fungicides and newer triazole products have been effective. After harvest, plow down and destroy crop residues, and remove volunteer plants. If possible, do not plant vegetable beet in close proximity to sugar beet fields. A disease forecasting system in the UK, based on the number of ground frosts in February and March, helps guide control strategies in sugar beet. These forecasts could be useful to vegetable growers, as they provide an indication of disease risk and inoculum pressure from sugar beet crops.

References

Asher, M. J. C. and Williams, G. E. 1991. Forecasting the national incidence of sugar beet powdery mildew from weather data in Britain. *Plant Pathology* 40:10–107.

Hills, F. J., Chiarappa, L., and Geng, S. 1980. Powdery mildew of sugar beet: Disease and crop loss assessment. *Phytopathology* 70:680–682.

Ruppel, E. G., Hills, F. J., and Mumford, D. L. 1975. Epidemiological observations on the sugarbeet powdery mildew epiphytotic in western U. S. A. in 1974. *Plant Disease Reporter* 59:283–286.

Peronospora farinosa f. sp. *betae*

DOWNY MILDEW

Introduction and significance

This downy mildew affects sugar beet, mangold, Swiss chard, wild sea beet, and table beet. Other forma speciales (f. sp.) affect different members of the Chenopodiaceae; for example, *P. farinosa* f. sp. *spinaciae* is the important downy mildew of spinach. Beet downy mildew is found in almost all beet producing areas and is most common in mild climates such as coastal California and Oregon in the USA and in parts of Europe. Downy mildew has been associated with malformed roots of table beet in the UK. When downy mildew causes death of the apical growing point, plant growth is severely hampered and losses of more than 50% can result.

Symptoms and diagnostic feature

Downy mildew can infect plants at any stage, and initial symptoms are chlorosis and distortion of the youngest leaves. The pathogen grows systemically within the young leaves and may invade the growing point, resulting in reduced growth and a rosette of spindly, deformed, chlorotic leaves, which curl downwards (**138**). Such leaves can be noticeably thickened and brittle. Under humid conditions, a dense, purple-gray growth appears on both upper and lower leaf surfaces. Individual downy mildew lesions can also develop on leaves. These lesions are chlorotic, irregular in shape, and later dry and turn brown. If the growing point has been invaded, a dark heart rot of the crown may develop. This heart rot can be difficult to recognize as a downy mildew symptom if the infected leaves have senesced and dropped off the plant. If older plants are infected, such plants can recover and produce additional leaves that are healthy. In seed crops, the main seed stalk is stunted and distorted, sepals and bracts are swollen, and the overall effect is a 'witch's broom' symptom.

Causal agent

Downy mildew is caused by the oomycete *P. farinosa* f. sp. *betae*. The pathogen produces dichotomously branched sporangiophores that emerge through stomata. The sporangia are produced at the tips of the sporangiophore branches, are hyaline to pale violet, and measure 20–28 x 17–24 µm. Oospores measure 26–36 µm and under cool moist conditions are produced in infected vegetative tissues and seed. The pathogen was previously known as *P. schachtii*.

Disease cycle

The pathogen can be seedborne and therefore spread from place to place in seed. About 1% of infested seeds give rise to infected plants. Oospores in the soil germinate to produce mycelium and then sporangia. Overwintering mycelia in seed crops, weed hosts, and volunteer plants also produce sporangia. Sporangia are airborne and are carried by wind to host plants. For severe epidemics to develop, cool, moist conditions (optimum temperature is 8° C) are required. Sporangia form at 5–22° C and 60–100% relative humidity (optimum 12° C, 85% relative humidity) and germinate at 1–30° C (optimum 4–10° C). At least 6 hours of leaf wetness and cool temperatures (optimum 7–15° C) are required for infection. There is little infection at temperatures above 20° C.

Control

Use seed that does not have significant levels of the pathogen. Use seed that has been produced in dry regions that discourage downy mildew development. Remove volunteer and weed plants that harbor the pathogen. Destroy crop residues after harvest. There is some disease resistance available in sugar beet. Use fungicides as required; protectant sprays of dithiocarbamates or phenylamides have given some control.

References

Byford, W. J. 1981. Downy mildews of beet and spinach. In: *The Downy Mildews* (Ed. D. M. Spooner) pp. 531–543. Academic Press London.

MacFarlane, J. W. 1968. Elimination of downy mildew as a major sugarbeet disease in the coastal valleys of California. *Plant Disease Reporter* 52:297–299.

138 Beet leaves infected with downy mildew.

Pleospora bjoerlingii (anamorph = *Phoma betae*)

BLACK LEG

Introduction and significance.

Black leg is an important disease of beet. The disease occurs in most beet producing areas and is especially a concern in Europe, North America, and Africa. The pathogen causes symptoms on both roots and foliage.

Symptoms and diagnostic features

Symptoms on young plants include pre-emergence damping-off, where seed and germinated seedlings die before growing above ground. Damping-off symptoms are indistinguishable from those caused by soilborne pathogens *Aphanomyces* and *Pythium*. Young plants that do emerge can develop black lesions on stem tissue that is in contact with soil. These lower stem symptoms are the reason for the common name of black leg. Foliar symptoms are pale brown leaf spots that can measure up to 2 cm in diameter and contain concentric rings of the fungus fruiting bodies, or pycnidia. These spots are mostly found on the older leaves and may be centered on rust pustules. In seed crops, the stems develop dark streaks and lesions with gray centers. Severe infection causes stems to break at the crown and roots to have a dry black rot. Pedicels on the flowering stems can develop lesions. Infected roots initially have water-soaked lesions which turn brown and develop into deep, sunken black lesions containing gray white mycelium and pycnidia. Severely affected roots become spongy and full of cavities (**139**). Root rot often develops during storage of table beet. Root rot symptoms may be confused with boron deficiency, which causes death of the growing point and allows secondary rots to develop.

Causal agent

Black leg is caused by the ascomycete fungus *Pleospora bjoerlingii*. The *Pleospora* form produces black pseudothecia that measure 200–500 µm in diameter and have one short apical papillum each. Pseudothecia develop on rotted roots at the end of the season and on crop residues that persist over the winter. In the spring the pseudothecia produce brown ascospores that are 18–25 x 7–10 µm. The pathogen is usually encountered in its asexual stage *Phoma betae*. Black globose pycnidia, which are immersed in host tissue, produce hyaline, unicellular conidia that measure 5–8 x 3–4.3 µm.

Disease cycle

Phoma betae is a seedborne pathogen. Optimum conditions for disease development are temperatures of 14–18° C and periods of high humidity. The pathogen spreads between seedlings if conidia from diseased seedlings are splashed by rain and sprinkler irrigation. Infected seedlings that survive and continue to grow may have systemic infections. Late in the season the pathogen can form the pseudothecia perfect stage and overwinter on crop residues. In the spring, pseudothecia produce airborne ascospores, which cause leaf spots in newly planted root crops and stem infections in seed crops. At harvest, if beet foliage is closely trimmed to the root or if roots are wounded, root rots can develop during storage. Postharvest disease is favored by temperatures above 15° C but can take place at much lower temperatures. If rain occurs close to seed crop harvest time, seedborne infection can be increased.

139 Root lesions on beets caused by the black leg pathogen, *Phoma betae*.

Control

Use seed that does not have significant levels of the pathogen. Seed treatments are a standard practice for sugar beet and should also be considered for vegetable beet. A thiram seed treatment is now often used on table and sugar beet seed. Hot water seed treatment is effective, but is not used commercially. Fungicide sprays are rarely used on crops grown for root production, but may be useful for seed crops. Implement cultural practices such as crop rotation, burying crop residues after harvest, and maintaining adequate levels of phosphate, potash, manganese, and boron.

References

Bugbee, W. M. 1979. *Pleospora bjoerlingii* in the USA. *Phytopathology* 69:277–278.

Bugbee, W. M. and Campbell, L. G. 1990. Combined resistance in sugar beet to *Rhizoctonia solani*, *Phoma betae*, and *Botrytis cinerea*. *Plant Disease* 74:353–355.

Byford, W. J. and Gambogi, P. 1985. *Phoma* and other fungi on sugar beet seed. *Transactions of the British Mycological Society* 84:21–28.

Heiberg, B. C. and Ramsey, G. B. 1948. Phoma rot of garden beets. *Phytopathology* 38:343–347.

Maude, R. B. and Bambridge, J. M. 1985. Effects of seed treatments and storage on the incidence of *Phoma betae* and the viability of infected red beet seeds. *Plant Pathology* 34:435–437.

Monte, E. and Garcia-Acha, I. 1988. Vegetative and reproductive structures of *Phoma betae* in vitro. *Transactions of the British Mycological Society* 90:233–245.

Ramularia beticola

RAMULARIA LEAF SPOT

Introduction and significance

This disease is widespread in western and central Europe and North America. Most research on this problem has involved sugar beet and fodder beet, but the problem also affects table beet, spinach beet, and Swiss chard. Ramularia leaf spot is important in seed crops, where it can cause severe defoliation in warm and wet weather. Yield losses may reach 10 to 20%, but such severity is unusual. The typical impact of the disease is to reduce the quality of the edible foliage.

Symptoms and diagnostic features

Symptoms are pale brown, circular or irregularly shaped leaf spots that range in size from 5 mm to more than 10 mm in diameter (**140, 141**). The spots are usually larger than those caused by *Cercospora beticola*, and *Ramularia* infections lack the distinctive red-brown margin of *Cercospora* lesions. In the UK, symptoms first occur in June and tend to worsen as crops near maturity in autumn. In seed crops, leaf spot severity increases as temperatures rise in the spring. Most leaf spots occur on the older leaves, and multiple lesions lead to early death of the leaf. The undersides of the lesions usually have a fine white, granular appearance because of the growth of the pathogen.

Causal agent

Ramularia leaf spot is caused by the fungus *Ramularia beticola*. The pathogen produces clusters of hyaline conidiophores that emerge from leaf stomata. Conidiophores bear hyaline, cylindrical, mainly two-celled conidia that measure 15 x 1.5 µm. Conidia form in short chains. There is no known teleomorph. Sporulation is usually present on leaf lesions and appears as tiny white tufts of fungal growth.

Disease cycle

The fungus sporulates when relative humidity is above 70% and temperatures are in the range of 5–20° C. Optimum temperatures for fungal growth are 16–17° C. After spores penetrate through leaf stomata, there is an incubation period of 16 days at 17° C before symptoms occur. This pathogen produces survival structures called pseudosclerotia that are found in crop debris and soil. The fungus can persist in the soil for more than 2 years. Sources of inoculum are buried roots, piles of culled beets, overwintered spinach beet plantings, and overwintered seed crops. The pathogen may also be seedborne.

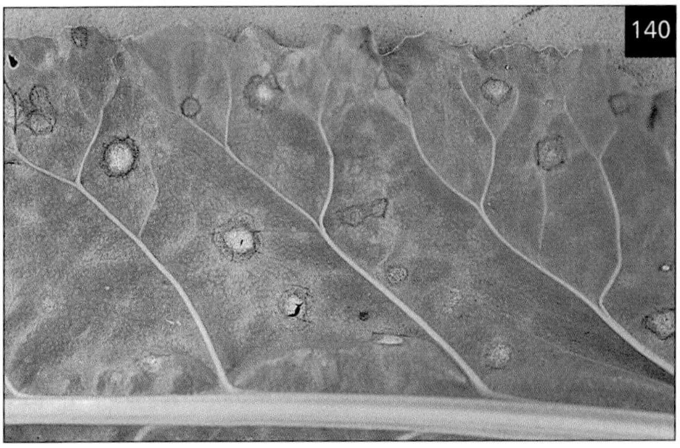

140 Ramularia leaf spot of beet.

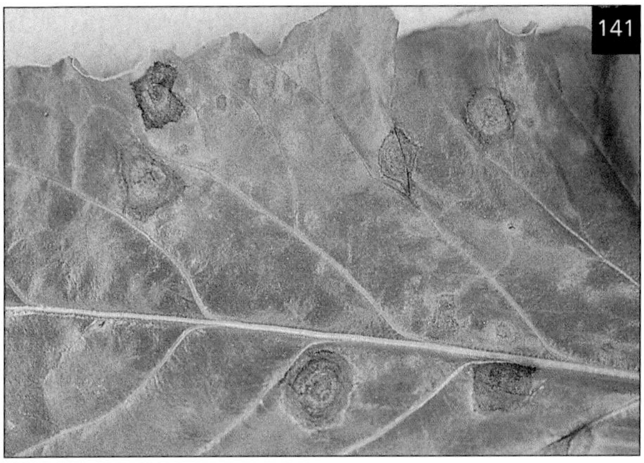

141 Ramularia leaf spot of beet.

Control

Resistant table beet cultivars have not been developed, but plant more tolerant selections if such can be identified. Remove volunteer and weed hosts. Bury crop residues after harvest. Do not plant production crops close to seed crops. Fungicides can provide some control, though these materials are generally more important for seed crops. Effective fungicides include dithiocarbamate and triazole products.

References

Byford, W. J. 1975. *Ramularia beticola* in sugar-beet seed crops in England. *Journal of Agricultural Science* 85:369–375.

Byford, W. J. 1976. Experiments with fungicide sprays to control *Ramularia beticola* in sugar-beet seed crops. *Annals of Applied Biology* 82:291–297.

Streptomyces spp.

SCAB

Introduction and significance

Scab lesions on table beet can cause considerable reductions in quality, resulting in loss of marketable yield.

Symptoms and diagnostic features

Scab causes irregularly shaped root lesions that vary in size. Lesions coalesce and affect large areas on the beet root. In some cases, the disease results in tumor-like outgrowths (**142**). Another disease, crown gall, caused by the bacterium *Agrobacterium tumefaciens*, produces similar symptoms consisting of large, rounded, swollen root galls having an irregular rough surface.

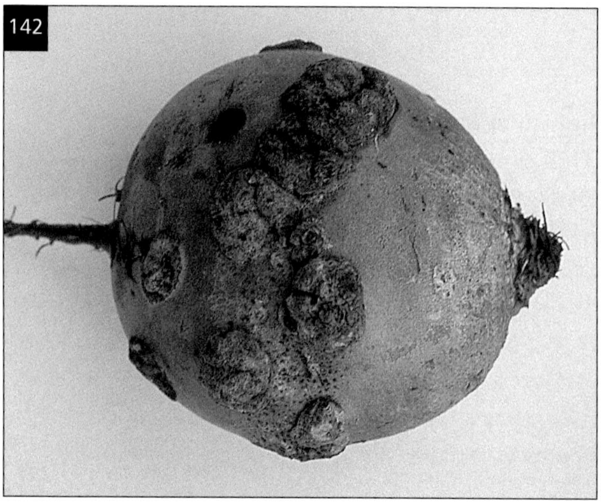

142 Raised lesions on beet caused by scab disease.

Causal agent

The pathogen is considered to be the actinomycete *Streptomyces scabies*. Actinomycetes are Gram-positive bacteria that can form branching filaments and which are common soil inhabitants. Recent research on potato has identified several species of *Streptomyces* that cause common scab, so the etiology of scab disease may be more complex. Potato and table beet are often grown in the same rotation, so the same pathogen possibly infects both crops. *Streptomyces* species survive for long periods in the absence of a host crop.

Disease cycle

Table beet is most susceptible to infection when the hypocotyl starts to swell (called decortication), which takes place about 5 weeks after planting. Infection occurs through wounds or lenticels near the soil surface, but sometimes roots and aerial parts are also infected. Beets are only susceptible for a 2 to 3 week period, after which tissues are suberized and become resistant to infection. Infection is favored by light, sandy soils, alkaline soil pH, warm temperatures (13–25° C), and soil water potentials below –40 kPa (= –0.4 bar).

Control

There are cultivar differences, so plant resistant or tolerant cultivars. Avoid planting susceptible crops in known infested fields. Do not apply lime to infested fields if beets will be planted, since disease severity is reduced in acid soils. Implement a crop rotation scheme that reduces the frequency of root crops. Maintain adequate soil moisture during decortication.

References

Adams, M. J. and Lapwood, D. H. 1978. The period of susceptibility of red beet to *Streptomyces* scab. *Plant Pathology* 27:97–98.

Adams, M. J., Lapwood, D. H., and Rankin, B. 1976. The growth of red beet and its infection by *Streptomyces* spp. *Plant Pathology* 25:147–151.

Adams, M. J., Lapwood, D. H., and Crisp, A. F. 1976. The susceptibility of red beet cultivars to *Streptomyces* scab. *Plant Pathology* 25:31–33.

Uromyces betae

RUST

Introduction and significance

Rust occurs worldwide and affects sugar beet, fodder beet, mangold, table beet, Swiss chard, and wild sea beet (*Beta maritima*). In parts of Europe and Asia the disease can be sufficiently severe to reduce yield. For many crops rust is a minor problem, although it has increased in prevalence in recent years.

Symptoms and diagnostic features

Symptoms consist of typical orange, raised pustules that measure 1–3 mm in diameter (**143**). Pustules develop on both top and bottom leaf surfaces and are more conspicuous on beet having green foliage than on red-leaved beet. Necrotic tissue forms around the larger pustules, especially when pustules develop in clusters or rings. With severe rust, the leaves support extensive numbers of pustules, are covered with the powdery orange spores, and begin to turn yellow and die. Later in the season, the pustules become dark brown, particularly on senescent foliage, as the brown teliospores are produced. Severe attacks cause premature senescence of foliage and reduce both yield and quality.

Causal agent.

Rust is caused by the basidiomycete fungus *Uromyces betae*. Like many rusts, there are multiple spore types for this autoecious rust. Orange-brown to golden-brown urediniospores are single-celled, oval in shape, echinulate, have two to three equatorial pores, and measure 26–33 x 19–24 µm. Dark brown teliospores are oval, smooth, with attached pedicels, measure 26–30 x 18–22 µm, with hyaline papilli covering the single apical germ pores. An aecidial stage occurs in the spring when temperatures are 10–13° C. Aecidiospores are slightly roughened (verruculose), measure 23–26 x 18–22 µm, and are produced in yellow, cup-shaped aecidia. Optimum temperature for aecidiospore germination is 15° C.

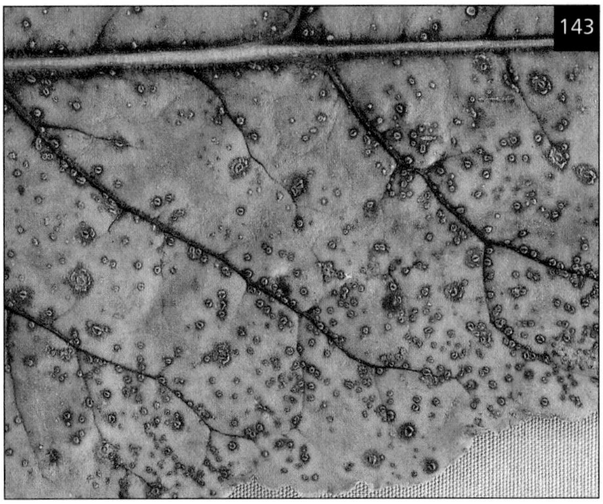

143 Rust pustules on beet.

Disease cycle

Inoculum consists of windborne spores from overwintered seed crops, volunteer plants, and weed hosts. The pathogen may be seedborne. Urediniospores are the primary inoculum responsible for spread of the disease. Optimal temperatures for urediniospore germination are in the range of 10–22° C. Teliospores are important as overwintering structures that can survive on crop residues and for up to a year on the soil surface. Aecidiospores are produced in the spring and are a third spore type capable of initiating infection, though they are seen less frequently than the other spore stages. Rust development is favored by warm (15–22° C), moist conditions and is inhibited by dry conditions at higher temperatures.

Control

While resistant cultivars are available for the sugar beet, no such varieties are yet developed for table beet. Remove diseased volunteer and weed hosts. Do not plant production fields in close proximity to seed crops. If available, apply fungicides.

References

Punithalingam, E. 1968. *CMI Descriptions of Pathogenic Fungi No. 177. Uromyces betae.*

Beet curly top virus

BEET CURLY TOP

144 **Curled leaves and stunted growth of sugar beet caused by *Beet curly top virus*.**

145 **Curled sugar beet leaves caused by *Beet curly top virus*.**

Introduction and significance

Virus diseases are of major economic importance in sugar beet and related vegetables in the Amaranthaceae. This plant family is subject to many virus pathogens, but only a few viruses are consistent concerns for vegetables such as table beet and Swiss chard. *Beet curly top virus* (BCTV) occurs in North and South America, Asia, the Middle East, and the Mediterranean region. This virus is an important pathogen of many crops such as pepper, tomato, and chenopodium plants.

Symptoms and diagnostic features

Symptoms consist of stunted plants and reduced foliage. Leaves are severely rolled upwards, crinkled (**144, 145**), and sometimes form enations which are small, swollen growths visible on the lower leaf surface. Taproots can become distorted and support proliferation of fine roots, resulting in a 'hairy root' symptom; this root proliferation symptom can be misdiagnosed as rhizomania. Phloem tissue can be necrotic when seen in dissected roots.

Causal agent

BCTV is a geminivirus with isometric particles that measure 18–22 nm in diameter and which occur singly or in pairs. The BCTV genome is a single-stranded circular DNA. BCTV is vectored in a persistent manner by the beet leafhopper (*Circulifer tenellus*). *Circulifer opacipennis* is a vector in the Mediterranean region. In the plant, BCTV is restricted to the phloem tissue. On a molecular level, researchers have compared strains of BCTV from North America and the Middle East and found them to be similar, providing evidence that these various BCTV strains share a common origin.

Disease cycle

This virus infects many weed and crop hosts. Recent research on curly top disease as it occurs in Amaranthaceae, Fabaceae, Solanaceae, and other crops indicates that the viral agent may differ depending upon the host being considered. Curly top as a disease may actually be caused by one of four different curly top virus species: *Beet curly top virus* (BCTV), *Beet mild curly top virus* (BMCTV), *Beet severe curly top virus* (BSCTV), and *Spinach curly top virus* (SCTV). Research is on-going to further determine the relationships of these various viruses.

Control

Follow general suggestions for managing virus diseases (see Part 1).

References

Briddon, R. W., Stenger, D. C., Bedford, I. D., Stanley, J., Izadpanah, K., Markham, P. G. 1998. Comparison of a beet curly top virus isolate originating from the old world with those from the new world. *European Journal of Plant Pathology* 104:77–84.

Duffus, J. E. 1983. Epidemiology and control of curly top diseases of sugarbeet and other crops. In: *Plant Virus Epidemiology*, R. T. Plumb and J. M. Thresh (eds.). Blackwell Scientific, Oxford, pp. 297–304.

Soto, M. J. and Gilbertson, R. L. 2003. Distribution and rate of movement of the curtovirus beet mild curly top virus (family Geminiviridae) in the beet leafhopper. *Phytopathology* 93:478–484.

Wang, H., Gurusinghe, P. de A., and Falk, B. W. 1999. Systemic insecticides and plant age affect beet curly top virus transmission to selected host plants. *Plant Disease* 83:351–355.

Beet leaf curl virus

BEET LEAF CURL

Introduction and significance

Beet leaf curl virus (BLCV) is a rhabdovirus that is restricted to central Europe, where its hosts are beet, spinach, and bean.

Symptoms and diagnostic features

Symptoms on beet are vein clearing of the youngest leaves, which curl as the veins fail to elongate. There is proliferation of small leaves so that the plant has a rosette-like appearance. Symptoms are more severe when temperatures are high. Root growth is also reduced.

Causal agent and disease cycle

Virus particles are bacilliform and measure 225 x 80 nm. The vector is the adult beet lace bug (*Piesma quadratum*) that can acquire the virus during its nymph stages, but transmits it in a persistent manner as an adult.

Control

Follow general suggestions for managing virus diseases (see Part 1). Control has been achieved by early planting of strips of beet around the edge of fields that attract lacebugs as they move into crops from hedgerows. The strips of beet are then plowed into the ground.

Beet mild yellowing virus

BEET MILD YELLOWING

Introduction and significance

The Polerovirus genus of viruses has three different species that affect these crops. *Beet mild yellowing virus* (BMYV) is the most important yellows virus on Amaranthaceae hosts in the UK and Western Europe. *Beet western yellows virus* (BWYV) is prevalent in the USA and elsewhere and infects many other plants in addition to beet, Swiss chard, and spinach. A third polerovirus pathogen is *Beet chlorosis virus* (BCHV).

Symptoms and diagnostic features

Infected leaves develop a bright yellow to orange-yellow color (**146, 147**) but lack the brown or red spotting associated with *Beet yellows virus* (BYV). Older leaves are thickened, brittle, and become extensively yellow. Leaf veins and tissues adjacent to the veins may remain green. This yellows symptom may be confused with various nutrient deficiency symptoms, notably those caused by nitrogen, iron, or magnesium deficiency.

Causal agent

Though BMYV and BWYV are serologically identical, these two distinct viruses do not infect all the same hosts, greatly complicating the etiology of this disease. These three viruses all infect spinach and various beet species. However, while some American isolates of BWYV infect lettuce, BMYV strains from Europe cannot infect this host.

Disease cycle

BMYV and BWYV may be found in numerous crop and weed plants, and the host list is extensive, including over 150 documented plant species. All three viruses are poleroviruses and are vectored by several aphid vectors, especially the green peach aphid (*Myzus persicae*), in a persistent manner. There are reports that low levels of seed transmission can occur.

Control

Follow general suggestions for managing virus diseases (see Part 1). Forecasts of yellows disease risk are released annually in the UK for sugar beet growers; these could be useful to vegetable growers as well.

146 Symptoms on beet caused by *Beet mild yellowing virus*.

147 Patches of beets infected with *Beet mild yellowing virus*.

References

Dewar, A. M. 1994. The Virus Yellows Warning Scheme – an Integrated Pest Management system for sugar beet in the UK. In: *Individuals, Populations, and Patterns in Ecology*. Leather, S. R., Watts, A. D., Mills, N. J., and Walters, K. F. A. (eds), pp. 173–185.

Dewar, A. M., Haylock, L. A., and Ecclestone, P. M. J. 1996. Strategies for controlling aphids and virus yellows in sugar beet. *Proceedings of the Brighton Crop Protection Conference – Pests & Diseases* 1996 1:185–190.

Duffus, J. E. 1973. The yellowing virus diseases of beet. *Advances in Virus Research* 18:347–386.

Duffus, J. E. and Russell, G. E. 1975. Serological relationships between beet western yellows virus and beet mild yellowing virus. *Phytopathology* 65:811–815.

Smith, H. G. 1989. Distribution and infectivity of yellowing viruses in field-grown sugar beet plants. *Annals of Applied Biology* 116:503–511.

Stevens, M., Smith, H. G., and Hallsworth, P. B. 1994. The host range of beet yellowing viruses among common arable weed species. *Plant Pathology* 43:579–588.

Beet mosaic virus

BEET MOSAIC

Introduction and significance

Beet mosaic virus (BtMV) is worldwide in distribution and is one of many viruses that can be found in Chenopodium plants.

Symptoms and diagnostic features

BtMV causes a pale green to yellow green mottle or mosaic that is similar to other mosaic disease symptoms. A backward curling of the leaf tip, when present, is a characteristic feature. On spinach, young leaves also curl backward. Spinach leaves develop bright yellow flecks that merge to produce large chlorotic areas. Plants are stunted and leaves become necrotic.

Causal agent and disease cycle

BtMV is a potyvirus and has flexuous filamentous particles measuring 690–770 x 13 nm. BtMV is transmitted in a nonpersistent manner by several aphid species, including *Myzus persicae* and *Aphis fabae*. The pathogen host range is broad and includes many species in diverse plant families. Several weeds are also important hosts.

Control

Follow general suggestions for managing virus diseases (see Part 1).

References

Cockbain, A. J., Gibbs, A. J., and Heathcote, G. D. 1963. Some factors affecting the transmission of sugar-beet mosaic virus and pea mosaic viruses by *Aphis fabae* and *Myzus persicae*. *Annals of Applied Biology* 52:133–143.

Rogov, V. V., Karasev, A. V., Agranovsky, A. A., and Gorbunova, N. I. 1991. Characterization of an isolate of beet mosaic virus from South Kazakhstan. *Plant Pathology* 40:515–523.

Beet necrotic yellow vein virus

BEET NECROTIC YELLOW VEIN, RHIZOMANIA

Introduction and significance

Beet necrotic yellow vein virus (BNYVV) is a damaging virus that affects many plants in the Amaranthaceae. The disease caused by BNYVV, rhizomania, occurs virtually wherever beet and related plants are grown.

Symptoms and diagnostic features

Leaves of infected plants show mild yellowing symptoms and an upright growth. There may be wilting of leaves, occasional veinal yellowing, and necrotic lesions. The taproot is usually stunted and shows constriction and rotting below soil level. The vascular rings in the root are brown and discolored. The name rhizomania means 'root madness,' or 'crazy root,' and refers to the extensive proliferation of lateral roots (**148, 149**).

148 Proliferation of small feeder roots on sugar beet affected by rhizomania.

149 Rhizomania disease on beet.

Causal agent

Different BNYVV strains have been documented, such as the A type, B type, and the virulent P strain (named after Pithiviers, France) that was discovered in France and recently detected in the UK. BNYVV is a furovirus and has rod-shaped particles in three size ranges (65–100, 270, or 390 x 20 nm).

Disease cycle

BNYVV is vectored by *Polymyxa betae*, a soil inhabiting plasmodiophorid. *Polymyxa betae* may sometimes cause direct damage to young beet plants and can be considered a minor root pathogen. Transmission is favored by warm soils with high moisture content, which favors vector activity and movement. Little infection occurs below 10° C and the optimum temperature for vector and virus multiplication is 25° C.

Control

Follow general suggestions for managing virus diseases (see Part 1). Do not plant susceptible hosts in fields having a history of this problem. Do not allow infested soil to be moved to uninfested fields.

References

Al musa, A. M. and Mink, G. I. 1981. Beet necrotic yellow vein virus in North America. *Phytopathology* 71:773–776.

Asher, M. J. C. 2002. The development of sugar-beet rhizomania and its control in the UK. *Proceedings of the BCPC Conference – Pests & Diseases 2002* 1:113–120.

Barr, K. J. and Asher, M. J. C. 1992. The host range of *Polymyxa betae* in Britain. *Plant Pathology* 41:64–68.

Blunt, S. J., Asher, M. J. C., and Gilligan C. A. 1991. Infection of sugar beet by *Polymyxa betae* in relation to soil temperature. *Plant Pathology* 40:257–267.

Gerik, J. S. and Duffus, J. E. 1988. Differences in vectoring ability and aggressiveness of isolates of *Polymyxa betae*. *Phytopathology* 78:1340–1343.

Hill, S. A. and Torrance, L. 1989. Rhizomania disease of sugar beet in England. *Plant Pathology* 38:114–122.

Lennefors, B.-L., Lindsten, K., and Koenig, R. 2000. First record of A and B type beet necrotic yellow vein virus in sugar beets in Sweden. *European Journal of Plant Pathology* 106:199–201.

Payne, P. A. and Asher, M. J. C. 1990. The incidence of *Polymyxa betae* and other fungal root parasites of sugar beet in Britain. *Plant Pathology* 39:443–451.

Tamada, T. and Baba, T. 1973. Beet necrotic yellow vein virus from 'rhizomania' affected sugar beet in Japan. *Annals of Phytopathological Society of Japan* 39:325–332.

Wisler, G. C., Lewellen, R. T., Sears, J. L., Liu, H.-Y., and Duffus, J. E. 1999. Specificity of TAS-ELISA for beet necrotic yellow vein virus and its application for determining rhizomania resistance in field-grown sugar beets. *Plant Disease* 83:864–870.

Beet pseudo-yellows virus

BEET PSEUDO-YELLOWS

Introduction and significance

Beet pseudo-yellows virus (BPYV) was first described in California on greenhouse crops. However, the pathogen is now worldwide in distribution.

Symptoms and diagnostic features

BPYV causes stunting and reduced growth. Mature leaves show bright yellow spots that measure 1–1.5 cm in diameter; such leaves later become thickened and brittle. The chlorosis caused by BPYV tends to be more uniform with less greening along leaf veins than with BYV or the poleroviruses.

Causal agent and disease cycle

BPYV is a closterovirus. This virus infects a wide range of crops including beet, carrot, cucumber, lettuce, pumpkin, spinach, strawberry, and numerous weeds. The only known vector is the greenhouse whitefly (*Trialeurodes vaporariorum*), which transmits BPYV in a semi-persistent manner.

Control

Follow general suggestions for managing virus diseases (see Part 1).

References

Duffus, J. E. 1965. Beet pseudo-yellows, transmitted by the greenhouse whitefly (*Trialeurodes vaporariorum*). *Phytopathology* 55:450–453.

Wintermantel, W. M. 2004. Pumpkin (*Cucurbita maxima* and *C. pepo*), a new host of beet pseudo-yellows virus in California. *Plant Disease* 88:82.

Beet western yellows virus

BEET WESTERN YELLOWS

Introduction and significance

The *Polerovirus* genus of viruses has three different species that affect these crops. *Beet western yellows virus* (BWYV) is prevalent in the USA and elsewhere and infects many other plants in addition to beet, Swiss chard, and spinach. *Beet mild yellowing virus* (BMYV) is the most important yellows virus on Amaranthaceae hosts in the UK and Western Europe. A third polerovirus pathogen is *Beet chlorosis virus* (BCHV).

Symptoms and diagnostic features

Infected leaves develop a bright yellow to orange-yellow color but lack the brown or red spotting associated with *Beet yellows virus* (BYV). Older leaves are thickened, brittle, and become extensively yellow. Leaf veins and tissues adjacent to the veins usually remain green (**150**, **151**). This yellows symptom may be confused with various nutrient deficiency symptoms, notably those caused by nitrogen, iron, or magnesium deficiency.

Causal agent

Though BWYV and BMYV are serologically identical, these two distinct viruses do not infect all the same hosts, greatly complicating the etiology of this disease. These three viruses all infect spinach and various beet species. However, while some American isolates of BWYV infect lettuce, BMYV strains from Europe cannot infect this host.

150 Interveinal yellowing and necrotric spotting caused by *Beet western yellows virus* on beet.

151 Interveinal yellowing caused by *Beet western yellows virus* on beet.

Disease cycle

BWYV and BMYV may be found in numerous crop and weed plants, and the host list is extensive, including over 150 documented plant species. All three viruses are poleroviruses and are vectored by several aphid vectors, especially the green peach aphid (*Myzus persicae*), in a persistent manner. There are reports that low levels of seed transmission can occur.

Control

Follow general suggestions for managing virus diseases (see Part 1).

References

Dewar, A. M. 1994, The Virus Yellows Warning Scheme – an Integrated Pest Management system for sugar beet in the UK. In: *Individuals, Populations, and Patterns in Ecology*. Leather, S. R., Watts, A. D., Mills, N. J., and Walters, K. F. A. (eds), pp. 173–185.

Dewar, A. M., Haylock, L. A., and Ecclestone, P. M. J. 1996. Strategies for controlling aphids and virus yellows in sugar beet. *Proceedings of the Brighton Crop Protection Conference – Pests & Diseases* 1996 1:185–190.

Duffus, J. E. 1973. The yellowing virus diseases of beet. *Advances in Virus Research* 18:347–386.

Duffus, J. E. and Russell, G. E. 1975. Serological relationships between beet western yellows virus and beet mild yellowing virus. *Phytopathology* 65:811–815.

Smith, H. G. 1989. Distribution and infectivity of yellowing viruses in field-grown sugar beet plants. *Annals of Applied Biology* 116: 503–511.

Stevens, M., Smith, H. G., and Hallsworth, P. B. 1994. The host range of beet yellowing viruses among common arable weed species. *Plant Pathology* 43:579–588.

Beet yellows virus

BEET YELLOWS

Introduction and significance

Beet yellows virus (BYV) occurs throughout the world on spinach and beets. In Europe, BYV and BMYV together can cause significant losses on sugar beet.

Symptoms and diagnostic features

Symptoms from BYV on table beet include mild yellowing, vein clearing, or vein yellowing of the younger leaves. Leaf necrosis can sometimes be associated with yellowing symptoms. However, BYV probably does not cause such necrosis; the necrosis is more likely due to decay caused by *Alternaria* and other secondary agents. Older leaves are thickened, brittle, and turn yellow (**152**). Such leaves later develop red to brown spots, which then gives the foliage a bronze color. Cold weather slows down the development of chlorosis, but increases necrotic symptoms. In spinach, young leaves show chlorotic spotting and vein clearing, and subsequently turn yellow, necrotic, and distorted. The plants are stunted and may die if the growing point becomes necrotic.

Causal agent and disease cycle

BYV is a closterovirus having flexuous filamentous particles that measure 1250 x 10 nm. At least 35 aphid species have been reported as vectors, but the most important are usually *Myzus persicae* and *Aphis fabae*. The virus is transmitted in a semi-persistent manner and is not retained after molting and passed to the progeny. BYV has a very wide host range, including many of the weed species infected by the polerovirus group.

Control

Follow general suggestions for managing virus diseases (see Part 1). Forecasts of yellows disease risk are released annually in the UK for sugar beet growers; these disease forecasts could be useful to vegetable growers as well.

152 ***Beet yellows virus* on beet.**

References

Duffus, J. E. 1973. The yellowing virus diseases of beet. *Advances in Virus Research* 18:347–386.

Smith, H. G. and Hallsworth, P. B. 1990. The effects of yellowing viruses on yield of sugar beet in field trials 1985–1987. *Annals of Applied Biology* 116:503–511.

Smith, H. G. and Hinckes, J. A. 1987. Studies of the distribution of yellowing viruses in the sugar beet root crop from 1981 to 1984. *Plant Pathology* 36:125–134.

Brassicaceae Mustard family

MEMBERS OF THE BRASSICACEAE (mustard family) are among the most widely grown vegetables and consist of over 300 genera. Brassicaceae plants are also commonly called crucifers because the flowers have four petals that extend opposite each other and form the shape of a square cross, hence are cruciform. The major vegetable group of Brassicaceae plants are the *Brassica oleracea* or cole crops. Cole crops include the following familiar plants: kale and collards (*B. oleracea* var. *acephala*), broccoli and cauliflower (both classified as *B. oleracea* var. *botrytis*, though some taxonomists separate broccoli and name it *B. oleracea* var. *italica*), cabbage (*B. oleracea* var. *capitata*), Brussels sprouts (*B. oleracea* var. *gemmifera*), kohlrabi (*B. oleracea* var. *gongylodes*). In Europe broccoli is also called calabrese.

Other *Brassica* species vegetables include turnip (*B. rapa*), rutabaga or swede (*B. napus* var. *napobrassica*), Chinese cabbage (*B. pekinensis*), broccoli raab or rappini (*B. rapa* subsp. *rapa*), and pak-choi or bok choi (*B. chinensis*). Various leafy mustards are produced and are classified as subspecies of *B. juncea, B. campestris, B. alba, B. napus*, and *B. nigra*.

Radish (*Rhaphanus sativus*) is a crucifer vegetable grown for the fleshy part of the plant that is the stem hypocotyl and upper root. There is a great diversity in cultivars, ranging from small, round varieties to very large, elongated types and colors that are red, white, green, and even black. Other cultivars, such as *R. sativus* cv. *longipinnatus*, produce much larger roots and are known as Chinese radish, mooli, or daikon.

Other Brassicaceae vegetables outside the *Brassica* genus include horseradish (*Armoracia rusticana*), watercress (*Nasturtium officinale*), cress (*Lepidium sativum*), and rocket salad or arugula (*Eruca sativa*). Many of these and other brassica vegetables are considered specialty items; diseases of these plants are described in the Specialty Crops chapter.

***Erwinia* spp., *Pseudomonas* spp.**

BACTERIAL HEAD/SPEAR ROT OF BROCCOLI

153 Early infection of bacterial head rot of broccoli.

Introduction and significance

This disease is highly dependent on weather and can cause extensive losses of broccoli in California, the southeast USA, and Europe (where broccoli is known as calabrese). The disease occurs sporadically wherever the crop is grown.

Symptoms and diagnostic features

Initial symptoms consist of a dark, water-soaked decay of the surfaces of small groups of florets in the broccoli head, or spear (153). Later, infected portions of the head

154 Advanced infection of bacterial head rot of broccoli.

turn dark brown to black (**154**). The tissue becomes soft, collapses, and emits a bad odor. Secondary molds and bacteria cause further decay. Bacterial head rot symptoms can resemble those of Alternaria head rot. With bacterial head rot, fungal sporulation is absent unless secondary molds colonize the diseased florets.

Causal agents

Bacterial head rot is caused by a complex of pathogenic bacteria including the following: *Erwinia carotovora* subsp. *carotovora*, *Pseudomonas fluorescens*, *P. marginalis*, *P. viridiflava*, and perhaps others. The roles of the different bacterial agents may differ. Some species, such as *P. fluorescens*, produce a surfactant-like chemical (viscosin) that allows bacteria to more readily penetrate the very waxy surface of broccoli florets. Other bacteria do not make such substances and cannot reach broccoli tissues on their own. If multiple species are present, the viscosin-producing species allow the non-producers to likewise penetrate the surface wax and infect the broccoli tissues.

Disease cycle

Severe disease is always associated with periods of cool temperatures and wet, foggy, or rainy weather when crops begin to form the early broccoli flower heads. Frequent overhead sprinkler irrigation can promote this disease. Infection is also tied with fluctuating day/night temperatures that allow water to condense on the heads. Excessive nitrogen applications increase disease severity. Secondary organisms are very often present on the diseased florets and this leads to rapid breakdown of tissues and difficulty in disease diagnosis.

Control

Some differences in cultivar susceptibility exist, based on architecture and exposure of the broccoli heads. Broccoli heads having domed, and not flat, head shapes are less likely to have severe head rot. Plant these less susceptible cultivars, if available. Avoid planting broccoli during the wet, rainy part of the year. Avoid using overhead sprinklers, or schedule irrigations to reduce the number of applications during head formation. Increasing the interval between irrigations from 2 to 8 days reduced head rotting by 50% in Oregon. Protectant copper sprays have been used with some success under UK conditions, but such have not proven useful in California.

References

Canaday, C. H. and Wyatt, J. E. 1992. Effects of nitrogen fertilization on bacterial soft rot in two broccoli cultivars, one resistant and one susceptible to the disease. *Plant Disease* 76:989–991.

Canaday, C. H., Wyatt, J. E., and Mullins, J. A. 1991. Resistance in broccoli to bacterial soft rot caused by *Pseudomonas marginalis* and fluorescent *Pseudomonas* species. *Plant Disease* 75:715–720.

Darling, D., Harling, R., Simpson, R. A., McRoberts, N., and Hunter, E. A. 2000. Susceptibility of broccoli cultivars to bacterial head rot: in vitro screening and the role of head morphology in resistance. *European Journal of Plant Pathology* 106:11–17.

Hernandez-Anguiano, A. M., Suslow, T. V., Leloup, L., and Kado, C. I. 2004. Biosurfactants produced by *Pseudomonas fluorescens* and soft-rotting of harvested florets of broccoli and cauliflower. *Plant Pathology* 53:596–601.

Hildebrand, P. D. 1989. Surfactant-like characteristics and identity of bacteria associated with broccoli head rot in Atlantic Canada. *Canadian Journal of Plant Pathology* 11:205–214.

Laycock, M. V., Hildebrand, P. D., Thibault, P., Walter, J. A., and Wright, J. L. C. 1991. Viscosin, a potent peptidolipid biosurfactant and phytopathogenic mediator produced by a pectolytic strain of *Pseudomonas fluorescens*. *Journal of Agricultural Food Chemistry* 39:483–489.

Ludy, R. L., Powelson, M. L., and Hemphill, D. D. 1997. Effect of sprinkler irrigation on bacterial soft rot and yield of broccoli. *Phytopathology* 81:614–618.

Robertson, S., Brokenshire, T., Kellock, L. J., Sutton, M., Chard, J., and Harling, R. 1993. Bacterial spear (head) rot of calabrese in Scotland: causal organisms, cultivar susceptibility and disease control. *Proceedings Crop Protection in Northern Britain* 1993:265–270.

Wimalajeewa, D. L. S., Hallam, N. D., Hayward, A. C., and Price, T. V. 1987. The etiology of head rot disease of broccoli. *Australian Journal of Agricultural Research* 38:735–742.

Wimalajeewa, D. L. S., Hayward, A. C., and Price, T. V. 1985. Head rot of broccoli in Victoria, Australia, caused by *Pseudomonas marginalis*. *Plant Disease* 69:177.

Pseudomonas syringae pv. *alisalensis*

BACTERIAL BLIGHT

Introduction and significance

In California this disease is most often found on broccoli raab (*Brassica rapa* subsp. *rapa*) and more recently on broccoli and arugula. The disease is very damaging to broccoli raab. This pathogen has also been detected in the eastern USA. This is a recently described disease and pathogen, and one that is gaining in importance on crops like broccoli raab.

Symptoms and diagnostic features

On broccoli raab (see chapter on Specialty Crops, page 389), small, angular, water-soaked specks occur on leaves. These flecks expand, turn brown, and become surrounded by bright yellow borders. As disease develops, the affected areas enlarge and coalesce, causing large sections of the leaf to turn yellow, then brown. With time the leaf can wilt and collapse, resulting in blight-like symptoms. On broccoli, the pathogen causes large, angular leaf spots that also begin as water-soaked lesions that later turn tan to brown (**155**).

Causal agent

Bacterial blight is caused by the bacterium *Pseudomonas syringae* pv. *alisalensis*.This pathogen appears to be a new pathovar in the *P. syringae* group. The pathogen is an aerobic, Gram-negative bacterium that can be isolated on standard microbiological media. Colonies are cream to light yellow in color, smooth, and have a notably sticky consistency. When cultured on King's medium B, this organism produces a diffusible pigment that fluoresces blue under ultraviolet light. There is evidence that this pathogen is seedborne.

Disease cycle

Little is known about the specifics of disease development. The bacterium is splash dispersed by rain and overhead sprinkler irrigation. Leaf symptoms always begin on the lower, protected leaves where shading keeps the conditions humid and cool. As disease develops, the symptoms gradually move up the broccoli raab stem and eventually reach the top shoots. Seedborne inoculum would account for the occurrence of this disease in broccoli raab fields that never were planted to this crop before.

Control

Reduce or eliminate the use of overhead sprinkler irrigation. Do not plant a subsequent broccoli raab crop in fields having undecomposed, infested crop residues. Copper applications provide only minimal protection against this pathogen.

References

Bull, C. T., Goldman, P., and Koike, S. T. 2004. Bacterial blight on arugula, a new disease caused by *Pseudomonas syringae* pv. *alisalensis* in California. *Plant Disease* 88:1384.

Cintas, N. A., Koike, S. T., and Bull, C. T. 2002. A new pathovar, *Pseudomonas syringae* pv. *alisalensis* pv. nov., proposed for the causal agent of bacterial blight of broccoli and broccoli raab. *Plant Disease* 86:992–998.

Koike, S. T., Cintas, N. A., and Bull, C. T. 2000. Bacterial blight, a new disease of broccoli caused by *Pseudomonas syringae* in California. *Plant Disease* 84:370.

Koike, S. T., Henderson, D. M., Azad, H. R., Cooksey, D. A., and Little, E. L. 1998. Bacterial blight of broccoli raab: a new disease caused by a pathovar of *Pseudomonas syringae*. *Plant Disease* 82:727–731.

155 Tan lesions of bacterial blight of broccoli.

Pseudomonas syringae pv. *maculicola*

BACTERIAL LEAF SPOT

Introduction and significance

This disease is occasionally important during wet summers in Europe and other production areas, but is usually of minor significance. Cabbage is reported to be the most commonly affected crop, but bacterial leaf spot also occurs on cauliflower, collards, Brussels sprout, radish, and turnip. In the USA, this disease is often most important as a seedling disease that reduces quality and vigor of greenhouse produced transplants.

Symptoms and diagnostic features

Infections start as dark, water-soaked specks on leaves that can be seen from both leaf surfaces (**156**). Leaf spots typically remain small and measure 3–4 mm in diameter, but enlarging and merging of spots occurs if conditions are favorable. Older leaf spots turn tan and may sometimes have a purple border surrounding them. Severe infection can cause blotching, leaf distortion and premature leaf fall. Infected cauliflower heads may be discolored. On transplants (**157**), bacterial leaf spot may resemble early downy mildew infections that are not sporulating. The small, dark leaf spot symptoms are the reason for the older disease name of pepper leaf spot.

Causal agent

The cause of bacterial leaf spot is *Pseudomonas syringae* pv. *maculicola*. The pathogen is an aerobic, Gram-negative bacterium that can be isolated on standard microbiological media. Colonies are cream to light yellow in color and smooth. When cultured on King's medium B, this organism produces a diffusible pigment that fluoresces blue under ultraviolet light. Strains of this pathogen are host specific to crucifers. The pathogen is seedborne.

Disease cycle

The pathogen appears to survive between crops on infested debris. In the field the bacterium is splash dispersed by rain and overhead sprinkler irrigation. Disease development can be particularly severe on greenhouse produced transplants due to the use of overhead sprinklers, humid protected conditions, and the practice of clipping and mowing transplants to strengthen transplant stems.

156 Leaf spots of bacterial leaf spot of cauliflower.

157 Cauliflower transplants infected with bacterial leaf spot.

Control

Reduce or eliminate the use of overhead sprinkler irrigation. Do not plant a subsequent crucifer crop in fields having undecomposed, infested crop residues. Copper applications provide only minimal protection against this pathogen. In greenhouses, schedule irrigations to allow for maximum drying of foliage. Remove transplant trays that show symptoms of the disease. Copper applications in the controlled setting of greenhouses are probably more effective than in field use. Sanitize old transplant trays and bench tops so the pathogen is not spread to new plantings. Mowing of transplants greatly increases the risk of spreading the disease. Use seed that does not have significant levels of the pathogen. Appropriate seed treatments can also contribute to the management of seedborne inoculum. Treat infested seed with hot water.

References

Cintas, N. A., Bull, C. T., Koike, S. T., and Bouzar, H. 2001. A new bacterial leaf spot disease of broccolini, caused by *Pseudomonas syringae* pv. *maculicola*, in California. *Plant Disease* 85:1207.

Koike, S. T., Smith, R. F., Van Buren, A. M., and Maddox, D. A. 1996. A new bacterial disease of arugula in California. *Plant Disease* 80:464.

Lewis-Ivey, M. L., Wright, S., and Miller, S. A. 2002. Report of bacterial leaf spot on collards and turnip leaves in Ohio. *Plant Disease* 86:186.

Peters, B. J., Asha, G. J., Cotherb, E. J., Hailstones, D. L., Nobleb, D. H., and Urwin, N. A. R. 2004. *Pseudomonas syringae* pv. *maculicola* in Australia: pathogenic, phenotypic and genetic diversity. *Plant Pathology* 53:73–79.

Shackleton, D. A. 1996. A bacterial leaf spot of cauliflower in New Zealand caused by *Pseudomonas syringae* pv. *maculicola*. *New Zealand Journal of Science* 9:872–877.

Wiebe, W. L. and Campbell, R. N. 1993. Characterization of *Pseudomonas syringae* pv. *maculicola* and comparison with *Pseudomonas syringae* pv. tomato. *Plant Disease* 77:414–419.

Zhao, Y. F., Damicone, J. P., and Bender, C. L. 2002. Detection, survival, and sources of inoculum for bacterial diseases of leafy crucifers in Oklahoma. *Plant Disease* 86:883–888.

Zhao, Y. F., Damicone, J. P., Demezas, D. H., Rangaswamy, V., and Bender, C. L. 2000. Bacterial leaf spot of leafy crucifers in Oklahoma caused by *Pseudomonas syringae* pv. *maculicola*. *Plant Disease* 84:1015–1020.

Xanthomonas campestris pv. *campestris*

BLACK ROT

Introduction and significance

Black rot is one of the most important diseases of crucifers. It is most damaging in tropical, subtropical, and other areas with warm, humid climates. For example, significant black rot occurs on cabbage grown in Africa. In cooler regions, losses are less severe, but the disease has become more prevalent in recent years, possibly because spread is favored by modern propagation systems. The seedborne nature of the pathogen has resulted in its distribution throughout the world.

Symptoms and diagnostic features

Black rot symptoms can vary greatly depending on the particular brassica host, age of host when infected, particular strain of the pathogen, and environmental conditions. Initial symptoms consist of small, water-soaked leaf spots. These then become brown specks with a chlorotic margin and can mimic other bacterial diseases (**158**). Other early symptoms are angular or V-shaped yellow lesions that often develop along the leaf edges.

158

156 Early infection of black rot of cauliflower.

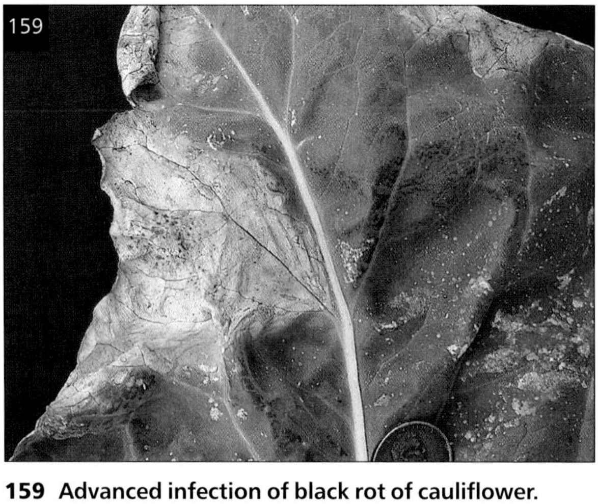

159 Advanced infection of black rot of cauliflower.

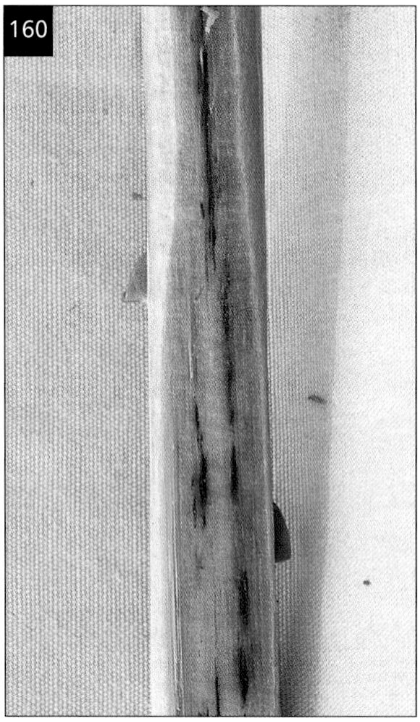

160 Vascular blackening of black rot of cauliflower.

These lesions later dry up and turn tan or brown (**159**). Black veins are sometimes seen within these tan lesions. Severely infected leaves may wither and drop off the plant. If systemic infection occurs, the vascular tissues in petioles and main stems turn black (**160**). Some strains produce symptoms known as 'blight,' which are characterized by a sudden collapse of interveinal tissues and a lack of veinal necrosis. Systemic lesions may develop in the center of the leaf when temperatures are high (20-28° C), but most marginal lesions occur at 16° C. However, if temperatures are cool, black rot symptoms may not be expressed.

Causal agent

The cause of black rot is the aerobic, Gram-negative bacterium *Xanthomonas campestris* pv. *campestris*. The pathogen can be isolated on standard microbiological media and produces yellow, mucoid, slow growing colonies typical of most xanthomonads. Various starch-based media aid in the isolation and identification of this organism. Strains of this pathogen are host specific to crucifers and infect *Brassica* species, radish, crucifer weeds, and ornamental crucifers such as stock (*Matthiola* spp.) and wallflower (*Cheiranthus cheiri*). This pathogen is seedborne, and different races have been documented. Also, the highly virulent variants that cause the blight and plant collapse symptoms can be differentiated from leaf spot causing strains via monoclonal antibody and molecular tests. Strains infecting horseradish are *X. campestris* pv. *amoraciae*; this organism is only a weak pathogen on cabbage and cauliflower.

Disease cycle

Initial inoculum comes from infested seed, diseased weeds and volunteer crucifers, and infested but undecomposed crop residues. Bacteria are dispersed by splashing rain or sprinkler irrigation from inoculum sources to susceptible plants. Cotyledons and leaves are infected through leaf openings called hydathodes, which are located where veins end at the leaf margin. Infection can also occur through wounds or roots. Symptoms only occur when sufficient inoculum builds up on the leaf and if weather conditions are conducive for infection and symptom expression. If conditions are humid and warm, and there is splashing water, disease can progress rapidly and cause extensive damage. In seed crops, infected plants may be symptomless, though the pathogen is present in the vascular system and is able to spread to pods and seed.

Control

Use seed that does not have significant levels of the pathogen. Treat seed with sodium hypochlorite or hot water. Produce seed in areas having a dry climate, and have such seed inspected and certified. Rotate crops with non-hosts for 2 to 3 years to avoid problems from infected crop residues. However, once crop residues are gone, the bacterium does not survive in the soil. Do not place seedbed or transplant areas adjacent to field production crops. Remove symptomatic transplants and surrounding plants and trays. Mowing of transplants greatly increases the risk of spreading the pathogen. Use resistant cultivars if available. The application of chemicals to foliage is not very effective for black rot control.

References

Alvarez, A. M., Benedict, A. A., Mizumoto, C. Y., Hunter, J. E., and Gabriel, D. W. 1994. Serological, pathological, and genetic diversity among strains of *Xanthomonas campestris* infecting crucifers. *Phytopathology* 84:1449–1457.

Alvarez, A. M. and Lou, K. 1985. Rapid identification of *Xanthomonas campestris* pv. *campestris* by ELISA. *Plant Disease* 69:1082–1086.

Babadoost, M., Derie, M. L., and Gabrielson, R. L. 1996. Efficacy of sodium hypochlorite treatments for control of *Xanthomonas campestris* pv. *campestris* in brassica seeds. *Seed Science and Technology* 24:7–15.

Chun, W. W. C., and Alvarez, A.M. 1983. A starch-methionine medium for isolation of *Xanthomonas campestris* pv. *campestris* from plant debris in soil. *Plant Disease* 67:632–635.

Claflin, L. E., Vidaver, A. K., and Sasser, M. 1987. MXP, a semi-selective medium for *Xanthomonas campestris* pv. *phaseoli*. *Phytopathology* 77:730–734.

Franken, A. A. J. M. 1992 . Comparison of immunofluorescence microscopy and dilution-plating for the detection of *Xanthomonas campestris* pv. *campestris* in crucifer seeds. *Netherlands Journal of Plant Pathology* 98:169–178.

Kocks, C. G., Ruissen, M. A., Zadoks, J. C., and Duijkers, M. G. 1998. Survival and extinction of *Xanthomonas campestris* pv. *campestris* in soil. *European Journal of Plant Pathology* 104:911–923.

Kocks, C .G., Zadoks, J. C., and Ruissen, M. A. 1999. Spatio-temporal development of black rot (*X. campestris* pv. *campestris*) in cabbage in relation to initial inoculum levels in field plots in The Netherlands. *Plant Pathology* 48:176–188.

Massomo, S. M. S., Nielsen, H., Mabagala, R. B., Mansfeld-Giese, K., Hockenhull, J., and Mortensen, C. N. 2003. Identification and characterisation of *Xanthomonas campestris* pv. *campestris* strains from Tanzania by pathogenicity tests, Biolog, rep-PCR and fatty acid methyl ester analysis. *European Journal of Plant Pathology* 109:775–789.

Poplawsky, A. R. and Chun, W. 1995. Strains of *Xanthomonas campestris* pv. *campestris* with atypical pigmentation isolated from commercial crucifer seed. *Plant Disease* 79:1021–1024.

Randhawa, P. S. and Schaad, N. W. 1984. Selective isolation of *Xanthomonas campestris* pv. *campestris* from crucifer seeds. *Phytopathology* 74:268–272.

Sahin, F. and Miller, S. A. 1997. A new pathotype of *Xanthomonas campestris* pv. *amoraciae* that causes bacterial leaf spot of radish. *Plant Disease* 81:1334.

Schaad, N. W. and Dianese, J. C. 1981. Crucifer weeds as sources of inoculum of *Xanthomonas campestris* in black rot of crucifers. *Phytopathology* 71:1215–1220.

Schultz, T. and Gabrielson, R. L. 1986. *Xanthomonas campestris* pv. *campestris* in western Washington crucifer seed fields: occurrence and survival. *Phytopathology* 76:1306–1309.

Shaw, J. J. and Kado, C. I. 1988. Whole plant wound inoculation for consistent reproduction of black rot of crucifers. *Phytopathology* 78:981–986.

Shigaki, T., Nelson, S. C., and Alavarez, A. M. 2000. Symptomless spread of blight inducing strains of *Xanthomonas campestris* pv. *campestris* on cabbage seedlings in misted seedbeds. *European Journal of Plant Pathology* 106: 339–346.

Tamura, K., Takikawa, Y., Tsuyumu, S., and Goto, M. 1994. Bacterial spot of crucifers caused by *Xanthomonas campestris* pv. *raphani*. *Annals of the Phytopathological Society of Japan* 60:281–287.

Vauterin, L., Rademaker, J., and Swings, J. 2000. Synopsis on the taxonomy of the genus *Xanthomonas*. *Plant Disease* 90:677–682.

Vicente, J. G., Conway, J., Roberts, S. J., and Taylor, J. D. 2001. Identification and origin of *Xanthomonas campestris* pv. *campetris* races and related pathovars. *Phytopathology* 91:492–499.

Williams, P. H. 1980. Black rot: a continuing threat to world crucifers. *Plant Disease* 64:736–742.

Zhao, Y., Damicone, J. P., Demezas, D. H., and Bender, C. L. 2000. Bacterial leaf spot diseases of leafy crucifers in Oklahoma caused by pathovars of *Xanthomonas campestris*. *Plant Disease* 84:1008–1014.

Albugo candida

WHITE RUST, WHITE BLISTER

Introduction and significance

This is a common disease of crucifer crops and related weeds and ornamentals. Affected vegetables include broccoli, Brussels sprouts, cauliflower, radish, horseradish, mustard, Chinese cabbage, rappini, and turnip. Yield reductions are rare, and the main impact of the disease is reduced quality due to the prominent signs of the pathogen.

Symptoms and diagnostic features

Early signs are small, raised, white or cream-colored blisters (sori) that measure 2–3 mm in diameter and form underneath the plant epidermis on leaves, stems, and flower stalks (**161**). The epidermis covering the pustule will later rupture, releasing powdery white sporangia. Later, larger blisters (2–3 cm diameter) may develop. The pathogen also causes distortion of leaf tissue, which may have a yellow or red discoloration. Severely infected leaves wither and die. Older pustules can become brown and rotten, making them difficult to recognize later in the season. Infection can develop systemically and cause twisted, deformed growth of the stem, leaves, or flowers. On flowering shoots, this condition is known as 'stagheads' because of the extensive twisting of infected tissue (**162**). Clubroot-like swellings may be produced on roots.

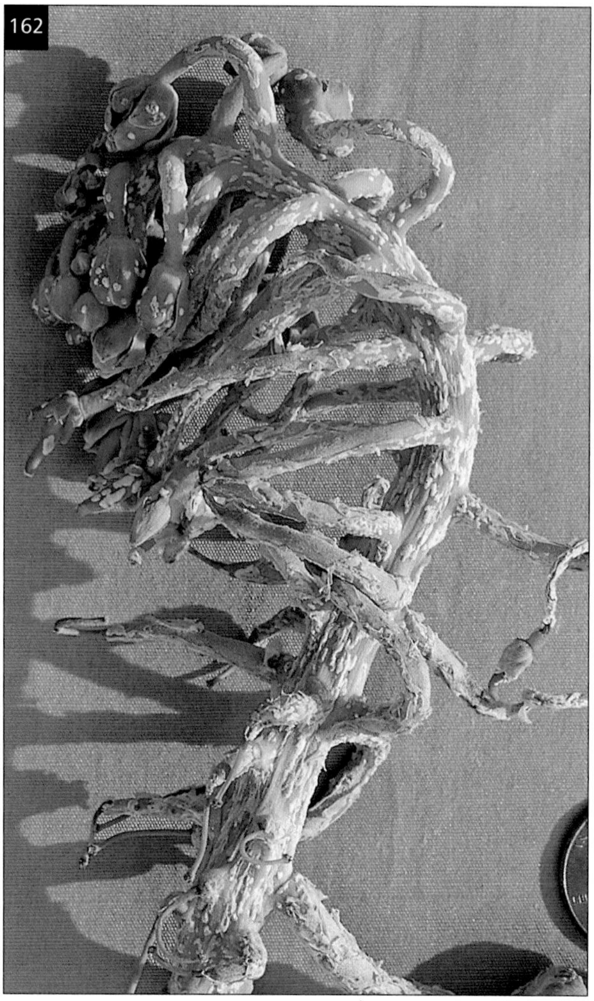

162 **Cauliflower seed stalk infection with white rust.**

161 **Leaf pustules of white rust of bok choy.**

163 **Pustules of white rust of radish.**

On radish, white rust signs include the characteristic raised white pustules on the leaves that break open to release white powdery spores (**163**). White rust also causes severe infection of flower shoots and swellings on the root. However, note that root galls on radish may also be caused by clubroot disease. Radish isolates are unlikely to affect other crucifers such as cabbage, cauliflower, and turnip.

Causal agent

White rust is caused by the oomycete *Albugo candida*. In infected tissues this organism produces globose, chocolate-brown oospores that measure 30–55 µm in diameter and have a thick ornamented spore wall. Within the raised pustules, club-shaped sporangiophores produce sporangia that are borne in basipetal chains. Sporangia are globose to oval, hyaline, with thin walls, and measure 12–18 µm in diameter. Downy mildew (*Peronospora parasitica*) sometimes grows on white rust pustules. There are many races of this pathogen, some of which have restricted host ranges and cannot infect all crucifers.

Disease cycle

This pathogen survives in debris and soil as resilient oospores. Oospores are therefore the main primary inoculum, which germinate to produce five to seven swimming zoospores. Zoospores swim to or are splashed onto susceptible hosts, resulting in infection and formation of the raised pustules. Airborne sporangia from the pustules result in pathogen spread within the crop. After a period of drying, sporangia also release zoospores on the surface of the host plant; these zoopsores encyst and penetrate the host directly. Leaf surface moisture from dew, fog, or rain is required for infection. Infection can occur in 4 to 6 hours under optimal temperatures. Germination occurs over the range of 0–18° C, but infection requires warmer temperatures of 15–25° C, with an optimum of 20° C. Greater disease severity occurs with increasing leaf-wetness duration. Symptoms appear about 10–14 days after infection.

Control

Bury crop residues after harvest to reduce spread of airborne spores to nearby crops. Use resistant cultivars when available and rotate with non-hosts where the disease occurs regularly. Some disease control effects have been reported for applications of P and K fertilizers (70–100 kg/ha). Apply protectant fungicides such as chlorothalonil, phenylamide, and strobilurin products.

References

Goyal, B. K., Verma, P. R., Spurr, D. T., and Reddy, M. S. 1996. *Albugo candida* staghead formation in *Brassica juncea* in relation to plant age, inoculation sites, and incubation conditions. *Plant Pathology* 45:787–794.

Liu, J. Q. and Rimmer, S. R. 1993. Production and germination of oospores of *Albugo candida*. *Canadian Journal of Plant Pathology*15:265–271.

Koike, S. T. 1996. Outbreak of white rust, caused by *Albugo candida*, on Japanese mustard and tah tsai in California. *Plant Disease* 80:1302.

Petrie, G. A. 1988. Races of *Albugo candida* (white rust and staghead) on cultivated Cruciferae in Saskatchewan. *Canadian Journal of Plant Pathology* 10:142–150.

Petrie, G. A. and Verma, P. R. 1974. A simple method for germinating oospores of *Albugo candida*. *Canadian Journal of Plant Science* 54:595–596.

Pidskalny, R. S. and Rimmer, S. R. 1985. Virulence of *Albugo candida* from turnip rape (*Brassica campestris*) and mustard (*Brassica juncea*) on various crucifers. *Canadian Journal of Plant Pathology* 7:283–286.

Scheck, H. J. and Koike, S. T. 1999. First occurrence of white rust of arugula, caused by *Albugo candida*. *Plant Disease* 83:877.

Verma, P. R. and Petrie, G. A. 1975. Germination of oospores of *Albugo candida*. *Canadian Journal of Botany* 53:836–842.

Williams, P. H. and Pound, G.S. 1963. Nature and inheritance of resistance to *Albugo candida* in radish. *Phytopathology* 53:1150–1154.

Alternaria brassicae, A. brassicicola

ALTERNARIA LEAF SPOT/ HEAD ROT, DARK LEAF SPOT

Introduction and significance

Alternaria pathogens are important on crucifer crops worldwide, causing foliar, pod, seed, and broccoli-head diseases. All the major brassica types (*B. campestris*, *B. oleracea*, *B. napus*, *B. rapa*) are susceptible. In many cases, these pathogens do not reduce the weight and size of the harvested commodity, but cause losses because of reduced quality and appearance. However, Chinese cabbage, bok choy, and leafy brassicas can be seriously damaged and rendered unmarketable.

Symptoms and diagnostic features

The main symptom of this disease is leaf spot. On leaves, small, dark specks measuring 1–2 mm in diameter first develop and later enlarge into circular, tan to brown spots that are visible from both sides of the leaf. These can enlarge to 5–25 mm in diameter and usually contain concentric rings, giving the spots a target-like appearance (**164**, **165**). Spots sometimes have yellow halos caused by a fungal toxin. If conditions are favorable, dark green conidia will be visible on the spots. Old leaf spots become papery in texture and may tear. Spots are most apparent on the oldest leaves. Leaf spots caused by *A. brassicicola* tend to be darker and may have a more irregular margin than those caused by *A. brassicae*. Spots on petioles tend to be brown and oval or elliptical in shape (**166**).

Susceptible crops such as turnip and Chinese cabbage can be almost defoliated by Alternaria leaf spot. In the UK, Alternaria leaf spot can significantly reduce Brussels sprout yields because of the low tolerance for spots and blemishes (**167**). Alternaria leaf spot is also important on stored white cabbage, turnip, and swede roots because disease development continues at low temperatures.

Alternaria brassicicola causes a disease of cauliflower and broccoli heads (**168**) when spores are splashed or blown onto the immature flower heads. Symptoms consist of a water-soaked decay of the crucifer head that later turns dark brown to black. The tissue becomes soft and mushy and the pathogen sporulates on the head. Secondary molds and bacteria cause further decay. Alternaria head rot can be severe during wet and cold weather in California. Alternaria head rot and bacterial head rot symptoms on broccoli are similar, so diagnosis needs to be done carefully. In seed crops, *Alternaria* species cause black or brown spots on the stems and pods, which results in premature ripening and splitting of the pods.

Causal agents

Alternaria leaf spot and related diseases are caused by *A. brassicae* and *A. brassicicola*. These pathogens can be isolated on standard microbiological media. Both species sporulate well in culture if incubated under 12-hour light/12-hour dark conditions. The two species are readily distinguished by spore size and morphology. *Alternaria brassicae* conidia from leaves are obclavate

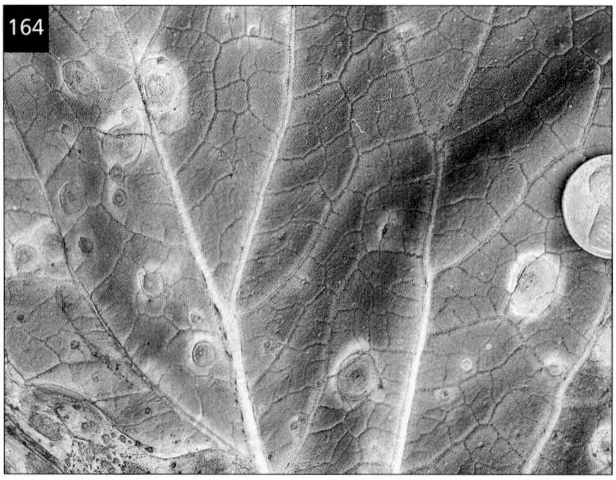

164 Symptoms of Alternaria leaf spot of Chinese cabbage.

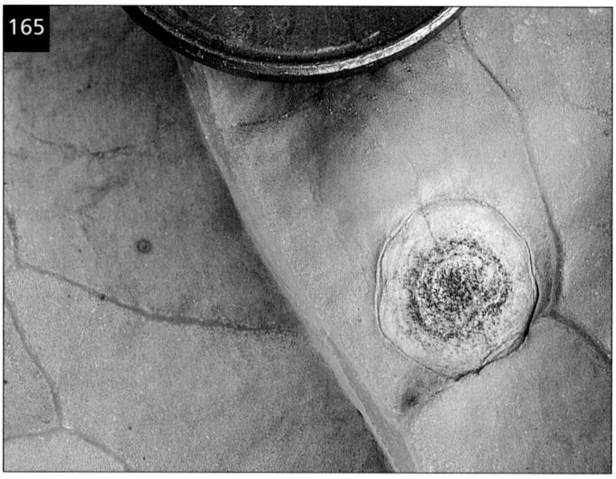

165 Close-up of Alternaria leaf spot of cauliflower.

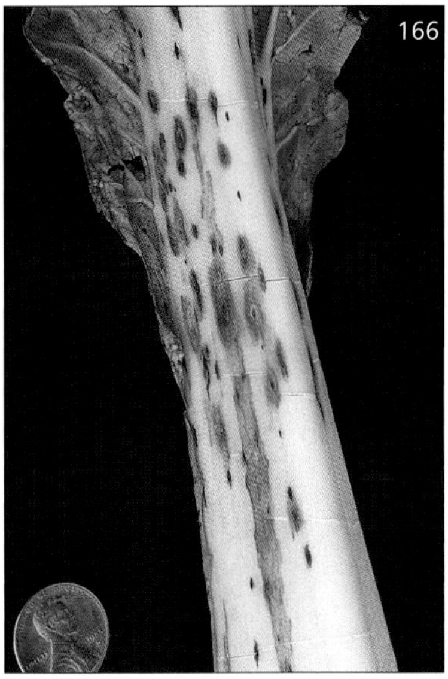

166 Petiole lesions of Alternaria leaf spot of bok choy.

in shape with one slender, unbranched beak that measures 55–75 µm long; beaks extend from the apical end of the spore body. The total conidial length, including the narrow beak, for *A. brassicae* is 75–350 µm. The width of the main spore body is 10–35 µm. Conidia have six to 19 transverse septa and one to eight longitudinal or oblique septa and are borne in short chains of up to four spores. *Alternaria brassicicola* conidia are borne in longer chains of 20 or more and are smaller, measuring 18–130 x 8–20 µm. Conidia are cylindrical and tapering near the apex. This species lacks the long beak, and has 1 to 10 transverse and few longitudinal septa.

The pathogen on radish (**169**) is yet another species, *A. raphani*. *Alternaria raphani* conidia are obclavate in shape with one short, unbranched beak. The total conidial length is 50–130 µm. The width of the main spore body is 14–30 µm. Conidia have three to seven transverse septa and several longitudinal or oblique septa and are borne in short chains of two to three spores.

Disease cycle

Seedborne inoculum can be both on the seed surface or internal in seed tissue. *A. brassicae* may survive for only a few months when seed storage temperatures are greater than 20° C, while *A. brassicicola* may survive for 12 years when there is internal seed infection. Seedlings produced from infected seed may have tiny necrotic spots or show stunting or death of cotyledons. Infected seed can be reduced in vigor and have lower

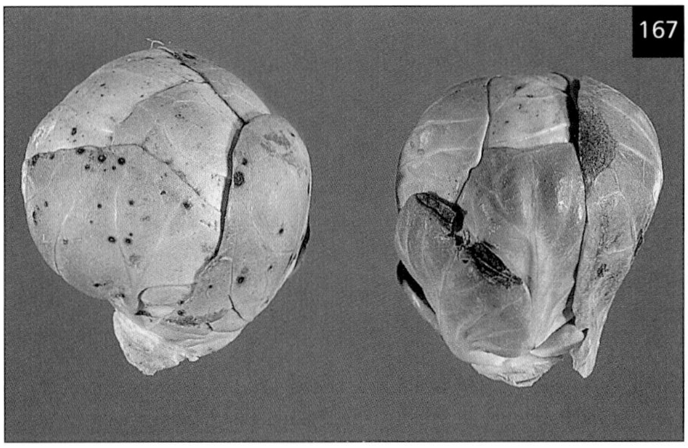

167 Alternaria leaf spot on Brussels sprouts caused by *A. brassicae*. Larger lesions are caused by the ring spot pathogen.

168 Broccoli infected with Alternaria head rot.

169 Leaf infections of Alternaria leaf spot of radish.

germination rates. At least 12 hours of 90% relative humidity are required for sporulation on plant tissues. Spore production is optimal at 18–24° C for *A. brassicae* and 20–30° C for *A. brassicicola.* Sporulation of these pathogens is favored by alternating light and dark periods. Conidia are released during periods of falling humidity and spore trap records reveal peaks in the afternoon when humidity is low. Spores are dispersed by splashing water and by wind. Both *Alternaria* species require free water for infection. A minimum of 6–8 hours leaf wetness is required for infection, and severe disease is associated with prolonged warm, wet weather. Older leaves and pods are more susceptible to infection and show increased disease as they senesce. These pathogens are able to survive on crop residues, and microsclerotia and chlamydospores enable *A. brassicae* to survive for longer periods. When oilseed rape or other crucifer seed crops are harvested, airborne inoculum can be dislodged and carry over to vegetable plantings. These two pathogens also affect many crucifer weeds and native *Brassica* species. Insect transmission has been reported for *A. brassicicola.*

Control

Use seed that does not have significant levels of the pathogen. Apply seed treatments if seed is infected. Seed treatments include fungicides such as thiram and iprodione, and hot water soaks at 50° C for 20 minutes. Destroy diseased crop residues after harvest to reduce inoculum sources. Do not plant vegetable brassicas close to field brassicas such as oilseed rape. Foliar fungicides can provide effective control. Such protectants include iprodione, triazoles such as difenoconazole and tebuconazole, and strobilurin products. Forecasting systems to predict infection are still being developed. Some differences in cultivar susceptibility may exist. For broccoli head rot, avoid using overhead sprinkler irrigation. Broccoli cultivars that are resistant to the head-rot phase of this disease have not been identified; however, those cultivars that form a more rounded, dome-shaped head generally have less severe head rot. Seed crops should be produced in arid climates and without the use of overhead sprinklers.

References

Babadoost, M. and Gabrielson, R. L. 1979. Pathogens causing Alternaria diseases of *Brassica* seed crops in Western Washington. *Plant Disease Reporter* 63:815–820.

Bains, P. S. and Tewari, J. P. 1987. Purification, chemical characterization and host-specificity of the toxin produced by *Alternaria brassicae. Physiological and Molecular Plant Pathology* 30:259–271.

Bassey, E. O. and Gabrielson, R. L. 1983. The effect of humidity, seed infection level, temperature, and nutrient stress on cabbage seedling disease caused by *Alternaria brassicicola. Seed Science and Technology* 11:403–410.

Bassey, E. O. and Gabrielson, R. L. 1983. Factors affecting accuracy of 2,4-D assays of crucifer seed for *Alternaria brassicicola* and relation of assays to seedling disease potential. *Seed Science and Technology* 11:411–420.

Cerkauskas, R. F., Stobbs, L. W., Lowery, D. T., Van Driel, L., Liu, W., and VanSchagen, J. 1998. Diseases, pests, and abiotic problems associated with oriental cruciferous vegetables in southern Ontario in 1993-1994. *Canadian Journal of Plant Pathology* 20:87–94.

Chen, L. Y., Price, T. V., and Park-Ng, Z. 2003. Conidial dispersal by *Alternaria brassicicola* on Chinese cabbage (*Brassica pekinensis*) in the field and under simulated conditions. *Plant Pathology* 52:536–545.

Dillard, H. R., Cobb, A. C., and Lamboy, J. S. 1998. Transmission of *Alternaria brassicicola* to cabbage by flea beetles. *Plant Disease* 82:153–157.

Guillemette, T., Iacomi-Vasilescu, B., and Simoneau, P. 2004. Conventional and real-time PCR-based assay for detecting pathogenic *Alternaria brassicae* in cruciferous seed. *Plant Disease* 88:490–496.

Hong, C. X. and Fitt, B. D. L. 1996. Factors affecting the incubation period of dark leaf and pod spot (*Alternaria brassicae*) on oilseed rape (*Brassica napus*). *European Journal of Plant Pathology* 102:545–553.

Huang, R. and Levy, Y. 1995. Characterization of iprodione-resistant isolates of *Alternaria brassicicola. Plant Disease* 79:828–833.

Humpherson-Jones, F.M. and Maude, R.B. 1982. Studies on the epidemiology of *Alternaria brassicicola* in *Brassica oleracea* seed production crops. *Annals of Applied Biology* 100:61-71.

Koike, S. T. 1996. Japanese mustard, tah tsai, and red mustard as hosts of *Alternaria brassicae. Plant Disease* 80:822.

Koike, S. T. 1997. Broccoli raab as a host of *Alternaria brassicae. Plant Disease* 81:552.

Koike, S. T. and Molinar, R. H. 1997. Daikon (*Raphanus sativus* L. cv. *longipinnatus*) as a host of *Alternaria brassicae* in California. *Plant Disease* 81:1094.

Tsuneda, A. and Skoropad, W. P. 1977. Formation of microsclerotia and chlamydospores from conidia of *Alternaria brassicae. Canadian Journal of Botany* 56:1333–1334.

Botrytis cinerea (teleomorph = *Botryotinia fuckeliana*)

GRAY MOLD

Introduction and significance

Disease caused by *Botrytis* usually has a minor impact on crucifers because infected, older leaves are removed during harvest. More severe problems can occur if crops are affected by other problems, such as chemical burn, nutrient deficiency, or weather damage, and are subsequently colonized by gray mold. Gray mold is an important problem in stored produce, particularly on white cabbage.

Symptoms and diagnostic features

Symptoms are brown blotches on leaves or stems that extend in size from 1–2 cm in diameter to very large sections of diseased tissue. Gray mold results in a soft, mushy rot in which tissues disintegrate if handled. Gray fungal growth is usually present in tufts near the center of foliar lesions or generally over cauliflower heads (**170**) or other organs. On stored cabbage (**171**) the outer wrapper leaves may bear extensive sporulation. Gray mold symptoms are almost always associated with senescence or physical damage to host tissues.

170 Gray mold on cauliflower head.

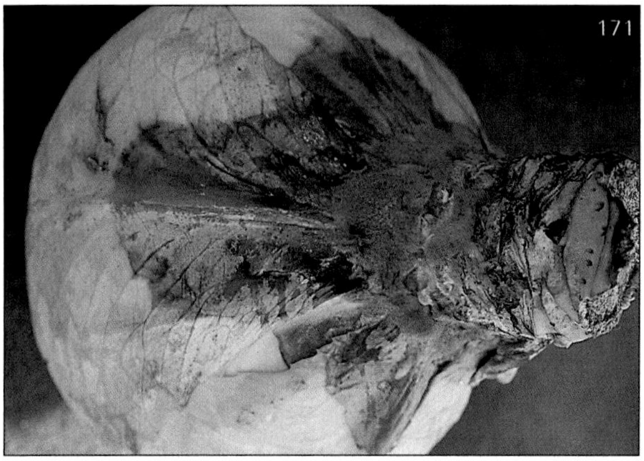

171 Gray mold on stored white cabbage.

Causal agent

Gray mold is caused by the ascomycete fungus *Botryotinia fuckeliana*. However, this stage is rarely observed on crucifers and the disease is associated with the anamorph *Botrytis cinerea*. Conidiophores are long, can measure up to 1–2 mm, become gray-brown with maturity, and branch irregularly near the apex. Conidia are clustered at the branch tips and are single-celled, pale brown, ellipsoid to obovoid, and measure 6–18 x 4–11 µm. If formed, sclerotia are black, oval or irregular in shape, and measure 4–15 mm. On host tissue the fungus produces characteristically profuse mycelial growth that is initially fluffy, white or pale gray, later becoming darker gray- brown when it sporulates.

Disease cycle

Botrytis cinerea survives in and around fields as a saprophyte on crop debris, as a pathogen on numerous crop and weed plants, and as sclerotia in the soil. Conidia develop from these sources and become windborne. When conidia land on senescent or damaged crucifer tissue, they will germinate within a few hours if free moisture is available and rapidly colonize this food base. Once established in the senescent tissue, the pathogen will progress into adjacent healthy stems and leaves, resulting in disease symptoms and the production of additional conidia. Cool temperatures, free moisture, and high humidity favor disease development. Factors that predispose crucifer crops to gray mold are physical damage due to passing farm equipment, burn from fertilizers and pesticides, pest infestations (especially from boring insects such as root maggots), drought stress, frost damage, overmaturity of the crop, and tipburn due to nutritional imbalances.

Control

Prevent or minimize injury to the crop. Provide sufficient fertilizers for the crop, and harvest the crop before the commodity becomes overmature. For cabbage, handle and harvest the crop under dry conditions and before the occurrence of frost. Postharvest fungicide dips are used on cabbage in some countries.

References

Davies, R. M. and Heale, J. B. 1985. *Botrytis cinerea* in stored cabbage: the use of germ tube growth on leaf discs as an indication of potential head rot. *Plant Pathology* 34:408–414.

Yoder, O. C. and Whalen, M. L. 1975. Factors affecting postharvest infection of stored cabbage tissue by *Botrytis cinerea*. *Canadian Journal of Botany* 53:691–699.

Erysiphe cruciferarum

POWDERY MILDEW

Introduction and significance

Crucifer crops are frequently affected by powdery mildew, which occurs in all production areas. There are large differences in susceptibility between different species and cultivars. Severe epidemics commonly occur in swede and turnip that result in reduced yield and quality. More commonly, the foliage and stems of brassicas may exhibit disease, though these infections have limited effects on quality and yield.

Symptoms and diagnostic features

Powdery mildew is initiated from airborne conidia and appears as small, radiating, diffuse colonies of superficial white mycelium that can measure from 5–15 mm in diameter. These ectophytic colonies are readily removed by rubbing the plant surface. Severe, advanced infections produce a dense white powdery covering of leaves (**172, 173**), stems, and seed crop pods. Severe attacks cause chlorosis, early defoliation, and necrosis of the tips of young leaves of cauliflower and cabbage. Powdery mildew colonies may be gray and restricted in size on resistant cultivars as the host reaction produces black speckling beneath the colony. Conspicuous gray or purple symptoms occur on the stems of Brussels sprout, while on sprouts there may be white colonies or fine black speckling in radiating lines.

Causal agent

Powdery mildew is caused by the ascomycete fungus *Erysiphe cruciferarum*. The asexual conidia measure 30–52 x 11–17 μm and are borne singly or in chains of only two spores. There appears to be physiological specialization within the pathogen; for example, isolates from turnip will not infect Brussels sprout, and vice versa. The sexual stage has only been reported occasionally and consists of spherical cleistothecia that are brown or black. Like all powdery mildews, *E. cruciferarum* is an obligate pathogen.

Disease cycle

The pathogen survives on overwintering crops and volunteers. The main period of powdery mildew activity in northern Europe is usually from July until October, but there is great seasonal variation for the disease. Conidia readily spread via winds and can travel significant distances. Powdery mildew conidia do not require water for germination, which can take place at low relative humidity. Rain or free water can actually inhibit

172 Severe powdery mildew of swede.

173 Close-up of powdery mildew on underside of Brussels sprout leaf.

powdery development on the upper leaf surface. Germ-tube growth occurs mainly at 98–100% relative humidity. In some areas of California, powdery mildew is most severe on greenhouse-grown vegetable crucifers. The exact role of cleistothecia in the disease cycle is not known, but this spore stage could provide ascospores as primary inoculum.

Control

Apply fungicides such as sulfur, triazole, and morpholine products. Less susceptible crops such as cauliflower and cabbage may not require fungicide treatment. For swede and turnip, delay planting dates as this can reduce disease severity. Plant resistant cultivars, particularly for Brussels sprout, cabbage, and swede. Two forms of resistance have been recognized: suppression of fungal development beyond the appressorial stage, and delayed spore production.

References

Dixon, G. R. 1974. Field studies of powdery mildew (*Erysiphe cruciferarum*) on Brussels sprouts. *Plant Pathology* 23:105–109.

Koike, S. T. and Saenz, G. S. 1997. First report of powdery mildew, caused by *Erysiphe cruciferarum*, on broccoli raab in California. *Plant Disease* 81:1093.

Rudyard, S. A. and Wheeler, B. E. J. 1985. The development of *Erysiphe cruciferarum* on field-grown Brussels sprouts and associated changes in soluble amino-acids in foliage leaves. *Plant Pathology* 34:616–625

Vakalounakis, D. J. 1993. First record of *Erysiphe cruciferarum* on *Lunaria biennis* in Greece. *Plant Pathology* 43:424–425.

Fusarium oxysporum* f. sp. *conglutinans*, *F. oxysporum* f. sp. *raphani

FUSARIUM WILT, FUSARIUM YELLOWS

Introduction and significance

This is sometimes a destructive disease of crucifers in temperate and tropical regions. In California, this disease affects primarily cabbage and radish.

Symptoms and diagnostic features

Initial symptoms consist of yellowing of the lower leaves, often on one side of the plant (**174**). These leaves later turn brown and drop off. A tan to brown discoloration of the xylem in the main stem, larger petioles, and larger roots is characteristic of this disease (**175**). With time, the entire plant may yellow, wilt, and collapse. This disease causes more severe symptoms on summer crops due to warmer soil temperatures; Fusarium develops most rapidly at temperatures ranging from 24–29° C, and little symptom development occurs below 16° C. Verticillium wilt (caused by *Verticillium* species) has similar vascular discoloration, but such coloration tends to be black and not brown. Black rot (caused by *Xanthomonas campestris*) also causes vascular discoloration, but the blackened xylem tends to be in the upper, less woody parts of the plant.

Fusarium yellows is also a damaging disease of radish and is particularly important in the USA. Initial symptoms consist of a dull, yellow-green chlorosis of

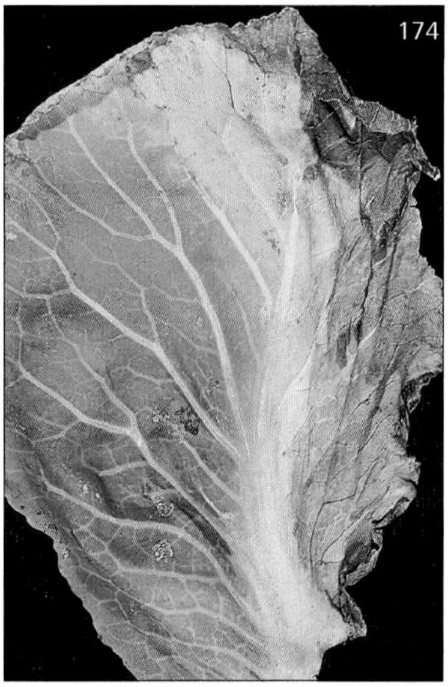

174 One-sided chlorosis and necrosis on cabbage caused by Fusarium wilt.

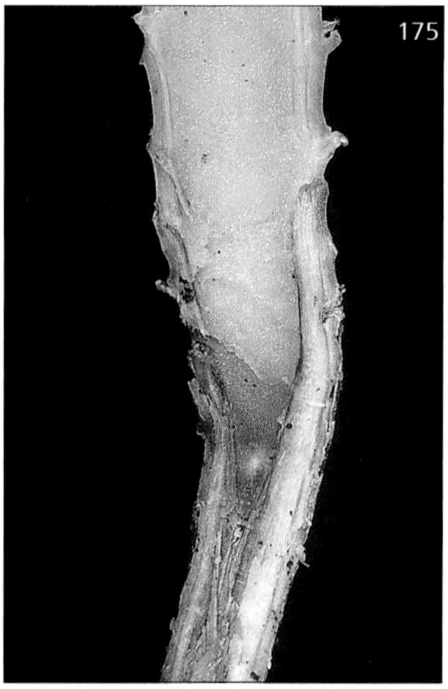

175 Vascular discoloration of Fusarium wilt of cabbage.

176 Stunted plants of Fusarium wilt of radish seedlings.

177 Stunted plants of Fusarium wilt of radish. Healthy plant on left.

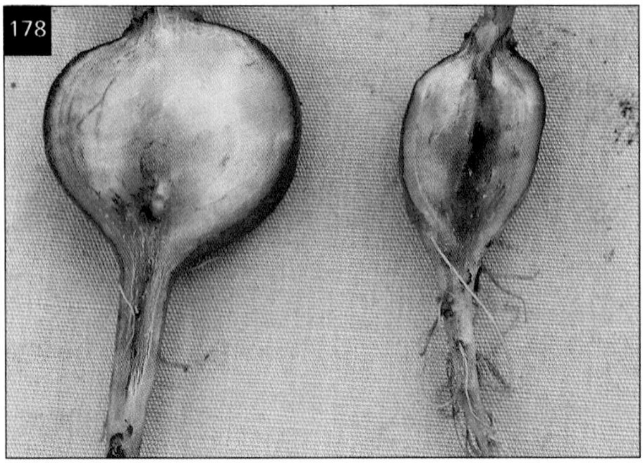

178 Vascular discoloration of Fusarium wilt of radish.

leaves that is followed by more general yellowing of the whole plant (**176**). Leaves may die prematurely and drop off the plant. Growth is stunted (**177**) and some plants die rapidly. Most characteristic is a brown to black vascular discoloration in the central core of the radish root (**178**). Symptom development is temperature dependent and is most severe at 17–35° C. At lower temperatures, plants may be infected, but remain symptomless.

Causal agents

Fusarium yellows on brassica hosts is caused by the fungus *Fusarium oxysporum* f. sp. *conglutinans*. The pathogen can be isolated on standard microbiological media. Semi-selective media such as Komada or FS can be useful for some samples or to purify isolates. The pathogen morphology and colony characteristics are similar to other *F. oxysporum* fungi. Microconidia are one- or two-celled and measure 6–15 x 2.5–4 µm. Macroconidia are two to four septate and measure 25–33 x 3.5–5.5 µm. This pathogen is host specific to crucifers, and can be further divided into various vegetative compatibility groups (VCG). Several races have been identified. Race 1 infects a wide range of crucifers, including radish. Race 2 is able to infect radish and some other brassicas, but is not able to cause disease on *B. oleracea* crops such as cabbage, cauliflower, and Brussels sprout. Some researchers suggest that this race 2 be designated as *F. oxysporum* f. sp. *raphani*.

Disease cycle

This pathogen is soilborne and produces chlamydospores that enable it to survive for many years in soil. It may also be seedborne. Infection occurs through the root system and the pathogen subsequently penetrates the vascular tissues, where growth and sporulation take place.

Control

Prevent the movement of infested soil to clean fields. Provide adequate fertilizers to the plants; in particular, it appears that potassium deficiency may increase disease severity. Plant resistant cultivars. Type A resistance is monogenic and prevents pathogen penetration of the vascular system except at high temperatures (>30° C). Type B resistance is polygenic and allows only limited vascular colonization at low temperatures, but loses this resistance as temperatures rise.

References

Beckman, C. H. 1967. Respiratory response of radish varieties resistant and susceptible to vascular infection by *Fusarium oxysporum* f. sp. *conglutinans*. *Phytopathology* 57:699–702.

Bosland, P. W. and Williams, P. H. 1987. An evaluation of *Fusarium oxysporum* from crucifers based on pathogenicity, isozyme polymorphism, vegetative compatibility, and geographic origin. *Canadian Journal of Botany* 65:2067–2073.

Bosland, P. W., Williams, P. H., and Morrison, R. H. 1988. Influence of soil temperature on expression of yellows and wilt of crucifers by *Fusarium oxysporum*. *Plant Disease* 72:777–780.

Farnham, M. W., Keinath, A. P., and Smith, J. P. 2001. Characterization of Fusarium yellows resistance in collard. *Plant Disease* 85:890–894.

Komada, H. 1975. Development of a selective medium for quantitative isolation of *Fusarium oxysporum* from natural soil. *Review of Plant Protection Research* 8:114–124.

Momol, E. A. and Kistler, H. C. 1992. Mitochondrial plasmids do not determine host range of crucifer-infecting strains of *Fusarium oxysporum*. *Plant Pathology* 41:103–112.

Ramirez-Villapadua, J., Endo, R. M., Bosland, P., and Williams, P. H. 1985. A new race of *Fusarium oxysporum* f. sp. *conglutinans* that attacks cabbage with type A resistance. *Plant Disease* 69: 612–613.

Leptosphaeria maculans (anamorph = *Phoma lingam*), *L. biglobosa* (anamorph = *P. lingam*)

BLACK LEG, PHOMA LEAF SPOT & CANKER

Introduction and significance

Black leg is probably one of the most important diseases of crucifer vegetables as well as forage and arable crucifers, particularly oilseed rape. Many regions, including North America, Australia, and Europe, experience significant crop loss from this problem. Black leg may be less important in China because the pathogen there is the less aggressive B-type. Black leg is the accepted and most common name for this disease, though Phoma leaf spot and canker is also used. On root crops, the disease is called dry rot.

Symptoms and diagnostic features

Foliar symptoms are leaf spots that can be green-brown, pale brown, tan, or almost white. Spots contain numerous dark, tiny fruiting bodies or pycnidia (**179**, **180**). Spots tend to be circular or oval in shape and range in size from 0.5-3.0 cm in diameter. There may be a chlorotic halo around the lesion, and leaf veins within and extending out of the lesion may be black. These symptoms are believed to be caused by A-type isolates. Symptoms caused by the B-type are usually smaller (1–3 mm) and darker spots, similar to those caused by *Alternaria* pathogens. These leaf spots have only a few pycnidia that are difficult to see without a dissecting microscope. Radish appears unaffected by B-type, but

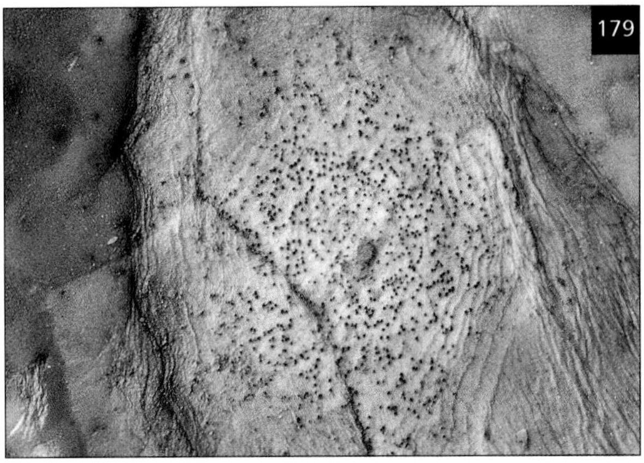

179 Leaf lesion with pycnidia of black leg of cauliflower.

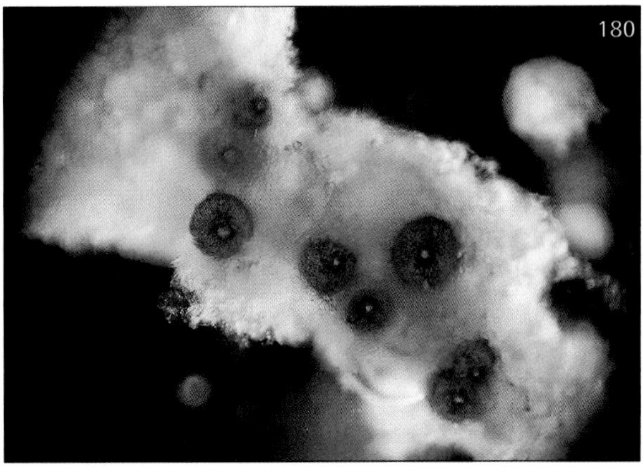

180 Microscopic view of black leg pycnidia.

develops leaf lesions after infection by A-type isolates. From leaf spots, the pathogen spreads via leaf veins to the petiole and eventually the main stem.

The main stem may then develop the black leg symptom where large, sunken, elongated, brown to black cankers form (**181**, **182**, **183**). These cankers have uneven borders and, as the tissue deteriorates, the stems may become girdled, resulting in stunting, wilting, and general poor growth. If cankers enlarge, the stem may rot and break, causing the plant to fall over. Cankers usually contain pycnidia. The woody tissues of the stem become black, which is why the disease is called black leg. Seedlings with cankers may fail to mature or die. Though cankers may come from the spread of the pathogen from leaves, cankers also form directly from inoculum making contact with the hyocotyl or stem, usually at the soil level.

On root crops such as swede and turnip, a dry rot of the bulb, or hypocotyl, develops. The upper neck and shoulder areas have large brown lesions with black margins. Pycnidia form in large numbers in the lesions. Sectioning the bulb usually reveals a large dry rot extending deep into the fleshy tissues (**184**, **185**). When lesions dry, severely affected bulbs become shrivelled and unmarketable. Secondary fungal and bacterial decay organisms often invade this dry rot tissue.

Causal agents

Black leg is caused by the ascomycete fungus *Leptosphaeria maculans*. The pathogen can be isolated on standard microbiological media. However, it probably is easiest to identify this pathogen by directly examining diseased tissue for the fruiting bodies of the *Phoma lingam* asexual stage. Look for globose, dark brown to black, ostiolate pycnidia that exude small (3–5 x 1–2 µm), single-celled, cylindrical conidia. The conidial exudate is deep pink, which is a useful diagnostic feature to distinguish *P. lingam* from other *Phoma* species such as *P. exigua* or *P. herbarum*. After a brief incubation in humid conditions, white mycelial growth usually forms around the edge of lesions or cankers, which is a useful diagnostic feature. The ascospores of *L. maculans* are large, yellow-brown, multicellular spores measuring 35–70 x 5–8 µm. They are produced in black, globose pseudothecia measuring 300–350 µm in diameter and forming on infected woody residues of crucifer hosts.

This pathogen is now considered to be a species complex. Many, but not all, isolates can be placed into one of two virulence groups (A and B). Each group can be further divided into phylogenetically distinct subgroups (A: PG2, PG3, PG4; B: NA1, NA2, NA3). The A-group is slow-growing in culture and does not produce brown pigment in culture though it does make sirodesmin toxins (isolates are designated as Tox+). The B-group grows more rapidly in culture and produces a brown pigment on potato dextrose agar, but it does not have sirodesmin toxins (Tox0). Molecular diagnostic tests are now available to distinguish these species. The aggressive or A-types are currently named *L. maculans*

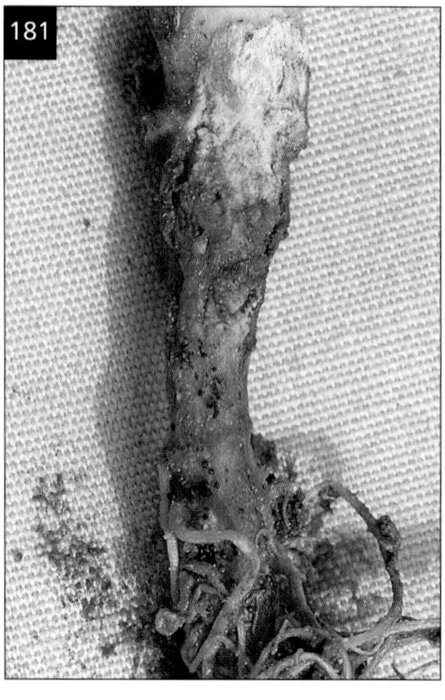

181 Stem infection of black leg of cauliflower.

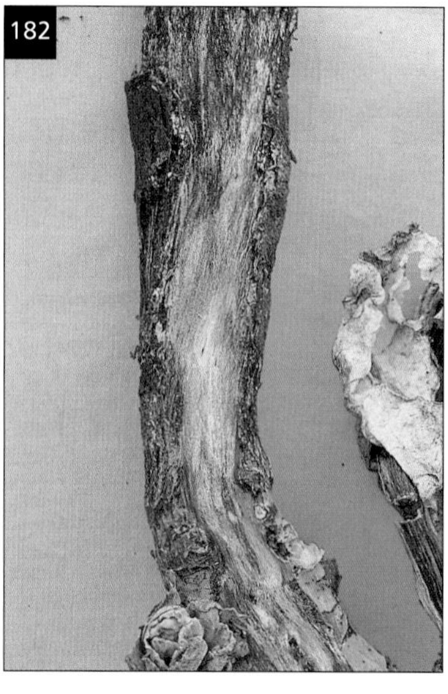

182 Internal discoloration of black leg of Brussels sprout.

and these cause the most severe black leg symptoms. The less aggressive B-type has recently been named *L. biglobosa*, which is differentiated by morphological and cultural characteristics. Interestingly, both *Leptosphaeria* species have a *P. lingam* asexual stage. Research and discussion are still ongoing regarding the taxonomy of this pathogen.

Disease cycle

The pathogen is commonly seedborne and low levels of infection (<0.1%) may initiate significant disease outbreaks. In some areas, however, low levels of seedborne inoculum are less important than airborne ascospores. High numbers of ascospores can be released from infected winter oilseed rape debris for several months each year. Spores can travel more than a kilometer to infect vegetable crucifers. Once the Phoma leaf spot stage is established, secondary spread takes place by water-dispersed conidia. The importance of conidia in the disease cycle appears to vary between different crops and geographic regions. Optimum temperatures for ascospore infection are 15–20° C. A minimum of 4 hours leaf wetness is required for infection, but maximum infection requires up to 48 hours. Symptoms appear in 3 to 4 days at 20° C or 14 days at 5° C. B-group ascospores produce long straight hyphae, which penetrate the leaf through stomata or wounds, while A-type ascospores produce shorter, curled hyphae that penetrate through the cuticle as well as through stomata and wounds.

184 Dry rot of swede caused by black leg.

Control

Use seed that does not have significant levels of the pathogen. Test seed using established seed-assay protocols. If seed are possibly infested, treat them with fungicides or hot water. Effective fungicides include thiram, iprodione, and fenpropimorph. Resistant cultivars have not been developed, though some *Brassica oleracea* crops show considerable tolerance to the leaf-infection phase of blackl eg. Apply foliar fungicides to protect susceptible cultivars. Avoid placing vegetable crucifers close to oilseed rape fields; close proximity is likely to increase disease pressure from airborne ascospores.

183 Close-up of Brussels sprout stem infection and pycnidia of black leg.

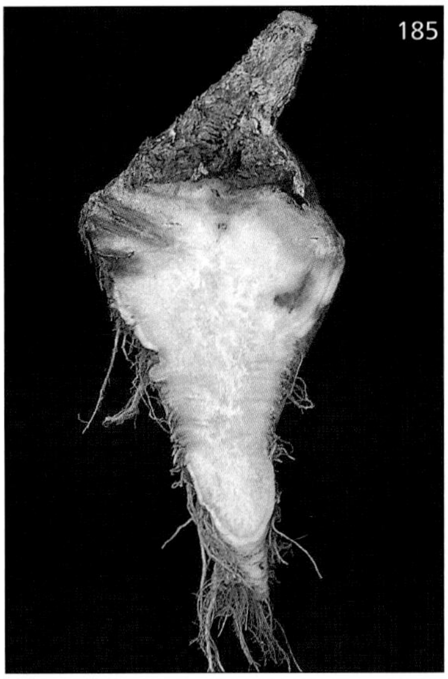

185 Section of swede bulb showing internal rotting due to black leg.

References

Bonman, J. M., Gabrielson, R. L., Williams, P. H., and Delwiche, P. A. 1981. Virulence of *Phoma lingam* to cabbage. *Plant Disease* 65:865–867.

Gabrielson, R. L. 1983. Blackleg disease of crucifers caused by *Leptosphaeria maculans* (*Phoma lingam*) and its control. *Seed Science and Technology* 11:749–780.

Gabrielson, L. 1988. Inoculum thresholds for seed-borne pathogens. *Phytopathology* 78:868–872.

Huang, H. J., Toscano-Underwood, C., Fitt, B. D. L., Hiu, X. J., and Hall, A. M. 2003. Effects of temperature on ascospore germination and penetration of oilseed rape (*Brassica napus*) leaves by A- or B-group *Leptosphaeria maculans* (Phoma stem canker). *Plant Pathology* 52:245–255.

Johnson, R. D. and Lewis, B. G. 1994. Variation in host range, systemic infection and epidemiology of *Leptosphaeria maculans*. *Plant Pathology* 43:417–424.

Maguire, J. D., Gabrielson, R. L., Mulanax, M. W., and Russell, T. S. 1978. Factors affecting sensitivity of 2,4-D assays of crucifer seed for *Phoma lingam*. *Seed Science and Technology* 6:915–924.

Maude, R. B., Humpherson Jones, F. M., and Shuring, C. G. 1984. Treatments to control *Phoma* and *Alternaria* infections of brassica seeds. *Plant Pathology* 33:525–535.

Mengistu, A., Rimmer, S. R., Koch, E., and Williams, P. H. 1991. Pathogenicity grouping of isolates of *Leptosphaeria maculans* on *Brassica napus* cultivars and their disease reaction profiles on rapid-cycling Brassicas. *Plant Disease* 75:1279–1282.

Pongam, P., Osborn, T. C., and Williams, P. H. 1999. Assessment of genetic variation among *Leptosphaeria maculans* isolates using pathogenicity data and AFLP analysis. *Plant Disease* 83:149–154.

Purwantara, A., Barrins, J. M., Cozijnsen, A. J., Ades, P. K., and Howlett, B. J. 2000. Genetic diversity of isolates of the *Leptosphaeria maculans* species complex from Australia, Europe, and North America using amplified fragment length polymorphism analysis. *Mycological Research* 104:772–781.

Shoemaker, R. A. and Brun, H. 2001. The teleomorph of the weakly aggressive segregate of *Leptosphaeria maculans*. *Canadian Journal of Botany* 79:412–419.

Thurwachter, F., Garbe, V., and Hoppe, H. H. 1999. Ascospore discharge, leaf infestation and variations in pathogenicity as criteria to predict impact of *Leptosphaeria maculans* on oilseed rape. *Journal of Phytopathology* 147:215–222.

Toscano-Underwood, C., West, J. S., Fitt, B. D. L., Todd, A. D., and Jedryczka, M. 2001. Development of Phoma lesions on oilseed rape leaves inoculated with ascospores of A- and B-group *Leptosphaeria maculans* (stem canker) at different temperatures and wetness durations. *Plant Pathology* 50:28–41.

Vanniasingham, V. M. and Gilligan, C. A. 1989. Effects of host, pathogen, and environmental factors on latent period and production of pycnidia of *Leptosphaeria maculans* on oilseed rape leaves in controlled environments. *Mycological Research* 93:167–174.

West, J. S., Balesdent, M.-H., Rouxel, T., Narcy, J. P., Huang, Y-J., Roux, J., Steed, J. M., Fitt, B. D. L., and Schmit, J. 2002. Colonization of winter oilseed rape tissues by A/Tox+ and B/Tox0 *Leptosphaeria maculans* (Phoma stem canker) in France and England. *Plant Pathology* 51:311–321.

West, J. S., Kharbanda, P. D., Barbetti, M.J., and Fitt, B. D. L. 2001. Epidemiology and management of *Leptosphaeria maculans* (Phoma stem canker) on oilseed rape in Australia, Canada and Europe. *Plant Pathology* 50:10–27.

Williams, P. H. 1992. Biology of *Leptosphaeria maculans*. *Canadian Journal of Plant Pathology*. 14:30–35.

Williams, R. H. and Fitt, B. D. L. 1999. Differentiating A and B groups of *Leptosphaeria maculans*, causal agent of stem canker (blackleg) of oilseed rape. *Plant Pathology* 48:161–175.

Mycosphaerella brassicicola
(anamorph = *Asteromella brassicae*)

RING SPOT

Introduction and significance

This disease is common in cool moist areas, particularly coastal regions of north-west Europe, California, New Zealand, and Australia. Severe problems are usually associated with intensive cropping and overlapping production of brassicas. It is important in winter cauliflower, Brussels sprout and cabbage in the UK, where it is the most important foliar disease on these crops. In California this disease is mostly found on Brussels sprouts.

Symptoms and diagnostic features

Symptoms occur on foliage and on the pods of seed crops. Initially spots are dark, measure 3–5 mm in diameter, and are green-brown or gray-black (**186**, **187**). Spots enlarge to 2–3 cm in diameter, appear gray when dry, but turn black when the foliage is wet. Leaf spots are visible on both sides of the leaf. Leaf spots are restricted by veins so that they often have an angular appearance; this is one feature that distinguishes ringspot from Alternaria leaf spot. Ring spot spots are often surrounded by a yellow halo, and when spotting is severe, there is extensive leaf yellowing (**188**) and early defoliation. Severe ring spot can result in significant yield loss. The larger leaf spots have concentric zones of tiny black fruiting bodies, hence the name ring spot, that form 3 to 4 weeks after infection.

The small size and high density of these fruiting bodies differs from those of Phoma leaf spot (black leg) which has widely-spaced, brown pycnidia. Symptoms are usually first found on the oldest leaves, but under high disease pressure new symptoms can occur on any leaf. Symptoms on sprouts of Brussels sprout are initially black spots that measure 3–5 mm in diameter. Lesions develop first on the lower sprouts. Mature lesions on sprouts reach 1–2 cm in diameter and are gray when dry but black when wet. Pod symptoms are similar to those on leaves and may be confused with those of Alternaria or white leaf spot diseases. Ring spot can also develop on stored cabbage heads and may penetrate through several leaf layers during storage.

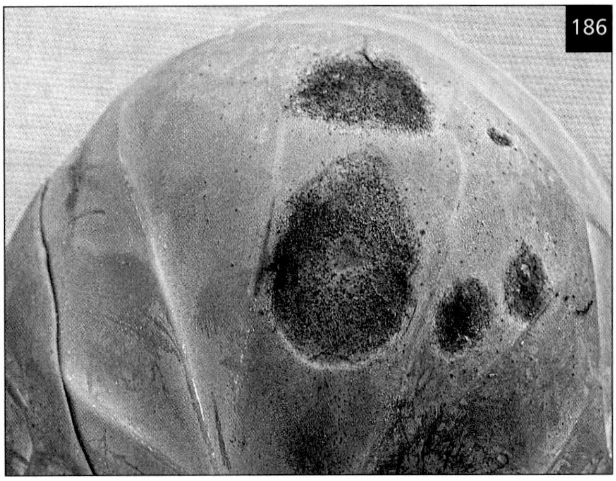

186 Ring spot lesions on Brussels sprout.

Causal agent

Ring spot is caused by the ascomycete fungus *Mycosphaerella brassicicola*. The pseudothecia of the sexual stage are usually present in older leaf lesions and appear as very small, black specks and structures arranged in concentric rings. Pseudothecia release ascospores that are hyaline, two-celled, with rounded ends, and measure 30–45 x 12–18 µm. The pathogen has an asexual stage named *Asteromella brassicae*. *Asteromella brassicae* also forms in the leaf spots, but spores are considered non-infective. Conidia are hyaline, cylindrical, measure 3–5 x 1 µm, and are produced in globular brown pycnidia. There is evidence that some isolates have restricted host ranges and infect only cabbage. The pathogen may be seedborne.

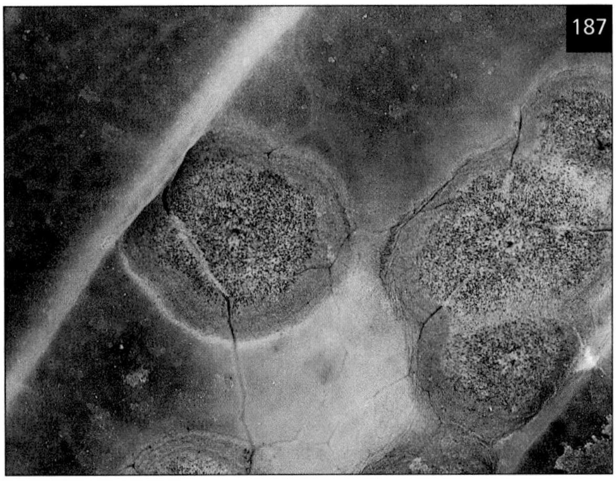

187 Ring spot on Brussels sprout.

Disease cycle

Ring spot is initiated by airborne ascospores, which enables the pathogen to spread from nearby crops and infected residues to new plantings. Survival on plant residues is thought to be limited. Once ring spot is established in a crop, secondary spread occurs by new ascospores produced in leaf lesions. Seedborne inoculum may be another source of the pathogen. The fungus requires briefer periods of leaf wetness than *Alternaria*, and conditions in the UK are rarely limiting for infection in crops like Brussels sprout. The optimum temperatures for infection are 16–20° C, and symptoms appear 10 to 14 days after infection. Mature pseudothecia appear after 4 consecutive days with 100% relative humidity, perhaps as soon as 3 weeks after initial infection. Severe disease is usually associated with prolonged periods (7 to 14 days) of wet weather.

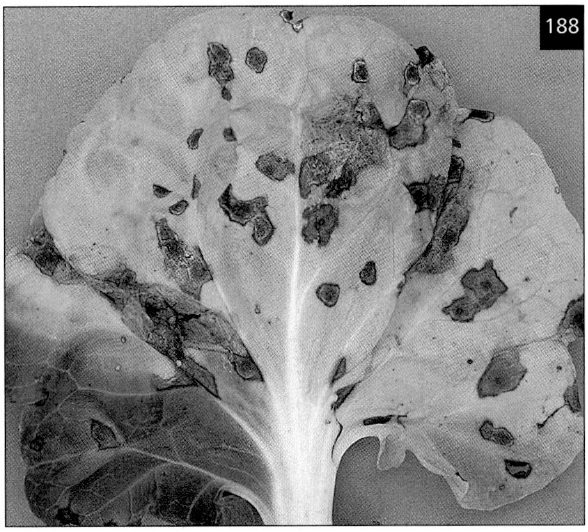

188 Ring spot on cabbage.

Control

Rotate with non-hosts and bury infected crop residues to reduce inoculum sources. Do not plant vegetable brassicas adjacent to field-grown brassicas, such as oilseed rape, or kale used for game cover in the UK. Plant resistant Brussels sprout and cauliflower cultivars. Apply fungicides such as chlorothalonil, triazole, and strobilurin products. Treat infected seed with hot water or a thiram soak. Disease forecasting and inoculum detection systems are being evaluated in the UK.

References

Davies, J. M. L., Ann, D. M., Dobson, S. C., Jones, O. W., Gladders, P., Melville, S. C., McPherson, G.M., Popham, M. D., Price, R. I., and Wafford, J.D. 1986. Fungicide spray timing on Brussels sprouts. *Aspects of Applied Biology* 12: 21–28.

Dring, D. M. 1961. Studies on *Mycosphaerella brassicicola*. *Transactions of the British Mycological Society* 44:253–264.

Hartill, W. F. T. and Sutton, P. G. 1980. Inhibition of germination of *Mycosphaerella brassicicola* ascospores on young cabbage and cauliflower leaves. *Annals of Applied Biology* 96:153–161.

Kennedy, R., Wakeham, A. J., and Cullington, J. E. 1999. Production and immunodetection of ascospores of *Mycosphaerella brassicicola* ringspot of vegetable crucifers. *Plant Pathology* 48:297–307.

Van den Ende, J. E. and Frinking, H. D. 1993. Comparison of inoculation methods with *Mycosphaerella brassicicola* on *Brassica oleracea* var. *capitata*: ascospores versus mycelial fragments. *Netherlands Journal of Plant Pathology* 99, Supplement 3:69–81.

Wafford, J. D., Gladders, P., and McPherson, G. M. 1986. The incidence and severity of Brussels sprout diseases and the influence of oilseed rape. *Aspects of Applied Biology* 12:1–12.

Wakeham, A. J. and Kennedy, R. 2002. Quantification of airborne inoculum using antibody based systems. The BCPC Conference – Pests & Diseases 1:409–416.

Wicks, T. J. and Vogelzang. B. 1988. Effect of fungicides applied after infection on the control of *Mycosphaerella brassicicola* on Brussels sprouts. *Australian Journal of Experimental Agriculture* 28:411–416.

Mycosphaerella capsellae
(anamorph = *Pseudocercosporella capsellae)*

WHITE LEAF SPOT

Introduction and significance

White leaf spot is a minor foliar disease of brassicas that may occasionally cause economic losses in parts of northern Europe, North America, and the southern hemisphere. Swede, turnip, various mustards, and Chinese cabbage are among the affected hosts, with fewer reports from *Brassica oleracea* vegetables. In the UK, this disease is most often reported on turnip and swede.

Symptoms and diagnostic features

The primary symptoms are leaf spots that first are small (2–3 mm in diameter), gray to gray-green, and circular. Later in disease development, lesions remain circular, become light tan to white, and can grow up to 10–12 mm in diameter (**189**). Dark streaks and speckling are sometimes apparent within these leaf spots. White conidial growth may be observed on the leaf spots, often on the leaf undersides (**190**). On seed crops, the most severe symptom is known as 'gray stem' from the color of the infected upper stems. Pods develop elongated to oval lesions that may resemble *Alternaria* infections but often have the diagnostic light tan color and dark reticulation.

Causal agent

White leaf spot is caused by the ascomycete fungus *Mycosphaerella capsellae*. The fungus forms perithecia and ascospores that are slightly curved, hyaline, two-celled, and measure 15–23 x 3–3.5 µm. However, the pathogen is mostly present on plants as its anamorph stage *Pseudocercosporella capsellae*. *Pseudocercosporella capsellae* can be isolated on standard microbiological media, though the fungus is very slow-growing and difficult to obtain in pure culture. After many days on potato dextrose agar (PDA), isolations can result in slow-growing, raised, black, stromatic colonies that produce few conidia. On 2% water agar, colony morphology is similar to that on PDA, but colonies also release a purple-pink pigment into the media. Confirmation of white leaf spot might better be accomplished by microscopic examination of leaf-spot tissue that has been just collected from the field or incubated under moist conditions. From plant tissue, conidiophores are solitary or grouped in fascicles. Conidia are hyaline, cylindrical, one to three septate, and measure 30–70 x 2–3 µm. Yet another spore stage, called spermatia, is produced in black spermogonia on leaves and pods and consists of very small spores that measure 3–4 x 1 µm; spermatia appear to be non-infective.

Disease cycle

Under UK conditions, disease development in winter oilseed rape suggests that ascospores are produced mainly in autumn (October to January) and spread via air to new spring crops. The sexual stage therefore enables the pathogen to survive between crops. Ascospore discharge is associated with wet conditions, and wet periods are required for infection. Symptoms appear in 6 to 8 days under optimum temperatures of 15–20° C, but may take 24 days to appear at 5° C. Conidia that grow on leaf spots are splash or wind dispersed only short distances to other leaves and plants. The pathogen does not persist in soil, but survives on weed hosts and volunteer brassicas.

Control

Control measures are rarely needed. Plow in infected crop residues after harvest. Do not plant vegetable brassicas adjacent to oilseed rape in regions where white leaf spot is common.

References

Campbell, R. and Greathead, A. S. 1978. *Pseudocercosporella* white spot of crucifers in California. *Plant Disease Reporter* 62:1066–1068.

Cerkauskas, R. F., Stobbs, L. W., Lowery, D. T., Van Driel, L., Liu, W., and VanSchagen, J. 1998. Diseases, pests, and abiotic problems associated with oriental cruciferous vegetables in southern Ontario in 1993-1994. *Canadian Journal of Plant Pathology* 20:87–94.

Inman, A. J., Fitt, B. D. L., and Evans, R. L. 1992. Epidemiology in relation to control of white leaf spot (*Mycosphaerella capsellae*) on oilseed rape. *Proceedings of the Brighton Crop Protection Conference – Pests and Diseases* 2:681–686.

Inman, A. J., Fitt, B. D. L., Todd, A. D., and Evans, R. L. 1999. Ascospores as primary inoculum for epidemics of white leaf spot (*Mycosphaerella capsellae*) in winter oilseed rape in the UK. *Plant Pathology* 48:308–319.

Inman, A. J., Sivanesan, A., Fitt, B. D. L., and Evans, R. L. 1991. The biology of *Mycosphaerella capsellae*, the teleomorph of *Pseudocercosporella capsellae*, cause of white leaf spot of oilseed rape. *Mycological Research* 95:1334–1342.

Koike, S. T. 1996. Red mustard, tah tsai, and Japanese mustard as hosts of *Pseudocercosporella capsellae* in California. *Plant Disease* 80:960.

Petrie, G. A. and Vanterpool, T. C. 1978. *Pseudocercosporella capsellae*, the cause of white leaf spot and grey stem of Cruciferae in western Canada. *Canadian Plant Disease Survey* 58:69–72.

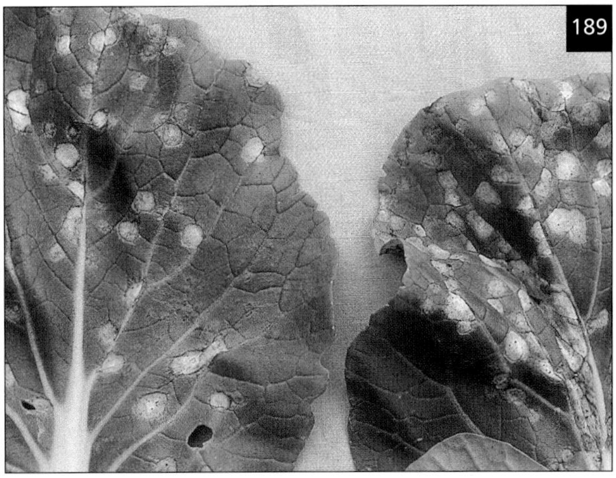

189 Symptoms of white leaf spot of bok choy.

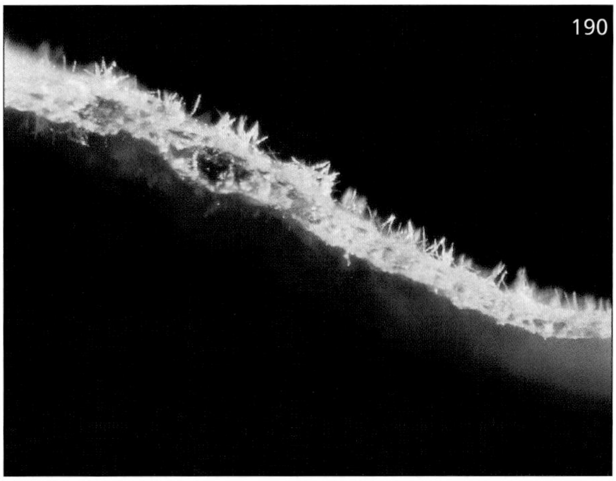

190 Sporulation of white leaf spot of Chinese cabbage.

Peronospora parasitica

DOWNY MILDEW

Introduction and significance

This common disease occurs worldwide and affects most vegetable, ornamental, and weed crucifers. It is particularly important at the seedling stage, but it can also cause problems on crops close to maturity. Severe attacks result in loss of seedlings. In California, downy mildew is rarely a concern on established, older plants.

Symptoms and diagnostic features

The first symptoms are irregular yellow patches on the cotyledons or young leaves with white fungal growth on the underside of the lesions. Under cool, humid conditions, sporulation can also occur on the upper lesion surface. With a hand lens, it is possible to see the erect fungal conidiophores bearing spores on 'tree-like' branches. This growth is distinct from the white radiating mycelium of powdery mildew. In established downy mildew lesions, there is usually irregular dark speckling and eventually the center of the lesion becomes bleached and papery. Downy mildew is often most severe on the cotyledons and first true leaves (**191**) and this can result in the death of seedlings, particularly if frost conditions occur.

On older plants, downy mildew causes variously sized yellow blotches that have an angular appearance because of the major leaf veins. If disease development is extensive, leaves may take on a blighted effect due to the numerous infection sites (**192**). On Brussels sprout, downy mildew causes either irregular shaped, yellow lesions with irregular dark speckling, or black lesions. Systemic infection results in a gray or black flecking and

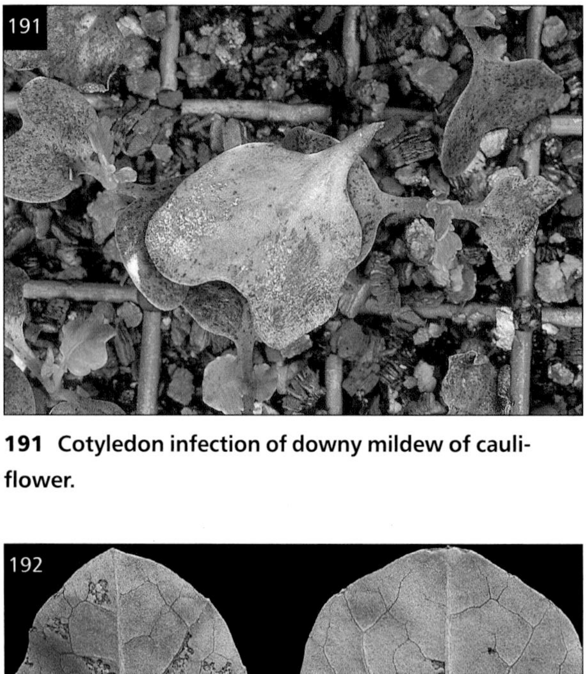

191 Cotyledon infection of downy mildew of cauliflower.

193 Systemic infection of downy mildew in broccoli stems.

192 Leaf infections of downy mildew of cauliflower.

194 Black discoloration and white sporulation of downy mildew on cauliflower florets.

streaking of internal tissues, notably in cauliflower and broccoli flower stalks, sprouts of Brussels sprout, and cabbage heads (**193**, **194**). Pod infection occurs in seed crops and in some species there is infection and distortion of the flower stem.

Downy mildew is an important problem on radish because the pathogen can infect the fleshy root and cause external black patches and lesions on the root shoulder and internal grey or black flecking and streaking in the hypocotyl and root (**195**, **196**, **197**). Either symptom renders the radish unmarketable. The surface of the root can show scarring and is prone to cracking or splitting. On radish leaves, downy mildew produces irregularly shaped, angular, yellow lesions that contain dark speckling (**198**, **199**). White sporulation of the pathogen occurs on the undersurface of the leaf.

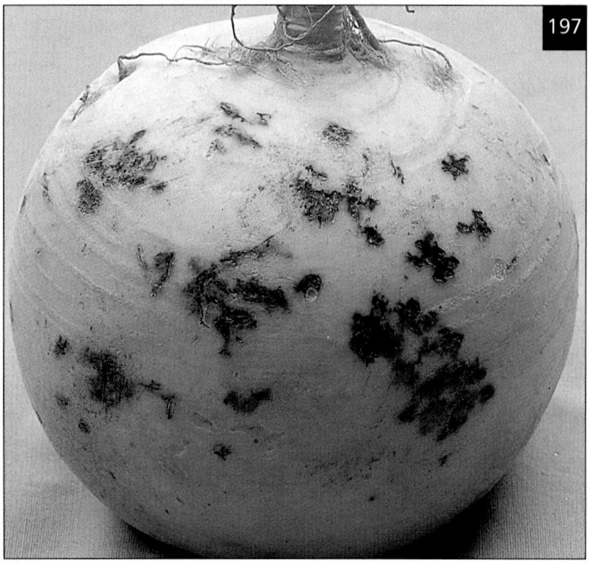
197 Black lesions of downy mildew of white radish.

195 Black lesions of downy mildew of red radish.

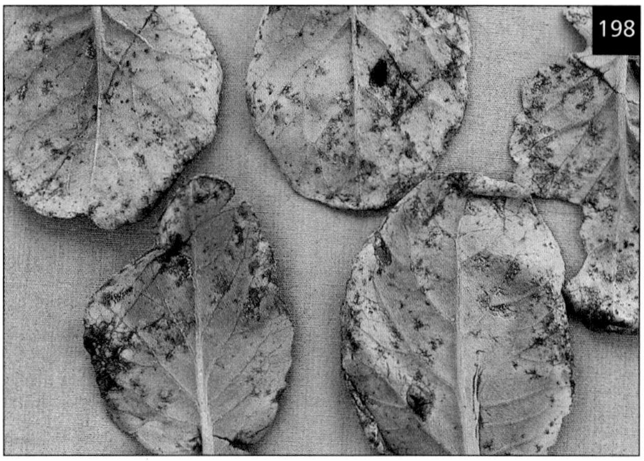
198 Leaf infections of downy mildew of radish.

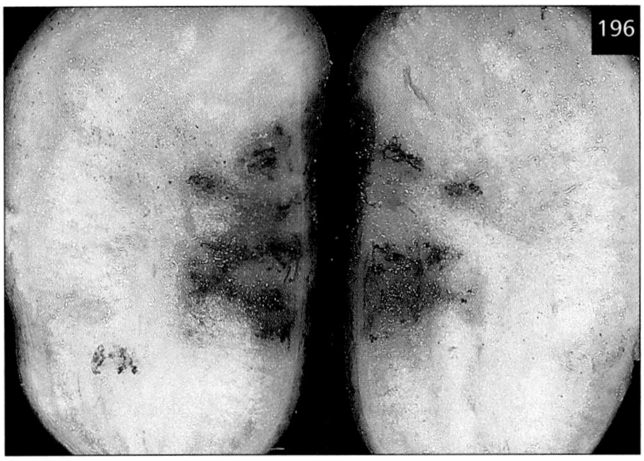
196 Internal root discoloration of downy mildew of radish.

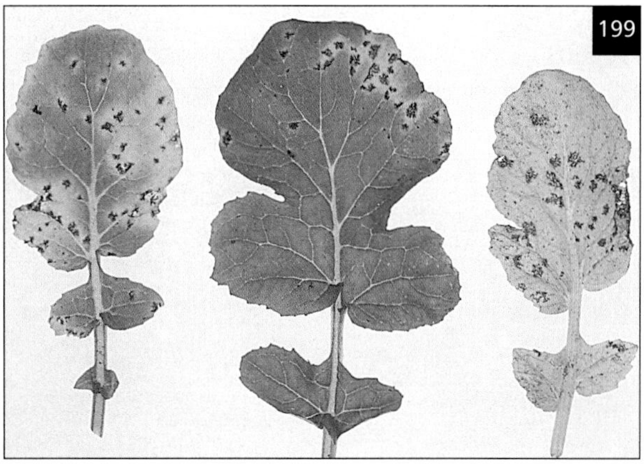
199 Leaf infections of downy mildew of radish.

Causal agent

Downy mildew is caused by the oomycete *Peronospora parasitica*. Due to recent taxonomic work, some researchers have renamed this organism *Hyaloperonospora parasitica*. This obligate pathogen forms mycelium and haustoria inside host tissues. The sporangiophores emerge through the stomata, are dichotomously branched, and terminate with slender tips. The sporangia are hyaline, elliptical, and measure 10–20 x 20–22 µm. Sexual oospores are globular, thick walled, and measure 26–43 µm. Some *P. parasitica* isolates and races form few or no oospores. There is some evidence that the pathogen may be seedborne. There is considerable host specialization within *P. parasitica* and strains affecting radish may be only weakly pathogenic on other brassicas.

Disease cycle

Cool (10–15° C), moist conditions favor downy mildew sporulation, germination of sporangia, and infection. Lesion expansion is most rapid at 20° C, and at this temperature the asexual cycle can be completed in only 3 or 4 days. Sporangia are dispersed via winds and splashing water and penetrate through the leaf cuticle or stomata. If oospores are produced, these may persist in crop residues and in soil and act as sources of initial inoculum.

Control

Manage downy mildew on transplants by improving air movement around the plants, irrigating early in the day so foliage can readily dry, and applying fungicides. Fungicides such as metalaxyl can be used as seed treatments for controlling early downy mildew on transplants, though metalaxyl resistance occurs in the UK and other regions. For field control, fungicides may also be necessary. Plant resistant or tolerant cultivars.

References

Cerkauskas, R. F., Stobbs, L. W., Lowery, D. T., Van Driel, L., Liu, W., and VanSchagen, J. 1998. Diseases, pests, and abiotic problems associated with oriental cruciferous vegetables in southern Ontario in 1993-1994. *Canadian Journal of Plant Pathology* 20:87–94.

Constantinescu, O. and Fatehi, J. 2002. *Peronospora*-like fungi (*Chromista, Peronosporales*) parasitic on Brassicaceae and related hosts. *Nova Hedwigia* 74:291–338.

Dickinson, C. H. and Greenhalgh, J. R. 1977. Host range and taxonomy of *Peronospora* on crucifers. *Transactions of the British Mycological Society* 69:111–116.

Dickson, M. H. and Petzoldt, R. 1993. Plant age and isolate source affect expression of downy mildew resistance in broccoli. *Horticultural Science* 28:730–731.

Jensen, B. D., Hockenhull, J., and Munk, L. 1999. Seedling and adult plant resistance to downy mildew (*Peronospora parasitica*) in cauliflower (*Brassica oleracea* convar. *botrytis* var. *botrytis*). *Plant Pathology* 48:604–612.

Kluczewski, S. M. and Lucas, J. A. 1982. Development and physiology of infection by the downy mildew fungus *Peronospora parasitica* (Pers. ex Fr.) Fr. in susceptible and resistant *Brassica* species. *Plant Pathology* 31:373–379.

Kluczewski, S. M. and Lucas, J. A. 1983. Host infection and oospore formation by *Peronospora parasitica* in agricultural and horticultural *Brassica* species. *Transactions of the British Mycological Society* 81:591–596.

Koike, S .T. 1998. Downy mildew of arugula, caused by *Peronospora parasitica*, in California. *Plant Disease* 82:1063.

McKay, A. G., Floyd, R. M., and Boyd, C. J. 1992. Phosphonic acid controls downy mildew (*Peronospora parasitica*) in cauliflower curds. *Australian Journal of Experimental Agriculture* 32:127–129.

Ramsey, G. B., Smith, M. A., and Wright, W. R. 1954. *Peronospora* on radish roots. *Phytopathology* 44:384–385.

Sherriff, C. and Lucas, J. A. 1989. Heterothallism and homothallism in *Peronospora parasitica. Mycological Research* 92:311–316.

Sherriff, C. and Lucas, J .A. 1990. The host range of isolates of downy mildew, *Peronospora parasitica*, from Brassica crop species. *Plant Pathology* 39:77–91.

Silue, D., Nashaat, N. I., and Tirilly, Y. 1996. Differential responses of *Brassica oleracea* and *B. rapa* accessions to seven isolates of *Peronospora parasitica* at the cotyledon stage. *Plant Disease* 80:142–144.

Thomas, C. E. and Jourdain, E. L. 1990. Evaluation of broccoli and cauliflower germplasm for resistance to Race 2 of *Peronospora parasitica. Horticultural Science* 25:1429–1431.

Wang, M., Farnham, M. W., and Thomas, C. E. 2001. Inheritance of true leaf stage downy mildew resistance in broccoli. *Journal of the American Society for Horticultural Science* 126:727–729.

Plasmodiophora brassicae

CLUBROOT

Introduction and significance

Clubroot is one of the most readily recognized diseases of crucifers. The disease may be increasing in importance as growers intensify rotations of brassicas and simultaneously try to reduce production costs by using less lime in the fields. Clubroot occurs wherever crucifers are grown and is believed to have originated in the Mediterranean region. Severely infested fields are often taken out of brassica production.

Symptoms and diagnostic features

Above-ground symptoms are severe stunting, yellowing, wilting, and other signs of a dysfunctional root system (**200**). Plant roots show extensive galling, swelling, and distortion (**201**, **202**). Large galls on the hypocotyl and main taproot are more damaging than galls on the smaller lateral roots. Fresh galls are firm in texture and retain the off-white color of root tissue. However, as disease progresses, the galls and swellings darken, become infested with secondary decay organisms, and are soft and rotted. Decay of roots releases inoculum into the soil. Disease tends to be more severe on summer crops because of warmer growing conditions, earlier infection of plants, and weather conditions that stress plants. Root galls caused by the turnip gall weevil (*Ceutorhynchus pleurostigma*) or small gall-like structures known as hybridization nodules may superficially resemble clubroot.

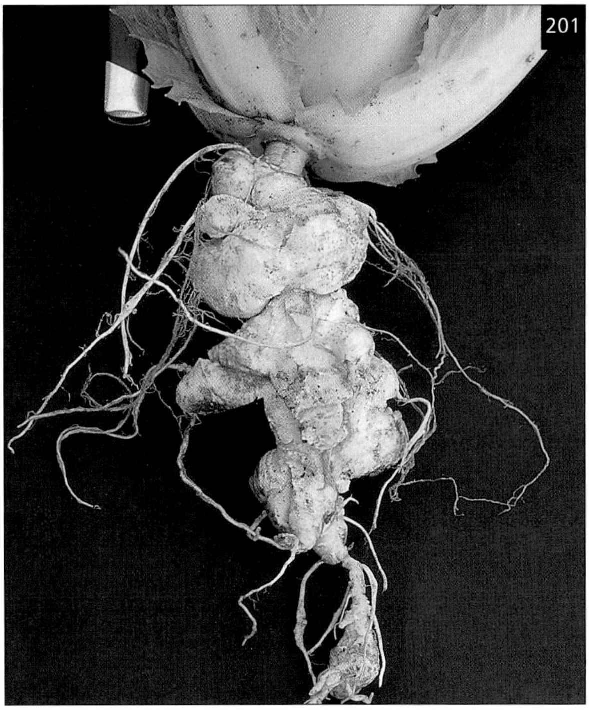

201 **Bok choy infected with clubroot disease.**

Causal agent

Clubroot is caused by the pathogen *Plasmodiophora brassicae*. This organism, previously considered a fungus, is currently considered more closely related to ciliate protozoans because of its reproductive structures and lack of a cell wall. The pathogen is placed in its own phylum, the *Plasmodiophoromycota*. The presence of clubbed roots containing resting spores (4 μm diameter)

200 **Stunted, declining cauliflower infected with clubroot disease.**

202 **Cauliflower transplants infected with clubroot disease.**

is diagnostic for this disease. The organism cannot be truly grown *in vitro*, though the amoebal stage has recently been cultured. *Plasmodiophora brassicae* has a complex life cycle that is still not fully understood. The pathogen can survive in soil for long periods, 20 years or longer, as thick-walled resting spores.

Many races of *P. brassicae* have been identified by using sets of cultivars with different genetic resistances. Depending on how an isolate infects one or more of the cultivars in the set, the isolate can be categorized as to a race or subset of the *P. brassicae* group. Several differential sets are used around the world: differential hosts by Williams; European clubroot differentials (ECD); CR F1 cultivars in Japan.

Disease cycle

The life cycle of clubroot is complex. Uninucleate thick-walled resting spores formed in the root galls are liberated into soil when the galls decay. Resting spores germinate in the presence of plant roots when temperatures are at least 16° C. Host infection is favored by temperatures above 18° C and disease severity increases with rising temperatures up to 26° C. Light intensity also influences clubroot development, with 600 Wh m^{-2} day^{-1} producing maximum clubroot severity on cabbage seedlings.

The resting spore germinates by releasing a single biflagellate zoospore that swims in soil water, finds a host root by chemotaxis, and attaches itself to a root hair. The attached zoospore forms a bullet-shaped structure (stachel) that allows it to forcibly inject zoospore contents into the root hair. Within the root hair, the pathogen becomes enclosed by the host cell plasmalemma and starts the second stage of its life cycle by forming a multi-nucleate plasmodium. The plasmodium produces 10 to 20 haploid secondary zoospores that are capable of infecting other root hairs or fuse in pairs to form larger binucleate cells, which infect cells of the root cortex. As the organism grows and divides, large multinucleate plasmodia form within the root. Infection leads to multiplication and enlargement of root cells due to changes in plant hormone (cytokinins and auxins) levels. The root cells accumulate starch when sugars are diverted to the affected roots. Galls develop within 3 to 4 weeks of infection.

Infection occurs over a wide range of both acid and alkaline soils, though acid soils favor the development of root symptoms. The pathogen can infect the root hairs of various non-crucifer plants including grasses and some weeds, but the clubroot stage only occurs in crucifers. On watercress, clubroot produces swellings on the shoot. Weed hosts, such as shepherd's purse (*Capsella bursa-pastoris*), may aid pathogen survival.

Control

Rotate crucifer crops with non-hosts on a 5 to 6 year cycle; this management step will not eliminate the pathogen from soil, but will prevent inoculum from continuing to increase. Maintain records of infested fields so that susceptible crops can be planted elsewhere. It is important to note that in California and other regions, mustard cover crops are planted in the fall and winter to add organic matter to the soil, increase soil microbial diversity, and possibly suppress some soilborne pathogens due to the glucosinolates released from decaying mustard plant residues. However, such mustards are susceptible to the clubroot pathogen and could sustain or increase *P. brassicae* inoculum levels.

For many fields, clubroot is successfully managed by applying lime and keeping soil pH above 7.2 to 7.3. Such soils are called responsive soils; the lime does not kill the pathogen, but creates conditions that prevent the formation of the root clubs. However, for other nonresponsive soils the liming treatment is much less effective. Calcium oxide acts rapidly and is generally considered more effective than calcium carbonate for clubroot control. A finely graded product is preferable to coarse material. There is evidence that the benefits of calcium salt applications are not solely due to effects on soil pH. The greatest reduction of symptoms has been obtained with calcium salts that raised both soil calcium concentration and soil pH.

Prevent the movement of infested soil and contaminated equipment to uninfested fields. Ensure that transplants and other materials do not harbor the pathogen. It is possible for transplants grown in soilless rooting medium to have clubroot due to contamination from infested dust and dirt. The resting spores can survive passage through the digestive system of cattle, so use caution in using manures as soil amendments. Resistant cultivars are available for various crucifer crops, but these do not generally provide complete control. Recently, resistance has broken down for clubroot resistant Chinese cabbage cultivars grown in Japan. Clubroot populations are highly variable and cultivar resistance will not be effective at all sites.

In the UK, soils are tested before planting to identify potential risks from clubroot. Such tests involve growing brassica seedlings in soil samples and take about 6 weeks to complete. Tests are available through the Advisory services. Soil fumigants have not proven to be economically viable for crucifer production.

References

Ann, D., Channon, A., Melville, S., and Antill, D. 1987. Clubroot control in cabbage and cauliflower by adding fungicide to the compost used for raising transplants in loose-filled cells. 1987. *BCPC Monograph No. 39 Application to Seeds and Soil*: 395–402.

Arnold, D. L., Blakesley, D., and Clarkson, J. M. 1996. Evidence for growth of *Plasmodiophora brassicae* in vitro. *Mycological Research* 100:535–540.

Buczacki, S.T., Ockendon, J. G., and Freeman, G. H. 1978. An analysis of some effects of light and soil temperature on clubroot disease. *Annals of Applied Biology* 88:229–238.

Campbell, R .N., Greathead, A. S., Myers, D. F., and deBoer, G. J. 1985. Factors related to control of clubroot of crucifers in the Salinas Valley, California. *Phytopathology* 75:665–670.

Castlebury, L. A., Maddox, J. V., and Glawe, D. A. 1994. A technique for the extraction and purification of viable *Plasmodiophora brassicae* resting spores from host root tissue. *Mycologia* 86:458–460.

Colhoun, J. 1958. Clubroot Disease of Crucifers Caused by *Plasmodiophora brassicae*: A Monograph. Phytopathological Paper No. 3. Commonwealth Mycological Institute.

Dixon, G. R .1996. Repression of the morphogenesis of *Plasmodiophora brassicae* Wor. by boron – A review. *Acta Horticulturae* 407:393–401.

Dixon, G. R. and Webster, M. A. 1988. Antagonistic effects of boron, calcium and pH on pathogenesis caused by *Plasmodiophora brassicae* Woronin (clubroot) – a review of recent work. *Crop Research* 28:83–95.

Dobson, R. L., Robak, J., and Gabrielson, R. L. 1983. Pathotypes of *Plasmodiophora brassicae* in Washington, Oregon, and California. *Plant Disease* 67:269–271.

Donald, E. C., Lawrence J. M., and Porter, I. J. 2004. Influence of particle size and application method on the efficacy of calcium cyanamide for control of clubroot of vegetable brassicas. *Crop Protection* 23:297–303.

Donald, E. C., Porter, I. J., and Lancaster, R. A. 2001. Band incorporation of fluazinam (Shirlan) into soil to control clubroot of vegetable brassica crops. *Australian Journal of Experimental Agriculture* 41:1223–1226.

Faggian, R., Bulman, S. R., Lawrie, A. C., and Porter, I. J. 1999. Specific polymerase chain reaction primers for the detection of *Plasmodiophora brassicae* in soil and water. *Phytopathology* 89:392–397.

Kuginuki, Y., Yoshikawa, H., and Hirai, M. 1999. Variation in virulence of *Plasmodiophora brassicae* in Japan tested with clubroot-resistant cultivars of Chinese cabbage (*Brassica rapa* L. ssp. *pekinensis*). *European European Journal of Plant Pathology* 105:327–332.

Jones, D. R., Ingram, D. S., and Dixon, G. R. 1982. Factors affecting tests for differential pathogenicity in populations of *Plasmodiophora brassicae*. *Plant Pathology* 31:229–238.

Manzanares-Dauleux, M. J., Divaret, I., Baron, F., and Thomas, G. 2001. Assessment of biological and molecular variability between and within field isolates of *Plasmodiophora brassicae*. *Plant Pathology* 50:165–173.

Murakami, H., Tsushima, S., Akimoto, T., Murakami, K., Goto, I., and Shishido, Y. 2000. Effects of growing leafy daikon (*Rhaphanus sativus*) on populations of *Plasmodiophora brasicae* (clubroot). *Plant Pathology* 49:584–589.

Myers, D. F. and Campbell, R. N. 1985. Lime and the control of clubroot of crucifers: effects of pH, calcium, magnesium, and their interactions. *Phytopathology* 75:670–673.

Robak, J. 1996. The effect of some crop rotations on decrease of clubroot, *Plasmodiophora brassicae,* in soils. *Proceedings of Brighton Crop Protection Conference – Pests & Diseases* 2:647–651.

Voorrips, R. E. 1996. Production, characterization and interaction of single-spore isolates of *Plasmodiophora brassicae*. *European Journal of Plant Pathology* 102:377–383.

Wallenhammar, A. C. and Arwidsson, O. 2001. Detection of *Plasmodiophora brassicae* by PCR in naturally infested soils. *European Journal of Plant Pathology* 107:313–321.

Webster, M. A. and Dixon, G. R. 1991. Calcium, pH, and inoculum concentration influencing colonization by *Plasmodiophora brassicae*. *Mycological Research* 95:64–73.

Webster, M. A. and Dixon, G. R. 1991. Boron, pH, and inoculum concentration influencing colonization by *Plasmodiophora brassicae*. *Mycological Research* 95:74–79.

Phytophthora brassicae

PHYTOPHTHORA STORAGE ROT

Introduction and significance

This disease is mainly important as a storage rot of white cabbage in Europe, where losses can reach 50%. Its prevalence as a foliar pathogen is probably underestimated.

Symptoms and diagnostic features

Symptoms consist of leaf spots that are grey or black, water-soaked, and measure 5–20 mm in diameter. These spots may resemble symptoms of gray mold or bacterial soft rot. Savoy cabbage (**203**) is especially susceptible because of the furrowed leaves that retain water for longer periods than smooth-leaved cabbage. For stored white cabbage, a firm brown rot extends from the stem base into the leaves. This disease also affects Chinese cabbage heads after short storage periods (**204**).

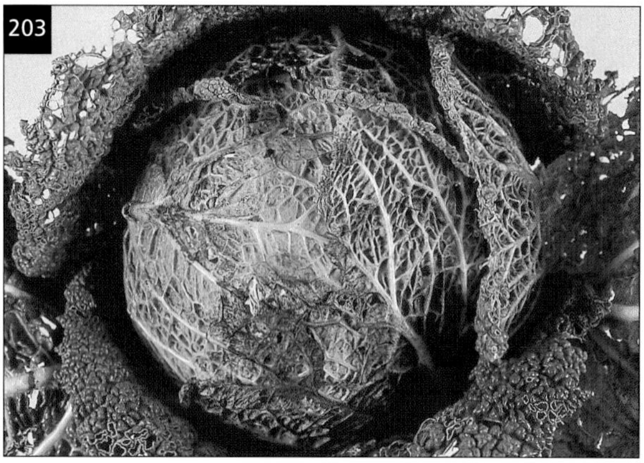

203 Storage rot symptoms on savoy cabbage.

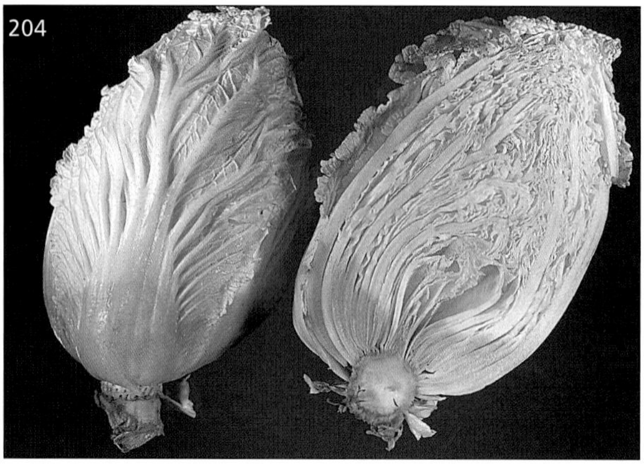

204 Chinese cabbage in storage affected by *Phytophthora brassicae*.

Causal agent

Phytophthora storage rot is caused by the oomycete *Phytophthora brassicae*. On solid media the pathogen produces abundant sporangia that have indistinct papilla and measure 40–74 x 25–48 µm. Few oogonia of the sexual stage are formed; oospores measure 30–65 µm in diameter. Though previously designated as *P. porri*, *P. brassicae* only infects brassica plants and is genetically and morphologically distinct from the allium pathogen.

Disease cycle

The precise epidemiology of this disease is not known. Oospores, sporangia, or zoospores in soil are probably splashed up onto cabbage heads and infect host tissues. Infection might also take place through handling and wounds that occur during harvest. Problems are often associated with particularly wet and muddy harvest conditions. Storage rot symptoms appear within 2 weeks at 20° C, but may take several weeks to appear at 0–2° C. There is usually little spread within storage bins. Leaf spot symptoms often occur in the field after heavy rain, again suggesting that soilborne inoculum is spread to plants by splashing water.

Control

Harvest cabbage under dry field conditions. Handle cabbage heads carefully and keep them free of soil and mud. Disinfect the cutting knives used for harvesting if they become dirty. Pre-storage dips with fungicides containing metalaxyl have given good control of storage rot.

References

Geeson, J. D. 1976. Storage rot of white cabbage caused by *Phytophthora porri*. *Plant Pathology* 25:115–116.

Hamm, P. B. and Koepsell, P. A. 1984. Phytophthora root rot of cabbage and cauliflower in Oregon. *Plant Disease* 68:533–535.

Man in t'Veld, W. A., de Cock, A .W. A. M., Ilieva, E., and Lévesque, C. A. 2002. Gene flow analysis of *Phytophthora porri* reveals a new species: *Phytophthora brassicae* sp. nov. *European Journal of Plant Pathology* 108:51–62.

Thompson, A. H., and Phillips, A. J. L. 1988. Root rot of cabbage caused by *Phytophthora drechsleri*. *Plant Pathology* 37:297–299.

Phytophthora megasperma

PHYTOPHTHORA ROOT ROT

Introduction and significance

This disease can damage vegetable brassicas, although it is usually of minor importance. *Phytophthora megasperma* is the most important and common species that causes root rot. In South Africa, *P. drechsleri* appears to be most prominent on cabbage. Some damping-off problems may be due to *P. cactorum*. Swede may occasionally develop a storage rot from *Phytophthora* infections.

Symptoms and diagnostic features

The external surfaces and internal tissues of infected roots are pale brown to black in color, rotted, and usually have a strong, unpleasant smell. Leaves, especially older ones, turn purple-red or yellow, and wilt. Later, the entire plant wilts and is unmarketable. At the soil line, the stem may develop dark, discrete lesions or turn black in general and become soft (**205**). Plants tend to fall over or be easily dislodged because the lateral roots have decayed (**206, 207, 208**). In the UK, *P. megasperma* is thought to have caused severe rotting of swede after several weeks of cold storage.

Causal agents

Phythophthora root rot is caused by the oomycete organism *P. megasperma*. The pathogen can be isolated on standard microbiological media. Semi-selective media such as PARP or PARPH can be useful for some samples or to clean up isolates. However, the hymexazol in PARPH may inhibit the growth of some *Phytophthora* species. The pathogen produces sporangia that are non-papillate and measure 35–50 x 25–35 µm. Examination of rotted root tissues will often

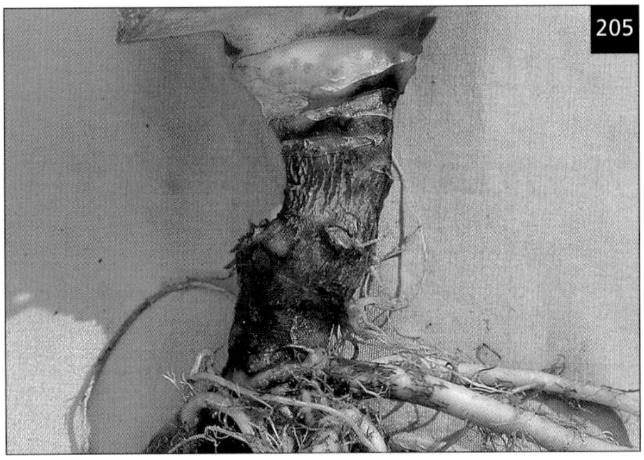

205 Necrotic cauliflower stem with Phytophthora root rot.

207 Collapsed cabbage with Phytophthora root rot.

206 Cabbage field with Phytophthora root rot.

208 Collapsed cauliflower with Phytophthora root rot.

reveal the sexual oospores of this organism, which are circular and thick walled. Floating pieces of symptomatic roots in water may readily induce sporangia to form. Other species, such as *P. drechsleri* and *P. cactorum*, also cause root rot diseases on crucifers.

Disease cycle

The pathogen survives in and relies on wet soil conditions. Plants become infected when soils are overly wet, over irrigated, or drain poorly. Hence, root rot is most often found at low spots in the field or at the end of irrigation runs. *Phytophthora megasperma* is a soil inhabitant and is able to affect a wide range of crop and weed species. This pathogen is therefore able to survive in the absence of susceptible brassica crops. In the soil, oospores germinate to produce sporangia, or sporangia are already present on organic matter or in the soil. Sporangia release zoospores that swim to roots and infect the host. Disease development is favored by soil temperatures in the range between 15-25° C.

Control

Prepare soils so that they are levelled, drain well, and are not compacted. Manage irrigation so that crops are not over watered. Avoid fields with a history of *Phytophthora* problems. Rotate crops so that crucifers are not planted back-to-back; however, crop rotations have limited effectiveness because *P. megasperma* is a soil inhabitant. Some cultivars of cauliflower may be tolerant to the disease. Fungicides are not usually effective.

References

Geeson, J. D., Browne, K. M., and McKeown, B. 1990. Storage rot of swede caused by *Phytophthora* sp. *Plant Pathology* 39:629–631.

Hamm, P. B. and Koepsell, P. A. 1984. Phytophthora root rot of cabbage and cauliflower in Oregon. *Plant Disease* 68:533–585.

Hansen, E .M., Brasier, C. M., Shaw, D. S., and Hamm, P. B. 1986. The taxonomic structure of *Phytophthora megasperma*: evidence for emerging biological species groups. *Transactions of the British Mycological Society* 87:557–573.

Kontaxis, D. G. and Rubatsky, V. E. 1983. Managing *Phytophthora* root rot in cauliflower. *California Agriculture* 37:12.

Thompson, A. H. and Phillips, A. J .L. 1988. Root rot of cabbage caused by *Phytophthora drechsleri*. *Plant Pathology* 37:297–299.

Pyrenopeziza brassicae
(anamorph = *Cylindrosporium concentricum*)

LIGHT LEAF SPOT

Introduction and significance

Light leaf spot occurs widely in northern Europe, the UK, Poland, and also in New Zealand and other areas with a cool, moist climate. This is the most important disease of winter oilseed rape in northern England and Scotland, and oilseed rape can be an important source of inoculum for vegetable brassicas. The main damage results from disease on leaves and petioles.

Symptoms and diagnostic features

This disease can be difficult to identify because symptoms vary and are influenced by weather conditions and host susceptibility. On very susceptible cultivars, leaves exhibit large white blotches that contain small green flecks. The pathogen grows under the leaf cuticle and disrupts the waxy surface as it sporulates, resulting in lifting of the leaf cuticle and appearance of bleached areas; this feature gives the disease the name light leaf spot. Blotches are usually surrounded by numerous dark fruiting bodies and associated white drops of spores that form in concentric rings (**209, 210**). In some cases the spore droplets do not appear until leaves are incubated under moist conditions for 1 or 2 days at 10–15° C. Sporulation can also occur on apparently healthy, green leaf tissue. Lesions can occur on

209 Severe light leaf spot on cauliflower with abundant white sporulation around lesions.

both upper and lower leaf surfaces, which causes the opposite leaf side to become yellow or red. Early infection of immature leaves or the growing point causes considerable leaf distortion and stunting. Leaf infection may also result in localized bubbling of the lamina, which is caused by the abnormal production of plant hormones.

On Brussels sprout, small groups of black spots occur in a 'thumb print' pattern, measuring 1–2 cm in diameter, on the leaf undersides and on the sprouts (**211**). This black spotting may appear similar to spotting caused by other pathogens such as *Alternaria* species, *Mycosphaerella brassicicola*, and *Peronospora parastica*. Severe light leaf spot infection of sprouts can cause early senescence and rotting of the outer leaves. Other symptoms include circular, red spots that measure 0.5–1.5 cm in diameter and occur singly on older leaves that are starting to turn yellow (**212**). These spots may not produce conidia after incubation. Senescing leaves may also have small 'green islands' which are associated with groups of black spots. On cauliflower, light leaf spot may cause pink-brown or black discoloration of petioles (**213**). When this occurs at the base of leaves attached to the cauliflower head, it adversely affects the appearance of the head. Similar symptoms are prevalent on the stems and pods of flowering brassica crops. Diseased pods can lead to seedborne infection, though the importance of this has not been established.

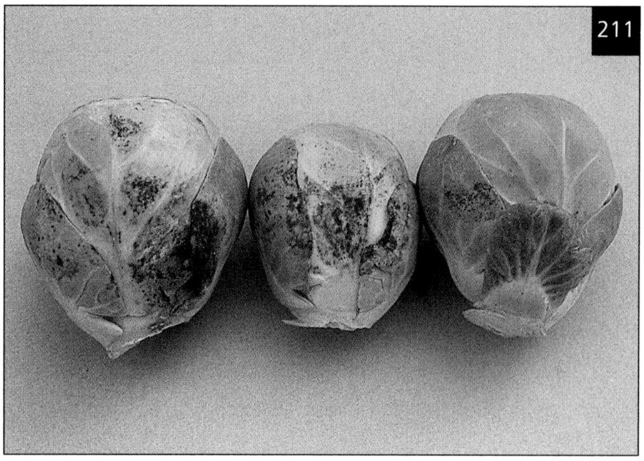

211 Light leaf spot on Brussels sprout buds.

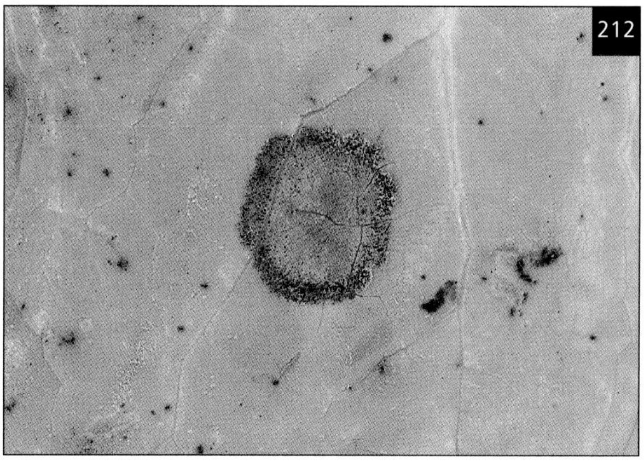

212 Close-up of light leaf spot watermark symptom on senescent Brussels sprout leaf.

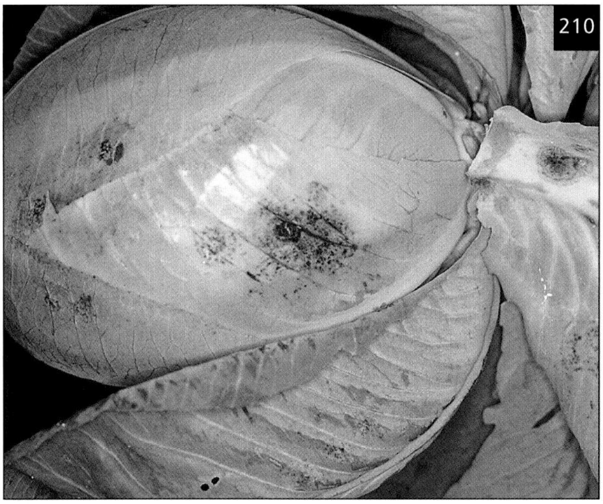

210 Light leaf spot lesion on cabbage head.

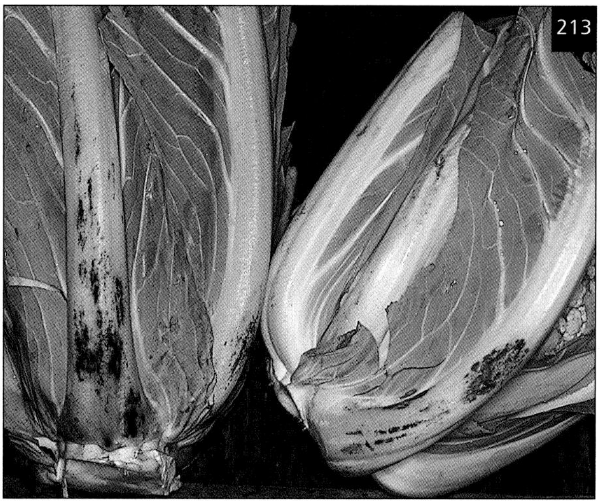

213 Blackening of cauliflower petioles infected with light leaf spot.

Causal agent

The cause of light leaf spot is the ascomycete fungus *Pyrenopeziza brassicae*. On affected hosts, the asexual stage *Cylindrosporium concentricum* is more usually encountered. This form of the pathogen makes a cushion-shaped mass of conidiophores called an acervulus. Acervuli break through the leaf epidermis and release white spore droplets. Conidia are hyaline, smooth, cylindrical, one-celled, and measure 10–16 x 3–4 µm. The fungus has two mating types that must both be present for the *P. brassicae* sexual stage to be formed. This stage produces small black fruiting structures called apothecia that are cup-shaped, sessile (without stalks), and measure 1–2 mm in diameter. Apothecia usually develop on the infected petioles of senescing or dead leaves and other plant residues. Airborne ascospores are released for a relatively short period, perhaps for only a few weeks. Ascospores are hyaline, cylindrical, measure 8–14 x 2.5–3.5 µm, and are zero to one septate.

Disease cycle

Ascospores become airborne and can either initiate new infections within the same crop or be dispersed to nearby crops. The formation of apothecia is temperature sensitive and is inhibited above 20° C. Ascospores and conidia have similar infection requirements, as studied on oilseed rape. On oilseed rape, 6 hours of leaf wetness are required for infection at 12–20° C. Much longer periods of leaf wetness are required for infection at lower temperatures. Following infection, symptoms appear after 14 days at 15° C and 28 days at 5° C. Disease severity is higher with longer periods of leaf wetness and cool temperatures. Optimum temperature for disease development is 15–16° C, but there is limited activity above 20° C. Once foliar symptoms develop, secondary inocula (conidia) appear and are dispersed to other leaves and plants by splashing water.

Control

Avoid planting susceptible vegetable brassicas in close proximity to oilseed rape. Note, however, that airborne ascospores can travel long distances in wind and still make contact with vegetable crops. Destroy infected crop residues after harvest to reduce airborne ascospores. Cultivars vary in their susceptibility to light leaf spot, so choose and plant more tolerant selections. Apply protectant fungicides if such are available. Disease forecasting systems have been developed for winter oilseed rape in the UK, and these and other programs may assist vegetable growers in timing fungicide applications.

References

Bradburne, R., Majer, D., Magrath, R., Werner, C. P., Lewis, B., and Mithen, R. 1999. Winter oilseed rape with high levels of resistance to *Pyrenopeziza brassicae* derived from wild *Brassica* species. *Plant Pathology* 48:550–558.

Cheah, L-H. and Hartill, W. F. T. 1985. Disease cycle in New Zealand of light leaf spot of brassicas, induced by *Pyrenopeziza brassicae*. *New Zealand Journal of Agricultural Research* 28:567–573.

Gilles, T., Fitt, B. D. L., and Jeger, M. J. 2000. Epidemiology in relation to methods for forecasting light leaf spot (*Pyrenopeziza brassicae*) severity on winter oilseed rape (*Brassica napus*) in the UK. *European Journal of Plant Pathology* 106:593–605.

Gilles, T., Fitt, B. D. L., Welham ,S .J., Evans, N., Steed, J.M., and Jeger, M. J. 2001. Modelling the effects of temperature and wetness duration on development of light leaf spot on oilseed rape leaves inoculated with *Pyrenopeziza brassicae* conidia. *Plant Pathology* 50:42–52.

Gilles, T., Fitt, B. D. L., Kennedy, R., Welham, S. J., and Jeger, M. J. 2000. Effects of temperature and wetness duration on conidia infection, latent period and asexual sporulation of *Pyrenopeziza brassicae* on leaves of oilseed rape. *Plant Pathology* 49:498–508.

Karolewskia, Z., Evans, N., Fitta, B. D. L., Baierl, A., Todd, A. D., and Foster, S. J. 2004. Comparative epidemiology of *Pyrenopeziza brassicae* (light leaf spot) ascospores and conidia from Polish and UK populations. *Plant Pathology* 53:29–37.

Mattock, S. E., Ingram, D. S., and Gilligan, C. A. 1981. Resistance of cultivated brassicas to *Pyrenopeziza brassicae*. *Transactions of the British Mycological Society* 77:153–159.

McCartney, H. A. and Lacey, M. 1990. The production and release of ascospores of *Pyrenopeziza brassicae* in oilseed rape crops. *Plant Pathology* 39:17–32.

Rawlinson, C. J., Sutton, B. C., and Muthyalu, G. 1978. Taxonomy and biology of *Pyrenopeziza brassicae* sp. nov. (*Cylindrosporium concentricum*), a pathogen of winter oilseed rape (*Brassica napus* ssp. *oleifera*). *Transactions of the British Mycological Society* 71:425–439.

Sclerotinia minor, S. sclerotiorum

WHITE MOLD, SCLEROTINIA STEM ROT, WATERY SOFT ROT

Introduction and significance

Sclerotinia diseases are usually of minor importance in brassica crops. The pathogens have wide host ranges and will cause problems if inoculum pressure is high or conditions favor their development. There can be significant losses in seed crops because flower petals are very susceptible and facilitate stem infections.

Symptoms and diagnostic features

Two species of *Sclerotinia* cause disease on brassicas. *Sclerotinia minor* (**214**) only infects stems or leaves in contact with the soil. Once infected, tissues become water-soaked, brown, and soft. Plants wilt and collapse when the stem is girdled by the lesion. White mycelium and numerous small (up to 3–5 mm), black sclerotia form on the outside and sometimes on the inside of stems.

Sclerotinia sclerotiorum can also infect lower leaves and stems, resulting in symptoms similar to those caused by *S. minor*. In addition, *S. sclerotiorum* produces ascospores that infect the upper parts of plants, resulting in a pale brown, watery, soft rot of these tissues (**215, 216, 217**). In the field, the apothecia that produce ascospores are often difficult to find.

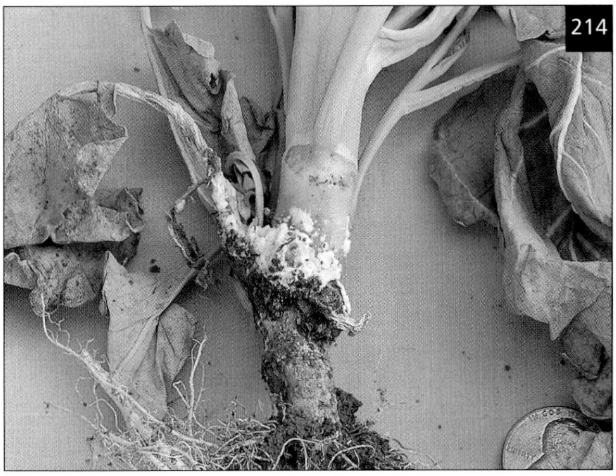

214 Broccoli seedling infected with *Sclerotinia minor.*

216 Diseased cauliflower infected with *Sclerotinia sclerotiorum.*

215 Cauliflower seedlings, in a transplant tray, infected by *Sclerotinia sclerotiorum*

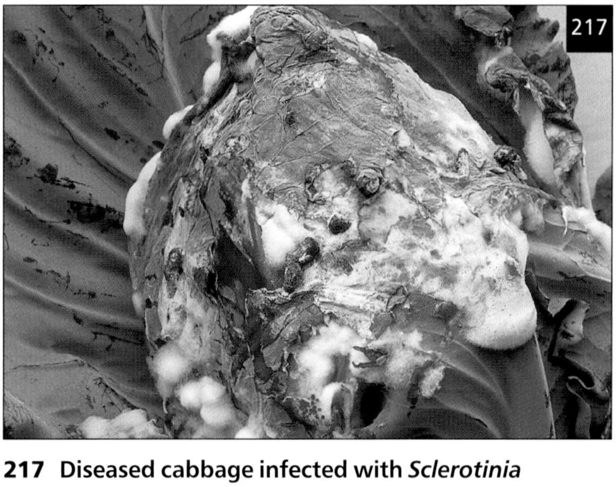

217 Diseased cabbage infected with *Sclerotinia sclerotiorum.*

218 Broccoli seed plant infected with *Sclerotinia sclerotiorum* and showing sclerotia inside seed stalks.

Along with the characteristic white mycelium, *S. sclerotiorum* forms sclerotia (**218**) that are significantly larger (5–10 mm long) than those of *S. minor*.

Causal agents and disease cycle

White mold is caused by two pathogens, *Sclerotinia sclerotiorum* and *S. minor*. These pathogens are ascomycete fungi and can be isolated on standard microbiological media. The two species are differentiated on the basis of sclerotia size, with *S. minor* producing smaller, more numerous sclerotia than *S. sclerotiorum*. Also, only *S. sclerotiorum* normally produces the sexual apothecia fruiting body and ascospores. For detailed descriptions of these pathogens and the disease cycle, see the bean white mold section in the chapter on legume diseases.

Control

Management steps for controlling white mold in vegetable brassicas is not usually required. For an integrated approach to Sclerotinia control, see the bean white mold section in the chapter on legume diseases.

References

Dillard, H. R. and Cobb, A. C. 1995. Relationship between leaf injury and colonization of cabbage by *Sclerotinia sclerotiorum*. *Crop Protection* 14:677–682.

Dillard, H. R. and Hunter, J. E. 1986. Association of common ragweed with Sclerotinia rot of cabbage in New York State. *Plant Disease* 70: 26–28.

Hao, J. J., Subbarao, K. V., and Koike, S. T. 2003. Effects of broccoli rotation on lettuce drop caused by *Sclerotinia minor* and on the population density of sclerotia in soil. *Plant Disease* 87:159–166.

Hims, M. J. 1979. Wild plants as a source of *Sclerotinia sclerotiorum* infecting oilseed rape. *Plant Pathology* 28: 197–198.

Hudyncia, J., Shew, H. D., Cody, B. R., and Cubeta, M. A. 2000. Evaluation of wounds as a factor to infection of cabbage by ascospores of *Sclerotinia sclerotiorum*. *Plant Disease* 84: 316–320.

Koike, S. T., Gonzales, T. G., Vidauri, M., and Subbarao, K. V. 1994. First report of *Sclerotinia minor* as a pathogen of cauliflower in California. *Plant Disease* 78:1216.

Turkington, T. K. and Morrall, R. A. A. 1993. Use of petal infestation to forecast Sclerotinia stem rot of canola: the influence of inoculum variation over the flowering period and canopy density. *Phytopathology* 83: 682–689.

Thanatephorus cucumeris
(anamorph = *Rhizoctonia solani*)

RHIZOCTONIA DISEASES: DAMPING-OFF, WIRESTEM, BOTTOM/ROOT ROT

Introduction and significance

On Brassicas, there are several *Rhizoctonia* diseases: damping-off, wirestem, bottom rot, root rot, and foliar blight. *Rhizoctonia* can be very damaging to seedlings just after planting, while black spot on swede can result in rejection of crops for market. In fields heavily cropped to susceptible plants, *Rhizoctonia* can cause significant stand reduction due to seedling (damping-off, wirestem) diseases. This pathogen is an important cause of root rot on radish.

Symptoms and diagnostic features

Damping-off and wirestem occur on newly emerged or young seedlings (**219**). For direct seeded crops, the fungus can attack seed or newly emerged seedlings and kill them prior to emergence (pre-emergence damping-off), or can attack roots and lower stem tissue (hypocotyl) shortly after the plant has emerged above ground (post-emergence damping-off). For recently emerged plants, the stem in contact with soil develops a dark, water-soaked to brown discoloration; these stems are often girdled and the plants fall over. Seedlings become progressively less susceptible as they get older and later infection is confined to the outer tissues of the hypocotyl and roots. Hypocotyl infections of older seedlings consist of browning and cracking of the epidermis, lesion formation, and outer stem tissue decay. When the outer stem deteriorates, only the fibrous inner xylem is intact and remains as a wiry tissue, hence the name wirestem (**220**). Affected plants wilt, turn purple (or blue in the case of cabbage), and remain stunted. Seedlings may break off at the soil line. Transplanted crucifers are subject only to this wirestem phase of seedling diseases. The pathogen can be identified in the field by its coarse mycelium that sometimes causes soil particles to adhere to and dangle from diseased stems (**221**). Cauliflower is more susceptible than cabbage or Brussels sprout .

220 Close-up of wirestem caused by *Rhizoctonia solani* on cauliflower. Healthy plant is on the right.

219 Wirestem caused by *Rhizoctonia solani* on cauliflower in the field.

221 Wirestem caused by *Rhizoctonia solani* on cauliflower. Mycelial strands dangle with soil particles.

222 Bottom rot caused by *Rhizoctonia solani* on Chinese cabbage.

223 *Rhizoctonia* causing yellowing of foliage on radish plants.

Bottom rot is a problem on cabbage, Chinese cabbage, and other head-forming brassicas (222). Once head formation begins, the petioles of lower leaves in contact with the soil become infected and develop dark brown, oval lesions. Secondary decay organisms may follow and make these lesions soft and watery. *Rhizoctonia solani* often produces a web of mycelial growth and sclerotia over the surface of the lesions. On root brassicas, root rot is a common problem, especially in the UK. Small (0.5–1.0 cm), sunken, black or brown lesions develop on roots and result in disfigured taproots. Sometimes roots show large brown lesions with concentric rings of fungal growth.

Rhizoctonia root rot reduces radish stands because seed or germinated seedlings are killed before emerging from the soil (pre-emergence damping-off) or seedlings die shortly after emerging above ground (post-emergence damping-off) (223). Plants that are attacked at later stages show grey or black streaking and lesions on the root surface (224). This external scarring contrasts with black root rot caused by *Aphanomyces raphani* in which there is blackening of internal tissues. Radishes have an uneven appearance when only one side of the root is affected.

Bok choy plants in Asia and cabbage in the USA have suffered from a foliar blight caused by *R. solani*. Circular to irregularly shaped spots measuring 1–3 mm in diameter occur on leaves. Spots become necrotic and sometimes have irregular borders.

Causal agent

Rhizoctonia diseases are caused by *R. solani*. The pathogen is isolated on standard microbiological media, and low nutrient media (water agar or acidified corn meal agar) are often useful because the characteristic hyphal structures are more readily seen. Look for brown hyphae that are up to 12 µm in diameter, have dolipore septa, approximately right angle branching, and a crosswall in the adjacent branch. Loose aggregations of hyphae form sclerotia in some isolates. This fungus does not produce asexual spores. If environmental conditions are suitable, the perfect stage (*Thanatephorus cucumeris*) of this basidiomycete pathogen may be observed as a white, thin, flat hymenial layer present on plant surfaces near the soil. This layer bears basidia and hyaline basidiospores that measure 6–14 x 4–8 µm. Specific molecular tests or reactions with reference isolates are required to identify the anastomosis groups (AG); because only compatible isolates fuse or mate with each other, mating studies reveal distinct AGs within *R. solani*. *Rhizoctonia solani* is not host specific to crucifers and will infect a wide range of other crops.

Disease cycle

Rhizoctonia solani survives saprophytically in soil as mycelium or sclerotia; these fungal structures are the inocula for infecting seedlings and older plants. Airborne basidiospores are produced by the *Thanatephorus* stage, but their importance in disease develop-

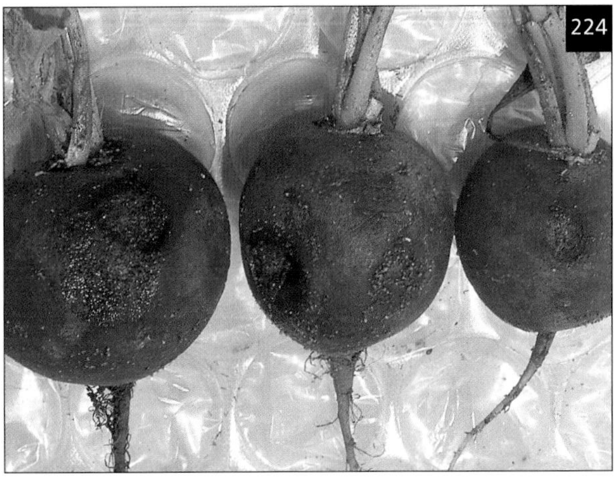

224 Root lesions of *Rhizoctonia solani* on radish root.

ment is not determined. *Rhizoctonia* is favored by warm soil conditions (25–30° C), but is capable of causing problems at much lower temperatures when seedling growth is slow. On brassicas, AG2-1 appears to be most important on young plants and at lower temperatures, while AG4 is associated with mature plant infection and warmer conditions. Both these AGs are also reported on radish. Fungal growth occurs at osmotic water potentials as low as -4 MPa (= -40 bars), but the optimum is -500 kPa (= -5 bars). Populations of *R. solani* in soil fluctuate in response to host substrates and activity of the soil microflora.

Control

Practice thorough sanitation at transplant nurseries to prevent contamination by *R. solani*. Clean and sanitize transplant trays and benches. Ensure that rooting media are not contaminated by infested soil or diseased plant residues. Successful sanitation steps at the transplant nursery eliminate the need to apply fungicides to the transplants during their production. When direct seeding in the field, use seed that has been treated with a fungicide. When placing transplants in the field, avoid planting them too deeply in the soil as the hypocotyl stem tissue is the most susceptible part of the plant. Practice crop rotation so that non-hosts are included in the rotation. Do not plant susceptible crucifers in fields having undecomposed crop residues. Fungicide treatments in the field have had limited effectiveness.

References

Abawi, G. S. and Martin, S. B. 1985. Rhizoctonia foliar blight of cabbage in New York State. *Plant Disease* 69:158–161.

Chet, I. and Baker, R. 1980. Induction of suppressiveness to *Rhizoctonia solani* in soil. *Phytopathology* 70:994–998.

Henis, Y., Ghaffar, A., and Baker, R.1978. Integrated control of *Rhizoctonia solani* damping-off of radish: Effect of successive plantings, PCNB, and *Trichoderma harzianum* on pathogen and disease. *Phytopathology* 68:900–907.

Humayden, H. S., Williams, P. H., Jacobsen, B. J., and Bissonnette, H. L. 1974. Resistance in radish to *Aphanomyces raphani* and *Rhizoctonia solani*. *Plant Disease Reporter* 60:156–160.

Kataria, H. R., Verma, P. R., and Rakow, G. 1993. Fungicidal control of damping-off and seedling root-rot in *Brassica* sp. caused by *Rhizoctonia solani* in the growth chamber. *Annals of Applied Biology* 123:247–256.

Keijer, J., Korsman, M. G., Dullemans, A. M., Houterman, P. M., de Bree, J., and van Silfhout, C.H. 1997. In vitro analysis of host plant specificity in *Rhizoctonia solani*. *Plant Pathology* 46:659–669.

Keinath, A. P. 1995. Relationships between inoculum density of *Rhizoctonia solani*, wirestem incidence and severity, and growth of cabbage. *Phytopathology* 85:1487–1492.

Keinath, A. P. and Farnham, M. W. 1997. Differential cultivars and criteria for evaluating resistance to *Rhizoctonia solani* in seedling *Brassica oleracea*. *Plant Disease* 81:946–952.

Keinath, A. P. and Farnham, M. W. 2001. Effect of wirestem severity on survival and head production of transplanted broccoli and cabbage. *Plant Disease* 85:639–643.

Keinath, A. P., Harrison, H. F., Marino, P. C., Jackson, D. M., and Pullaro, T. C. 2003. Increase in populations of *Rhizoctonia solani* and wirestem of collard with velvet bean cover crop mulch. *Plant Disease* 87:719–725.

Roy, S. K., Das, B. C., and Bora, L. C. 1998. Non-pesticidal management of damping-off of cabbage caused by *Rhizoctonia solani*. *Journal of the Agricultural Science Society of North East India* 11:127–130.

Thornton, C. R., O'Neill, T. M., Hilton, G., and Gilligan, C. A. 1999. Detection and recovery of *Rhizoctonia solani* in naturally infested glasshouse soils using a combined baiting, double monoclonal antibody ELISA. *Plant Pathology* 48:627–634.

Yang, G. H., Chen, X. Q., Chen, H. R., Naito, S., Ogoshi, A., and Zhao, J. F. 2004. First report of foliar blight in *Brassica rapa* ssp. *chinensis* caused by *Rhizoctonia solani* AG-4. *Plant Pathology* 53:260.

Verticillium dahliae, V. longisporum

VERTICILLIUM WILT

Introduction and significance

Verticillium wilt is an important problem in various crucifer vegetables in California. In France, Germany, and Sweden, Verticillium wilt is a significant problem on oilseed rape, which could pose a threat to vegetable brassicas. In the UK, the last report of Verticillium wilt in brassicas was on Brussels sprout in 1961. Verticillium wilt is a problem on horseradish.

225 Collapsed and dying foliage of cauliflower infected with Verticillium wilt.

Symptoms and diagnostic features

Symptoms can appear 3 to 4 weeks after transplanting for cauliflower in California. The older, lower leaves turn yellow and wilt (**225**), with one side of the leaf more severely affected than the other. These leaves later turn brown and drop off the stem, often when plants approach maturity and form flower buds. The xylem of stems and roots becomes black and microsclerotia can be detected in vascular tissues (**226**, **227**). Overall growth of the plant is stunted (**228**). Verticillium wilt can be a significant problem for cabbage, cauliflower, Brussels sprouts, Chinese cabbage, and bok choy. Horseradish has also been reported as a host. It is interesting that broccoli, a close relative of cauliflower, is not susceptible to this pathogen. In California, the disease is most prevalent in summer and autumn crops

Causal agents

Verticillium wilt is caused by imperfect fungus *Verticillium*. The pathogen can be isolated on standard microbiological media. Semi-selective media such as NP-10 can be useful for some samples or to purify isolates. On general media, the pathogen forms the expected verticillate conidiophores, hyaline single-celled conidia, and black microsclerotia. However, the species involved in causing Verticillium wilt of brassica is not clear. *Verticillium dahliae* is a key pathogen of worldwide importance on many crops, and has been listed as the cause of Verticillium wilt of crucifers. However, in 1997 a second species, *V. longisporum*, was proposed as the distinct pathogen that causes Verticillium wilt of crucifers. *Verticillium longisporum* has been distinguished from *V. dahliae* by its elongated rather than spherical microsclerotia, longer spores (7.1–8.8 µm compared with 3.5–5.5 µm for *V. dahliae*) and fewer phialides per node on conidiophores (3 v. 4 to 5). There is uncertainty about the validity of the name *V. longisporum* because not all *Verticillium* isolates from crucifers have these features, *V. dahliae* isolates from other crops can infect brassicas, and molecular data indicates the existence of at least three distinct types of *Verticillium* isolates from crucifer. Therefore this new species name has yet to be generally accepted. The pathogen may be seedborne.

226 Discolored vascular tissue of Chinese cabbage infected with Verticillium wilt.

227 Discolored vascular ring of cauliflower infected with Verticillium wilt.

228 Stunted Brussels sprouts infected with Verticillium wilt.

Disease cycle

The pathogen is a true soilborne organism and forms microsclerotia that enable it to survive in soil for a decade or longer. Plants can be infected within a few weeks after planting. Microsclerotia germinate, enter plants via roots, and grow systemically within the host. Disease symptoms often are expressed when plants mature and form reproductive structures such as flower buds. Soil populations of only one microsclerotia per gram of soil can result in 5% infection. This pathogen is becoming more widespread in Europe and may be introduced or distributed via contaminated soil moved by machinery, footwear, plant material (such as potato tubers), or by wind-blown soil particles.

Control

Maintain records so that infested fields are not used for susceptible crucifers. Avoid moving infested soil and equipment to clean fields. Plant non-susceptible crucifers such as broccoli. Resistant cultivars are not available, though for cauliflower there are vigorous hybrids that show some tolerance to the disease. If it is necessary to use infested fields, schedule plantings of susceptible crops like cauliflower for winter periods; cold soil temperatures inhibit the activity of the pathogen and can suppress development of disease. Interestingly, researchers find that incorporating the biomass from brassica cover crops or crops such as broccoli can also reduce soil inoculum of *V. dahliae* and subsequent disease on cauliflower.

References

Barbara, D. J. and Clewes, E. 2003. Plant pathogenic *Verticillium* species: how many of them are there? *Molecular Plant Pathology* 4:297–305.

Bhat, R. G. and Subbarao, K. V. 2001. Reaction of broccoli to isolates of *Verticillium dahliae* from various hosts. *Plant Disease* 85:141–146.

Collins, A., Okoli, C. A. N., Morton, A., Parry, D., Edwards, S. G., and Barbara, D. J. 2003. Isolates of *Verticillium dahliae* pathogenic to crucifers are of at least three distinct molecular types. *Phytopathology* 93:364–376.

Karapapa, V. K., Bainbridge, B. W., and Heale, J. B. 1997. Morphological and molecular characterization of *Verticillium longisporum* comb. nov. pathogenic to oilseed rape. *Mycological Research* 101:1281–1294.

Koike, S .T., Subbarao, K. V., Davis, R. M., Gordon, T. R., and Hubbard, J. C. 1994. Verticillium wilt of cauliflower in California. *Plant Disease* 78:1116–1121.

Sorensen, L. H., Schneider, A. T., and Davis, J. R. 1991. Influence of sodium polygalacturonate sources and improved recovery of *Verticillium* species from soil. (Abstract). *Phytopathology* 81:1347.

Subbarao, K. V., Chassot, A., Gordon, T. R., Hubbard, J. C., Bonello, P., Mullin, R., Okamoto, D., Davis, R. M. and Koike, S. T. 1995. Genetic relationships and cross pathogenicities of *Verticillium dahliae* isolates from cauliflower and other crops. *Phytopathology* 85:1105–1112.

Subbarao, K. V., Hubbard, J. C., and Koike, S. T. 1999. Evaluation of broccoli residue incorporation into field soil for Verticillium wilt control in cauliflower. *Plant Disease* 83:124–129.

Xiao, C. L. and Subbarao, K. V. 1998. Relationships between *Verticillium dahliae* inoculum density and wilt incidence, severity, and growth on cauliflower. *Phytopathology* 88:1108–1115.

Xiao, C ., Subbarao, K. V., Schulbach, K. F., and Koike, S. T. 1998. Effects of crop rotation and irrigation on *Verticillium dahliae* microsclerotia in soil and wilt on cauliflower. *Phytopathology* 88:1046–1055.

Beet western yellows virus

BEET WESTERN YELLOWS

Introduction and significance

Beet western yellows virus (BWYV) is widespread in winter oilseed rape in Europe. Yield reductions of 10–30% have been reported in oilseed rape and growth may be reduced in vegetable brassicas. In other areas such as coastal California, BWYV is rarely found in brassicas but is occasionally detected in lettuce, spinach, and endive.

Symptoms and diagnostic features

Symptoms include reddening of the foliage, though this is very similar to nutrient or stress factors and therefore is not diagnostic. Infected plants of *B. napus* and other vegetable brassicas can remain symptomless. Occasionally, distinctive red or yellow coloration may develop in cauliflower (**229**). In the UK, tipburn symptoms in stored white cabbage have been attributed to BWYV, though not all infected heads developed symptoms.

Causal agent and disease cycle

BWYV is a polerovirus and infects many hosts, including crucifers, beet, lettuce, and other crop and weed plants. BWYV has isometric particles with diameters of 25 nm and single-stranded, linear RNA genomes. Some isolates or strains of this virus have different abilities to infect certain plants; thus, not all strains of BWYV may be able to infect all known hosts, greatly complicating the etiology of this disease for crucifers. BWYV is transmitted by aphids in a persistent manner (for up to 50 days). *Myzus persicae* is likely to be the main vector on brassicas.

Control

Follow general suggestions for managing virus diseases (see Part 1).

229

229 Yellowing of cauliflower foliage caused by *Beet western yellows virus.*

References

Hardwick, N .V., Davies, J. M.L., and Wright, D. M. 1994. The incidence of three virus diseases of winter oilseed rape in England and Wales in the 1991/92 and 1992/3 growing season. *Plant Pathology* 43:1045–1049.

Hunter, P. J., Jones, J. E., and Walsh, J. A. 2002. Involvement of beet western yellows virus, cauliflower mosaic virus, and turnip mosaic virus in internal disorders of stored white cabbage. *Phytopathology* 92:816–826.

Pallett, D. W., Thurston, M. I., Cortina-Borja, M., Edwards, M-L., Alexander, M., Mitchell, E., Raybould, A. F., and Cooper, J. I. 2002. The incidence of viruses in wild *Brassica rapa* ssp. *sylvestris* in southern England. *Annals of Applied Biology* 141:163–170.

Smith, H. G. and Hinckes, J. A. 1985. Studies on beet western yellows virus in oilseed rape (*Brassica napus* ssp. *oleifera*) and sugar beet (*Beta vulgaris*). *Annals of Applied Biology* 107:473–484.

Cauliflower mosaic virus

CAULIFLOWER MOSAIC

Introduction and significance

Cauliflower mosaic virus (CaMV) is common in temperate regions worldwide. It is often found causing only mild symptoms in crucifers, but yield losses of 20–50% can occasionally occur.

Symptoms and diagnostic features

Symptoms consist of leaf mosaics or mottles that are similar to those caused by other viruses. Infection of young plants can result in some stunting and leaf vein clearing, often near the base of the leaf. Leaf enations are sometimes produced. On mature cauliflower there can be vein banding with dark green areas near the veins and chlorotic patches in the interveinal leaf tissue. The mosaic of light and dark green regions continues to develop and waxy bloom is lost in interveinal areas. On cabbage, the virus causes a diffuse mottle with distinct vein clearing at high temperatures (28° C) or chlorotic vein banding at lower temperatures (20° C). Brussels sprouts show fairly mild symptoms and plants can be distinguished by their paler green color (**230**). In Chinese cabbage, vein clearing occurs over the whole leaf, which can senesce prematurely. Inner leaves show obvious light and dark green mottling between the veins. CaMV is not implicated directly in storage disorders of cabbage, but it can exacerbate problems caused by other viruses. In seed crops, the flowering shoots of infected plants may be stunted, twisted, and develop black spotting on stems and pods.

Causal agent and disease cycle

CaMV is the type member of the caulimovirus group and has isometric particles that measure 50 nm, and double stranded DNA. CaMV is transmitted in a semi-persistent manner by many aphid species, particularly *Brevicoryne brassicae* and *Myzus persicae*, which both breed on crucifers. Aphids can acquire the virus in only1–2 minutes and transmit it immediately.

Control

Follow general suggestions for managing virus diseases (see Part 1).

230 Mild symptoms on Brussels sprouts caused by *Cauliflower mosaic virus*.

References

Hunter, P. J., Jones, J. E., and Walsh, J. A. 2002. Involvement of beet western yellows virus, cauliflower mosaic virus, and turnip mosaic virus in internal disorders of stored white cabbage. *Phytopathology* 92:816–826.

Shepherd, R. J., Wakeman, R. J., and Romanko, R. R. 1968. DNA in cauliflower mosaic virus. *Virology* 36:150–152.

Turnip mosaic virus

TURNIP MOSAIC

Introduction and significance

Turnip mosaic virus (TuMV) is the most severe of the common viruses of crucifers and occurs worldwide. It affects most cultivated brassicas, horseradish, and watercress. Other vegetables, such as lettuce, and ornamentals are hosts. Rhubarb is also affected in Europe.

Symptoms and diagnostic features

TuMV causes chlorotic or necrotic spots and rings, or a general mosaic. Some cultivars develop systemic necrosis and mosaics. Early infection causes severe crinkling and stunting of leaves (**231**) and, in some cases, necrosis and death of plants. On cabbage, some sectors of the leaves may be more severely affected by black spotting than others (**232**). In stored cabbage, inner leaves of heads affected by TuMV may develop large (5-10 mm) black lesions (sometimes referred to as cigar burn) that can result in crop rejection.

Causal agent and disease cycle

TuMV is a member of the potyvirus group and has filamentous particles measuring 750 x 12 nm. There are several strains of the virus and these differ in host range and symptomatology. TuMV is transmitted by many aphid species in a nonpersistent manner. *Brevicoryne brassicae* and *Myzus persicae* are important vectors.

Control

Follow general suggestions for managing virus diseases (see Part 1). Watercress should be raised annually from seed. Dig out and remove old, possibly infected rhubarb plants, and replant with new, virus-free stocks.

References

Edwardson, J. R. and Purcifull, D. E. 1970. Turnip mosaic virus-induced inclusions. *Phytopathology* 60: 85–88.

Hughes, S. L., Green, S. K., Lydiate, D. J., and Walsh, J. A. 2002. Resistance to Turnip mosaic virus in *Brassica rapa* and *B. napus* and the analysis of genetic inheritance in selected lines. *Plant Pathology* 51: 567–573.

Jenner, C. E. and Walsh, J. A. 1996. Pathotype variation in turnip mosaic virus with special reference to European isolates. *Plant Pathology* 45: 848–856.

231 Bok choy infected with *Turnip mosaic virus*. Healthy plant is on the right.

232 Necrotic spotting on cabbage caused by *Turnip mosaic virus*.

Capsicum Pepper

PEPPER (*Capsicum annuum*, *C. frutescens*, and other species) is an important Solanaceae (nightshade family) crop that is grown for its flavorful, pungent fruit. This group of plants most likely originated in the tropical and subtropical regions of the Americas. Pepper fruit are used in different ways as fresh fruit that are raw or cooked, dehydrated whole fruit, or ground into spices and used in diverse processed products. Peppers are used in medicines, cosmetics, and grown as ornamental plants. Peppers are categorized by their degree of sweetness or hotness. Differing levels of capsaicin, a phenolic amide, give pepper its hotness.

Xanthomonas campestris pv. *vesicatoria*

BACTERIAL SPOT

Introduction and significance

Bacterial spot occurs throughout the world but is most serious in tropical and subtropical pepper–growing areas where it is favored by high humidity and rainfall.

Symptoms and diagnostic features

Initial foliar symptoms consist of irregularly shaped, water-soaked spots on leaves (**233**). These spots later turn dark brown to black-brown and usually remain smaller than 5 mm in diameter (**234**). As disease progresses, the spots may coalesce and result in leaves having large necrotic areas. Severely diseased foliage, especially lower leaves, can defoliate from the plant. While infected leaves may show some chlorosis, the individual spots usually are not surrounded by yellow halos. Dark streaks and patches may develop on petioles and stems. Infections on fruit appear as irregularly shaped, brown, raised rough scabs that measure 2 to 5 mm in diameter (**235**). Fruit spots are often clustered near the stem end of the fruit where water and water-splashed inoculum collect.

Causal agent

Bacterial spot is caused by the aerobic, Gram-negative bacterium *Xanthomonas campestris* pv. *vesicatoria*. The pathogen can be isolated on standard microbiological media and produces yellow, mucoid, slow growing colonies typical of most xanthomonads. However, *X. campestris* pv. *vesicatoria* only weakly hydrolyzes starch, so starch-based semi-selective media

233 **Leaf spots of bacterial spot on pepper transplants.**

234 **Leaf spots of bacterial spot of pepper.**

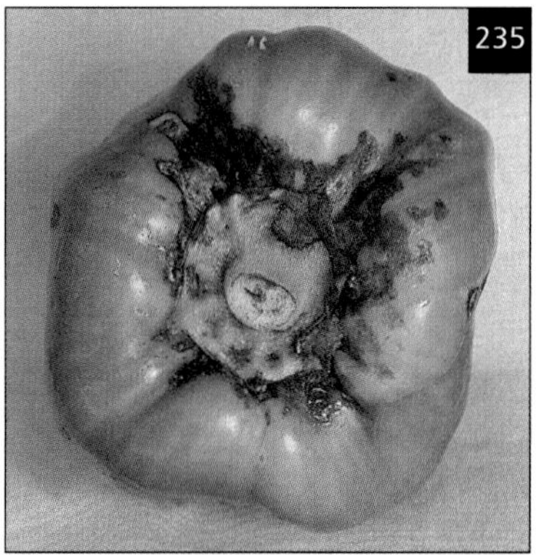

235 **Fruit lesions of bacterial spot of pepper.**

such as SX or MXP media are not diagnostic for identifying this pathogen. Tween medium is helpful because this bacterium forms characteristic white calcium salt crystals when growing on it.

Xanthomonad pathogens from pepper and tomato hosts are a complex group of organisms. At least eight pepper strains of this pathogen are host specific to pepper and are designated as P1, P2, P3, etc. Other strains are host specific to tomato. Finally, other strains are pathogenic on both of these hosts; these pepper-tomato strains are designated as PT strains. Researchers further find that xanthomonad pathogens from pepper and tomato can be divided into various groups (A, B, C, and D) based on genetic and biochemical parameters. The assignment of these groups to various species and pathovars is still being debated.

This pathogen is seedborne. Another bacterial disease is caused by *Pseudomonas syringae* pv. *syringae* and causes symptoms that closely resemble those of bacterial spot.

Disease cycle

Primary inoculum can come from infested seed, plant debris in soil, or reservoir plant hosts. Infested seed is a particularly important inoculum source because if such seed are used to produce transplants, these plant materials are often grown under conditions that can encourage disease development and spread. The practice of using overhead sprinkler irrigation in greenhouses can significantly spread the pathogen. Once diseased transplants are in the field, the pathogen can be spread plant-to-plant via splashing water, contaminated tools and implements, and workers' hands. Disease development is favored by high humidity and warm temperatures in the 24–30° C range. The pathogen is not a true soilborne organism, but it can survive in the soil on infested plant residues. The pathogen can also overwinter on volunteer pepper plants and on weeds such as black nightshade (*Solanum nigrum*) and ground cherry (*Physalis minima*).

Control

Carefully monitor pepper seed fields so that bacterial spot problems can be identified and managed from an early stage. Use seed that does not have significant levels of the pathogen. Appropriate seed treatments can also contribute to the management of seedborne inoculum. Treat seed with hydrochloric acid, calcium hypochlorite, hot water, or other materials. Seed health testing and certification programs help regulate the availability and cleanliness of such seed. Seed tests usually involve the washing of a 10,000 seed sample and subsequent plating of the liquid onto semi-selective medium. Discard heavily infested seed.

Inspect transplants and remove suspect plants and surrounding transplant trays. Sanitize benches that hold transplants, transplant trays, and equipment that comes in contact with plants. Consider applying preventative copper sprays for protecting transplants. Avoid using overhead sprinkler irrigation in the field. With an appropriate disinfectant, periodically and regularly sanitize tools such as clippers and pruning shears. Do not allow equipment or workers to pass through fields when foliage is wet. In the field, copper and copper-maneb sprays may be helpful depending on weather conditions and disease pressure. Once the pepper crop is finished, incorporate the residues to enhance plant decomposition and the dissipation of bacteria. Rotate to a non-host crop before returning to pepper and do not allow volunteer peppers or weed hosts to survive.

References

Abbasi, P. A., Soltani, N., Cuppels, D. A., and Lazarovits, G. 2002. Reduction of bacterial spot disease severity on tomato and pepper plants with foliar applications of ammonium lignosulfonate and potassium phosphate. *Plant Disease* 86:1232–1236.

Bashan, Y., Diab, S., and Okon, Y. 1982. Survival of *Xanthomonas campestris* pv. *vesicatoria* in pepper seeds and roots in symptomless and dry leaves in non-host plants and in the soil. *Plant and Soil* 68:161–170.

Bashan, Y., Okon, Y., and Henis, Y. 1982. Long-term survival of *Pseudomonas syringae* pv. *tomato* and *Xanthomonas campestris* pv. *vesicatoria* in tomato and pepper seeds. *Phytopathology* 72:1143–1144.

Bouzar, H., Jones, J. B., Stall, R. E., Hodge, N. C., Minsavage, G. V., Benedict, A. A., and Alvarez, A. M. 1994. Physiological, chemical, serological, and pathogenic analyses of a worldwide collection of *Xanthomonas campestris* pv. *vesicatoria* strains. *Phytopathology* 84:663–671.

Buonaurio, R., Stravato, V. M., and Scortichini, M. 1994. Characterization of *Xanthomonas campestris* pv. *vesicatoria* from *Capsicum annuum* in Italy. *Plant Disease* 78:296–299.

Conover, R. A. and Gerhold, N. R. 1981. Mixtures of copper and maneb or mancozeb for control of bacterial spot of tomato and their compatibility for control of fungus diseases. *Proceedings Florida State Horticultural Society* 94:154–156.

Gitaitis, R. D., Chang, C. J., Sijam, K., and Dowler, C. C. 1991. A differential medium for semiselective isolation of *Xanthomonas campestris* pv. *vesicatoria* and other cellulolytic Xanthomonads from various natural sources. *Plant Disease* 75:1274–1278.

Jones, J. B., Stall, R. E., and Bouzar, H. 1998. Diversity among Xanthomonads pathogenic on pepper and tomato. *Annual Review of Phytopathology* 36:41–58.

Kousik, C. S. and Ritchie, D. F. 1995. Isolation of pepper races 4 and 5 of *Xanthomonas campestris* pv. *vesicatoria* from diseased peppers in southeastern U. S. fields. *Plant Disease* 79:540.

Kousik, C. S. and Ritchie, D. F. 1998. Response of bell pepper cultivars to bacterial spot pathogen races that individually overcome major resistance genes. *Plant Disease* 82:181–186.

Martin, H. L., Hamilton, V. A., and Kopittke, R. A. 2004. Copper tolerance in Australian populations of *Xanthomonas campestris* pv. *vesicatoria* contributes to poor field control of bacterial spot of pepper. *Plant Disease* 88:921–924.

McGuire, R. G., Jones, J. B., and Sasser, M. 1986. Tween media for semiselective isolation of *Xanthomonas campestris* pv. *vesicatoria* from soil and plant material. *Plant Disease* 70:887–891.

Obradovic, A., Mavridis, A., Rudolph, K., Janse, J. D., Arsenijevic, M., Jones, J. B., Minsavage, G. V., Wang, J.-F. 2004. Characterization and PCR-based typing of *Xanthomonas campestris* pv. *vesicatoria* from peppers and tomatoes in Serbia. *European Journal of Plant Pathology*110:285–292.

Rowell, B., Jones, R. T., Nesmith, W., Satanek, A., and Snyder, J. C. 2001. Bacterial spot resistance, yield, and quality of bell and specialty peppers. *HortTechnology* 11:648–657.

Sijam, K., Chang, C. J., and Gitaitis, R. D. 1992. A medium for differentiating tomato and pepper strains of *Xanthomonas campestris* pv. *vesicatoria*. *Canadian Journal of Plant Pathology* 14:182–184.

Stahl, R. E., Beaulieu, C., Egel, D., Hodge, N. C., Leite, R. P., Minsavage, G. V., Bouzar, H., Jones, J. B., Alvarez, A. M., and Benedict, A. A. 1994. Two genetically diverse groups of strains are included in *Xanthomonas campestris* pv. *vesicatoria*. *International Journal of Systematic Bacteriology* 44:47–53.

Tegegn, T. 1985. A review of bacterial leaf spot of peppers (*Capsicum annuum*) caused by *Xanthomonas campestris* pv. *vesicatoria* and some methods of its control. *Acta Horticulturae* 158:369–376.

Athelia rolfsii (anamorph = *Sclerotium rolfsii)*

SOUTHERN BLIGHT

Introduction and significance

Southern blight, or Sclerotium stem rot, occurs on a large number of vegetable and ornamental plants.

Symptoms and diagnostic features

On pepper, the early symptoms consist of a water-soaked lesion on crown and lower stem tissue that is in contact with the soil. These infection sites turn light to dark brown and can rapidly girdle the entire crown. Above-ground symptoms consist of wilting and a quick collapse of all foliage (**236**). If soil moisture conditions are suitable, the pathogen will form a thick, white mycelial mat or layer on crown, lower stem, and even

236 Collapsed pepper plant affected by southern blight.

237 Pepper stem exhibiting white mycelial growth from *Sclerotium rolfsii*.

on the soil surrounding the crown (237). Small (1–2 mm in diameter), spherical, tan to brown sclerotia form profusely on and in this white growth. Sclerotia are characterized by having an outer differentiated, pigmented rind.

Causal agent

Southern blight is caused by *Sclerotium rolfsii*, which is an imperfect fungus in the *Mycelia Sterilia* category and produces no asexual spores. *Sclerotium rolfsii* has a broad host range and forms a basidiomycete perfect stage (*Athelia rolfsii*), though it is unknown whether this stage is involved in the disease.

Disease cycle

Because of its resilient sclerotia, the pathogen can survive in the soil and in crop debris for many years. When susceptible crops are planted into infested soil, the sclerotia germinate and infect the plant. The fungus is favored by high temperatures above 30° C.

Control

Rotate with non-host plants so that soil inoculum levels are reduced; however, because of the ability of *S. rolfsii* to survive in soil, crop rotations will not eliminate the disease. Deep plowing of fields prior to planting, which inverts the soil profile, may help reduce inoculum levels. Pre-plant treatment of soil with effective fumigants will provide some control but will not eradicate the pathogen and are expensive to use.

References

Brown, J. E., Stevens, C., Osborn, M. C., and Bryce, H. M. 1989. Black plastic mulch and spunbonded polyester row cover as method of southern blight control in bell pepper. *Plant Disease* 73:931–932.

Jenkins, S. F. and Averre, C. W. 1986. Problems and progress in integrated control of southern blight of vegetables. *Plant Disease* 70:614–619.

Punja, Z. K. 1985. The biology, ecology, and control of *Sclerotium rolfsii*. *Annual Review of Phytopathology* 23:97–127.

Ristaino, J. B., Perry, K. B., and Lumsden, R. D. 1996. Soil solarization and *Gliocladium virens* reduced the incidence of southern blight (*Sclerotium rolfsii*) in bell pepper in the field. *Biocontrol Science & Technology* 6:583–593.

Botrytis cinerea (teleomorph = *Botryotinia fuckeliana*)

GRAY MOLD

Introduction and significance

Gray mold of pepper is found in all pepper producing regions of the world. The disease can occasionally be damaging, especially if humid conditions prevail. The disease can often be more prevalent in greenhouse settings.

Symptoms and diagnostic features

Petioles and stems become infected and develop tan to light brown lesions (238). The developing lesion can eventually girdle the entire petiole or stem and show concentric rings due to the sporulation of the pathogen and coloration of the lesion. Leaves may also have brown lesions and sporulation. The gray mold pathogen often infects petioles, stems, and leaves that are damaged or senescing. Flower petals are also subject to gray mold infections. Infections on green or ripe fruit result in a soft, decayed, brown to gray rot that can eventually envelop the entire fruit (**239**). The fungus usually sporulates on the fruit calyx or in the center of the fruit lesion where the epidermis has split. Pepper transplants that are damaged during the planting process can develop stem infections and die. Gray mold can be particularly damaging in greenhouses due to the elevated humidity in such environments.

Causal agent

Gray mold is caused by *Botrytis cinerea*. The sexual stage, *Botryotinia fuckeliana*, is rarely found on the crop. Conidiophores of *B. cinerea* are long (1–2 mm), become gray-brown with maturity, and branch irregularly near the apex. Conidia are clustered at the branch tips and are single-celled, pale brown, ellipsoid to obovoid, and measure 6–18 x 4–11 µm. The pathogen can be isolated on standard microbiological media. Some isolates sporulate poorly in culture unless incubated under lights (12 h light/12 h dark). The pathogen forms survival structures (sclerotia) that are black, oblong or dome-shaped, and measure 4–10 mm. The fungus grows best at 18–23° C but is inhibited at warm temperatures above 32° C. On host tissue the fungus produces characteristically profuse sporulation that is dense, velvety, and grayish brown in color.

Disease cycle

Botrytis cinerea survives in and around fields as a saprophyte on crop debris, as a pathogen on numerous crops and weed plants, and as sclerotia in the soil. Conidia develop from these sources and become windborne. When conidia land on senescent or damaged pepper tissue, they will germinate if free moisture is available and rapidly colonize this food base. Once established, the pathogen will grow into adjacent healthy stems and leaves, resulting in disease symptoms and the production of additional conidia. Cool temperatures, free moisture, and high humidity favor the development of the disease. Pepper tissues that are damaged from other diseases can also become colonized by *B. cinerea* acting as a secondary decay organism.

Control

Reduce plant wetness by avoiding or reducing sprinkler irrigation. To reduce overall humidity, adequately ventilate or heat greenhouses. Fungicides may be useful in protecting fruit from gray mold.

238 Stem lesion of gray mold of pepper.

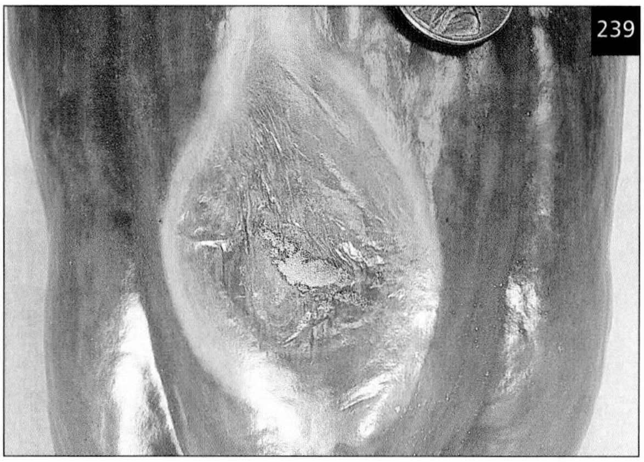

239 Fruit lesion of gray mold of pepper.

References

Elad, Y., Yunis, H., and Volpin, H. 1993. Effect of nutrition on susceptibility of cucumber, eggplant, and pepper crops to *Botrytis cinerea*. *Canadian Journal of Botany* 71:602–608.

Park, S. E., Lee, J. T., Chung, S.O., Kim, H. E., Park, S. H., Lee, J. T., Chung, S. O., and Kim, H. K. 1999. Forecasting the pepper grey mould rot to predict the initial infection by *Botrytis cinerea* in greenhouse conditions. *Plant Pathology Journal* 15:158–161.

Colletotrichum capsici, C. coccodes, C. gloeosporioides

ANTHRACNOSE

Introduction and significance

Pepper anthracnose occurs throughout many of the pepper growing regions of the world. The disease has greatest impact on the fruit. Both immature and ripe fruit are subject to infection, though ripe fruit are more susceptible.

Symptoms and diagnostic features

Initial symptoms on pepper fruit consist of round to oval, tan lesions. As disease develops, the fruit lesions expand, remain circular in shape, and show concentric rings of black, orange, and tan colors (**240, 241**). The black color comes from the fungal structures (*microsclerotia* and *acervuli*) that form in the rings. If humid, wet weather occurs, the fruiting bodies in the lesions will release orange to pink-colored spore masses. Lesions can enlarge and become quite large (30 mm or more in diameter) and sunken. Leaves and stems are also susceptible to anthracnose and develop irregularly shaped, tan to brown lesions that often have darker brown borders.

Causal agents

Anthracnose is caused by several species of the fungus *Colletotrichum*: *C. capsici*, *C. coccodes*, *C. gloeosporioides*. All species form minute (approximately 0.3 mm in diameter), cup-shaped acervuli fruiting bodies that are usually present in fruit lesions. Acervuli release single-celled, cyndrical hyaline conidia. These measure 19–28 x 2.7–5 µm (*C. capsici*), 16–24 x 2–5 µm (*C. coccodes*), or 11–19 x 2.7–5 µm (*C. gloeosporioides*). Some species produce long, brown, septate setae that are present in the acervuli. *C. coccodes* and *C. gloeosporioides* form small (0.2–0.4 mm in diameter), irregularly shaped survival structures called microsclerotia.

Disease cycle

The fungus survives in soil in the form of microsclerotia or as acervuli and microsclerotia in dried plant residue. The pathogen can be splashed up onto pepper foliage and fruit and initiate infections. In addition, fruit that are in contact with the soil become infected by soilborne inoculum. Ripe fruit are particularly susceptible to infection. Optimum temperatures for disease development are 20–24° C. With wet, humid conditions, conidia are produced in a gelatinous material in acervuli. Dispersal of conidia requires splashing water from rain or sprinklers. The pathogen can be seedborne in pepper.

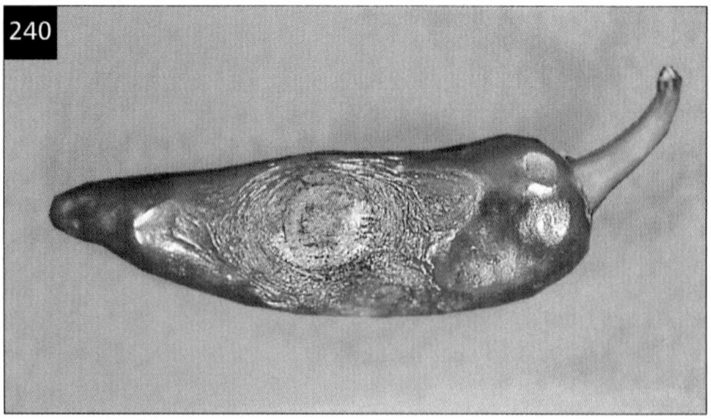

240 Fruit lesions on Jalapeno pepper caused by anthracnose.

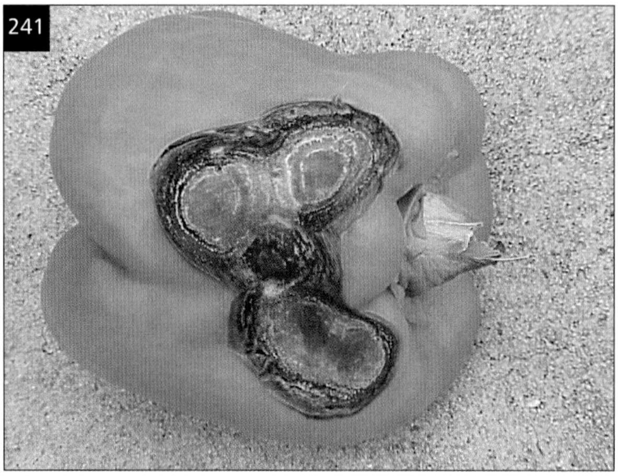

241 Fruit lesions on bell pepper caused by anthracnose.

Control

Rotate crops so that non-hosts are grown every other year. Many weeds can support the pathogen, so practise good weed control. Use seed that does not have significant levels of the pathogen. Appropriate seed treatments can also contribute to the management of seedborne inoculum. Because infested seed can result in diseased transplants, inspect transplants and discard diseased plants. Avoid sprinkler irrigation, as splashing water spreads conidia. Apply fungicides as necessary.

References

Farley, J. D. 1976. Survival of *Colletotrichum coccodes* in soil. *Phytopathology* 66:640–641.

Freeman, S., Katan, T., and Shabi, E. 1998. Characterization of *Colleototrichum* species responsible for anthracnose diseases of various fruits. *Plant Disease* 82:596–605.

Lewis Ivey, M. L., Nava-Diaz, C., and Miller, S. A. 2004. Identification and management of *Colletotrichum acutatum* on immature bell peppers. *Plant Disease* 88:1198–1204.

Manandhar, J. B., Hartman, G. L., and Wang, T. C. 1995. Conidial germination and appressorial formation of *Colleotrichum capsici* and *C. gloeosporioides* isolates from pepper. *Plant Disease* 79:361–366.

Manandhar, J. B., Hartman, G. L., and Wang, T. C. 1995. Anthracnose development on pepper fruits inoculated with *Colleotrichum gloeosporioides*. *Plant Disease* 79:380–383.

Raid, R. N. and Pennypacker, S. P. 1987. Weeds as hosts for *Colletotrichum coccodes*. *Plant Disease* 71:643–646.

242 Powdery mildew sporulation on underside of pepper leaf.

Leveillula taurica (anamorph = *Oidiopsis taurica*)

POWDERY MILDEW

Introduction and significance

Powdery mildew is widespread on peppers throughout the world and sometimes can cause severe damage and crop loss.

Symptoms and diagnostic features

The disease initially causes light green, irregularly shaped patches on leaves. These patches can later turn chlorotic and have brown, necrotic centers. Careful examination of the undersides of these leaves reveals the white powdery growth of this pathogen (**242**). The edges of severely infected leaves will curl upward. Such leaves will fall off the plant and result in defoliation of older foliage. Significant leaf drop will cause the fruit to be exposed and become sunburned (**243**). Fruit damage can therefore indirectly result from powdery mildew. This pathogen only infects the older leaves, with the younger leaves escaping infection until they mature.

243 Defoliation and resulting sunburned fruit caused by powdery mildew of pepper.

Causal agent

Powdery mildew is caused by *Leveillula taurica* (anamorph = *Oidiopsis taurica*). This mildew species does not form epiphytic mycelium and all conidiophores develop from endophytic mycelium and emerge through stomata in the lower leaf epidermis. Conidiophores can be branched and carry one or sometimes two conidia. Conidia are hyaline, single-celled, and dimorphic. Primary (terminal) conidia are lanceolate with distinct apical points. Secondary conidia are ellipsoid–cylindric and lack the apical point. Conidial dimensions for both types vary according to the host plant but generally are 50–70 x 16–24 µm. For both conidial types, length-to-width ratios are greater than three. Though not reported to occur on pepper, this fungus can form a sexual stage consisting of globose cleistothecia having numerous hypha-like appendages and containing up to 20 asci. *Leveillula taurica* appears to have a broad host range of numerous crops and weeds. However, research indicates that there may be distinct subpopulations that have more restricted host ranges.

Disease cycle

Leveillula taurica is an obligate pathogen and survives on overwintering peppers, alternate hosts, or possibly as cleistothecia on other crops. Conidia are dispersed by winds. Powdery mildew development is favored by mild temperatures below 30°C.

Control

Apply fungicides, such as sulfur, if the disease becomes severe. Some pepper types (nonpungent cultivars) are tolerant to powdery mildew, and researchers are developing resistant lines for other pepper types.

References

Correll, J. C., Gordon, T. R., and Elliott, V. J. 1987. Host range, specificity, and biometrical measurements of *Leveillula taurica* in California. *Plant Disease* 71:248–251.

de Souza, V. L. and Café-Filho, A. C. 2003. Resistance to *Leveillula taurica* in the genus *Capsicum*. *Plant Pathology* 52:613–619.

Palti, J. 1971. Biological characteristics, distribution, and control of *Leveillula taurica*. *Phytopathologia Mediterranea* 10:139–153.

Reuveni, R. and Rotem, J. 1973. Epidemics of *Leveillula taurica* on tomatoes and peppers as affected by conditions of humidity. *Phytopathologische Zeitschrift* 76:153–157.

Shifress, C., Pilowsky, M., and Zacks, J. M. 1992. Resistance to *Leveillula taurica* mildew in *Capsicum annuum*. *Phytoparasitica* 20:279–283.

Phytophthora capsici, P. nicotianae var. *parasitica*

PHYTOPHTHORA BLIGHT/ ROOT AND CROWN ROTS

Introduction and significance

There are two important soilborne *Phytophthora* species that cause several diseases on pepper. *Phytophthora* diseases take the forms of seed decays, seedling damping-off, transplant stem rots, root and crown rots, foliar blights and fruit rots. This section will focus on root and crown rots and foliar blights. In recent years this pathogen has increased in importance on various vegetable crops in the USA.

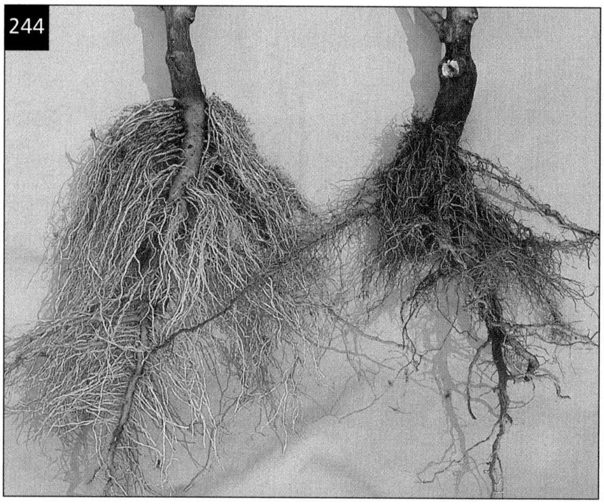

244 Root rot of Phytophthora root rot of pepper. Healthy plant is on the left.

245 Wilting of plant affected by Phytophthora root rot (on left). Healthy plant is on right.

246 Pepper field affected by Phytophthora root rot.

Symptoms and diagnostic features

Symptoms of Phytophthora root rot initially consist of water-soaked root lesions that later turn dark gray to chocolate-brown (**244**). The discoloration can occur on both the fine feeder and larger taproots. As lesions expand, individual roots become girdled or entirely rotted. The discoloration will affect both vascular and stele tissues of the root and can move up the main taproot and into the plant crown and lower main stem. In advanced stages, the roots will be soft and decayed. The plant crown can show both surface and internal discoloration. Above ground symptoms consist of foliage that first turns dull gray-green, then later wilts (**245**). The entire plant canopy can rapidly collapse and die (**246**).

Phytophthora blight of pepper results in above-ground infections. Irregularly shaped, black lesions develop on stems and in axils of branches. Leaves can have gray to brown, circular to oval, water-soaked spots that can enlarge and cover large portions of the leaf. This disease can also cause fruit infections that result in rotted fruit tissues that support white sporulation of the pathogen.

Causal agents

Phytophthora root and crown rots are caused by *P. capsici* and *P. parasitica* (= *P. nicotianae* var. *parasitica*). Phytophthora blight is caused by *P. capsici*. Both species are oomycete organisms, soil inhabitants, and can persist in soils for extended periods of time. *Phytophthora capsici* forms irregularly shaped sporangia that can be spherical, ovoid, elongated, or have more than one apex. Sporangia are papillate, deciduous, have pedicels that are 10 or more µm in length, and measure 30–60 x 25–35 µm. *Phytophthora parasitica* sporangia vary greatly and can be ellipsoidal, ovoid, pyriform, or spherical with distinct papilla. Sporangia are not deciduous and measure 11–60 x 20–45 µm.

Disease cycle

The root and crown diseases require the presence of wet soils and free water. Compacted, finely textured, and poorly draining soils create conditions favorable for root rot. In addition to pepper, these *Phytophthora* species can infect tomato, cucurbits, and other hosts. The blight disease occurs if inoculum is splashed from the soil and onto lower foliage and stems. Later in disease development, *P. capsici* forms sporangia on infected leaves, stems, or fruit. These sporangia are carried by winds or splashing water onto other plants, resulting in spread of the disease. Infection conditions are optimum if free moisture is present and temperatures are 24–33° C.

Control

Plant peppers in soils that are prepared properly and drain well. Carefully manage irrigation so that excess soil water is avoided. Keep bed tops dry by using subsurface drip irrigation. Fungicides may help manage both root and fruit infections. Some new pepper cultivars have some resistance to these pathogens.

References

Cafe-Filho, A. C. and Duniway, J. M. 1996. Effect of location of drip irrigation emitters and position of *Phytophthora capsici* infections in roots on Phytophthora root rot of pepper. *Phytopathology* 86:1364–1369.

Cafe-Filho, A. C. and Duniway, J. M. 1995. Dispersal of *Phytophthora capsici* and *Phytophthora parasitica* in furrow-irrigated rows of bell pepper, tomato, and squash. *Plant Pathology* 44:1025–1032.

Cafe-Filho, A. C. and Duniway, J. M. 1995. Effects of furrow irrigation schedules and host genotype on Phytophthora root rot of pepper. *Plant Disease* 79:39–43.

Hausbeck, M. K. and Lamour, K. H. 2004. *Phytophthora capsici* on vegetable crops: research progress and management challenges. *Plant Disease* 88:1292–1303.

Hwang, B.-K. and Kim, C.-H. 1995. Phytophthora blight of pepper and its control in Korea. *Plant Disease* 79:221–227.

Lamour, K. H. and Hausbeck, M. K. 2003. Effect of crop rotation on the survival of *Phytophthora capsici* in Michigan. *Plant Disease* 87:841–845.

Larkin, R. P., Ristaino, J. B., and Campbell, C. L. 1995. Detection and quantification of *Phytophthora capsici* in soil. *Phytopathology* 85:1057–1063.

Oelke, L. M., Bosland, P. W., and Steiner, R. 2003. Differentiation of race specific resistance to Phytophthora root rot and foliar blight in *Capsicum annuum*. *Journal of the American Society for Horticultural Science* 128:213–218.

Parra, G. and Ristaino, J. B. 2001. Resistance to mefenoxam and metalaxyl among field isolates of *Phytophthora capsici* causing Phytophthora blight of bell pepper. *Plant Disease* 85:1069–1075.

Ristaino, J. B. 1991. Influence of rainfall, drip irrigation, and inoculum density on the development of Phytopthora root and crown rot epidemics and yield in bell pepper. *Phytopathology* 81:922–929.

Ristaino, J. B. and Gumpertz, M. L. 2000. New frontiers in the study of dispersal and spatial analysis of epidemics caused by species in the genus *Phytophthora*. *Annual Review of Phytopathology* 38:541–576.

Ristaino, J. B. and Johnston, S. A. 1999. Ecologically based approaches to management of Phytophthora blight of bell pepper. *Plant Disease* 83:1080–1089.

Sujkowski, L. S., Parra, G. R., Gumpertz, M. L., and Ristaino, J. B. 2000. Temporal dynamics of Phytophthora blight on bell pepper in relation to the mechanisms of dispersal of primary inoculum of *Phytophthora capsici* in soil. *Phytopathology* 90:148–156.

***Phytophthora* spp., *Pythium* spp., *Rhizoctonia solani* (teleomorph = *Thanatephorus cucumis*)**

DAMPING-OFF, PYTHIUM ROOT ROT

Introduction and significance

Several soilborne pathogens cause seed, seedling, and transplant diseases of pepper. Pathogens include species of *Phytophthora*, *Pythium*, and *Rhizoctonia*. If soil conditions favor the pathogen, plant stands can be significantly reduced.

Symptoms and diagnostic features

These diseases have various phases. Pepper seeds can be infected prior to germination and result in seed death. Newly germinated seedlings can be infected to such a degree that plants do not emerge above the soil (pre-emergence damping-off). Pepper seedlings might emerge from the ground but become diseased after emergence (post-emergence damping-off). Finally, transplants placed in the ground can develop root rots or stem lesions from these same pathogens. Initial symptoms of post-emergence damping-off occur on seedling stems in contact with the soil. These symptoms consist of shriveled stems that have discolored, tan to dark brown lesions. With time the lower stem collapses, roots decay, and the cotyledons and leaves will wilt. Such plants sometimes fall over. Damping-off diseases often result in death of the seedling and subsequent reduction of plant stands. However, even if plants do not succumb to these pathogens, the surviving plant may be stunted, delayed in development, and less productive (**247**).

These three pathogens can also infect roots and crowns of older plants. Such infections are usually restricted to the small feeder roots which turn brown and become rotted. Affected plants have reduced vigor and foliage may turn chlorotic.

Causal agents

Several species of *Phytophthora* and *Pythium* can cause damping-off in pepper. These pathogens are in the oomycete group of organisms. These organisms survive in the soil as saprophytes and are favored by wet soil conditions. With the exception of *P. ultimum*, these pathogens usually produce zoospores that swim to and infect susceptible tissues. Sexual structures (antheridia, oogonia, and oospores) are produced by all species. In addition to pepper, these pathogens can infect numerous other plants.

Rhizoctonia solani is a soilborne fungus with a very broad host range. *R. solani* has no asexual fruiting structures or spores, but produces characteristically coarse, brown, approximately right-angled branching hyphae. The hyphae are distinctly constricted at branch points, and cross walls with dolipore septa are deposited just after the branching. Hyphal cells are multi-nucleate. Small, tan to brown loosely aggregated clumps of mycelia function as sclerotia. This fungus can survive by infecting and thriving on a great number of plant hosts, besides pepper, and can also persist in the soil as a saprophyte. The teleomorph stage, *Thanatephorus cucumis*, is not commonly observed.

Disease cycle

Most of these soilborne pathogens survive in fields for indefinite amounts of time. Wet soil conditions and cool temperatures (15–20° C) generally favor these organisms and their ability to grow and infect hosts. However, for pathogens such as *P. aphanidermatum* and *P. myriotylum*, warmer soil conditions (32–37° C) are favorable. *Rhizoctonia solani* can also be favored by warmer soils. Seedlings are most susceptible to infection, though these pathogens can infect the feeder roots of mature plants.

Control

Damping-off is primarily controlled by creating conditions unfavorable for the pathogens. Plant on raised beds and in soils that drain well so that overly wet soil conditions are avoided. Carefully manage irrigation so that excess water is not applied. Plant seed that has been treated with fungicides, and avoid planting seed too deeply, which delays seedling emergence and increases the chance of infection. Post-plant fungicides may provide control for some of these pathogens. Avoid planting too soon into fields that still have extensive crop residue in the soil. Rotate crops because consecutive pepper plantings will increase the populations of the soilborne pathogens. For transplants, prepare good quality beds and do not place transplants too deep into the beds.

247 Poor growth of pepper plants infected by *Pythium* (plant on the right). Healthy plant is on the left.

References

Chellemi, D. O., Mitchell, D. J., Kannwischer-Mitchell, M. E., Rayside, P. A., and Rosskopf, E. N. 2000. *Pythium* spp. associated with bell pepper production in Florida. *Plant Disease* 84:1271–1274.

Mao, W., Lewis, J. A., Lumsden, R. D., and Hebbar, K. P. 1998. Biocontrol of selected soilborne diseases of tomato and pepper plants. *Crop Protection* 17:535–542.

Sclerotinia minor, S. sclerotiorum

WHITE MOLD, SCLEROTINIA ROT

Introduction and significance

White mold, or Sclerotinia rot, is an occasional problem on peppers. The types of symptoms seen on pepper depend on which species of *Sclerotinia* is involved and which stage of the pathogen is present. White mold caused by *S. sclerotiorum* is generally the more important disease.

Symptoms and diagnostic features

Sclerotinia minor only infects the pepper stem tissue that is in contact with the soil (**248**). *Sclerotinia minor* causes a water-soaked lesion to develop at the crown and lower stem. The lesion enlarges and can girdle the plant, resulting in the collapse of the canopy and foliage. With time the crown and stem lesions turn light tan to off-white in color. White mycelium and small (3–5 mm), black, irregularly shaped sclerotia form around and within the decayed crown.

Sclerotinia sclerotiorum is the other species that can attack pepper. Airborne ascospores can cause infections throughout the pepper canopy (**249**). These above-ground infections usually occur on damaged stems or petioles, or where a nutrient source, such as a senescent flower petal, falls onto stems or petioles. These infections are water-soaked lesions that gradually enlarge and encircle the stems. Older infections turn off-white, white-gray, or tan in color and show zonate rings.

248 Pepper crowns infected by *Sclerotinia minor*.

249 Upper canopy infections of *Sclerotinia sclerotiorum* of pepper.

White mycelium can be observed on infected lesions if conditions are favorable. The large, black sclerotia can grow on the outer surface or in the central cavity of stems (**250**). *Sclerotinia sclerotiorum* can also infect pepper crowns and lower stems (**251**). Though not always observed, fruit can become infected and develop a water-soaked, dull green lesion that later turns into a soft, watery rot. This pathogen produces white mycelium and black, oblong or dome-shaped sclerotia that are 5–10 mm long.

Causal agents and disease cycle

White mold is caused by two species of *Sclerotinia*, *S. minor* and *S. sclerotiorum*. The two pathogens are distinguished primarily by the size of sclerotia. *Sclerotinia minor* sclerotia are significantly smaller than those of *S. sclerotiorum*. In addition, *S. sclerotiorum* produces apothecia, while *S. minor* does not generally do so in the field. For detailed descriptions of these pathogens and the disease cycles, see the bean white mold section in the chapter on legume diseases.

Control

For an integrated approach to *Sclerotinia* control, see the bean white mold section in the chapter on legume diseases. Against *S. sclerotiorum*, fungicides may only be partially effective due to the sporadic release of infective ascospores during the course of a pepper crop. If peppers are grown in the greenhouse, reduce humidity levels by ventilating the structures.

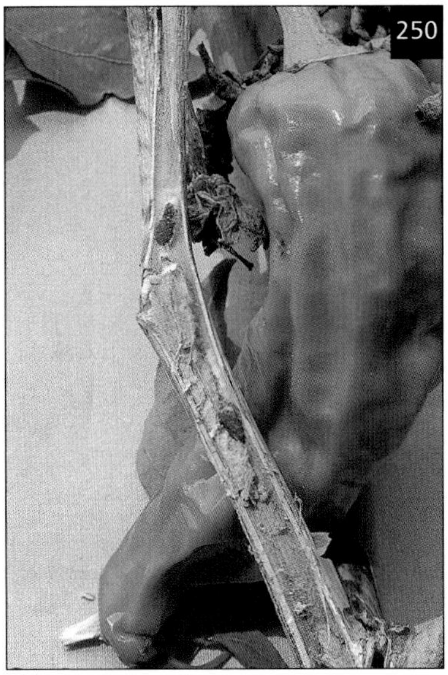

250 Mycelium and sclerotia of *Sclerotinia sclerotiorum* inside a pepper stem.

251 Pepper stems exhibiting mycelium of *Sclerotinia sclerotiorum*.

References

Abawi, G. S. and Grogan, R. G. 1979. Epidemiology of diseases caused by *Sclerotinia* species. *Phytopathology* 69:899–904.

Gonzalez, T. G., Henderson, D. M., and Koike, S. T. 1998. First report of bell pepper (*Capsicum annuum*) as a host of *Sclerotinia minor*. *Plant Disease* 82:832.

Kohn, L. M. 1979. Delimitation of the economically important plant pathogenic *Sclerotinia* species. *Phytopathology* 69:881–886.

Purdy, L. H. 1979. *Sclerotinia sclerotiorum*: history, diseases and symptomatology, host range, geographic distribution, and impact. *Phytopathology* 69:875–880.

Yanar, Y. and Miller, S. A. 2003. Resistance of pepper cultivars and accessions of *Capsicum spp.* to *Sclerotinia sclerotiorum*. *Plant Disease* 87:303–307.

Stemphylium solani, S. lycopersici

GRAY LEAF SPOT

Introduction and significance

Gray leaf spot disease is found in many pepper-producing areas and can commonly be found on both pepper and tomato.

Symptoms and diagnostic features

The main symptoms occur on leaves and initially consist of small (1–2 mm in diameter) reddish-brown spots. These spots enlarge into circular to oval spots that measure 3–5 mm in diameter and consist of white centers with red to brown margins. In severe cases, leaves having many spots can turn chlorotic and drop off the plant. Young pepper plants at the seedling stage are most sensitive to infection. Spots can also occur on petioles and stems but are not found on pepper fruit. Stem lesions are irregular in shape, white in color, and have red to brown margins. Stem lesions can weaken the plant and result in broken stems.

Causal agents

Gray leaf spot is caused by two species of the fungus *Stemphylium*. On V–8 juice agar incubated under lights, these pathogens produce dark green-brown mycelium and abundant conidia after approximately 10 days. For *S. solani*, conidiophores are mostly unbranched and 4–7 µm wide with distinctly swollen apical cells (8–11 µm wide) having darkly pigmented bands. Conidia dimensions are mostly 33–55 x 18–28 µm. Conidia are multicelled, pale to golden brown, smooth to finely verruculose, having three to six transverse and several longitudinal septa. The spore has a pointed apex and is usually constricted at the middle transverse septum.

For *S. lycopersici*, conidiophores are mostly unbranched and 6–7 µm wide with distinctly swollen apical cells (8–10 µm wide) having darkly pigmented bands. Conidia dimensions are mostly 50–74 x 16–23 µm. Conidia are multicelled, pale to medium brown, smooth to finely verruculose, having one to eight transverse and several longitudinal septa. The spore has a conical apex and is usually constricted at three transverse septa. The taxonomy of these species is still in need of clarification. *Stemphylium botryosum* f. sp. *lycopersici* and *S. floridanum* are other *Stemphylium* fungi or pathogen synonyms that are pathogenic on

Solanaceous hosts. The perfect stages of these pathogens, which are various species of *Pleospora*, are rarely found on this crop.

Disease cycle

These two *Stemphylium* pathogens survive as saprophytes on crop residues, or as pathogens on volunteer or wild Solanaceae plants. Winds and splashing water spread conidia to pepper hosts. Free moisture is required for infection to take place. The disease is most severe if conditions are humid and overcast, and if foliage is wet with dew or rain. These pathogens can be seedborne.

Control

Use seed that does not have significant levels of the pathogen. Appropriate seed treatments can also contribute to the management of seedborne inoculum. Inspect transplants and remove symptomatic plants. Rotate crops so that infected crop residues and volunteer plants are not present. Use fungicides if necessary.

References

Blazquez, C. H. 1969. Occurrence of gray leaf spot on peppers in Florida. *Plant Disease Reporter* 53:756.

Braverman, S. W. 1968. A new leaf spot of pepper incited by *Stemphylium botryosum* f. sp. *capsicum*. *Phytopathology* 58:1164–1167.

Verticillium dahliae

VERTICILLIUM WILT

Introduction and significance

Verticillium wilt is a well-known disease that affects hundreds of different plant species and is an important problem on pepper throughout the world. The closely related tomato and aubergine (eggplant) are also subject to this disease.

Symptoms and diagnostic features

On pepper, early symptoms consist of the slight chlorosis of lower leaves. The chlorosis can progress until leaves are bright yellow (**252**); such leaves will wilt and eventually fall off the plant. Plant shoots and the overall foliage will wilt, especially during the warmer times of the day. Internal vascular tissue has a tan to light brown discoloration (**253**). This coloring is most evident in the main stems closer to the crown; such discoloration may not be evident in the upper, smaller stems. Disease symptoms can be accentuated if the infected plant is bearing a heavy load of fruit or is stressed by some other factor. Even if infected plants do not die completely, plant growth and yields can be significantly reduced (**254**).

252 Chlorosis of pepper caused by Verticillium wilt.

Causal agent

Verticillium wilt is caused by the fungus *Verticillium dahliae*. The pathogen can be isolated on standard microbiological media, though semi-selective media such as NP–10 can be useful for isolation. On general purpose media, the pathogen forms the characteristic hyaline, verticillate conidiophores bearing 3–4 phialides at each node, and hyaline, single-celled, ellipsoidal conidia that measure 2–8 x 1–3 µm. Older cultures form dark brown to black torulose microsclerotia that consist of groups of swollen cells formed by repeated budding. Microsclerotia size varies greatly and is in the range of 15 to 100 µm in diameter. Microsclerotia enable the pathogen to survive in the soil for extended periods of time (up to 8 to 10 years). *Verticillium dahliae* is listed as having an extensive host range of crops and weeds. However, research has indicated that the *V. dahliae* isolates from various crops have distinct host ranges and may therefore exist as different strains. For example, isolates from bell pepper can infect many other non-pepper hosts; however, isolates from chile pepper infect only pepper and aubergine (eggplant).

When *V. dahliae* isolates from tomato are inoculated onto pepper, some will infect pepper while others cannot cause disease. Therefore, the precise etiology of the various Verticillium wilt diseases from different crops appears to be quite complex.

Disease cycle

The pathogen survives in the soil as dormant microsclerotia or as epiphytes on non-host roots. Cool to moderate weather conditions favor the pathogen, and disease is enhanced at temperatures between 20–24° C. The microsclerotia germinate and infective hyphae enter host roots through wounds.

Control

In general, Verticillium wilt is best controlled by planting resistant or tolerant cultivars. However, presently there are no commercially available pepper cultivars with sufficient resistance. Pre-plant treatment of soil with effective fumigants will give some control but will not eradicate the pathogen. For greenhouse production, steaming of soil can also provide some control. Rotate crops so that pepper is not planted in fields having a history of the problem. Rotation with non-host crops, such as small grains and corn, can lower inoculum levels but will not eradicate the pathogen. Minimize spread of infested soil to uninfested areas.

References

Bhat, R. G. and Subbarao, K. V. 1999. Host range specificity in *Verticillium dahliae*. *Phytopathology* 89:1218–1225.

Bhat, R. G., Smith, R. F., Koike, S. T., Wu, B. M., and Subbarao, K. V. 2003. Characterization of *Verticillium dahliae* isolates and wilt epidemics of pepper. *Plant Disease* 87:789–797.

Bletsos, F. A., Thanassoulopoulos, C. C., and Roupakias, D. G. 1999. Water stress and Verticillium wilt severity on eggplant (*Solanum melongena* L.). *Journal of Phytopathology* 147:243–248.

Evans, G. and McKeen, C. D. 1975. A strain of *Verticillium dahliae* pathogenic to sweet pepper in southwestern Ontario. *Canadian Journal of Plant Science* 55:857–859.

Nagao, H., Arai, H., Oshima, S., Koike, M., and Iijima, T. 1998. Vegetative compatibility of an isolate of *Verticillium dahliae* pathogenic to both tomato and pepper. *Mycoscience* 39:37–42.

Riley, M. K. and Bosland, P. W. 1998. Commercial planting media effective in screening for Verticillium wilt of *Capsicum annuum*. *HortScience* 33:285–286.

Sorensen, L. H., Schneider, A. T., and Davis, J. R. 1991. Influence of sodium polygalacturonate sources and improved recovery of Verticillium species from soil. (Abstract) *Phytopathology* 81:1347.

253 Vascular discoloration of pepper caused by Verticillium wilt.

254 Advanced symptoms and plant collapse of pepper caused by Verticillium wilt.

Beet curly top virus

BEET CURLY TOP

Introduction and significance

Pepper is susceptible to a large number of virus pathogens. Over 70 such agents have been documented to some degree, and other virus-like diseases have yet to be fully characterized. Many of these diseases are economically important worldwide, while certain others are only significant in particular pepper producing areas. *Beet curly top virus* (BCTV) occurs in North and South America, Asia, the Middle East, and the Mediterranean region. It is an important pathogen of many crops such as beet, tomato, and pepper.

Symptoms and diagnostic features

BCTV can result in severely stunted, chlorotic plants if peppers are infected early in development. Such plants have dull to bright yellow leaves that are brittle, thickened, and rolled (255, 256). Plants that are infected later show some degree of reduced growth, chlorosis, and rolled brittle leaves. Fruit set can be reduced.

Causal agent and disease cycle

BCTV is a geminivirus with isometric particles that measure 18–22 nm in diameter and which occur singly or in pairs. The BCTV genome is a single stranded circular DNA. BCTV is vectored by either the beet leafhopper (*Circulifer tenellus*) in the United States or *C. opacipennis* in the Mediterranean region. In the plant, BCTV is restricted to the phloem tissue. This virus has many plant hosts. Curly top as a disease may actually be caused by one of four different curly top virus species: *Beet curly top virus* (BCTV), *Beet mild curly top virus* (BMCTV), *Beet severe curly top virus* (BSCTV), and *Spinach curly top virus* (SCTV). Research is on-going to further determine the relationships of these various viruses. On a molecular level, researchers have compared strains of BCTV from North America and the Middle East and found them to be similar, providing evidence that these various BCTV strains share a common origin.

Control

Follow general suggestions for managing virus diseases (see Part 1).

255 Leaf rolling symptom of *Beet curly top virus* of pepper.

256 Stunted and chlorotic pepper infected with *Beet curly top virus*.

References

Abdalla, O. A., Desjardins, P. R., and Dodds, J. A. 1991. Identification, disease incidence, and distribution of viruses infecting peppers in California. *Plant Disease* 75:1019–1023.

Briddon, R. W., Stenger, D. C., Bedford, I. D., Stanley, J., Izadpanah, K., and Markham, P. G. 1998. Comparison of a beet curly top virus isolate originating from the old world with those from the new world. *European Journal of Plant Pathology*104:77–84.

Soto, M. J. and Gilbertson, R. L. 2003. Distribution and rate of movement of the curtovirus Beet mild curly top virus (family *Geminiviridae*) in the beet leafhopper. *Phytopathology* 93:478–484.

Stenger, D. C., Duffus, J. E., and Villalon, B. 1990. Biological and genomic properties of a geminivirus isolated from pepper. *Phytopathology* 80:704–709.

Stenger, D. C. and McMahon, C. L. 1997. Genotypic diversity of beet curly top virus populations in the western United States. *Phytopathology* 87:737–744

Wang, H., Gurusinghe, P. de A., and Falk, B. W. 1999. Systemic insecticides and plant age affect beet curly top virus transmission to selected host plants. *Plant Disease* 83:351–355.

Cucumber mosaic virus

CUCUMBER MOSAIC

Introduction and significance

Cucumber mosaic virus (CMV) is commonly found throughout the world and can cause disease on perhaps over 1,000 plant hosts, including pepper and many other vegetable crops.

Symptoms and diagnostic features

CMV symptoms on pepper are extremely variable and can range from mild mosaics, diffuse chlorotic lesions, necrotic specks, ring spots, and many other manifestations (**257, 258, 259**). Infected plants may not produce as many fruit, or fruit that do develop are small or have ring spot or other markings.

Causal agent and disease cycle

CMV is a cucumovirus with virions that are isometric (28–29 nm in diameter) and contain three single stranded RNAs. A number of CMV strains have been documented. CMV is transmitted by many aphid vectors, and in pepper the virus can be seedborne.

Control

Follow general suggestions for managing virus diseases (see Part 1). CMV is seedborne in pepper. Use seed that does not have significant levels of the pathogen.

References

Abdalla, O. A., Desjardins, P. R., and Dodds, J. A. 1991. Identification, disease incidence, and distribution of viruses infecting peppers in California. *Plant Disease* 75:1019–1023.

Gallitelli, D. 2000. The ecology of cucumber mosaic virus and sustainable agriculture. *Virus Research* 71:9–21.

Rodriguez-Alvarado, G., Kurath, G., and Dodds, J. A. 1995. Heterogeneity in pepper isolates of cucumber mosaic virus. *Plant Disease* 79:450–455.

257 **Necrotic growing point of pepper caused by *Cucumber mosaic virus*.**

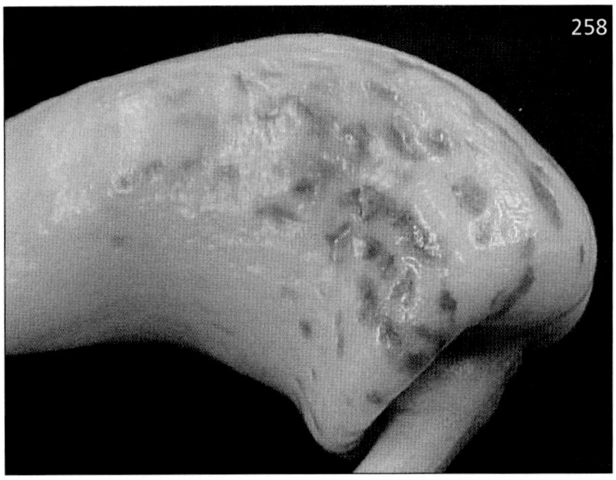

258 **Necrotic spotting on pepper fruit infected with *Cucumber mosaic virus*.**

259 **Necrotic veins of pepper leaf caused by *Cucumber mosaic virus*.**

Pepper mild mottle virus

PEPPER MILD MOTTLE

Introduction and significance

Pepper mild mottle virus (PMMV) is a serious pathogen worldwide.

Symptoms and diagnostic features

Foliar symptoms are usually slight and consist of mild mottles, mosaics, or chlorosis (**260**). Fruit, however, can be severely malformed, small, and have necrotic and off-colored skin (**261**).

Causal agent and disease cycle

PMMV is a tobamovirus with virions that are rod shaped (312 x 18 nm) and contain an RNA genome. Four PMMV strains have been documented. This virus does not have a known vector, but is easily transmitted mechanically and is seedborne.

Control

Follow general suggestions for managing virus diseases (see Part 1). PMMV is seedborne in pepper. Use seed that does not have significant levels of the pathogen.

References

Alonso, E., Garcia-Luque, I., Avila-Rincon, M. J., Wicke, B., Serra, M. T., and Diaz-Ruiz, J. R. 1989. A tobamovirus causing heavy losses in protected pepper crops in Spain. *Journal of Phytopathology* 125:67–76.

Alonso, E., Garcia-Luque, I., de la Cruz, A., Wicke, B., Avila-Rincon, M. J., Serra, M. T., Castresana, C., and Diaz-Ruiz, J. R. 1991. Nucleotide sequence of the genomic RNA of pepper mild mottle virus, a resistance-breaking tobamovirus in pepper. *Journal of General Virology* 72:2875–2884.

Wetter, C. 1984. Serological identification of four tobamoviruses infecting pepper. *Plant Disease* 68:597–599.

260

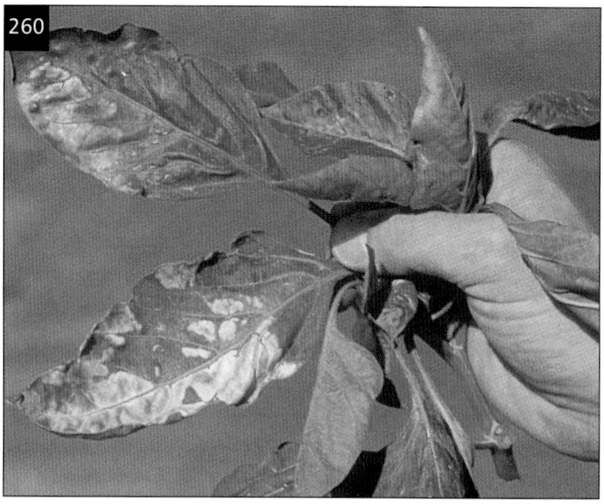

260 **Chlorotic leaf symptoms on pepper infected with *Pepper mild mottle virus*.**

261

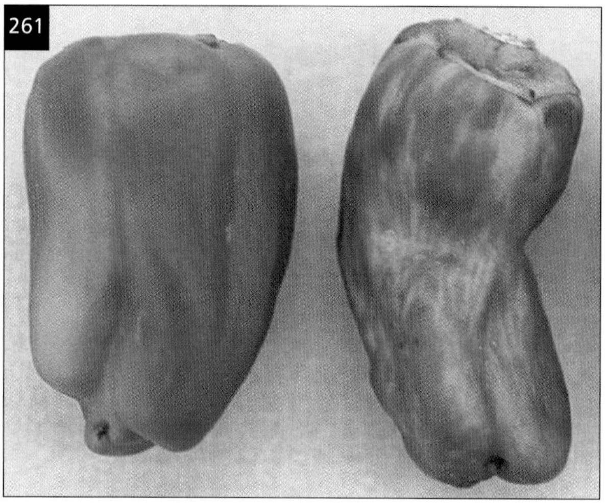

261 **Chlorotic blotches on pepper fruit infected with *Pepper mild mottle virus*.**

Potato virus Y

POTATO VIRUS Y

Introduction and significance

Potato virus Y (PVY) is a pathogen of solanaceous plants around the world.

Symptoms and diagnostic features

Symptoms can vary greatly but mostly consists of vein clearing that later turns into dark green vein banding. Mottle and mosaic patterns can also occur. Necrosis can develop on petioles and stems. Fruit symptoms, when present, appear as mosaic patterns and necrotic spotting on fruit skin.

Causal agent and disease cycle

PVY is a potyvirus with long (730 x 11 nm) flexuous rods that contain single stranded RNA. PVY is vectored by aphids and is also readily transmitted mechanically.

Control

Follow general suggestions for managing virus diseases (see Part 1).

References

Llave, C., Martínez, B., Díaz-Ruíz, J. R., and López-Abella, D. 1999. Serological analysis and coat protein sequence determination of *potato virus Y* (PVY) pepper pathotypes and differentiation from other PVY strains. *European Journal of Plant Pathology* 105:847–857.

Sinaloa tomato leaf curl virus

SINALOA TOMATO LEAF CURL

Introduction and significance

This virus is an important pathogen in North and Central America. This pathogen appears to infect only plants in the Solanaceae and Malvaceae plant families.

Symptoms and diagnostic features

Sinaloa tomato leaf curl virus (STLCV) causes interveinal chlorosis, mosaic patterns, and curling of leaves. Plants can be stunted and form fruit that are small and have necrotic tissue on the blossom end.

Causal agent and disease cycle

STLCV is a begomovirus with isometric particles that contain a DNA genome. STLCV is vectored by the *Bemisia tabaci* whitefly.

Control

Follow general suggestions for managing virus diseases (see Part 1).

References

Brown, J. K., Idris, A. M., and Fletcher, D. C. 1993. *Sinaloa tomato leaf curl virus*, a newly discovered geminivirus of tomato and pepper in west coast Mexico. *Plant Disease* 77:1262.

Idris, A. M., and Brown, J. K. 1998. Sinaloa tomato leaf curl geminivirus: biological and molecular evidence for a new subgroup III virus. *Phytopathology* 88:648–657.

Tobacco etch virus

TOBACCO ETCH

Introduction and significance

Tobacco etch virus (TEV) can be a serious virus pathogen in parts of North and South America.

Symptoms and diagnostic features

Leaf mosaics, mottles, and distortions are typical symptoms. Plants can be stunted and produce fruit with severe mosaics and deformities.

Causal agent and disease cycle

TEV is a potyvirus with long (730 x 12 nm) flexuous rods that contain single stranded RNA. TEV is vectored by aphids. Several TEV strains have been documented.

Control

Follow general suggestions for managing virus diseases (see Part 1).

References

Ariyaratne, I., Hobbs, H. A., Valverde, R. A., Black, L. L., and Dufresne, D. J. 1996. Resistance of *Capsicum* spp. genotypes to tobacco etch potyvirus isolates from the Western hemisphere. *Plant Disease* 80:1257–1261.

Tobacco mosaic virus, Tomato mosaic virus

TOBACCO MOSAIC, TOMATO MOSAIC

Introduction and significance

Tobacco mosaic (TMV) and *Tomato mosaic* (ToMV) *viruses* are two closely related virus pathogens. Both viruses can infect pepper and other solanaceous plants.

Symptoms and diagnostic features

TMV and ToMV both cause light and dark green mosaics and distortions of leaves. Necrotic flecks and spots can also appear on leaves. Fruit can be small, deformed, and show chlorosis and necrosis in the skin.

Causal agents and disease cycle

TMV and ToMV are viruses in the tobamovirus group. These viruses have straight rod particles that measure 300 x 18 nm and contain single stranded, linear RNA genomes. TMV has no known vector, is readily transmitted mechanically, and is possibly seedborne in pepper. ToMV also lacks a known vector, is transmitted mechanically, and can be seedborne in pepper and tomato.

Control

Follow general suggestions for managing virus diseases (see Part 1). Both ToMV and TMV are seedborne in pepper. Use seed that does not have significant levels of the pathogen.

References

Broadbent, L. 1976. Epidemiology and control of tomato mosaic virus. *Annual Review of Phytopathology* 14:75–96.

Tomato spotted wilt virus

TOMATO SPOTTED WILT

Introduction and significance

This virus can be very common on pepper and tomato crops, as well as on neighboring crops of other types. In some regions, such as California, greenhouse produced peppers can be especially hard hit by this disease.

Symptoms and diagnostic features

Tomato spotted wilt virus (TSWV) causes leaves to develop irregularly shaped to circular, black, small spots; chlorotic and necrotic ring spots can also occur. Stems and shoots may have black streaks or lesions on the epidermis (**262**). Severely affected plants may wilt or be stunted. Symptomatic fruit will develop chlorotic rings, patches, and lesions (**263**).

Causal agent

TSWV is a tospovirus and has isometric particles, measuring approximately 80–110 nm, which are surrounded with membranes. TSWV has a genome consisting of three linear single-stranded RNAs and is vectored by several species of thrips; at least eight species are found to be vectors. The western flower (*Frankliniella occidentalis*) and tobacco (*F. fusca*) thrips are probably the most important vectors.

Disease cycle

TSWV has one of the most extensive host ranges of any known plant virus and can infect over 900 different cultivated and weedy plant species. Therefore initial inoculum can come from any number of landscape plants, weeds, and other plants. Thrips vector the virus to the pepper crops. It is well documented that thrips insects can only acquire the virus as larvae that feed on infected plants; after acquiring the virus, the insects carry the virus for the rest of their lives. The virus is not seedborne in pepper.

Control

Follow general suggestions for managing virus diseases (see Part 1). The broad host range of TSWV and the difficulty in controlling thrips makes this disease particularly difficult to manage. Some sweet and hot pepper cultivars have the Tsw gene that confers resistance to TSWV. In Italy, however, some resistance breaking (RB) strains have overcome this resistance.

262 **Stem and leaf lesions caused by *Tomato spotted wilt virus* on pepper.**

263 **Severe symptoms of *Tomato spotted wilt virus* on fruit of jalapeno pepper. Healthy fruit is on the right.**

References

Best, R. J. 1968. *Tomato spotted wilt virus. Advances in Virus Research* 13:66–146.

Cho, J. J., Mau, R. F. L., Gonsalves, D., and Mitchell, W. C. 1986. Reservoir weed hosts of *tomato spotted wilt virus. Plant Disease* 70:1014–1017.

Greenough, D. R., Black, L. L., and Bond, W. P. 1990. Aluminum-surfaced mulch: an approach to the control of *tomato spotted wilt virus* in solanaceous crops. *Plant Disease* 74:805–808.

Groves, R. L., Walgenbach, J. F., Moyer, J. W., and Kennedy, G. G. 2002. The role of weed hosts and tobacco thrips, *Frankliniella fusca*, in the epidemiology of *tomato spotted wilt virus. Plant Disease* 86:573–582.

Hobbs, H. A., Black, L. L., Story, R. N., Valverde, R. A., Bond, W. P., Gatti, J. M. Jr., Schaeffer, D. O., and Johnson, R. R. 1993. Transmission of *tomato spotted wilt virus* from pepper and three weed hosts by *Frankliniella fusca. Plant Disease* 77:797–799.

Maris, P. C., Joosten, N. N., Goldbach, R. W., and Peters, D. 2003. Restricted spread of *Tomato spotted wilt virus* in thrips-resistant pepper. *Phytopathology* 93:1223–1227.

Maris, P. C., Joosten, N. N., Goldbach, R. W., and Peters, D. 2004. Tomato spotted wilt virus infection improves host suitability for its vector *Frankliniella occidentalis. Phytopathology* 94:706–711.

Maris, P. C., Jootsen, N. N., Peters, D., and Goldbach, R. W. 2003. Thrips resistance in pepper and its consequences for the acquisition and inoculation of *Tomato spotted wilt virus* by the western flower thrips. *Phytopathology* 93:96–101.

Roggero, P., Masenga, V., and Tavella, L. 2002. Field isolates of *tomato spotted wilt virus* overcoming resistance in pepper and their spread to other hosts in Italy. *Plant Disease* 86:950–954.

Cucurbitaceae Gourd family

The Cucurbitaceae (gourd family) includes a wide variety of vegetable crops that are commercially grown primarily for their edible fruits and sometimes immature blossoms. Cultivated cucurbits are classified into two groups, the Cucurbiteae and Sicyoideae. Notable commercial species include the following: cucumber (*Cucumis sativus*); muskmelon and other melons (*Cucumis melo*); cantaloupe (*C. melo* var. *cantalupensis*); watermelon (*Citrullus lanatus*); summer and winter squashes, gourds, or courgette (*Cucurbita pepo*); vegetable marrow (*C. pepo* var. *medullosa*); pumpkin (*Cucurbita maxima* and *C. pepo*); Chinese winter melon (*Benincasa hispida*); bitter melon (*Momordica cochinchinensis*). It is believed that many of these species originated from tropical regions in Africa, the Americas, and Asia.

Acidovorax avenae subsp. *citrulli*

BACTERIAL FRUIT BLOTCH

Introduction and significance

Compared to most cucurbit diseases, bacterial fruit blotch is a recently described problem, being first observed and characterized in 1988–1989. Watermelon is the primary host. However, honeydew and musk melons, cantaloupe, pumpkin, citron, and squash are also susceptible.

Symptoms and diagnostic features

Symptoms on leaves are not particularly striking and may resemble other diseases. Leaf spots are small, water-soaked to brown, irregularly shaped, and with angular edges. Such spots may sometimes be surrounded by chlorotic borders (**264**, **265**). The pathogen is seedborne, resulting in seedling infections where the cotyledons develop irregularly shaped water-soaked lesions (**266**). Lesions can expand and run along the cotyledon midrib and become brown to red-brown and necrotic. Hypocotyls on young seedlings can develop lesions and subsequently collapse and die.

The most important impact of this disease is on watermelon fruits. The upper portion develops small (less than 1 cm) irregularly shaped lesions that rapidly expand into large blotches that can cover most or all of the fruit (**267**). Early in development, blotches appear either water-soaked, dull gray-green, or dark green in color. In time the older, central area of lesions can turn brown to red-brown and necrotic, with the epidermis cracking and an amber-colored exudate oozing out of

264 **Watermelon leaf showing leaf spots caused by *Acidovorax avenae* subsp. *citrulli*.**

265 **Watermelon leaf showing necrotic lesion caused by *Acidovorax avenae* subsp. *citrulli*.**

266 Watermelon cotyledons infected with the bacterial fruit blotch pathogen.

267 Fruit lesions of bacterial fruit blotch on watermelon.

the central blotch area. Blotches are shallow infections that usually do not penetrate into the watermelon flesh. Secondary decay organisms that invade fruit blotch lesions are mostly responsible for fruit breakdown, rot, and collapse. Triploid (seedless) cultivars of watermelon may be less susceptible.

Causal agent
Bacterial fruit blotch is caused by *Acidovorax avenae* subsp. *citrulli* (formerly named *Pseudomonas pseudoalcaligenes* subsp. *citrulli*). This bacterium is a Gram-negative rod having a single polar flagellum, is nonfluorescent in culture, and forms white colonies. When inoculated into tobacco leaves, this pathogen generally causes a hypersensitive reaction. However, some strains are not able to cause the tobacco hypersensitivity reaction, and infect watermelon seedlings but not fruit.

Disease cycle
The pathogen is seedborne in many cucurbit species and can therefore be present when seedlings germinate in the field or in transplant greenhouses. Warm temperatures and high humidity favor disease development. Splashing water from rain and sprinkler irrigation spread the bacteria from plant to plant. Environmental conditions inside transplant greenhouses are particularly favorable for this disease. The bacterium infects immature fruit through stomata, and young fruit apparently are most susceptible to infection. It appears that the bacterium can infect fruit pulp and seeds if it enters through blossoms early in fruit development. The pathogen survives in infested crop residues and fruit rinds, and on diseased volunteer and weed cucurbits. Apparently this disease does not readily spread in storage from diseased to healthy watermelon fruit if proper postharvest conditions are maintained.

Control
Use seed that does not have significant levels of the pathogen. Seed treatments such as bleach, other bactericides, and fermentation plus seed-drying procedures may be helpful in cleaning up infested seed. Remove cucurbit weeds and volunteer reservoir hosts. Complete resistance is not yet available for watermelon. Rotate out of cucurbit crops for at least 3 years. Reduce or eliminate the use of sprinkler irrigation. For transplant greenhouses, practice good sanitation, reduce humidity levels, reduce sprinkler irrigation, and remove symptomatic plants and surrounding trays. Accept and plant transplants that do not show symptoms of the disease. Copper sprays provide some control.

References

Frankle, W. G., Hopkins, D. L., and Stall, R. E. 1993. Ingress of watermelon fruit blotch bacterium into fruit. *Plant Disease* 77:1090–1092.

Hopkins, D. L. and Thompson, C. M. 2002. Seed transmission of *Acidovorax avenae* subsp. *citrulli* in cucurbits. *HortScience* 37:924–926.

Isakeit, T., Black, M. C., Barnes, L. W., and Jones, J. B. 1997. First report of infection of honeydew with *Acidovorax avenae* subsp. *citrulli*. *Plant Disease* 81:694.

Latin, R. X. and Hopkins, D. L. 1995. Bacterial fruit blotch of watermelon: the hypothetical exam question becomes reality. *Plant Disease* 79:761–765.

Latin, R., Tikhonova, I., and Rane, K. 1995. Factors affecting the survival and spread of *Acidovorax avenae* subsp. *citrulli* in watermelon transplant production facilities. *Phytopathology* 85:1413–1417.

Lovic, B. R. and Hopkins, D. L. 2003. Production steps to reduce seed contamination by pathogens of cucurbits. *HortTechnology* 13:50–54.

Rane, K. K. and Latin, R. X. 1992. Bacterial fruit blotch of watermelon: association of the pathogen with seed. *Plant Disease* 76:509–512.

Rushing, J. W., Keinath, A. P., and Cook, W. P. 1999. Postharvest development and transmission of watermelon fruit blotch. *HortTechnology* 9:217–219.

Somodi, G. C., Jones, J. B., Hopkins, D. L., Stall, R. E., Kucharek, T. A., Hodge, N. C., and Watterson, J. C. 1991. Occurrence of a bacterial watermelon fruit blotch in Florida. *Plant Disease* 75:1053–1056.

Sowell, G., Jr. and Schaad, N. W. 1979. *Pseudomonas pseudoalcaligenes* subsp. *citrulli* on watermelon: seed transmission and resistance of plant introductions. *Plant Disease Reporter* 63:437–441.

Walcott, R. R., Gitaitis, R. D., and Castro, A. C. 2003. Role of blossoms in watermelon seed infestation by *Acidovorax avenae* subsp. *citrulli*. *Phytopathology* 93:528–534.

Willems, A., Goor, M., Thielemans, S., Gillis, M., Kersters, K., and De Ley, J. 1992. Transfer of several phytopathogenic *Pseudomonas* species to *Acidovorax* as *Acidovorax avenae* subsp. *avenae* subsp. *nov*., comb. *nov*., *Acidovorax avenae* subsp. *citrulli*, *Acidovorax avenae* subsp. *cattleyae*, and *Acidovorax konjaci*. *International Journal of Systematic Bacteriology* 42:107–119.

Erwinia tracheiphila

BACTERIAL WILT

Introduction and significance

Bacterial wilt is a serious disease of cucurbits such as cucumbers and muskmelons, while it is less of a concern on various other cucurbit crops. The disease is more important in North America than in Europe.

Symptoms and diagnostic features

Initial symptoms consist of the wilting of a few to several leaves along a stem. As disease develops, the foliage will exhibit extensive wilting. The wilting can occur on individual runners or stems, or throughout an entire plant's foliage. Wilting foliage takes on a dark green color, but later can become chlorotic and then necrotic. In advanced stages of the disease the entire plant will collapse and die (**268**). Sticky strands of bacterial ooze can be observed when the cut ends of symptomatic stems are slowly drawn apart (**269**). Squash fruit can be infected and show small, irregular water-soaked areas on the fruit surface.

268

268 Collapsing oriental melon vines caused by bacterial wilt.

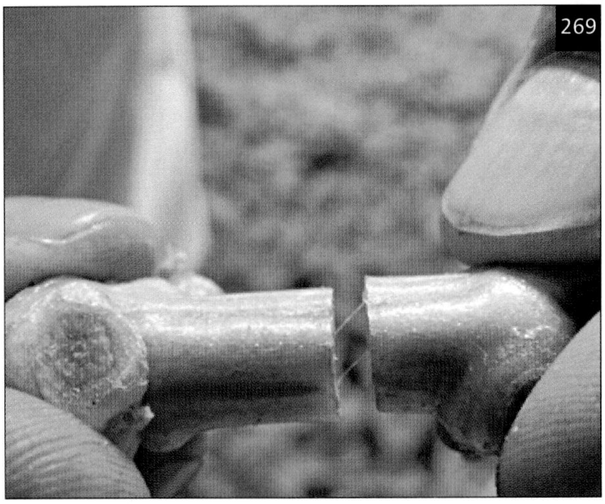

269 Bacterial strands extending from cut ends of melon stem infected with bacterial wilt.

Causal agent

The cause of bacterial wilt is the Gram-negative bacterium *Erwinia tracheiphila*, which is classified in the *Erwinia amylovora* taxonomic group. This pathogen differs from most *Erwinia* species in that *E. tracheiphila* does not liquefy gelatin, shows weak action in milk, does not reduce nitrates, and does not metabolize inorganic nitrogen sources. This pathogen is host specific to the Cucurbitaceae. *Erwinia tracheiphila* is mechanically transmitted by a number of insects, primarily the striped cucumber beetle (*Acalymma vittatum*) and the spotted cucumber beetle (*Diabrotica undecimpunctata*). Insects that are in contact with and feed on diseased plant tissue become contaminated with the pathogen. These insects move to other plants and inoculate them when their infested mouthparts feed on the plants. The bacteria enter the xylem via these feeding wounds and become systemic within the plant. Wilting occurs when the multiplying bacteria and resulting resins plug the vascular tissue of the host plant. This disease develops most rapidly in young, succulent cucurbit plants.

Disease cycle

The complete epidemiology of bacterial wilt has not yet been determined. The pathogen is apparently not seedborne, nor does it survive in soil beyond 2 or 3 months. In contrast to earlier studies, it now appears that the pathogen does not persist within the insect vector's intestinal tract. The initial inoculum of bacterial wilt, therefore, is not known. Infected weed or volunteer cucurbit hosts may be the source of inoculum from which insects become infested. The pathogen may survive in dried plant tissues for a limited time.

Control

Plant resistant cultivars as they become available. Control the vector insects by applying systemic insecticides to the soil or contact insecticides to host plants.

References

Leach, J. G. 1964. Observations on cucumber beetles as vectors of curcurbit wilt. *Phytopathology* 54:606–607.

Main, C. E. and Walker, J. C. 1971. Physiological responses of susceptible and resistant cucumber to *Erwinia tracheiphila*. *Phytopathology* 61:518–522.

Pseudomonas syringae pv. *lachrymans*

ANGULAR LEAF SPOT

Introduction and significance

Angular leaf spot occurs on cucurbits throughout the world and is particularly serious in more humid, warmer regions. The disease can be severe on cucumber grown outside in fields, but is less commonly found on cucurbits grown inside greenhouses.

Symptoms and diagnostic features

Initial symptoms occur on leaves and consist of small, irregularly shaped, water-soaked to gray-colored lesions. As these lesions expand, they become angular in shape as the lesion edges become delimited by leaf veins (**270**). On some cultivars the lesions can be surrounded by chlorotic borders. As lesions age they turn tan to gray and become dry. The dried tissue often tears and falls out, resulting in a 'shot hole' or tattered appearance (**271**). If conditions are wet or humid, bacterial exudates can ooze onto lesion surfaces and dry into a white residue. Lesions and bacterial ooze can also develop on petioles and stems.

Early symptoms of fruit infections consist of typically oval to circular lesions that are small (1 to 5 mm in diameter) and water-soaked. The water-soaked appearance can later turn into a brown discoloration of fruit tissues. Exudates sometimes collect on these fruit spots. Lesions can later develop into deep internal rots; such infections, along with the activity of secondary decay organisms, result in unmarketable fruit.

Causal agent

The pathogen is the aerobic, Gram-negative bacterium *Pseudomonas syringae* pv. *lachrymans*, strains of which are host specific to cucurbit plants. When cultured on Kings medium B, this organism produces a diffusible pigment that fluoresces blue under ultraviolet light.

Other *Pseudomonas* pathogens affect the cucurbit group of plants, though in many reports it is not clear exactly which species and pathovars are involved. Recently a *Pseudomonas* disease has been reported on cantaloupe in Europe. The pathogen is a pathovar of *P. syringae* and causes necrotic leaf spots with water-soaked margins, stem and petiole cankers, and sunken fruit lesions. Fruit infection can cause immature fruit to drop and more mature fruit to develop dry rot cavities. The bacterium is also pathogenic to sugar beet and therefore may be *P. syringae* pv. *aptata*. Another bacterium, *P. syringae* pv. *syringae*, is reported to also cause disease on cantaloupe and squash in the USA.

Disease cycle

This pathogen is internally seedborne, which results in infection of the seedling cotyledons upon germination. Bacteria are splashed from plant to plant by splashing water from rain and sprinkler irrigation. Insects and physical contact from passing equipment and harvesters can also spread the pathogen. Warm, humid conditions favor disease development. Optimum temperatures for disease development are 24–28° C, though the pathogen can survive at higher (36° C) temperatures. The pathogen can also survive in the soil on infested plant residues for perhaps up to 2 years.

270 Leaf lesions of angular leaf spot on cucumber.

271 Advanced symptoms of angular leaf spot on cucumber.

Control

Use seed that does not have significant levels of the pathogen. Seed treatments for this disease usually consist of hot water that has been acidified, or heated zinc or manganese sulfate solutions. Rotate crops so that seed and fruit production plantings are placed in fields that have not had cucurbits for 2 years. Do not use overhead sprinkler irrigation. Conduct production and harvesting procedures only when the foliage is dry. Use resistant cultivars if they are available. The application of copper sprays may also be useful. Do not over fertilize the crop with high nitrogen fertilizers, as excessive nitrogen may increase disease severity, especially on younger leaves.

References

Grogan, R. G., Lucas, L. T., and Kimble, K. A. 1971. Angular leaf spot of cucumber in California. *Plant Disease Reporter* 55:3–6.

Hopkins, D. L. and Schenck, N. C. 1972. Bacterial leaf spot of watermelon caused by *Pseudomonas lachrymans*. *Phytopathology* 62:542–545.

Kritzman, G. and Zutra, D. 1983. Survival of *Pseudomonas syringae* pv. *lachrymans* in soil plant debris, and the rhizosphere of nonhost plants. *Phytoparasitica* 11:99–108.

Langston, D. B., Jr., Sanders, F. H., Brock, J. H., Gitaitis, R. D., Flanders, J. T., and Beard, G. H. 2003. First report of a field outbreak of a bacterial leaf spot of cantaloupe and squash caused by *Pseudomonas syringae* pv. *syringae* in Georgia. *Plant Disease* 87:600.

Morris, C. E., Glaux, C., Latour, X., Gardan, L., Samson, R., and Pitrat, M. 2000. The relationship of host range, physiology, and genotype to virulence on cantaloupe in *Pseudomonas syringae* from cantaloupe blight epidemics in France. *Phytopathology* 90:636–646.

Serratia marcescens

CUCURBIT YELLOW VINE DISEASE

Introduction and significance

A relatively newly discovered disease, cucurbit yellow vine disease (CYVD) was first seen on squash and pumpkin plantings in the USA (Oklahoma, Texas) in 1988. The disease has since been documented in other states in the southeastern USA as well as in Massachusetts. Levels of damage vary, but CYVD can cause extensive crop loss to cantaloupe, pumpkin, squash, and watermelon.

Symptoms and diagnostic features

Initial symptoms usually occur 10 to 15 days before fruit maturity. Foliage rapidly turns a lime green, then bright yellow. Plants show stunted growth and eventual decline. Older leaves can be blighted or burned in appearance, and younger leaves on vine tips can curl and not properly expand. A characteristic symptom is the gold to honey-brown-colored phloem in the main roots and crowns of diseased plants. In advanced stages of CYVD, the root systems will rot; however, this is thought to be due to secondary decay organisms in the soil. In Oklahoma, watermelon and cantaloupe fields that are planted before 15 June show the greatest disease severity.

In some cases, at the time of flowering and fruit set, plants will not turn chlorotic but instead will wilt and collapse in a day or two. Fruit appear normal in shape, but in the case of watermelon the fruit lose their green color and become unmarketable. Because this bacterium is vectored by an insect, symptoms often appear in aggregated small patches or along the edges of fields; such patterns are related to the movement of the insect.

Causal agent

The pathogen is the aerobic, rod shaped, Gram-negative bacterium *Serratia marcescens* that inhabits the phloem tissue of cucurbit hosts. The pathogen can be isolated by grinding phloem tissue in buffer and then streaking the resulting suspension onto potato dextrose agar, nutrient agar, or purple agar. The bacterium can be maintained on nutrient agar. This bacterium is vectored by the squash bug (*Anasa tristis*).

Serratia marcescens is interesting in that strains of this bacterium are found in a number of diverse environments. On plants, *S. marcescens* is a non-pathogenic endophyte in cotton and rice, a crown rot pathogen of alfalfa and sainfoin (*Onobrychis vicifolia*), and now an important cucurbit pathogen. Other strains cause disease in insects, are opportunistic pathogens on humans, and inhabit water and soil ecosystems. Unlike many other *S. marcescens* strains, cucurbit strains do not produce the red prodigiosin pigment. While genetic analysis shows that the cucurbit strains are *S. marcescens*, other experiments show that the cucurbit strains form a distinct group and differ from other non-CYVD strains in biochemical and physiological aspects.

The host range of this cucurbit pathogen has not yet been determined. Researchers found, however, that the *S. marcescens* endophyte strains from rice were not pathogenic to cucurbits.

Disease cycle

The disease epidemiology of CYVD has not been determined. There is no evidence that the pathogen is seedborne. It appears that the squash bug vector is the most important factor in pathogen survival and distribution. Overwintering populations of this insect can transmit the pathogen to new crops in the spring.

Control

Manage the squash bug vector by spraying insecticides, plowing down old cucurbit plantings, and removing vegetation habitats that might help the insect overwinter.

References

Avila, F. J., Bruton, B. D., Fletcher, J., Sherwood, J. L., Pair, S. D., and Melcher, U. 1998. Polymerase chain reaction detection and phylogenetic characterization of an agent associated with yellow vine disease of cucurbits. *Phytopathology* 88:428–436.

Bruton, B. D., Fletcher, J., Pair, S. D., Shaw, M., and Sittertz-Bhatkar, H. 1998. Association of a phloem-limited bacterium with yellow vine disease in cucurbits. *Plant Disease* 82:512–520.

Bruton, B. D., Mitchell, F., Fletcher, J., Pair, S. D., Wayadande, A., Melcher, U., Brady, J., Bextine, B., and Popham, T. W. 2003. *Serratia marcescens*, a phloem-colonizing, squash bug-transmitted bacrerium: causal agent of cucurbit yellow vine disease. *Plant Disease* 87:937–944.

Lukezic, F. L., Hildebrand, D. C., Schroth, M. N., and Shinde, P. A. 1982. Association of *Serratia marcescens* with crown rot of alfalfa in Pennsylvania. *Phytopathology* 72:714–718.

Pair, S. D., Bruton, B. D., Mitchell, F., Fletcher, J., Wayadande, A., and Melcher, U. 2004. Overwintering squash bugs harbor and transmit the causal agent of cucurbit yellow vine disease. *Journal of Economic Entomology* 97:74–78.

Rascoe, J., Berg, M., Melcher, U., Mitchell, F. L., Bruton, B. D., Pair, S. D., and Fletcher, J. 2003. Identification, phylogenetic analysis, and biological characterization of *Serratia marcescens* strains causing cucurbit yellow vine disease. *Phytopathology* 93:1233–1229.

Acremonium cucurbitacearum*, *Fusarium* spp., *Phytophthora* spp., *Pythium* spp., *Rhizoctonia solani*, *Rhizopycnis vagum

DAMPING-OFF, ROOT ROTS

Introduction and significance

A number of soilborne pathogens cause seed and seedling diseases of cucurbits. Pathogens include species of *Acremonium*, *Fusarium*, *Phytophthora*, *Pythium*, *Rhizoctonia*, and *Rhizopycnis*. Some of these pathogens are also implicated in the vine decline complex that affects mature plants.

Symptoms and diagnostic features

These diseases have various phases. Cucurbit seeds can be infected prior to germination and result in seed death. Newly germinated seedlings can be infected to such a degree that plants do not emerge above the soil (pre-emergence damping-off). Finally, seedlings might emerge from the ground but become diseased after soil emergence (post-emergence damping-off) (**272**). Initial symptoms of post-emergence damping-off generally consist of yellow or light tan to dark brown lesions of the root and hypocotyl tissues (**273**). With time the hypocotyl shrivels, roots decay further, and the cotyledons and leaves wilt and collapse. Some pathogens initially attack the tissue where the seed coat is in contact with the hypocotyl after germination. Damping-off diseases often result in death of the seedling and subsequent reduction of plant stands (**274**). However, even if plants do not succumb to these pathogens, the surviving plant may be stunted and delayed in development.

Older, mature plants can also be attacked by these seedling pathogens. The feeder roots of mature plants will discolor and turn a water-soaked, tan to brown color. The root tissue becomes soft and can slough away. Upper taproot and lower stem tissue in contact with the soil can become first gray-green, then brown to red-brown, soft in texture, and rotted. Fruit in contact with the soil develop irregularly shaped, water-soaked to brown lesions. These fungi, along with *Verticillium dahliae*, are implicated in the vine decline disease that is caused by a complex of fungi.

Causal agents

Acremonium cucurbitacearum produces conidiophores that are simple, awl-shaped, erect monophialides measuring 13–54 µm long. Conidia collect in slimy aggregations and are usually one-celled, hyaline, oblong to cylindrical, and measure 5–7.5 x 2–4 µm. This pathogen produces chlamydospores that allow it to survive in the soil. Optimum disease development occurs at 24–27° C, though infection takes place over a broader temperature range (12-30° C).

Fusarium equiseti and *F. solani* can cause seedling diseases of cucurbits. However, even Fusarium wilt pathogens (various forma speciales of *F. oxysporum*), may sometimes cause damping-off. *Fusarium* species have branched or unbranched conidiophores that have phialides as conidiogenous cells. Phialides produce either small, one- to two-celled microconidia or larger, fusoid, curved, multicelled macroconidia. Macroconidiophores grow in clustered groups called sporodochia. Resilient chlamydospores enable *Fusarium* species to survive for long periods of time in the soil.

Phytophthora and *Pythium* species belong in the oomycete group of organisms. *Phytophthora drechsleri* has been reported to cause seedling damping-off. A number of *Pythium* species cause damping-off: *P. aphanidermatum*, *P. irregulare*, *P. myriotylum*, *P. ultimum*. All these pathogens survive in the soil as saprophytes and are favored by wet soil conditions. With the exception of *P. ultimum*, these organisms usually produce zoospores that swim to and infect susceptible tissues. Sexual structures (antheridia, oogonia, and oospores) are produced by all species. In addition to cucurbits, these pathogens can infect numerous other plants.

Rhizoctonia solani is a soilborne fungus with a very broad host range. *R. solani* has no asexual fruiting structures or spores, but produces characteristically coarse, brown, right-angle branching hyphae. The hyphae are distinctly constricted at branch points, and cross walls with dolipore septa are deposited just after the branching. Hyphal cells are multi-nucleate. Small, tan to brown loosely aggregated clumps of mycelia function as sclerotia. This fungus can survive by infecting and thriving on a great number of plant hosts, besides cucurbits, and can also persist in the soil as a saprophyte. The teleomorph stage, *Thanatephorus cucumis*, is not commonly observed.

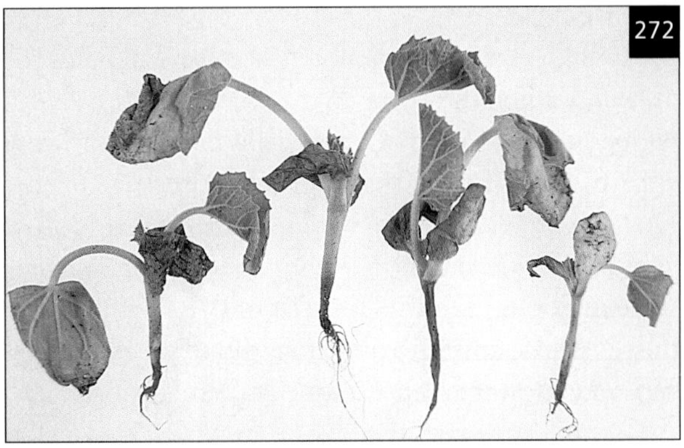

272 Muskmelon seedlings affected by *Pythium* damping-off.

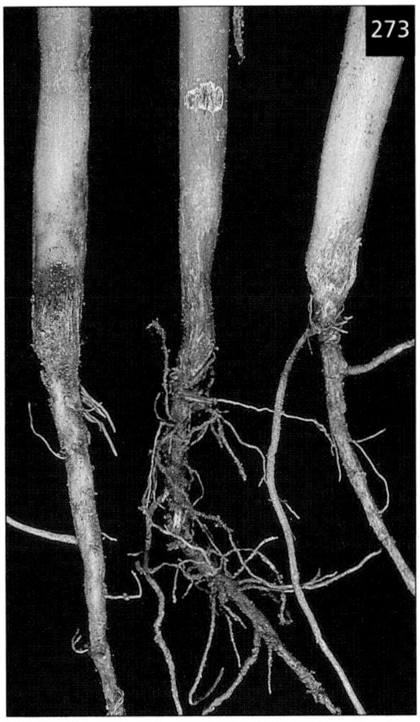

273 *Rhizoctonia* root rot on young watermelon seedlings.

274 Watermelon seedlings affected by *Pythium* damping-off.

Disease cycle

Most of these soilborne pathogens survive in fields for indefinite amounts of time. Wet soil conditions and cool temperatures (15–20° C) generally favor these fungi and their ability to grow and infect host cucurbits. However, for pathogens such as *Fusarium*, *Pythium aphanidermatum* and *Pythium myriotylum*, warmer soil conditions (32–37° C) are favorable. Seedlings are the most susceptible to infection, though these pathogens can infect the feeder roots of mature plants.

Control

Control damping-off by creating conditions unfavorable for the pathogens. For transplant production, practice good sanitation (keep growing areas and benches clean, use either new or thoroughly disinfected trays, do not reuse rooting media). Plant on raised beds and in well draining soils so that overly wet soil conditions are avoided. Carefully irrigate so that excess water is not applied. Plant fungicide-treated seed, and avoid planting seed too deeply, which delays seedling emergence and increases the chance of infection. Post-plant fungicides may provide control for some of these pathogens. Avoid planting too soon into fields that still have extensive crop residue in the soil. Some rotation crops can exacerbate the damping-off problems for cucurbits. If melons are planted after cotton, root rot caused by *F. equiseti* can be severe; melons following alfalfa might have increased *Pythium* problems.

References

Adams, G. C., Gubler, W. D., and Grogan, R. G. 1987. Seedling disease of muskmelon and mixed melons in California caused by *Fusarium equiseti*. *Plant Disease* 71:370–374.

Aegerter, B. J., Gordon, T. R., and Davis, R. M. 2000. Occurrence and pathogenicity of fungi associated with melon root rot and vine decline in California. *Plant Disease* 84:224–230.

Armengol, J., Vicent, A., Martínez-Culebras, P., Bruton, B. D., and García-Jiménez, J. 2003. Identification, occurrence and pathogenicity of *Rhizopycnis vagum* on muskmelon in Spain. *Plant Pathology* 52:68–73.

Garcia-Jimenez, J., Velacruz, M. T., Jorda, C., and Alfaro-Garcia, A. 1994. *Acremonium* species as the causal agent of muskmelon collapse in Spain. *Plant Disease* 78:416–419.

Cladosporium cucumerinum

SCAB

Introduction and significance

Scab can be an important foliar and fruit disease in some regions of the world, and losses can sometimes exceed 50%. The disease affects most cucurbit crops, although resistant cultivars have made scab less important for cucumber production.

Symptoms and diagnostic features

Initial symptoms on leaves consist of water-soaked, pale green lesions that are irregularly shaped. As the disease develops, lesions enlarge, turn gray to brown in color (**275**). Eventually the inner tissue of the lesion breaks and falls out, resulting in irregularly shaped holes and tears in the leaf (**276**). Leaf lesions may also have chlorotic borders around them. If newly expanded leaves are infected with numerous lesions, the leaves may twist and become deformed. Lesions on petioles and stems tend to be more elongated in shape but have the same initially water-soaked, then brown, color. Under humid conditions, the leaf, petiole, and stem lesions can be covered with the powdery, dark green fungal growth of the pathogen. Severe infection and subsequent profuse production of conidia can result in allergic reactions in sensitive individuals.

The greatest economic impact of scab is on the fruit (**277, 278**). Fruit infections begin as minute (2–4 mm), greasy looking, sunken specks on the fruit surface. These enlarge, become circular to oval in shape, gray in color, and remain sunken. Dark green sporulation may appear on the lesions, and such fruit lesions often ooze sticky exudates, hence the use of the additional name of 'gummosis' for this disease. In severe cases, the lesions will coalesce and form large sunken craters and cavities. Secondary decay organisms will colonize infected fruit tissue and contribute further to fruit rot.

Causal agent

Scab is caused by the fungus *Cladosporium cucumerinum*. Cultures appear olivaceous brown to greenish black at maturity. Conidia form in long, acropetal chains and are branched and olive-brown in color. Lower conidia are larger, have one or two septa, and are called ramoconidia. Upper conidia are smaller, one-celled, ellipsoidal to cylindrical to fusiform, and measure 4–9 x 3–5 µm. Another *Cladosporium* species,

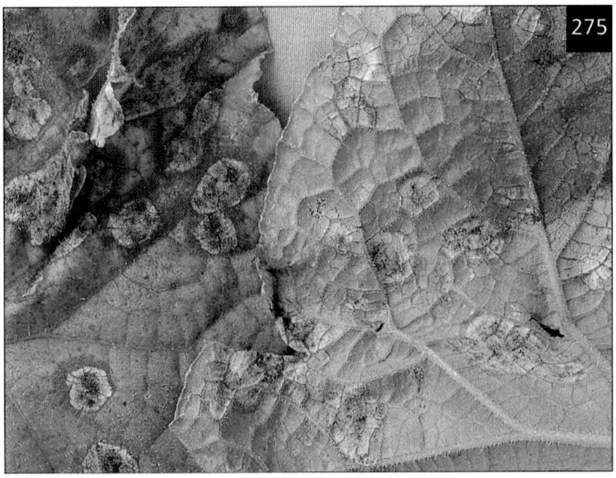

275 Leaf spots of scab on squash.

276 Tattered leaf symptoms of scab on squash.

C. tenuissimum, has been associated with small (3 mm in diameter), circular swellings on cucumber fruit grown in Israel.

Disease cycle

The scab pathogen can survive on old crop residues and other organic matter in the soil. Survival in some cases can reach up to 3 years. Conidia are dispersed long distances in the air, land on susceptible tissues, and germinate and infect the host if conditions are favorable. The pathogen requires 100% relative humidity or free moisture for infection. Optimum temperatures for the fungus are between 17–27° C, though growth can occur in the range of 5–30° C. The pathogen may be seedborne.

Control

Plant resistant cultivars if available. Rotate crops so that 2 or more years pass between susceptible cucurbit crops. Use seed that does not have significant levels of the pathogen. Avoid using sprinkler irrigation. Some fungicides may provide control if applied prior to fruit development.

References

Batta, Y. A. 2004. *Cladosporium tenuissimum* Cooke (Deuteromycotina: Hyphomycetes) as a causal organism of new disease on cucumber fruits. *European Journal of Plant Pathology* 110:1003–1009.

Crossan, D. F. and Sasser, J. M. 1969. Effect of rotation with corn on cucurbit scab. *Plant Disease Reporter* 53:452–453.

McKemy, J. M. and Morgan-Jones, G. 1992. Studies in the genus *Cladosporium sensu lato*. VII. Concerning *Cladosporium cucumerinum*, causal organism of crown blight and scab or gummosis of cucurbits. *Mycotaxon* 43:163–170.

277 Fruit lesions of scab on yellow squash.

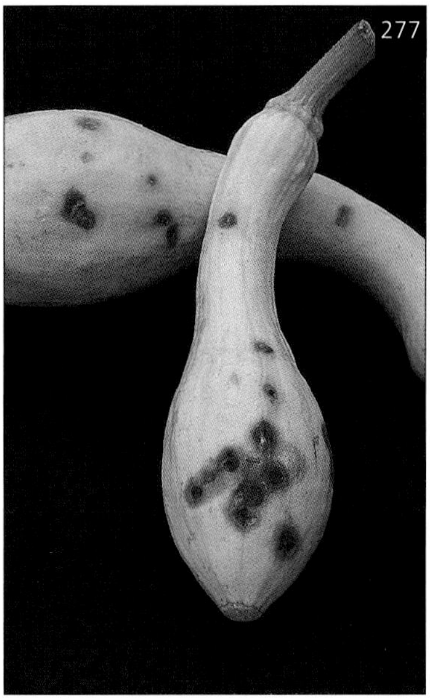

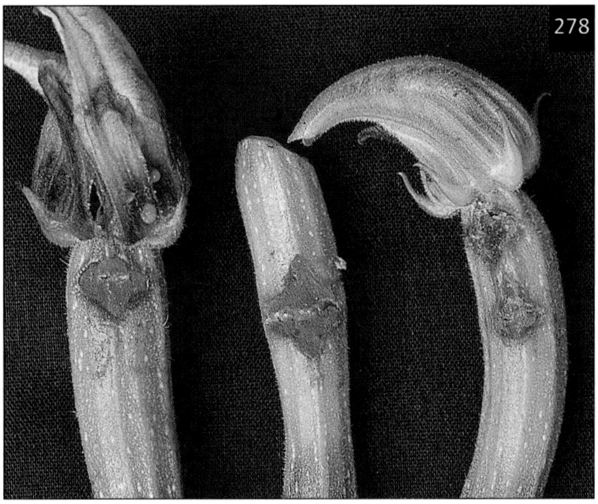

278 Fruit lesions of scab on immature zucchini squash.

Didymella bryoniae
(anamorph = *Phoma cucurbitacearum*)

GUMMY STEM BLIGHT, BLACK ROT

Introduction and significance

Gummy stem blight is one of the most important diseases of cucurbits and affects leaves, stems, and fruits of most of these species. If environmental conditions favor disease development, significant damage can occur to susceptible crops. Greenhouse-grown cucumbers are also affected.

Symptoms and diagnostic features

Early leaf symptoms consist of circular, water-soaked to gray green spots on cotyledons and true leaves. Such spots enlarge quickly, can become irregular in shape, and turn tan to dark brown. When spots coalesce, the entire leaf can become blighted. Leaf spots can be colonized with the small, dark fruiting bodies (pycnidia) of the fungus perfect stage (**279**). Stem infections consist of oblong water-soaked lesions that later turn brown to dark brown (**280**). As disease develops, such infections turn into cankers that can girdle the entire stem and result in wilting of the attached foliage and death of the seedling or plant. A brown, sticky exudate often oozes from stem cankers. Stem infections also support the development of pycnidia. On occasion, leaf and stem lesions contain the structures of the perfect stage (pseudothecia) of the pathogen. In greenhouses, cucumber plants are often first infected on the cut stubs left where fruit were harvested or shoots were pruned. Lesions on these cut stubs can expand and girdle the entire stem.

On fruit (**281**, **282**), the disease is called black rot and initially appears as water-soaked, dark green circular spots. Spots expand, darken, and have pycnidia and pseudothecia in the lesion centers. Depending on the crop species, the fruit rot may eventually take on a black color. If stored under humid conditions, fruit lesions may develop a white, cottony mycelial growth. Older fruit lesions can become leathery in texture and cracked. Infected, rotted fruit enables the pathogen to become seedborne. Black rot can be severe on Hubbard pumpkin and butternut, acorn, and buttercup squashes.

Causal agent

Gummy stem blight is caused by the fungus *Didymella bryoniae*. In older references the fungus is referred to as *Mycosphaerella melonis*. *Didymella bryoniae* forms dark, ostiolate pseudothecia that have bitunicate asci. Each ascus contains eight ascospores that are hyaline, oval, one-septate, and measure 14–18 x 4–7 µm. Ascospores have constrictions at the septa, and the upper cell is wider than the lower cell. The anamorph, *Phoma cucurbitacearum*, is usually more commonly found in the diseased tissue. Pycnidia of *P. cucurbitacearum* produce conidia that are hyaline, cylindrical, are often one-septate, and measure 6–13 x 2–4 µm. Researchers have found that some isolates are more virulent than others, showing that there is some diversity within pathogen populations.

Disease cycle

The fungus can survive in the soil and on crop residue. Studies indicate that the pathogen can persist in fields for 2 years without a host. Rains and winds blow ascospores or conidia from inoculum sources onto new crops. The fungus is most damaging with warm, moist weather, and optimum conditions for infection include free moisture and temperatures in the 20–24° C range. The pathogen may be seedborne, resulting in diseased seedlings and transplants. Fruit can rot 2–3 days after first becoming infected.

Control

To reduce soilborne inoculum, rotate crops on a 2-year cycle without cucurbit hosts. Apply protectant fungicides; such chemicals are especially needed to protect wounds that function as infection sites. Note that there are reports of the apparent failure of a strobilurin fungicide to control the disease, indicating that resistant isolates are present. In the USA, a disease forecasting program (Melcast) has been used to schedule fungicide applications for controlling gummy stem blight and anthracnose of melons.

Use seed that does not have significant levels of the pathogen. Carefully examine transplants and remove diseased plants. Greenhouse production of transplants or crops should include increased ventilation and reduced sprinkler irrigation. Avoid injuring fruit before or during harvest, as wounds enable the pathogen to enter fruit during storage.

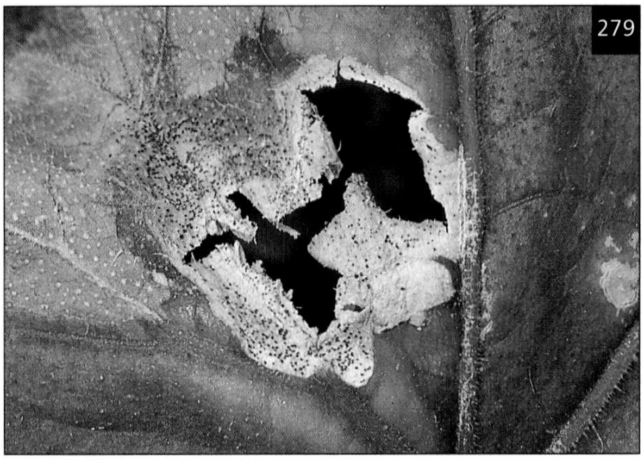

279 **Leaf spot of gummy stem blight on cucumber.**

280 **Stem lesion of gummy stem blight on cucumber.**

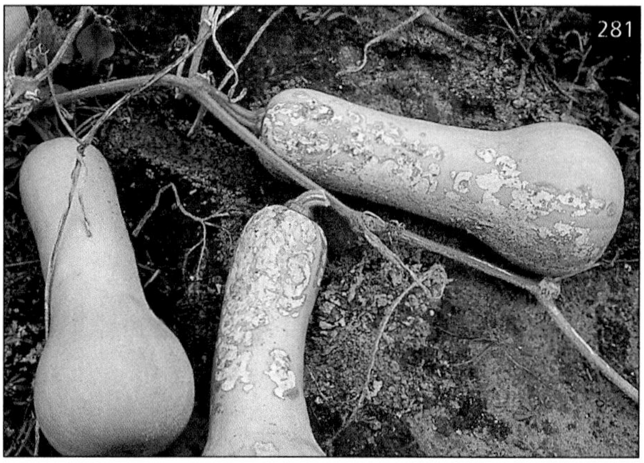

281 **Black rot fruit lesions on butternut squash.**

282 **Cut cantaloupe showing black rot lesions.**

References

Arny, C. J. and Rowe, R. C. 1991. Effects of temperature and duration of surface wetness on spore production and infection of cucumbers by *Didymella bryoniae*. *Phytopathology* 81:206–209.

Gleason, M. L., Parker, S. K., Pitblado, R. E., Latin, R. X., Speranzini, D., Hazzard, R. V., Maletta, M. J., Cowgill, W. P., and Biederstedt, D. L. 1997. Validation of a commercial system for remote estimation of wetness duration. *Plant Disease* 81:825–829.

Keinath, A. P. and DuBose, V. B. 2004. Evaluation of fungicides for prevention and management of powdery mildew on watermelon. *Crop Protection* 23:35–42.

Keinath, A. P., Farnham, M. W., and Zitter, T. A. 1995. Morphological, pathological, and genetic differentiation of *Didymella bryoniae* and *Phoma* sp. isolated from cucurbits. *Phytopathology* 85:364–369.

Kothera, R. T., Keinath, A. P., Dean, R. A., and Farnham, M. W. 2003. AFLP analysis of a worldwide collection of *Didymella bryoniae*. *Mycological Research* 107:297–304.

St. Amand, P. C. and Wehner, T. C. 1995. Eight isolates of *Didymella bryoniae* from geographically diverse areas exhibit variation in virulence but no isolate by cultivar interaction on *Cucumis sativus*. *Plant Disease* 79:1136–1139.

Somai, B. M., Dean, R. A., Farnham, M. W., Zitter, T. A., and Keinath, A. P. 2002. Internal transcribed spacer regions 1 and 2 and random amplified polymorphic DNA analysis of *Didymella bryoniae* and related *Phoma* species isolated from cucurbits. *Phytopathology* 92:997–1004.

Svedelius, G. 1990. Effects of environmental factors and leaf age on growth and infectivity of *Didymella bryoniae*. *Mycological Research* 94:885–889.

Utkhede, R. S. and Koch, C. A. 2002. Chemical and biological treatments for control of gummy stem blight of greenhouse cucumbers. *European Journal of Plant Pathology* 108:443–448.

van Steekelenburg, N. A. M. 1982. Factors influencing external fruit rot of cucumber caused by *Didymella bryoniae*. *Netherlands Journal of Plant Pathology* 88:47–56.

van Steekelenburg, N. A. M. 1985. Influence of humidity on incidence of *Didymella bryoniae* on cucumber leaves and growing tips under controlled conditions. *Netherlands Journal of Plant Pathology* 91:277–283.

Zhang, J. X., Bruton, B. D., Miller, M. E., and Isakeit, T. 1999. Relationship of developmental stages of cantaloupe fruit to black rot susceptibility and enzyme production by *Didymella bryoniae*. *Plant Disease* 83:1025–1032.

Erysiphe cichoracearum (= Golovinomyces cichoracearum), Sphaerotheca fuliginea (= Podosphaera xanthii), Leveillula taurica (anamorph = Oidiopsis taurica)

POWDERY MILDEW

Introduction and significance

Most of the various cucurbit crops are susceptible to powdery mildew, which is a prevalent foliar disease worldwide. If not managed, this disease can cause early plant senescence and reduction in yields. Several powdery mildew fungi infect the cucurbits.

Symptoms and diagnostic features

The first sign of the pathogen is the white, powdery growth of the fungus that occurs on both top and bottom leaf surfaces and on petioles and stems (**283**). Older, lower leaves are the first leaves to show infection (**284**). Powdery mildew leaf colonization can cause the underlying leaf tissue to turn chlorotic and later necrotic. In severe cases, the entire leaf can be covered with the white growth, turn brown, and die. Severely infected plants can be stunted, have smaller fruit and reduced yields, and senesce early. Loss of foliage can result in sunburned fruit because of exposure to direct sunlight. Fruit are rarely colonized.

Causal agents

There are two main powdery mildew pathogens that infect cucurbits: *Sphaerotheca fuliginea* and *Erysiphe cichoracearum*. Both of these species produce a whitish, epiphytic mycelium that grows superficially on host surfaces. Conidiophores of both species are borne on this surface mycelium. For *E. cichoracearum*, conidia are produced in long chains, are hyaline, ellipsoid to barrel-shaped, and measure 25–45 x 14–25 µm. Edge line shapes of immature conidial chains are sinulate, and conidia lack fibrosin bodies. *Erysiphe cichoracearum* conidia germinate with straight germ tubes and unlobed appressoria. Cleistothecia are not commonly observed but are globose, have numerous hypha-like appendages, and contain up to 10–25 unitunicate asci. Asci contain two ascospores that are ovoid to ellipsoid. Recently researchers have reclassified *E. cichoracearum* and recommend the name *Golovinomyces cichoracearum*.

The conidia of *S. fuliginea* are produced in long chains, are hyaline, ellipsoid to barrel-shaped, and measure 25–37 x 14–25 µm. Edge line shapes are crenate, and freshly produced conidia contain fibrosin bodies. *Sphaerotheca fuliginea* conidia germinate by producing forked germ tubes but no appressoria. Cleistothecia are rare and are globose, have numerous hypha-like appendages, and contain a single unitunicate ascus. Asci contain eight ascospores that are ovoid to ellipsoid. *Sphaerotheca fuliginea* appears to be more prevalent worldwide on cucurbits. This pathogen has recently been reclassified as *Podosphaera xanthii*.

283 Powdery mildew sporulation on squash leaf. Healthy leaf on right.

284 Older squash foliage infected by powdery mildew.

While both *S. fuliginea* and *E. cichoracearum* are reported as having broad host ranges, these are complex organisms, and sub-populations within each species are probably host specific to various plant groups. Therefore the *S. fuliginea* and *E. cichoracearum* isolates that infect cucurbits most likely do not infect non-cucurbit plants, and vice versa. On melons, the *S. fuliginea* pathogen has five described races, and *E. cichoracearum* has two described races.

Some cucurbit crops (such as cucumber) are also susceptible to a third powdery mildew fungus, *Leveillula taurica* (anamorph: *Oidiopsis taurica*). However, this particular fungus, which is an endophytic powdery mildew species, is of minor economic importance for cucurbit production. The fungus causes chlorotic, vein-delimited leaf lesions; profuse white sporulation develops on the undersides of these angular spots.

Disease cycle

As obligate pathogens, all powdery mildew species overwinter on volunteer or weed host plants that then serve as sources of initial inoculum. Greenhouse grown cucurbits or cucurbit crops grown beyond the production region may also be sources of initial inoculum. Conidia are wind blown for long distances. High relative humidity favors infection, though conidia can germinate and infect plants at 50% or lower relative humidity. Disease development is optimum at temperatures of 20–27° C. *Sphaerotheca* development can be quite rapid, with germ tubes appearing within 2 hours, conidiophores developing within 4 days after infection, and cleistothecia being produced after several weeks.

Control

Plant resistant cultivars and apply protectant fungicides such as sulfur. Resistance to systemic fungicides has been widespread for *S. fuliginea*. In some cases, breeding programs did not recognize that cucurbit powdery mildew was caused by different pathogens; therefore, some early information regarding cultivar resistance may not be reliable.

References

El-Ammari, S. S. and Khan, M. W. 1983. *Leveillula taurica* on greenhouse cucumber in Libya. *Plant Disease* 67:553–555.

Epinat, C., Pitrat, M., and Bertrand, F. 1993. Genetic analysis of resistance of five melon lines to powdery mildews. *Euphytica* 65:135–144.

Keinath, A. P. and DuBose, V. B. 2004. Evaluation of fungicides for prevention and management of powdery mildew on watermelon. *Crop Protection* 23:35–42.

Kiss, L. and Szentivanyi, O. 2001. Infection of bean with cucumber powdery mildew, *Podosphaera fusca*. *Plant Pathology* 50:411.

Kobori, R. F., Suzuki, O., Wierzbicki, R., Della Vecchia, P. T., and Camargo, L. E. A. 2004. Occurrence of *Podosphaera xanthii* Race 2 on *Cucumis melo* in Brazil. *Plant Disease* 88:1161.

McGrath, M. T. 2001. Fungicide resistance in cucurbit powdery mildew: experiences and challenges. *Plant Disease* 85:236–245.

McGrath, M. T. and Shishkoff, N. 2003. First report of the cucurbit powdery mildew fungus (*Podosphaera xanthii*) resistant to strobilurin fungicides in the United States. *Plant Disease* 87:1007.

McGrath, M. T. and Staniszewska, H. 1996. Management of powdery mildew in summer squash with host resistance, disease threshold-based fungicide programs, or an integrated program. *Plant Disease* 80:1044–1052.

McGrath, M. T., Staniszewska, H., and Shishkoff, N. 1996. Fungicide sensitivity of *Sphaerotheca fuliginea* populations in the United States. *Plant Disease* 80:697–703.

McGrath, M. T., Staniszewska, H., Shishkoff, N., and Casella, G. 1996. Distribution of mating types of *Sphaerotheca fuliginea* in the United States. *Plant Disease* 80:1098–1102.

Menzies, J., Bowen, P., Ehret, D., and Glass, A. D. M. 1992. Foliar applications of potassium silicate reduce severity of powdery mildew on cucumber, muskmelon, and zucchini squash. *Journal of the American Society for Horticultural Science* 117:902–905.

Romero, D., Rivera, M. E., Cazorla, F. M., De Vicente, A., and Pérez-García, A. 2003. Effect of mycoparasitic fungi on the development of *Sphaerotheca fusca* in melon leaves. *Mycological Research* 107:64–71.

Reuveni, M., Agapov, V. and Reuveni, R. 1995. Suppression of cucumber powdery mildew (*Sphaerotheca fuligena*) by foliar sprays of phosphate and potassium salts. *Plant Pathology* 44:31–39.

Sowell, G., Jr. 1982. Population shift of *Sphaerotheca fuliginea* on muskmelon from race 2 to race 1 in the southeastern United States. *Plant Disease* 66:130–131.

Vakalounakis, D. J. and Klironomou, E. 1995. Race and mating type identification of powdery mildew on cucurbits in Greece. *Plant Pathology* 44:1033–1038.

Vakalounakis, D. J. and Klironomou, E. 2001. Taxonomy of *Golovinomyces* on cucurbits. *Mycotaxon* 80:489–491.

Fusarium oxysporum f. sp. *cucumerinum*, f. sp. *melonis*, f. sp. *niveum*, f. sp. *radicis-cucumerinum*

FUSARIUM WILT, FUSARIUM ROOT/STEM ROT

Introduction and significance

Cucurbits are susceptible to several vascular wilt diseases caused by different formae speciales of the *Fusarium oxysporum* fungus. Cucumber, melon (or muskmelon), and watermelon are each subject to their own particular *F. oxysporum* pathogen. Squash apparently does not have its own specific *F. oxysporum* pathogen, but can be infected by the watermelon pathogen.

285 Wilting of watermelon vines caused by Fusarium wilt.

286 Advanced symptoms of Fusarium wilt of watermelon.

Symptoms and diagnostic features

These three wilt diseases show a similar range of symptoms, though particular expressions of disease vary depending upon the host, age of the plant, environmental conditions, abundance of the pathogen in soil, and virulence of the prevalent strains.

Seedlings can develop damping-off and lower stem infections due to Fusarium wilt pathogens; high damping-off incidence has been observed if watermelon is planted during warm weather and fields have high inoculum levels. However, the typical problem occurs on older seedlings and further advanced plants.

Initial symptoms include dull green to gray-green coloring of foliage, yellowing of older leaves, and temporary wilting of shoot tips and young leaves during the warmer time of the day. Symptoms initially may be one sided and only affect vines on a particular side of the plant. As disease develops, a wilting symptom may affect all leaves and shoots (**285**). Plants might be stunted, show a stiff appearance, and suffer from poor growth and reduced fruit set, while leaves become necrotic, brown, and dried up. Fruit that are present might become sunburned due to loss of functional foliage that covers the fruit. The vines will subsequently collapse and die (**286**). Symptom expression may be initiated, and most severe, when plants are stressed or when fruit set and development are taking place. An important diagnostic feature is the light to dark brown to red-brown discoloration of the xylem tissue in the taproot, crown, and lower stem; however, on some cultivars the vascular discoloration may be subtle or absent. On occasion a longitudinal, necrotic lesion occurs on the external surface of lower stems (**287**); such lesions might be more common on muskmelon.

Causal agents

The pathogen morphology and colony characteristics of these pathogens are similar to other *F. oxysporum* fungi. The fungus forms one- or two-celled, oval to kidney shaped microconidia on monophialides, and three- to five-celled, fusiform, curved macroconidia. Macroconidia are usually produced in cushion shaped structures called sporodochia. Resilient chlamydospores are also formed. The pathogen is usually readily isolated from symptomatic vascular tissue, though semi-selective media like Komada's medium can be helpful to reduce interference from secondary rot organisms.

287 Stem symptoms caused by Fusarium wilt on cucumber.

Fusarium wilt of cucumber (*Cucumis sativus*) is caused by *F. oxysporum* f. sp. *cucumerinum*. This pathogen primarily infects cucumber, though some isolates can infect watermelon and melon. Presently three races have been documented. Melon and muskmelon (both *Cucumis melo* plants) Fusarium wilt is caused by *F. oxysporum* f. sp. *melonis*. Four races of this pathogen are currently recognized and are designated as 0, 1, 2, and 1–2. *Fusarium oxysporum* f. sp. *niveum* is the causal agent of Fusarium wilt of watermelon (*Citrullus lanatus*). Three races (races 0, 1, and 2) are documented. While this pathogen is generally considered host specific to watermelon, there are several reports of isolates being able to infect cucumber, melon, and summer squash (*Cucurbita pepo*).

A different *F. oxysporum* pathogen causes a distinct disease called Fusarium root and stem rot. The pathogen, *F. oxysporum* f. sp. *radicis-cucumerinum*, causes tan to brown lesions at the plant crown and upper root areas. Affected plants grow poorly and can later collapse. This pathogen affects cucumber, muskmelon, squash, and other cucurbits. This disease is similar to crown and root rot of tomato.

Disease cycle

All wilt pathogens infecting cucurbits share very similar epidemiologies. The pathogens are true soilborne organisms and can persist in the soil for long periods of time as saprobes. Chlamydospores also enable these fungi to survive in the soil without a host. The pathogens can be seedborne. Fusarium wilt fungi enter the host via roots, colonize the xylem, and become systemically distributed. Fusarium wilt is more severe if cucurbits are planted in sandy, porous soils that have high nitrogen content and acidic pH values (pH 5.0–6.0). Optimum soil temperatures for disease development are between 20-27° C. *Fusarium oxysporum* f. sp. *melonis* occurs early in the season in cooler soils (18–22° C). After an infected plant dies or is incorporated into the soil, the fungus produces chlamydospores that are returned to the soil and increase the pathogen population. Fungal propagules are distributed whenever infested, contaminated soil is moved by equipment, vehicles, and water. Fusarium wilt severity increases when the host plant is stressed due to unfavorable growing conditions such as temperature extremes and heavy fruit loads.

Control

Use resistant cultivars, when available. Note that the performance of a resistant cultivar is influenced by the inoculum level in the soil. Grafting to resistant rootstocks (e.g. melon to *Benincasa cerifera* roots) has proven successful in combating Fusarium wilt. Use seed that does not have significant levels of the pathogen. Practice crop rotations so that pathogen populations are allowed to decline. In California, a 3-year rotation out of susceptible cucurbits is sufficient to avoid economic damage of the crop. Some soils show natural suppressiveness to these *Fusarium* pathogens. Fumigation of field soils can effectively lower soil populations of the pathogen, though such soils may be re-invaded. Researchers found that *F. oxysporum* f. sp. *melonis* rapidly recolonized soils treated with various fumigants.

References

Gerlagh, M. and Blok, W. 1988. *Fusarium oxysporum* f. sp. *cucurbitacearum* n. f. embracing all formae speciales of *F. oxysporum* attacking cucurbitaceous crops. *Netherlands Journal of Plant Pathology* 94:17–31.

Gordon, T. R., Okamoto, D., and Jacobson, D. J. 1989. Colonization of muskmelon and nonsusceptible crops by *Fusarium oxysporum* f. sp. *melonis* and other species of *Fusarium*. *Phytopathology* 79:1095–1100.

Gordon, T. R. and Okamoto, D. 1990. Colonization of crop residue by *Fusarium oxysporum* f. sp. *melonis* and other species of *Fusarium*. *Phytopathology* 80:381–386.

Gubler, W. D. and Grogan, R. G. 1976. Fusarium wilt of muskmelon in the San Joaquin Valley, California. *Plant Disease Reporter* 60:742–744.

Jenkins, S. F., Jr. and Wehner, T. C. 1983. Occurrence of *Fusarium oxysporum* f. sp. *cucumerinum* on greenhouse-grown *Cucumis sativus* seed stocks in North Carolina. *Plant Disease* 67:1024–1025.

Kim, D. H., Martyn, R. D., and Magill, C. W. 1993. Mitochondrial DNA (mtDNA)—relatedness among formae speciales of *Fusarium oxysporum* in the Cucurbitaceae. *Phytopathology* 83:91–97.

Larkin, R. P., Hopkins, D. L., and Martin, F. N. 1990. Vegetative compatibility within *Fusarium oxysporum* f. sp. *niveum* and its relationship to virulence, aggressiveness, and race. *Canadian Journal of Microbiology* 36:352–358.

Leary, J. V. and Wilbur, W. D. 1976. Identification of the races of *Fusarium oxysporum* f. sp. *melonis* causing wilt of muskmelon in California. *Phytopathology* 66:15–16.

Louvet, J., Alabouvette, C., and Rouxel, F. 1981. Microbiological suppressiveness of some soils to Fusarium wilts. In: *Fusarium: Disease, Biology and Taxonomy* (Eds R.E. Nelson, T.A. Tousoun, and R.J. Cooke) pp. 262–275. Pennsylvania State University Press.

Marois, J. J., Dunn, M. T., and Papavizas, G. C. 1983. Reinvasion of fumigated soils by *Fusarium oxysporum* f. sp. *melonis*. *Phytopathology* 73:680–684.

Martyn, R. D. and Bruton, B. D. 1989. An initial survey of the United States for races of *Fusarium oxysporum* f. sp. *niveum*. *HortScience* 24:696–698.

Martyn, R. D. and McLaughlin, R. J. 1983. Susceptibility of summer squash to watermelon wilt pathogen (*Fusarium oxysporum* f. sp. *niveum*). *Plant Disease* 67:263–266.

Martyn, R. D. and McLaughlin, R. J. 1983. Effects of inoculum concentration on the apparent resistance of watermelon to *Fusarium oxysporum* f. sp. *niveum*. *Plant Disease* 67:493–495.

McKeen, C. D. and Wensley, R. N. 1962. Cultural and pathogenic variation in the muskmelon wilt fungus, *Fusarium oxysporum* f. sp. *melonis*. *Canadian Journal of Microbiology* 8:769–784.

Punja, Z. K. and Parker, M. 2000. Development of Fusarium root and stem rot, a new disease on greenhouse grown cucumbers in British Columbia caused by *Fusarium oxysporum* f. sp. *radicis-cucumerinum*. *Canadian Journal of Plant Pathology* 22:349–363.

Risser, G., Banihashemi, Z., and Davis, D. W. 1976. A proposed nomenclature of *Fusarium oxysporum* f. sp. *melonis* races and resistance genes in *Cucumis melo*. *Phytopathology* 66:1105–1106.

Rose, S. and Punja, Z. K. 2004. Greenhouse cucumber cultivars differ in susceptibility to Fusarium root and stem rot. *HortTechnology* 14:240–242.

Vakalounakis, D. J. 1996. Root and stem rot of cucumber caused by *Fusarium oxysporum* f. sp. *radicis-cucumerinum* f. sp. *nov*. *Plant Disease* 80:313–316.

Wensley, R. N. and McKeen, C. D. 1962. Rapid test for pathogenicity of soil isolates of *Fusarium oxysporum* f. sp. *melonis*. *Canadian Journal of Microbiology* 8:818–819.

Zink, F. W. and Gubler, W. D. 1986. Inheritance of resistance to races 0 and 2 of *Fusarium oxysporum* f. sp. *melonis* in a gynoecious muskmelon. *Plant Disease* 70:676–678.

Fusarium solani* f. sp. *cucurbitae
(teleomorph = *Nectria haematococca*)

FUSARIUM CROWN & FOOT ROT

Introduction and significance

Fusarium crown and foot rot is primarily a squash and pumpkin problem, though other cucurbits can become infected. Quality of the squash or pumpkin fruit can be severely affected.

Symptoms and diagnostic features

Initial symptoms occur on the crown and upper root tissues of seedlings as well as older plants. Crown and upper root infections appear as water-soaked lesions near or beneath the soil line. With time, the lesions become darker and above-ground parts of the plants can wilt and eventually die. Lesions may girdle the plant and result in a sunken, shriveled appearance. Crowns in this condition can easily break off. White to pink mycelial growth may develop on symptomatic tissues. The pathogen generally does not infect upper stem or lower root tissues. The surfaces of squash and pumpkin fruit that are in contact with the soil can develop circular to oblong, tan to brown, firm, dry, sunken lesions. Such lesions may have concentric rings within them (**288, 289**). These spots remain firm unless secondary decay organisms enter the infected area and cause soft, wet rots.

Causal agent

The pathogen is *Fusarium solani* f. sp. *cucurbitae*. The fungus forms one- or two-celled, ovoid to oblong microconidia on elongated monophialides, and three- to five-celled, fusiform, slightly curved macroconidia on multi-branched polyphialides. Macroconidia are usually produced in cushion-shaped structures called sporodochia. Chlamydospores are also formed. Two races of this pathogen have been identified. Race 1 infects all plant parts and is distributed worldwide, while race 2 only infects the fruit and is apparently limited to California and Ohio in the USA. The teleomorph, *Nectria haematococca*, has not been observed in the field on infected cucurbits.

288 **Fruit lesions on pumpkin caused by *Fusarium solani*.**

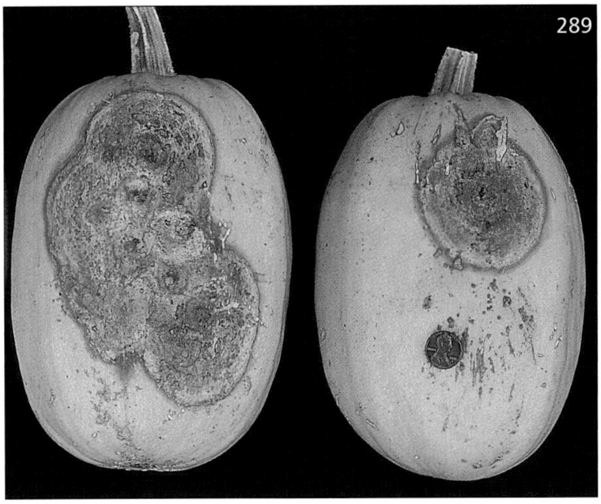

289 **Fruit lesions on spaghetti squash caused by *Fusarium solani*.**

Disease cycle

The pathogen can survive in the soil for a maximum of 2 to 3 years. For fruit infections to occur, the fruit must be in contact with moist soil that harbors inoculum. *Fusarium solani* f. sp. *cucurbitae* can also be seedborne.

Control

Rotate crops to non-cucurbits for 3 or more years. Use seed that does not have significant levels of the pathogen. Treat infested seed with hot water (15 minute soak at 55° C) or fungicides.

References

Champaco, E. R., Martyn, R. D., and Miller, M. E. 1993. Comparison of *Fusarium solani* and *F. oxysporum* as causal agents of fruit rot and root rot of muskmelon. *HortScience* 28:1174–1177.

Nagao, H., Sato, K., and Ogiwara, S. 1994. Susceptibility of *Cucurbita* spp. to the cucurbit root-rot fungus, *Fusarium solani* f. sp. *cucurbitae* race 1. *Agronomie* 2:95–102.

Nash, S. M. and Alexander, J. V. 1965. Comparative survival of *Fusarium solani* f. sp. *cucurbitae* and *F. solani* f. sp. *phaseoli* in soil. *Phytopathology* 55:963–966.

Sumner, D. R. 1976. Etiology and control of root rot of summer squash in Georgia. *Plant Disease Reporter* 60:923–927.

Toussoun, T. A. and Snyder, W. C. 1961. The pathogenicity, distribution, and control of two races of *Fusarium* (*Hypomyces*) *solani* f. sp. *cucurbitae*. *Phytopathology* 51:17–22.

Glomerella lagenarium
(anamorph = *Colletotrichum orbiculare*)

ANTHRACNOSE

Introduction and significance

Anthracnose disease occurs worldwide on a number of cucurbit crops and some wild cucurbit species. The disease is usually of minor importance in arid regions, but can be very damaging in high humidity cucurbit production areas such as the southeast USA. Anthracnose is widespread as far west as Oklahoma.

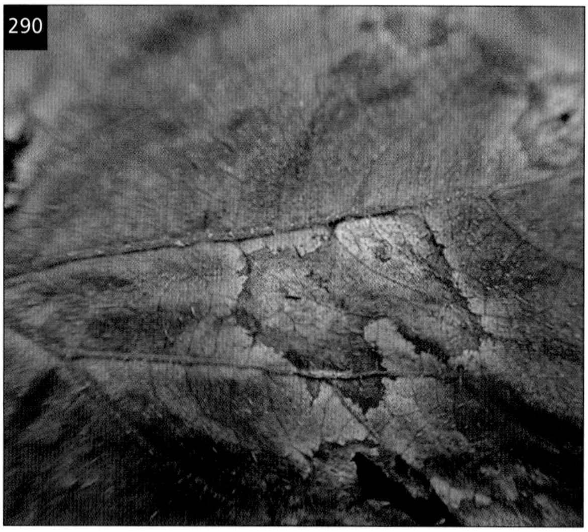

290 Leaf spot caused by the anthracnose pathogen on cucurbits.

291 Leaf spots of anthracnose on cucumber.

Symptoms and diagnostic features

Initial symptoms appear as irregularly shaped water-soaked leaf lesions that soon turn chlorotic. Lesions can expand, become greater than 1 cm in diameter, and turn tan to brown in color (**290**). On watermelon, such lesions are darker and appear brown to black. Extensive lesion development will cause the leaf to buckle and become distorted. Lesion tissue dries, cracks, and falls out (**291**). Lesions can also occur on petioles and stems and start out as elongated, oval to diamond- shaped, water-soaked lesions that later turn tan to brown. Stem lesions can be covered with the minute black fungal structures (acervuli) and pink spore masses that are exuded by these fruiting bodies. Clear, brown exudates can also collect on these stem lesions. Fruit symptoms are circular water-soaked areas that later become dark brown to black sunken lesions (**292**, **293**). If high humidity and water are present, acervuli of the pathogen will grow in the fruit lesions and exude gelatinous pink spore masses.

Causal agent

Anthracnose is caused by the fungus *Colletotrichum orbiculare*. Older references use the name *Colletotrichum lagenarium*. *Colletotrichum orbiculare* produces microscopic, black, cup-shaped structures (acervuli) that contain conidiophores and conidia. Conidia are hyaline, straight, cylindrical, nonseptate, with blunt ends, and measure 4–6 x 13–19 µm. Collectively, conidia are produced in pink gelatinous droplets. Black setae are usually present in the acervuli. The teleomorph *Glomerella lagenarium* is rarely found in association with cucurbit hosts. Distinct races of *C. orbiculare* have been identified.

Disease cycle

The pathogen overwinters on infected crop residues, volunteer cucurbit plants, and wild hosts. Conidia from these sources move to uninfected plants by splashing water, contact with contaminated tools and workers, and perhaps by insects. To germinate and infect the host, conidia require 24 hours of 100% relative humidity and optimum temperatures of 22–27° C. Spores can take up to 72 hours to infect the host, and symptoms first appear 4 days after infection. Subsequent disease development is dependent on warm, wet weather, and spread of secondary inoculum is primarily achieved via splashing rain or irrigation water.

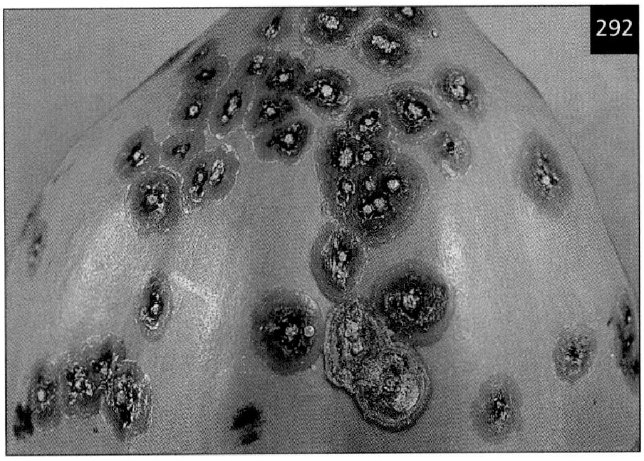

292 **Fruit lesions caused by *Colletotrichum orbiculare* on acorn squash.**

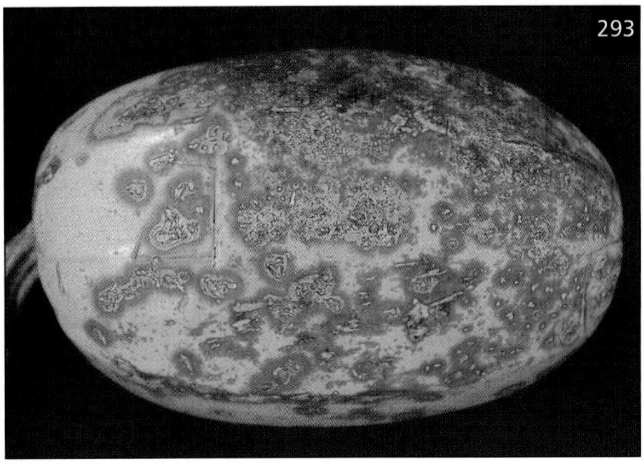

293 **Fruit lesions on spaghetti squash caused by anthracnose.**

The pathogen can also be seedborne and result in early infections on germinating seedlings or transplants placed into production fields. The pathogen infects several wild cucurbit species, which in turn can act as inoculum sources for production cucurbits.

Control

Use resistant cultivars as they are available. Use seed that does not have significant levels of the pathogen. Practice crop rotations in which cucurbits are not planted for 2 years. Apply protectant fungicides. In the USA, a disease forecasting program (Melcast) has been used to schedule fungicide applications for controlling anthracnose and gummy stem blight of melons. Avoid irrigating with overhead sprinklers.

References

Gleason, M. L., Parker, S. K., Pitblado, R. E., Latin, R. X., Speranzini, D., Hazzard, R. V., Maletta, M. J., Cowgill, W. P., and Biederstedt, D. L. 1997. Validation of a commercial system for remote estimation of wetness duration. *Plant Disease* 81:825–829.

Koike, S. T., Tidwell, T. E., Fogle, D. G., and Patterson, C. L. 1991. Anthracnose of greenhouse-grown watermelon transplants caused by *Colletotrichum orbiculare* in California. *Plant Disease* 75:644.

Leben, C. and Daft, G. C. 1968. Cucumber anthracnose: influence of nightly wetting of leaves on numbers of lesions. *Phytopathology* 58:264–265.

Monroe, J. S., Santini, J. B., and Latin, R. 1997. A model defining the relationship between temperature and leaf wetness duration, and infection of watermelon by *Colletotrichum orbiculare*. *Plant Disease* 81:739–742.

Thompson, D. C. and Jenkins, S. F. 1985. Effects of temperature, moisture, and cucumber cultivar resistance on lesion size increase and conidial production by *Colletotrichum lagenarium*. *Phytopathology* 75:828–832.

Thompson, D. C. and Jenkins, S. F. 1985. Influence of cultivar resistance, initial disease, environment, and fungicide concentration and timing on anthracnose development and yield loss in pickling cucumbers. *Phytopathology* 75:1422–1427.

Wasilwa, L. A., Correll, J. C., Morelock, T. E., and McNew, R. E. 1993. Reexamination of races of the cucurbit anthracnose pathogen *Colletotrichum orbiculare*. *Phytopathology* 83:1190–1198.

Macrophomina phaseolina

CHARCOAL ROT

Introduction and significance

Charcoal rot is usually most serious on melon hosts. The disease is usually of minor importance, but can be confused with other diseases.

Symptoms and diagnostic features

Initial symptoms consist of chlorosis, wilting, and necrosis of lower, older leaves near the crown. The crown and lower stem will develop a dark green, water-soaked lesion (**294**) that can eventually cover the entire crown and extend up to 15–20 cm above the ground. Brown exudates will collect on the lesion surfaces. Vines can be stunted or wilt, collapse, and die. In later stages of the disease (**295**), the stem and crown lesion will dry, crack, and support the development of minute, black microsclerotia of the pathogen. Other structures (pycnidia) might also form in the lesions. Roots can be extensively colonized. Fruit in contact with soil can also develop a firm decay.

Causal agent

Charcoal rot is caused by the fungus *Macrophomina phaseolina*, which is a soilborne fungus that has a wide host range but occurs mainly in hot climates where soil temperatures reach at least 28° C. The fungus produces spherical, dark brown, ostiolate pycnidia and single-celled conidia that are hyaline, ellipsoidal to ovoid, and measure 14–30 x 5–10 µm. In diseased tissues, profuse microsclerotia are produced that are irregular in shape, black, and range in length from 100–1000 µm.

Disease cycle

The pathogen persists in the soil and crop residue as microsclerotia. Roots become infected first and the fungus later progresses into the plant crown.

Control

Rotate with non-host crops so that soil inoculum does not build up to high levels. Some cultivars may have tolerance or resistance to this pathogen. Minimize plant stress by irrigating and fertilizing properly.

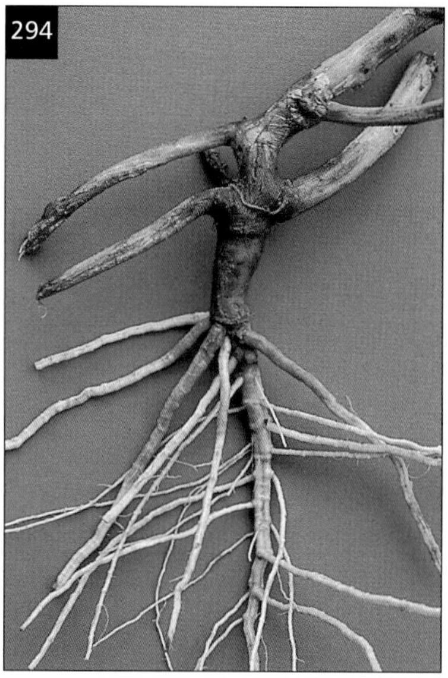

294 Early crown symptoms of charcoal rot on cantaloupe.

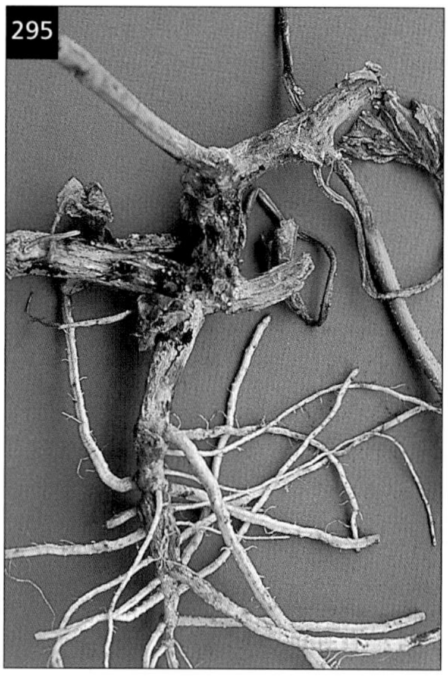

295 Advanced cracking and bleeding of charcoal rot on cantaloupe.

References

Bruton, B. D., Jeger, M. J., and Reuveni, R. 1987. *Macrophomina phaseolina* infection and vine decline in cantaloupe in relation to planting date, soil environment, and plant maturation. *Plant Disease* 71:259–263.

Bruton, B. D. and Reuveni, R. 1985. Vertical distribution of microsclerotia of *Macrophomina phaseolina* under various soil types and host crops. *Agriculture, Ecosystems, & Environment* 12:165–169.

Reuveni, R., Krikum, J., Machimias, A., and Shelvin, E. 1982. The role of *Macrophomina phaseolina* in a collapse of melon plants in Israel. *Phytoparasitica* 10:51–56.

Monosporascus cannonballus

MONOSPORASCUS ROOT ROT AND VINE DECLINE

Introduction and significance

Monosporascus root rot and vine decline affects a number of cucurbit crops, but is most severe on melons and watermelons. The disease has been detected in India, Israel, Japan, Spain, Taiwan, Tunisia, and the USA (in the southwest part of the country, including the states of California, Arizona, and Texas). A similar disease, reportedly caused by *M. eutypoides*, is documented in Israel.

Symptoms and diagnostic features

Initial symptoms include the stunting and poor growth of young plants. The older, lower leaves then begin to turn chlorotic, wilt, and collapse (**296**). These leaf symptoms progress up the vines until many or most of the leaves wilt and die (**297, 298**). Fruit on affected vines may be smaller, or cracked and sunburned due to exposure. Symptoms on roots include tan to red-brown lesions (**299**). These necrotic spots can expand and result in the death of small feeder roots. In severe cases the taproot and entire root system may rot and the plant dies. An important sign is the development of minute, black erumpent fungal structures (perithecia) in the brown root lesions. With a hand lens, clusters of the round, black spores of the fungus can be observed at the perithecial openings.

Causal agent

The disease is caused by the fungus *Monosporascus cannonballus*. This pathogen is a soilborne ascomycete that forms globose, brown perithecia. Perithecia contain clavate to pyriform, unitunicate asci that each

296 Chlorosis and decline of old cantaloupe leaves caused by *Monosporascus*.

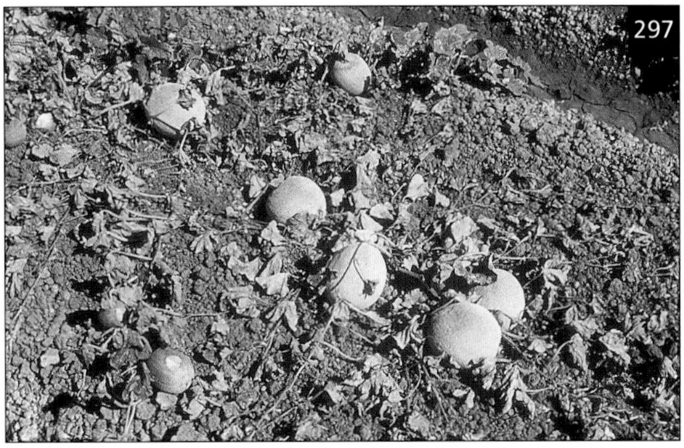

297 Dead cantaloupe vines killed by *Monosporascus*.

298 Foliage collapse of cantaloupe caused by *Monosporascus*.

299 Root lesions of root rot and vine decline of cantaloupe.

hold a single, one-celled, thick-walled, spherical, black ascospore that measures 35–50 µm in diameter (**300**, **301**). A dark, periapical ring surrounds the perithecial opening. An anamorph stage has not been identified. The fungus grows well under warm conditions and optimum *in vitro* growth takes place at 30–35° C.

Disease cycle

The complete epidemiology of Monosporascus root rot and vine decline is not yet documented. Mycelium or ascospores in soil presumably infect roots early in the season, with disease development and fungal growth encouraged by warmer temperatures later in the summer. Roots in the upper soil profile are affected first. Plant stress may make disease symptoms more severe. Ascospore production takes place in infected roots of declining and senescent plants. *Monosporascus cannonballus* is distributed wherever infested, contaminated soil is moved by equipment, vehicles, and water. Non-cucurbits such as wheat, corn, and sorghum may also be hosts to this pathogen.

Control

The fungus appears to survive in the soil for extended periods, so avoid planting melons and watermelons in known infested fields. Enhance plant growth by employing good production practices. Avoid excessive irrigation. In Israel, experiments demonstrated that reduced, less frequent irrigations helped delay plant collapse, though yields and fruit quality were also reduced. Recent research finds that pathogen inoculum production is reduced if old crop roots are dried out by cultivating fields after harvest or by applying fumigants (metam sodium) immediately after harvest.

300 Ascospores of *Monosporascus* on infected root.

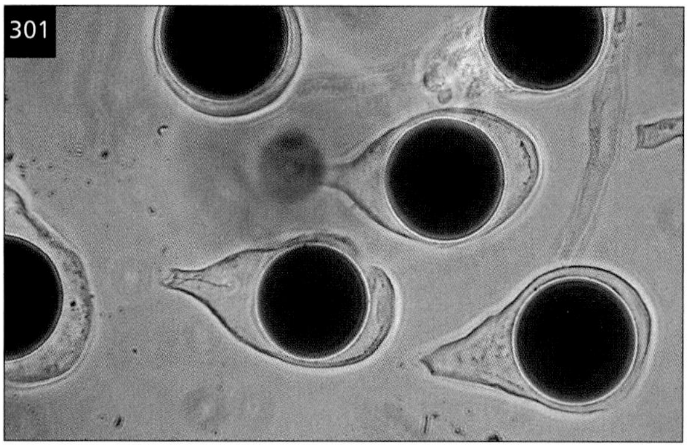

301 Asci and single ascospores of *Monosporascus*.

References

Martyn, R. D. and Miller, M. E. 1996. Monosporascus root rot and vine decline: an emerging disease of melons worldwide. *Plant Disease* 80:716–725.

Mertely, J. C., Martyn, R. D., Miller, M. E., and Bruton, B. D. 1991. Role of *Monosporascus cannonballus* and other fungi in a root rot/vine decline disease of muskmelon. *Plant Disease* 75:1133–1137.

Mertely, J. C., Martyn, R. D., Miller, M. E., and Bruton, B. D. 1993. An expanded host range for the muskmelon pathogen *Monosporascus cannonballus*. *Plant Disease* 77:667–673.

Mertely, J. C., Martyn, R. D., Miller, M. E., and Bruton, B. D. 1993. Quantification of *Monosporascus cannonballus* ascospores in three commercial muskmelon fields in south Texas. *Plant Disease* 77:766–771.

Radewald, K. C., Ferrin, D. M., and Stanghellini, M. E. 2004. Sanitation practices that inhibit reproduction of *Monosporascus cannonballus* in melon roots left in the field after crop termination. *Plant Pathology* 53:660–668.

Pivonia, S., Cohen, R., Cohen, S., Kigel, J., Levita, R., and Katan, J. 2004. Effect of irrigation regimes on disease expression in melon plants infected with *Monosporascus cannonballus*. *European Journal of Plant Pathology* 110:155–161.

Pivonia, S., Cohen, R., Kigel, J., and Katan, J. 2002. Effect of soil temperature on disease development in melon plants infected by *Monosporascus cannonballus*. *Plant Pathology* 51:472–479.

Pollack, F. G. and Uecker, F. A. 1974. *Monosporascus cannonballus*, an unusual ascomycete in cantaloupe roots. *Mycologia* 66:346–349.

Stanghellini, M. E., Ferrin, D. M, Kim, D. H., Waugh, M. M., Radewald, K. C., Sims, J. J., Ohr, H. D., Mayberry, K. S., Turini, T., and McCaslin, M. A. 2003. Application of preplant fumigants via drip irrigation systems for the management of root rot of melons caused by *Monosporascus cannonballus*. *Plant Disease* 87:1176–1178.

Stanghellini, M. E., Kim, D. H., and Rasmussen, S. L. 1996. Ascospores of *Monosporascus cannonballus*: germination and distribution in cultivated and desert soils in Arizona. *Phytopathology* 86:509–514.

Phytophthora capsici

PHYTOPHTHORA CROWN AND ROOT ROT

Introduction and significance

Phytophthora crown and root rot affects most cucurbit crops, with squash perhaps being the most sensitive. In several states in the USA, this pathogen has increased in importance in recent years.

Symptoms and diagnostic features

Initial symptoms consist of wilting of young shoots and leaves, followed by wilting and collapse of all foliage (**302**). The lower stem and crown turns tan to brown in color and is soft and rotted. The root system turns tan to brown and is rotted. When dug out of the ground, the small feeder roots break off and outer tissue of the larger roots sloughs off. The crown and root rot pathogen can also cause seedling damping-off, foliar blight, and fruit rot diseases (**303, 304, 305**).

Causal agent

The main pathogen is the oomycete *Phytophthora capsici*. Optimum growth occurs at 25–28° C. *Phytophthora capsici* produces sporangia that release zoospores that swim to and infect susceptible tissues. Also, sporangia are deciduous and can be aerially dispersed. As a heterothallic species, both mating types (A1 and A2) must be present for oospore production. In addition to cucurbits, *P. capsici* can infect aubergine (eggplant), pepper, tomato, and other crops.

302 Phytophthora crown and root rot in squash field.

303 *Phytophthora capsici* causing aerial blight symptoms on hard squash.

304 *Phytophthora capsici* infecting cucumber fruit.

305 *Phytophthora capsici* infecting pumpkin fruit.

Other *Phytophthora* species, such as *P. drechsleri*, also have been reported to cause crown and root rot of cucurbits.

Disease cycle

Phytophthora capsici survives in the soil as a saprophyte and pathogen of several host plants. If wet soil conditions persist, zoospores move to, and infect, the crown and root tissues of the host. Oospores are important sources of initial inoculum because they can survive as dormant propagules in soil for extended periods of time.

Control

Do not plant cucurbits in fields that have poorly draining, heavy textured soils. Prepare seed and transplant beds so that drainage is enhanced. Schedule irrigations so that excess water is not applied and fields drain properly. Rotate crops so that cucurbits and other hosts are not planted within 3 years of the previous susceptible planting. Some fungicides may also be helpful in controlling this disease, though resistance to phenylamide fungicides has been found.

References

Cafe-Filho, A. C., Duniway, J. M., and Davis, R. M. 1995. Effects of the frequency of furrow irrigation on root and fruit rots of squash caused by *Phytophthora capsici*. *Plant Disease* 79:44–48.

Hausbeck, M. K. and Lamour, K. H. 2004. *Phytophthora capsici* on vegetable crops: research progress and management challenges. *Plant Disease* 88:1292–1303.

Ho, H. H., Lu, J., and Gong, L. 1984. *Phytophthora drechsleri* causing blight of *Cucumis* species in China. *Mycologia* 76:115–121.

Lamour, K. H. and Hausbeck, M. K. 2003. Susceptibility of mefenoxam-treated cucurbits to isolates of *Phytophthora capsici* sensitive and insensitive to mefenoxam. *Plant Disease* 87:920–922.

Lamour, K. H. and Hausbeck, M. K. 2003. Effect of crop rotation on the survival of *Phytophthora capsici* in Michigan. *Plant Disease* 87:841–845.

Polach, F. J. and Webster, R. K. 1972. Identification of strains and inheritance of pathogenicity in *Phytophthora capsici*. *Phytopathology* 62:20–26.

Ristaino, J. B. 1990. Intraspecific variation among isolates of *Phytophthora capsici* from pepper and cucurbit fields in North Carolina. *Phytopathology* 80:1253–1259.

Tian, D. and Babadoost, M. 2004. Host range of *Phytophthora capsici* from pumpkin and pathogenicity of isolates. *Plant Disease* 88:485–489.

Pseudoperonospora cubensis

DOWNY MILDEW

Introduction and significance

Downy mildew is a major foliar disease of the various cucurbit crops and is particularly severe in warm temperate or tropical environments. Watermelon and some *Cucurbita* species (*C. maxima* and *C. moschata*) may less susceptible.

Symptoms and diagnostic features

Initial symptoms are found on the top surfaces of leaves and consist of small, pale green to greasy looking, angular or rectangular spots that are delimited by leaf veins (**306, 307**). These patches later turn slightly chlorotic to bright yellow. The early symptoms can give the leaf a mosaic or mottled appearance. As downy mildew lesions age, the tissue turns brown and necrotic. Lesions can expand and coalesce, resulting in the shriveling and death of large areas of leaf surface.

If environmental conditions are favorable, the lower leaf surfaces of infected areas will support the growth of light gray to dark purple sporulation that will also be delimited by leaf veins and therefore result in angular or rectangular patterns (**308**). Downy mildew can cause stunting or even death if young plants are infected early and severely. Fruit maturation and production might be prevented and fruit flavor and sugar content could also be affected.

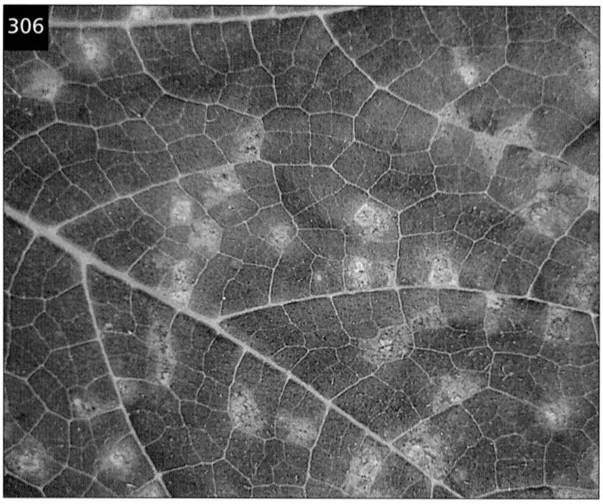

306 Downy mildew on summer squash. Upper surface of leaf showing angular lesions of downy mildew.

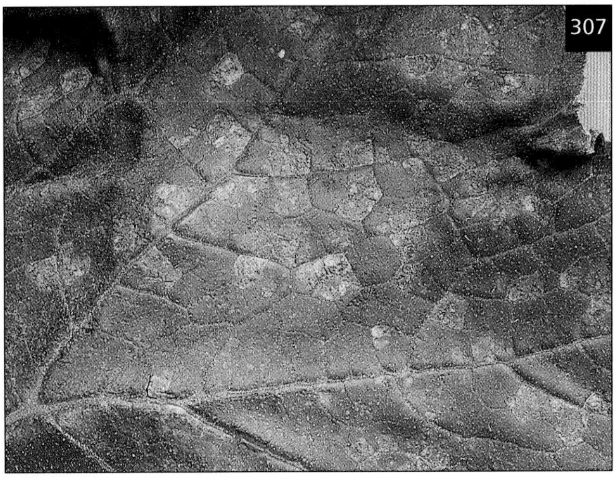

307 **Angular lesions of downy mildew of squash.**

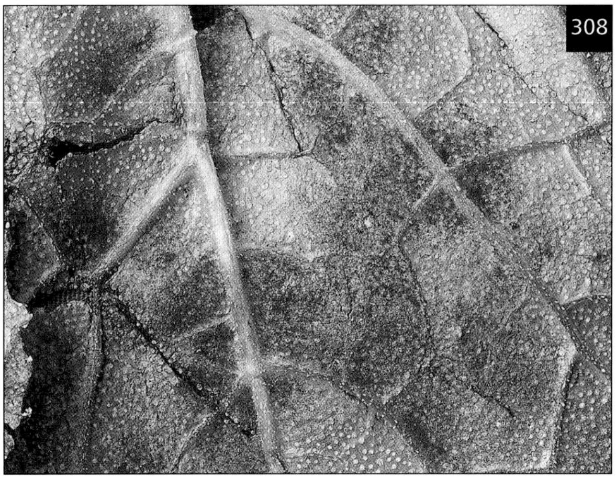

308 **Downy mildew sporulation on leaf underside of cucumber.**

Causal agent

Downy mildew is caused by the oomycete *Pseudoperonospora cubensis*, which is an obligate pathogen that is host specific to cucurbits. Sporangiophores emerge from leaf stomates, branch dichotomously, and terminate in slender curved tips. The sporangial tips bear a single, gray to purple, ovoid to ellipsoidal sporangium that measures 20–40 x 14–25 µm. Sporangia have a papilla on the distal end. When produced, oospores measure 22–42 µm in diameter; however, these sexual structures are rarely found. Five distinct physiological races or pathotypes have been reported for *P. cubensis*, each of which is host specific to particular cucurbit plants.

Disease cycle

As an obligate pathogen, *P. cubensis* overwinters on volunteer or weedy cucurbit plants that then serve as sources of initial inoculum. Greenhouse-grown cucurbits or cucurbit crops grown far beyond the production region may also be sources of initial inoculum. Sporangia are produced if humidity is at 100% for 6 hours and temperatures are between 15–20° C; these spores are wind blown or splashed onto susceptible host tissue and germinate if free moisture is present for at least 2 hours at 20–25° C. Sporangia germinate by releasing 2–15 motile zoospores that swim towards, and encyst on, leaf stomates. Cysts germinate with germ tubes that penetrate the stomates and initiate colonization of the leaf. The incubation period prior to symptom expression is 3–12 days depending on temperature, relative humidity, and inoculum concentration.

Control

Plant resistant cultivars, though these are currently only available for cucumber and melon. Apply protectant fungicides; however, some resistance to such materials has been documented. Avoid using overhead sprinkler irrigation. Ventilate greenhouses to reduce humidity levels.

References

Cohen, Y. 1977. The combined effects of temperature, leaf wetness, and inoculum concentration on infection of cucumber with *Pseudoperonospora cubensis*. *Canadian Journal of Botany* 55:1478–1487.

Katan, T. and Bashi, E. 1981. Resistance to metalaxyl in isolates of *Pseudoperonospora cubensis*, the downy mildew pathogen of cucurbits. *Plant Disease* 65:798–800.

Palti, J. and Cohen, Y. 1980. Downy mildew of cucurbits. *Phytoparasitica* 8:109–147.

Thomas, C. E., Inaba, T., and Cohen, Y. 1987. Physiological specialization in *Pseudoperonospora cubensis*. *Phytopathology* 77:1621–1624.

Verticillium dahliae

VERTICILLIUM WILT

Introduction and significance

Verticillium wilt affects most species in this crop group. The disease is found throughout the world.

Symptoms and diagnostic features

Initial symptoms typically occur after plants are established and have set fruit. Lower leaves will become off green and wilt during the warmer part of the day. As disease progresses, the leaves become chlorotic, withered, and later necrotic and dry. Such symptoms progress and affect all leaves and the tips of plant runners. Wilt symptoms are sometimes one sided and affect only the vines on one side of the plant. Plants eventually collapse and die (**309, 310**). An important symptom is the tan to brown discoloration of the xylem tissues in roots and stems. *Verticillium dahliae*, along with a series of other soilborne pathogens, is implicated in the vine decline disease that is caused by a complex of fungi. This pathogen has been found to cause an internal fruit decay on Chinese winter melon (*Benincasa hispida*).

Causal agent

The causal agent is *Verticillium dahliae*. The pathogen can be isolated on standard microbiological media, though semi-selective media such as NP-10 can be useful for isolation and for purifying cultures. On general purpose media, the pathogen forms the characteristic hyaline, verticillate conidiophores bearing three to four phialides at each node, and hyaline, single-celled, ellipsoidal conidia that measure 2–8 x 1–3 µm. Older cultures form dark brown to black torulose microsclerotia that consist of groups of swollen cells formed by repeated budding. Microsclerotia size varies greatly and ranges from 15 to 100 µm in diameter. Microsclerotia enable the pathogen to survive in the soil for extended periods of time (up to 8 to 10 years). The pathogen has a very wide host range.

309 Declining vines of Verticillium wilt of watermelon.

Disease cycle

Roots of susceptible plants grow near microsclerotia that are dormant in the soil. Microsclerotia are stimulated to germinate and infect the roots. The pathogen subsequently invades xylem tissue and becomes systemic in the plant. Disease progresses most rapidly during relatively cool temperatures (20–23° C) and when plants are undergoing stress, such as during flower and fruit formation. Diseased plants become colonized with microsclerotia, which are returned to the soil during crop residue incorporation.

Control

Plant resistant or tolerant cultivars as they become available. Rotate crops so that cucurbits are not planted in fields having a history of the problem. Soil tests have been developed that can estimate the numbers of microsclerotia per gram of soil; such tests may be useful for giving indications of potential risk of this disease. Minimize spread of infested soil to uninfested areas.

References

Aegerter, B. J., Gordon, T. R., and Davis, R. M. 2000. Occurrence and pathogenicity of fungi associated with melon root rot and vine decline in California. *Plant Disease* 84:224–230.

Gubler, W. D. and Bernhardt, E. A. 1992. Cavity rot of winter melon caused by *Verticillium dahliae*. *Plant Disease* 76:416–417.

Sorensen, L. H., Schneider, A. T., and Davis, J. R. 1991. Influence of sodium polygalacturonate sources and improved recovery of *Verticillium* species from soil. (Abstract) *Phytopathology* 81:1347.

310 **Collapsed plants of Verticillium wilt of cucumber.**

Cucumber mosaic virus

CUCUMBER MOSAIC

Introduction and significance

Cucumber mosaic virus (CMV) is commonly found throughout the world and infects over 800 plant (crop and weed) hosts, including cucurbits.

Symptoms and diagnostic features

CMV severely stunts plant growth and causes leaves to be distorted, reduced in size, curled or rolled, and show a yellow mosaic or mottle pattern with chlorotic patches (**311**). Young leaves at the growing point may form a rosette. Flowers can be distorted and may have green petals. Fruit can be small, malformed, and show various discolorations.

Causal agent and disease cycle

CMV is a cucumovirus with virions that are isometric (29 nm in diameter) and contain three single-stranded RNAs. A number of CMV strains have been documented. CMV is transmitted in a nonpersistent manner by a number of aphid vectors and is probably the most commonly encountered virus on cucurbit crops. CMV occurs in many weed hosts and can also be seedborne.

Control

Follow general suggestions for managing virus diseases (see Part 1). Virus resistant, transgenic squash cultivars are available. These plants derive their resistance from genes of the virus pathogen itself.

References

Alonso-Prados, J. L., Luis-Arteaga, M., Alvarez, J. M., Moriones, E., Batlle, A., Laviña, A., García-Arenal, F., and Fraile, A. 2003. Epidemics of aphid-transmitted viruses in melon crops in Spain. *European Journal of Plant Pathology* 109:129–138.

Arce-Ochoa, J. P., Dainello, F., Pike, L. M., and Drews, D. 1995. Field performance comparison of two transgenic summer squash hybrids to their parental hybrid line. *HortScience* 30:492–493.

Dodds, J. A., Lee, J. G., Nameth, S. T., and Laemmlen, F. F. 1984. Aphid- and whitefly-transmitted cucurbit viruses in the Imperial Valley, California. *Phytopathology* 74:221–225.

Fuchs, M., Tricoli, D. M., Carney, K. J., Schesser, M., McFerson, J. R., and Gonsalves, D. 1998. Comparative virus resistance and fruit yield of transgenic squash with single and multiple coat protein genes. *Plant Disease* 82:1350–1356.

Karchi, Z., Cohen, S., and Govers, A. 1975. Inheritance of resistance to cucumber mosaic virus in melons. *Phytopathology* 65:479–481.

Nameth, S. T., Dodds, J. A., Paulus, A. O., and Laemmlen, F. F. 1986. Cucurbit viruses of California. *Plant Disease* 70:8–12.

Provvidenti, R. 1993. Resistance to viral diseases of cucurbits. Pages 8–43 in: *Resistance to Viral Diseases of Vegetables: Genetics and Breeding*. M. M. Kyle, editor. Timber Press, Portland, Oregon.

311 ***Cucumber mosaic virus*** **affecting garden marrow.**

Papaya ringspot virus type W

PAPAYA RINGSPOT

Introduction and significance

This virus can result in significant crop loss in temperate as well as subtropical and tropical regions.

Symptoms and diagnostic features

Papaya ringspot virus type W (PRSV-W) causes severe stunting of the plant. Extremely malformed leaves have reduced lamina and irregular, dark green blisters (**312**, **313**). Fruit are distorted, knobby, and show color breaks.

312 Distorted pumpkin leaves affected by papaya ringspot.

313 Foliar symptoms of papaya ringspot on squash.

Causal agent and disease cycle

PRSV-W was previously named *Watermelon mosaic virus 1* (WMV-1). It is a potyvirus, with filamentous virions measuring 760–800 x 12 nm and containing single-stranded RNA. The host range of PRSV-W appears to be limited to cucurbits and does not include papaya. The papaya pathotype PRSV-P is serologically identical to PRSV-W but has the ability to infect cucurbits and papaya. PRSV-P is less important economically for cucurbit production. PRSV-W is vectored by aphid species.

Control

Follow general suggestions in Part 1.

References

Dodds, J. A., Lee, J. G., Nameth, S. T., and Laemmlen, F. F. 1984. Aphid- and whitefly-transmitted cucurbit viruses in the Imperial Valley, California. *Phytopathology* 74:221–225.

Wang, Y. J., Provvidenti, R., and Robinson, R. W. 1984. Inheritance of resistance to watermelon mosaic virus 1 in cucumber. *HortScience* 19:587–588.

Webb, R. E. and Scott, H. A. 1965. Isolation and identification of watermelon mosaic viruses 1 and 2. *Phytopathology* 55:895–900.

Squash leaf curl virus

SQUASH LEAF CURL

Introduction and significance

Squash leaf curl virus (SLCV) was first seen in the 1970s in the southwest region of the USA. Presently this virus is found in various cucurbit growing areas in the USA, Mexico, and Central America.

Symptoms and diagnostic features

SLCV causes leaves to take on a bright yellow mosaic or mottle symptom. Leaves can be curled or slightly distorted. Plants are stunted. Flowers produced on diseased plants may be late in opening or fall off. Fruit may not mature or can be malformed, bumpy, and have a break in color.

Causal agent and disease cycle

SLCV is a geminivirus with isometric particles that measure 20 x 30 nm and contain a single stranded, circular DNA genome. It is vectored by various bio types in the *Bemisia tabaci* whitefly group and appears to primarily infect plants in the cucurbit family.

Control

Follow general suggestions in Part 1.

References

Cohen, S., Duffus, J. E., Larsen, R. C., Liu, H. Y., and Flock, R. A. 1983. Purification, serology, and vector relationships of squash leaf curl virus, a whitefly-transmitted geminivirus. *Phytopathology* 73:1669–1673.

Dodds, J. A., Lee, J. G., Nameth, S. T., and Laemmlen, F. F. 1984. Aphid- and whitefly-transmitted cucurbit viruses in the Imperial Valley, California. *Phytopathology* 74:221–225.

Flock, R. A. and Mayhew, D. E. 1981. Squash leaf curl, a new disease of cucurbits in California. *Plant Disease* 65:75–76.

Nameth, S. T., Dodds, J. A., Paulus, A. O., and Laemmlen, F. F. 1986. Cucurbit viruses of California. *Plant Disease* 70:8–12.

Polston, J. E., Dodds, J. A., and Perring, T. M. 1989. Nucleic acid probes for detection and strain discrimination of cucurbit geminiviruses. *Phytopathology* 79:1123–1127.

Provvidenti, R. 1993. Resistance to viral diseases of cucurbits. Pages 8–43 in: *Resistance to Viral Diseases of Vegetables: Genetics and Breeding*. M. M. Kyle, editor. Timber Press, Portland, Oregon.

Squash mosaic virus

SQUASH MOSAIC

Introduction and significance

This disease has been known since the early 1900s. Its importance in many countries has decreased due to the use of virus-free seed. However, some cucurbit producing regions still experience losses from this virus.

Symptoms and diagnostic features

Squash mosaic virus (SqMV) symptoms are extremely variable. Leaves show a variety of mosaic, mottle, dark green vein banding, ring spot, and other virus-like foliar symptoms. Leaf enations can also develop. Plants are stunted and fruit are malformed and have color breaks.

Causal agent

SqMV is a comovirus that has isometric particles measuring 28–30 nm in diameter and a genome of two single-stranded linear RNAs. Distinct strains exist and cause different symptoms on different cucurbits.

Disease cycle

The virus is vectored by the western striped cucumber beetle (*Acalymma trivittatum*) and the spotted cucumber beetle (*Diabrotica undecimpunctata*). As a seedborne pathogen, an important means of spread is by the use of infested seed. The virus is very stable and can be mechanically transmitted through production and harvest procedures. This virus naturally infects plants in the cucurbit and chenopodium families.

Control

Follow general suggestions in Part 1. Use seed that has been tested and certified to be free of the virus.

References

Alvarez, M. and Campbell, R. N. 1978. Transmission and distribution of squash mosaic virus in seeds of cantaloupe. *Phytopathology* 68:257–263.

Nelson, M. R. and Knuhtsen, H. K. 1973. Squash mosaic virus variability: review and serological comparisons of six biotypes. *Phytopathology* 63:920–926.

Nolan, P. A. and Campbell, R. N. 1984. Squash mosaic virus detection in individual seeds and seed lots of cucurbits by enzyme-linked immunosorbent assay. *Plant Disease* 68:971–975.

Tobacco ringspot virus

TOBACCO RINGSPOT

Introduction and significance

Tobacco ringspot virus (TRSV) is found throughout the USA and has also been reported around the world.

Symptoms and diagnostic features

TRSV causes a very bright chlorosis on young foliage that takes the form of mosaics, mottles, and ring spots. Fruit will either abort, not grow to full size, or develop bumps, ring spots, mottles, and other distortions.

Causal agent and disease cycle

TRSV is a nepovirus that has isometric particles measuring 25–29 nm and a genome of two single-stranded RNAs. Various strains have been differentiated. This pathogen has a broad host range that includes plants in over 20 families and is vectored primarily by the soilborne dagger nematode (*Xiphinema americanum*). TRSV has been reported to be vectored non-specifically by aphids and mites; it can also be seedborne and pollen borne.

Control

Follow general suggestions in Part 1. Do not plant susceptible crops in fields having a history of this disease. Use seed that does not have significant levels of the pathogen.

Watermelon mosaic virus

WATERMELON MOSAIC VIRUS

Introduction and significance

Watermelon mosaic virus (WMV) was previously named *Watermelon mosaic virus 2* (WMV-2) and is found in cucurbit producing areas throughout the world.

Symptoms and diagnostic features

WMV causes leaves to develop mosaics, vein banding, rings, light green patches, and other symptoms (**314**). Leaves are sometimes crinkled. Fruit can be distorted or have color breaks (**315**).

314 Deformed squash leaf infected with *Watermelon mosaic virus*.

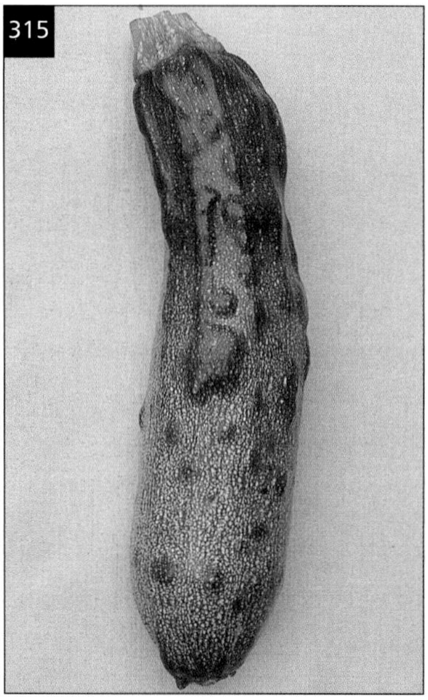

315 Deformed squash fruit infected with *Watermelon mosaic virus*.

Causal agent and disease cycle

WMV is a potyvirus that has filamentous particles measuring 730–765 nm long and single stranded linear RNA genomes. WMV infects cucurbits, legumes, and other plants and has a host range of over 150 species. This pathogen is vectored by aphids.

Control

Follow general suggestions for managing virus diseases (see Part 1). Virus resistant, transgenic squash cultivars are available. These plants derive their resistance from genes of the virus pathogen itself.

References

Alonso-Prados, J. L., Luis-Arteaga, M., Alvarez, J. M., Moriones, E., Batlle, A., Laviña, A., García-Arenal, F., and Fraile, A. 2003. Epidemics of aphid-transmitted viruses in melon crops in Spain. *European Journal of Plant Pathology* 109:129–138.

Arce-Ochoa, J. P., Dainello, F., Pike, L. M., and Drews, D. 1995. Field performance comparison of two transgenic summer squash hybrids to their parental hybrid line. *HortScience* 30:492–493.

Fuchs, M., Tricoli, D. M., Carney, K. J., Schesser, M., McFerson, J. R., and Gonsalves, D. 1998. Comparative virus resistance and fruit yield of transgenic squash with single and multiple coat protein genes. *Plant Disease* 82:1350–1356.

Mclean, G. D., Burt, J. R., Thomas, D. W., and Sproul, A. N. 1982. The use of reflective mulch to reduce the incidence of watermelon mosaic virus in Western Australia. *Crop Protection* 1:491–496.

Zucchini yellow mosaic virus

ZUCCHINI YELLOW MOSAIC

Introduction and significance

Zucchini yellow mosaic virus (ZYMV) is an important cucurbit virus throughout the world and can at times cause tremendous crop loss. The disease first appeared in northern Italy and southern France in the late 1970s, spread world-wide during the 1980s, and reached the UK in 1987.

Symptoms and diagnostic features

Leaves are severely distorted and develop dark green blisters, strip-like lamina (called laciniate or filiform), serrated edges, necrosis, enations, and other malformations (**316**). Plants can be extremely stunted and stem internodes are very short. In severe cases, most of the leaves have the laciniate or filiform distortion; in this case overall plant canopy and leaf surface area are significantly reduced. Fruit can be greatly distorted and have knobs, swellings, and cracks (**317**).

Causal agent and disease cycle

ZYMV is a potyvirus that has filamentous virions measuring 750 x 11 nm and single-stranded linear RNA genomes. Some strains have been differentiated. Information is lacking on the natural host range of this pathogen. ZYMV is vectored by aphids, can be seedborne within the seed coat, and may be spread on cutting knives during harvesting.

Control

Follow general suggestions for managing virus diseases (seePart 1). Resistant cultivars are available and mild-strain inoculation of seedlings has been shown to protect plants against severe strains. Virus resistant, transgenic squash cultivars are also available. These plants derive their resistance from genes of the virus pathogen itself.

316 Leaf symptoms of cucumber infected with *Zucchini yellow mosaic virus*.

317 Cucumber fruit deformed by *Zucchini yellow mosaic virus*.

References

Arce-Ochoa, J. P., Dainello, F., Pike, L. M., and Drews, D. 1995. Field performance comparison of two transgenic summer squash hybrids to their parental hybrid line. *HortScience* 30:492–493.

Blua, M. J. and Perring, T. M. 1989. Effect of zucchini yellow mosaic virus on development and yield of cantaloupe. *Plant Disease* 73:317–320.

Desbiez, C. and Lecoq, H. 1997. Zucchini yellow mosaic virus. *Plant Pathology* 46:809–829.

Fuchs, M., Tricoli, D. M., Carney, K. J., Schesser, M., McFerson, J. R., and Gonsalves, D. 1998. Comparative virus resistance and fruit yield of transgenic squash with single and multiple coat protein genes. *Plant Disease* 82:1350–1356.

Lisa, V., Boccardo, G., D'Agostino, G., Dellavalle, G., and d'Aquilio, M. 1981. Characterization of a potyvirus that causes zucchini yellow mosaic. *Phytopathology* 71:667–672.

Nameth, S. T., Dodds, J. A., Paulus, A. O., and Kishaba, A. 1985. Zucchini yellow mosaic virus associated with severe diseases of melon and watermelon in Southern California desert valleys. *Plant Disease* 69:785–788.

Nameth, S. T., Dodds, J. A., Paulus, A. O., and Laemmlen, F. F. 1986. Cucurbit viruses of California. *Plant Disease* 70:8–12.

Provvidenti, R. 1991. Inheritance of resistance to the Florida strain of zucchini yellow mosaic virus in watermelon. *HortScience* 26:407–408.

Provvidenti, R. 1993. Resistance to viral diseases of cucurbits. Pages 8–43 in: *Resistance to Viral Diseases of Vegetables: Genetics and Breeding*. M. M. Kyle, editor. Timber Press, Portland, Oregon.

Purcifull, D. E., Adlertz, W. C., Simone, G. W., Hiebert, E., and Christie, S. R. 1984. Serological relationships and partial characterization of zucchini yellow mosaic virus isolated from squash in Florida. *Plant Disease* 68:230–233.

Walkey, D. G. A., Lecoq, H., Collier, R., and Dobson, S. 1992. Studies on the control of zucchini yellow mosaic virus in courgettes by mild strain protection. *Plant Pathology* 41:762–771.

Fabaceae Pea family

The Fabaceae (pea family) is the second largest of the flowering plant familes, with over 16,000 species in 750 genera. Commonly known as legumes or pulses, they are second only to cereals in their economic and nutritional importance in the human diet. This chapter has individual sections on the diseases of broad bean, pea, and bean. The bean group is diverse and the genus *Phaseolus* comprises over 150 species. Common or snap bean (*P. vulgaris*) is primarily grown for its green pods that are used for fresh markets or for freezing and canning. Common bean is generally divided into several categories: French bean (both pods and under-developed seeds are consumed); haricot filet bean (only immature pods are eaten because fully mature pods are stringy); haricot bean (only fresh seed are eaten); dry field bean (dry seed are used for consumption, and the pods are not eaten). Scarlet runner bean (*Phaseolus coccineus*) is probably native to Central American and Mexico. This species is sometimes used as an ornamental plant, but is also grown for its young pods that are used in similar ways as common bean. The lima bean (*P. lunatus*) is grown for its large seeds, and the pods are not consumed. Lima beans are used for canned, frozen, and fresh market products.

Pea (*Pisum sativum*) is a familiar vegetable commodity that is used fresh, frozen, canned, and dried. Some cultivars are grown for their immature, edible pods and under-developed seed. Other cultivars are grown for the seed, as the pods are not consumed.

Broad bean (*Vicia faba*), also known as faba bean, horsebean, or English bean, is grown commercially for the very large seeds that are shelled and used for frozen, fresh, and dried beans. Broad bean is an important vegetable in the UK and other European countries. In the USA, in addition to being considered a specialty vegetable, broad bean is used as a cover crop for adding organic matter to the soil.

PHASEOLUS SPECIES (BEANS)

Pseudomonas syringae pv. *phaseolicola*

HALO BLIGHT

Introduction and significance

This seedborne bacterial disease is an important and widespread problem in both common and scarlet runner bean. It may be the most important disease of *Phaseolus* species in Europe. Halo blight is also damaging in cooler, high altitude climates in the tropics.

Symptoms and diagnostic features

Early symptoms appear as greasy looking, water-soaked, angular leaf lesions (**318**). The young lesions expand and coalesce into larger lesions that are brown and can reach 2–3 cm in diameter. Old lesions dry and become red to brown. Developing and older leaf spots usually are surrounded by a distinct yellow halo, which is a conspicuous feature of this disease (**319**). In severe cases, the bean plant can show systemic chlorosis in

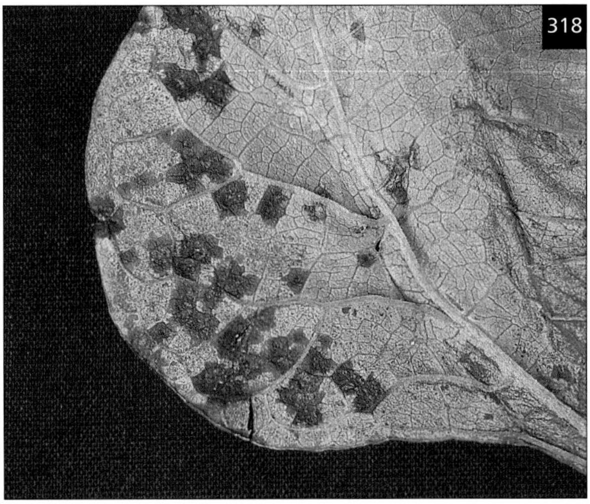

318 **Leaf lesions of halo blight of bean.**

319 **Chlorosis or 'halo' of halo blight of bean.**

which much of the foliage turns yellow. Seedlings grown from infected seed may be stunted and have a 'snake's head' symptom because leaf tissue (plumules) has been destroyed. Water-soaked or yellow-brown lesions also develop at the primary leaf node, which girdle and weaken the stem so that plants eventually snap off at this node. The disease may cause a red discoloration in the interveinal leaf tissue and elongated red streaks on stems. Leaf veins may also take on a red discoloration. Pods develop water-soaked, round to oval spots on sides of pods, and elongated lesions along the sutures. Early pod infection may cause seeds to rot or to remain immature. A cream or silver-colored bacterial ooze may be exuded from pod lesions.

Causal agent

Halo blight is caused by the bacterium *Pseudomonas syringae* pv. *phaseolicola*. This pathogen is an aerobic, Gram-negative, rod-shaped bacterium. The pathogen can be isolated on standard microbiological media and produces cream-colored colonies typical of most pseudomonads. When cultured on Kings medium B, it produces a diffusible pigment that fluoresces blue under ultraviolet light. Strains of this pathogen are host specific to bean, hence the pathovar (pv.) designation. This pathogen is seedborne. There are several races of this pathogen; the particular strain of the pathogen and environmental conditions influence the development and degree of the yellow halo, which is due to the production of the bacterial toxin phaseolotoxin.

Some strains of a different *Pseudomonas* pathogen, *Pseudomonas syringae* pv. *syringae*, are pathogenic to bean and cause bacterial brown spot disease. Symptoms of bacterial brown spot can be somewhat similar to those of halo blight. Spots tend to be more circular, are brown in color, and a yellow halo can also surround the lesion.

Disease cycle

Infected seed is the most important source of inoculum, as there appears to be limited survival on crop residues or in soil. Weed hosts may be important in some regions. The pathogen is spread by splashing water from rain or sprinkler irrigation. In field crops, only one infected plant in 10,000 is sufficient to cause an epidemic. The bacterium invades through stomata or wounds and then spreads through the vascular system. It is able to break out of the xylem and into the parenchyma where it causes tissue collapse. The disease cycle takes less than 7 days under optimum conditions. Disease development is most pronounced at cooler temperatures, 16–20° C; however, the production of toxin and appearance of halo symptoms develop at warmer temperatures: 20–23° C. Seeds are infected through the vascular tissues or by direct infection through pod walls.

Control

Use seed that does not have significant levels of the pathogen. Apply copper sprays to foliage, though such treatments may be only marginally effective. Choose resistant cultivars. Irrigate with furrow or drip irrigation, and avoid the use of overhead sprinklers. Eliminate volunteer and other hosts.

References

Cheng, G. Y., Legard, D. E., Hunter, J. E., and Burr, T. E. 1989. Modified bean pod assay to detect strains of *Pseudomonas syringae* pv. *syringae* that cause bacterial brown spot of snap bean. *Plant Disease* 73:419–423.

Hirano, S. S., Rouse, D. I., Clayton, M. K., and Upper, C. D. 1995. *Pseudomonas syringae* pv. *syringae* and bacterial brown spot of snap bean: a study of epiphytic phytopathogenic bacteria and associated disease. *Plant Disease* 79:1085–1093.

Legard, D. E. and Schwartz, H. F. 1987. Sources and management of *Pseudomonas syringae* pv. *phaseolicola* and *Pseudomonas syringae* pv. *syringae* epiphytes on dry beans in Colorado. *Phytopathology* 77:1503–1509.

Lindemann, J., Arny, D. C., and Upper, C. D. 1984. Epiphytic population of *Pseudomonas syringae* pv. *syringae* on snap beans and nonhost plants and the incidence of bacterial brown spot. *Phytopathology*. 74:1329–1333.

Rico, A., Lopez, R., Asenio, C., Aizpun, M. T., Asensio-S.-Manzanera, M. L., and Murillo, J. 2003. Nontoxigenic strains of *Pseudomonas syringae* pv. *phaseolicola* are a main cause of halo blight of beans in Spain and escape current detection methods. *Phytopathology* 93:1553–1559.

Schaad, N. W., Azad, H., Peet, R. C., and Panopoulos, N. J.1989. Identification of *Pseudomonas syringae* pv. *phaseolicola* by a DNA hybridization probe. *Phytopathology* 79: 903–907.

Serfontein, J. J. 1994. Occurrence of bacterial brown spot of dry beans in Transvaal province of South Africa. *Plant Pathology* 43: 597–599.

Taylor, J. D. 1972. Field studies on halo-blight of beans (*Pseudomonas phaseolicola*) and its control by foliar sprays. *Annals of Applied Biology* 70:191–197.

Taylor, J. D., Dudley, C. L., and Presly, L. 1979. Studies of halo-blight infected seed and disease transmission in dwarf beans. *Annals of Applied Biology* 93: 267–277.

Taylor, J. D., Phelps, K., and Dudley, C. L. 1979. Epidemiology and strategy for the control of halo-blight of beans. *Annals of Applied Biology* 93:167–172.

Taylor, J. D., Teverson, D. M., Allen, D. J., and Pastor–Corrales, M. A. 1996. Identification and origin of races of *Pseudomonas syringae* pv. *phaseolicola* from Africa and other bean growing areas. *Plant Pathology* 45: 469–478.

Taylor, J. D., Teverson, D. M., and Davis, J. H. C. 1996. Sources of resistance to *Pseudomonas syringae* pv. *phaseolicola* races in *Phaseolus vulgaris*. *Plant Pathology* 45:479–485.

Webster, D. M., Atkin, J. D., and Cross, J. E. 1983. Bacterial blights of snap beans and their control. *Plant Disease* 67:935–940.

Yessad-Carreau, S., Manceau, C., and Luisetti, J. 1994. Occurrence of specific reactions induced by *Pseudomonas syringae* pv. *syringae* on bean pods, lilac and pear plants. *Plant Pathology* 43:528–536.

Xanthomonas campestris pv. *phaseoli*

COMMON BACTERIAL BLIGHT

Introduction and significance

Like halo blight, this seedborne bacterial disease is a very important and widespread problem on bean. The disease is damaging in humid environments and is found in most bean production areas. This disease is actually caused by two very closely related, but distinct bacterial pathogens.

Symptoms and diagnostic features

Early symptoms are water-soaked, angular leaf lesions. The young lesions expand and coalesce into larger lesions that are gray-brown and irregular in shape. Leaf spots are usually surrounded by a narrow yellow halo (**320**). Pods develop water-soaked, round to oval spots on sides of pods (**321**). Pod lesions later turn brown to red-brown. Bacterial ooze may be exuded from pod lesions.

Causal agents

Common bacterial blight is caused by the bacterium *Xanthomonas campestris* pv. *phaseoli*. This pathogen is an aerobic, Gram-negative, rod-shaped bacterium. The pathogen can be isolated on standard microbiological media and produces yellow, mucoid, slow growing colonies typical of most xanthomonads. This bacterium hydrolyzes starch, so starch-based semi-selective media such as SX and MXP media are useful for isolating and identifying this pathogen. Strains of this pathogen are restricted to hosts in the Fabaceae family. This pathogen is seedborne.

320 **Angular lesions and surrounding yellow borders on bean leaves caused by common bacterial blight.**

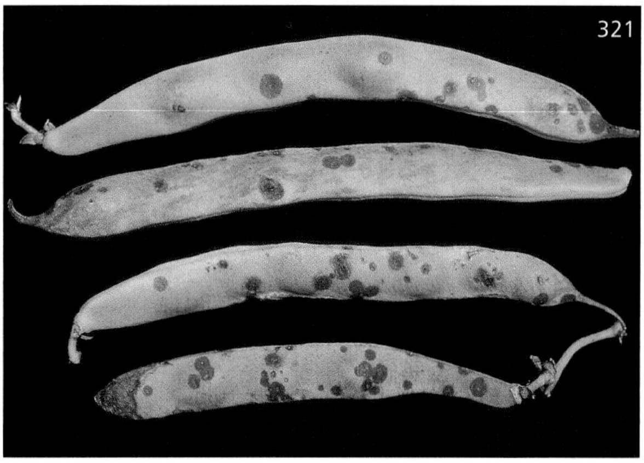

321 **Bean pod lesions caused by the common bacterial blight pathogen.**

Common bacterial blight is also caused by a variant of *X. campestris* pv. *phaseoli*. This variant causes the same disease symptoms, and has the same host range and similar biochemical profile. However, in culture this variant produces a brown pigment on tyrosine medium. The pigment is called fuscous, meaning dusky brown in color. This pathogen is named *X. campestris* pv. *phaseoli* var. *fuscans*. There is genetic evidence (in the form of low DNA–DNA homology) that these two variants are genetically distinct from each other.

Disease cycle

Infected seed is an important source of inoculum. Weed hosts may be important in some regions. In temperate climates, the pathogen may survive in crop residue and provide another source of inoculum. The pathogen is spread by splashing water from rain or sprinkler irrigation. Disease development is most pronounced at warmer temperatures of 28–32° C with high humidity.

Control

Use seed that does not have significant levels of the pathogen. Rotate crops so that beans are not planted into fields having infested crop residues. Apply copper sprays to foliage, though such treatments may be only marginally effective. Choose resistant cultivars. Irrigate with furrow or drip irrigation, and avoid the use of overhead sprinklers. Eliminate volunteer and other hosts.

References

Audy, P., Laroche, A., Saindon, G., Huang, H. C., and Gilbertson, R. L. 1994. Detection of bean common blight bacteria, *Xanthomonas campestris* pv. *phaseoli* and *Xanthomonas campestris* pv. *phaseoli* var. *fuscans*, using the polymerase chain reaction. *Phytopathology* 84:1185–1192.

Birch, P. R. J., Hyman, L. J., Taylor, R., Opio, A. F., Bragard, C., and Toth, I. K. 1997. RAPD PCR-based differentiation of *Xanthomonas campestris* pv. *phaseoli* and *Xanthomonas campestris* pv. *phaseoli* var. *fuscans*. *European Journal of Plant Pathology* 103:809–814.

Chan, J. W.Y.F. and Goodwin, P. H. 1999. Differentiation of *Xanthomonas campestris* pv. *phaseoli* from *Xanthomonas campestris* pv. *phaseoli* var. *fuscans* by PFGE and RFLP. *European Journal of Plant Pathology* 105:867–878.

Claflin, L. E., Vidaver, A. K., and Sasser, M. 1987. MXP, semiselective medium for *Xanthomonas campestris* pv. *phaseoli*. *Phytopathology* 77:730–734.

Gilbertson, R. L., Maxwell, D. P., Hagedorn, D. J., and Leong, S. A. 1989. Development and application of plasmid DNA probe for detection of bacteria causing common bacterial blight of bean. *Phytopathology* 79:518–525.

Gilbertson, R. L., Rand, R. E., and Hagedorn, D. J. 1990. Survival of *Xanthomonas campestris* pv. *phaseoli* and pectolytic strains of *X. campestris* in bean debris. *Plant Disease* 74:322–327.

Malin, E. M., Roth, D. A., and Belden, E. L. 1983. Indirect immunofluorescent staining for detection and identification of *Xanthomonas campestris* pv. *phaseoli* in naturally infected bean seed. *Plant Disease* 67:645–647.

Mkandawire, A. B. C., Mabagala, R. B., Guzman, P., Gepts, P., and Gilbertson, R. L. 2004. Genetic diversity and pathogenic variation of common blight bacteria (*Xanthomonas campestris* pv. *phaseoli* and *Xanthomonas campestris* pv. *phaseoli* var. *fuscans*) suggests pathogen coevolution with the common bean. *Phytopathology* 94:593–603.

Aphanomyces euteiches, Fusarium oxysporum* f. sp. *phaseoli, Phoma medicaginis* var. *pinodella, Pythium* spp., *Rhizoctonia solani, Thielaviopsis basicola

ROOT/FOOT ROT COMPLEX

Introduction and significance

Foot rot disease is extremely common, and symptoms of this problem can be found in most crops. The foot rot complex of pathogens is probably one of the most important disease problems in legumes.

Symptoms and diagnostic features

This group of pathogens causes damping-off of seeds and seedlings, and brown to red-brown, sunken lesions on plant hypocotyls and taproots below the soil surface (**322**, **323**, **324**, **325**, **326**). Lesions may continue to develop until the stem base and roots are severely discolored, shriveled, and nonfunctional. Affected plants may be stunted, unproductive, and in some cases will wither and die.

Causal agents

The main pathogens include *Fusarium solani* f. sp. *phaseoli*, *Rhizoctonia solani*, *Phoma medicaginis* var. *pinodella* (formerly known as *Ascochyta pinodella*), *Thielaviopsis basicola* (cause of black root rot), *Aphanomyces euteiches*, and *Pythium* species. The *Fusarium* and *Rhizoctonia* pathogens are particularly common in California. For *R. solani* a number of different isolates, or anastomosis groups (AGs) can infect bean. See the pea section on foot rot complex for more details (page 272).

Disease cycle

Inoculum resides in the soil in the form of various survival structures, spores, or mycelium. Nutrients and exudates released from germinating bean seeds and roots stimulate pathogen activity and germination of chlamydospores, oospores, and spores. Disease development occurs at a wide range of temperatures. Overly wet or poorly draining soils enhance disease severity and incidence. Other factors that increase the problem include compacted soils, acid soils, low soil fertility, and frequent plantings of legumes. Bean hypocotyls become less susceptible to *R. solani* approximately 2 weeks after planting; this increase in resistance is associated with calcification of the middle lamella between plant cells.

Control

Rotate crops and avoid over planting legumes. A minimum of 4 years between such crops is needed to help reduce disease pressure. Avoid growing beans in fields having histories of severe foot rot. Correct poor drainage and soil compaction problems. Irrigate fields so that excess water is not applied. Fungicide seed treatments are useful in reducing damping-off and improving crop establishment; however, these materials have little effect on the foot rot phase of the disease.

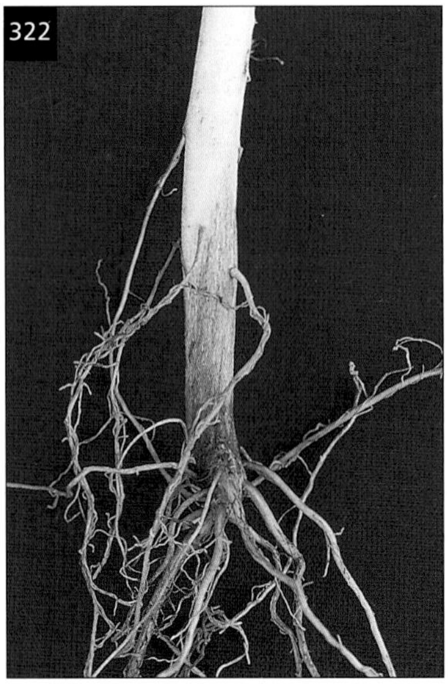

322 Hypocotyl lesions due to Fusarium root rot of bean.

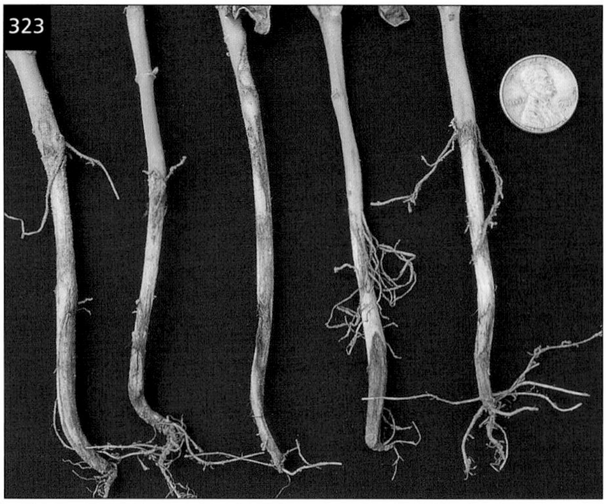

323 Hypocotyl lesions due to Rhizoctonia root rot of bean.

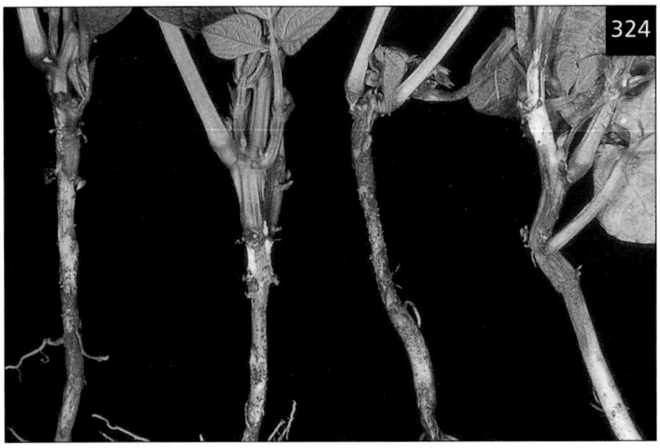

324 Infected hypocotyls of Pythium root rot of bean.

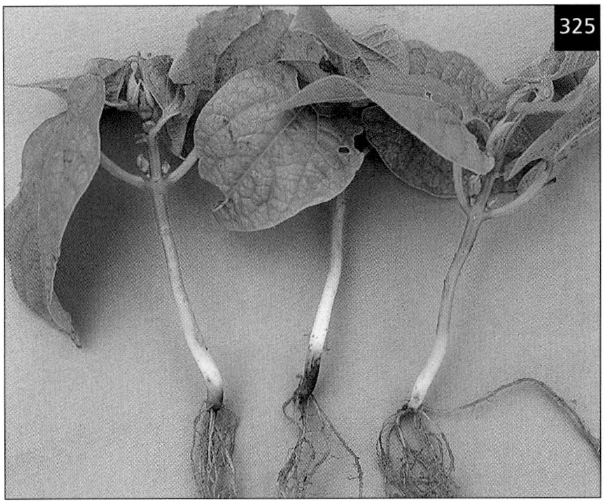

325 Blackening at stem base of bean caused by *Thielaviopsis basicola.*

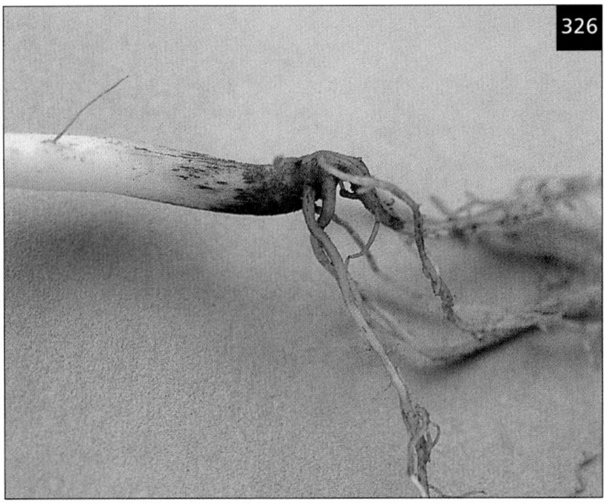

326 Lower stem lesion of bean caused by *Thielaviopsis basicola.*

References

Alves-Santos, F. M., *et al* 2002. Pathogenicity and race characterization of *Fusarium oxysporum* f. sp. *phaseoli* isolates from Spain and Greece. *Plant Pathology* 51:605–611.

Burke, D. W. and Miller, D. E. 1983. Control of Fusarium root rot with resistant beans and cultural management. *Plant Disease* 67:1312–1317.

Datnoff, L. E. and Sinclair, J. B. 1988. Interaction of *Fusarium oxysporum* and *Rhizoctonia solani* in causing root rot of soybeans. *Phytopathology* 78:771–777.

France, R. A. and Abawi, G. S. 1994. Interaction between *Meloidogyne incognita* and *Fusarium oxysporum* f. sp. *phaseoli* on selected bean genotypes. *Journal of Nematology* 26:467–474.

Godoy-Lutz, G., Steadman, J. R., Higgins, B., and Powers, K. 2003. Genetic variation among isolates of the web blight pathogen of common bean based on PCR-RFLP of its ITS-rDNA region. *Plant Disease* 87:766–771.

Hall, R. and Phillips, L. G. 1992. Effects of crop sequence and rainfall on population dynamics of *Fusarium oxysporum* f. sp. *phaseoli* in soil. *Canadian Journal of Botany* 70:2005–2008.

Henis, Y. and Ben-Yephet, Y. 1970. Effect of propagule size of *Rhizoctonia solani* on saprophytic growth, infectivity, and virulence on bean seedlings. *Phytopathology* 60:1351–1356.

Lloyd, A. B. and Lockwood, J. L. 1963. Effect of soil temperature, host variety, and fungus strain on Thielaviopsis root rot of peas. *Phytopathology* 53:329–331.

Maier, C. R. 1961. Black root-rot development on pinto beans, incited by *Thielaviopsis basicola* isolates, as influenced by different soil temperatures. *Plant Disease Reporter* 45:804–807.

Miller, D.E. and Burke, D. W. 1986. Reduction of Fusarium root rot and Sclerotinia wilt in beans with irrigation, tillage, and bean genotype. *Plant Disease* 70:163–166.

Muyolo, N. G., Lipps, P. E., and Schmitthenner, A. F. 1993. Reactions of dry bean, lima bean, and soybean cultivars to Rhizoctonia root and hypocotyl rot and web blight. *Plant Disease* 77:234–238.

Papavizas, G. C. and Adams, P. B. 1969. Survival of root-infecting fungi in soil: XII. Germination and survival of endoconidia and chlamydospores of *Thielaviopsis basicola* in fallow soil and in soil adjacent to germinating bean seed. *Phytopathology* 59:371–378.

Papavizas, G. C., Adams, P. B., Lumsden, R. D., Lewis, J. A., Dow, R. L., Ayers, W. A., and Kantzes, J. G. 1975. Ecology and epidemiology of *Rhizoctonia solani*. *Phytopathology* 65:871–877.

Paulus, A. O., Brendler, R. A., Nelson, J., and Otto, H. W. 1985. Rhizoctonia stem canker on beans. *California Agriculture* 39 (11/12):13–14.

Pfender, W. F. and Hagedorn, D. 1982. *Aphanomyces euteiches* f. sp. *phaseoli*, a causal agent of bean root and hypocotyl rot. *Phytopathology* 72:306–310.

Silbernagel, M. J. and Mills, L. J. 1990. Genetic and cultural control of Fusarium root rot in bush snap beans. *Plant Disease* 74:61–66.

Snapp, S., Kirk, W., Roman-Aviles, B., and Kelly, J. 2003. Root traits play a role in integrated management of Fusarium root rot in snap beans. *HortScience* 38:187–191.

Botrytis cinerea (teleomorph = *Botryotinia fuckeliana*)

GRAY MOLD

Introduction and significance

Gray mold is frequently damaging on common bean grown under cool temperate conditions. *Botrytis cinerea* is a weak pathogen, and problems are usually associated with damaged tissues. Crop loss occurs from diseased pods and also when it proves impractical for processors to pick out and remove infected pods.

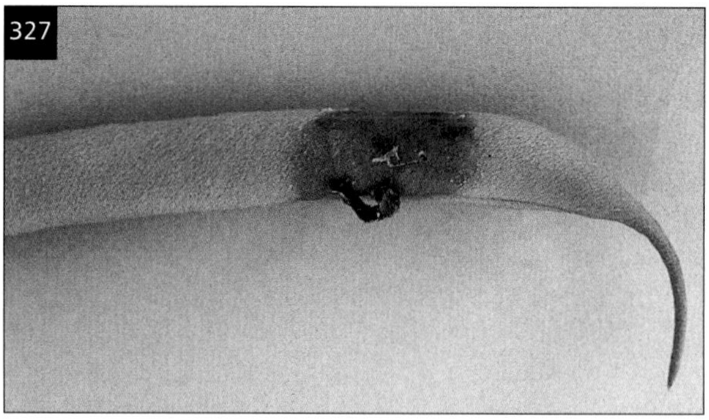

327 Bean pod lesion caused by *Botrytis cinerea* and associated with adhering petal.

Symptoms and diagnostic features

Young plants that are damaged during crop management operations or from environmental extremes are subject to infection by *Botrytis*, which can infect and girdle stems and other compromised tissues. Such plants can die. Leaves are subject to gray mold if they are damaged by winds and other abrasion factors, or if fallen petals adhere to leaves. The pathogen colonizes damaged leaf tissue, or grows on the nutrient-rich petal tissue and then bridges into healthy leaf tissue. In either case, the result is a water-soaked lesion that rapidly turns into a brown lesion that supports the gray growth of the fungus. However, the most serious problem involves infection of flowers and developing bean pods. Pod lesions often develop where floral tissues adhere to the base or tips of developing pods (**327**). Once pods have been colonized, there is rapid enlargement of water-soaked lesions and the development of the characteristic gray fungal growth (**328**). Direct contact between infected and healthy pods enables secondary spread to take place.

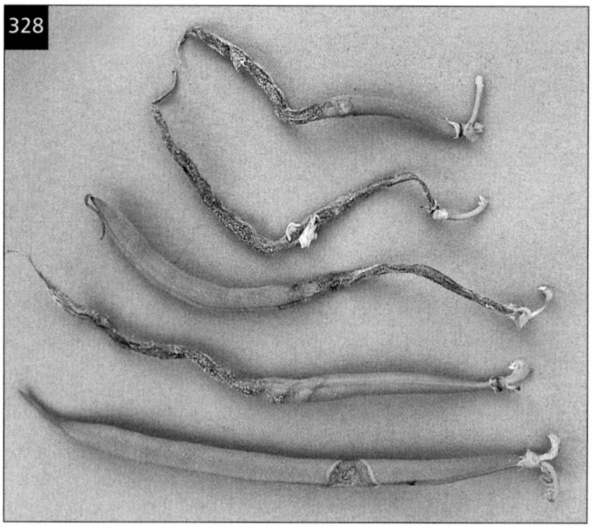

328 Bean pods infected with *Botrytis cinerea*.

Causal agent

Gray mold is caused by the ascomycete fungus *Botryotinia fuckeliana*. The disease is more commonly associated with the asexual form *Botrytis cinerea*. Conidiophores of *B. cinerea* are long (1–2 mm), become gray-brown with maturity, and branch irregularly near the apex. Conidia are clustered at the branch tips and are single-celled, pale brown, ellipsoid to obovoid, and measure 6–18 x 4–11 µm. The pathogen can be isolated on standard microbiological media. Some isolates sporulate poorly in culture unless incubated under lights (12 h light/12 h dark). If formed, sclerotia are black, oblong or dome-shaped, and measure 4–10 mm. The fungus grows best at 18–23° C but is inhibited at warm temperatures above 32° C. On host tissue the fungus produces characteristically profuse sporulation that is dense, velvety, and grayish brown in color.

Disease cycle

Under moderate (15–20° C) and wet conditions with high humidity (90–95% relative humidity), disease development can be rapid. *Botrytis* is also important on bean in storage and in transit as even a low level of infection can result in crop rejection. Under dry and warm conditions, the pathogen is inhibited and lesions dry out and become bleached. *Botrytis* requires wounded tissue or a food base, such as detached petals, to incite disease.

Control

Plant cultivars that have some tolerance to *Botrytis*; such differences in susceptibility have been seen in common bean. Cultivars that bear pods high off the ground may be less prone to pod infection. Manage the crop to avoid damage to the plants. Select sites that have good air movement that allows foliage to dry, and plant crop rows parallel to prevailing winds. Apply fungicides at early flowering; however, *B. cinerea* is widely known to develop resistance to many fungicides.

References

Johnson, K. B. and Powelson, M. L. 1983. Influence of prebloom disease established by *Botrytis cinerea* and environmental and host factors on gray mold pod rot of snap bean. *Plant Disease* 67:1198–1202.

Polach, F. J. and Abawi, G. S. 1975. The occurrence and biology of *Botryotinia fuckeliana* on beans in New York. *Phytopathology* 65:657–660.

Vulsteke, G. and Meeus, P. 1982. Chemical control of *Botrytis cinerea* and *Sclerotinia sclerotiorum* on dwarf snap beans. *Netherlands Journal of Plant Pathology* 88:79–85.

329 Powdery mildew of bean.

Erysiphe polygoni

POWDERY MILDEW

Introduction and significance

Powdery mildew has a worldwide distribution on bean and in some areas is a common problem. However, significant yield losses due to powdery mildew are rare.

Symptoms and diagnostic features

Early symptoms are darkened, discolored areas on the upper surfaces of the oldest leaves. Such spots become colonized with diffuse, white radiating fungal colonies (**329**). Severe powdery mildew growth may cause foliage to twist, become distorted, and turn yellow. The fungus colonizes stems and pods, causing pods to be small and deformed.

Causal agents

Powdery mildew is caused by the obligate fungus *Erysiphe polygoni*. Conidia are ellipsoidal, typically measure 26–52 x 15–23 µm, and are produced in chains. The perfect stage appears as small black fruiting bodies (cleistothecia) but is rarely formed on green foliage of bean. Distinct races may exist.

Recently a second powdery mildew was reported on bean in Hungary. Symptoms and signs of this mildew are similar to those of *E. polygoni*. This second powdery mildew pathogen is *Podosphaera fusca* (previously named *Sphaerotheca fuliginea* or *S. fusca*).

Disease cycle

Disease is favored by warm, dry days and cooler nights, leading to dew formation. The fungus overwinters on diseased hosts, and sometimes as cleistothecia on plant debris. Conidia are air dispersed. Powdery mildew usually occurs on bean late in the production season.

Control

Plant resistant cultivars or cultivars that are less severely affected by powdery mildew. Early planting and irrigating with overhead sprinklers can lower disease severity. Apply sulfur and other fungicides when early stages of disease are observed.

References

Kiss, L. and Szentivanyi, O. 2001. Infection of bean with cucumber powdery mildew, *Podosphaera fusca*. *Plant Pathology* 50:411.

Fusarium oxysporum f. sp. *phaseoli*

FUSARIUM WILT, FUSARIUM YELLOWS

Introduction and significance

Fusarium wilt of bean is found in North America, South America, Africa, and Europe. The disease can be severe in parts of Africa, Spain, Brazil, Colombia, Panama, and Costa Rica.

Symptoms and diagnostic features

Early symptoms include chlorosis of leaves and premature senescence of lower foliage. As the disease progresses, foliage can become bright yellow. Vascular tissue of main taproot, crown, and lower stem turns red-brown in color.

Causal agent

Fusarium wilt is caused by the fungus *Fusarium oxysporum* f. sp. *phaseoli*. From diseased vascular tissue in stems, the pathogen can be isolated onto standard agar media such as potato dextrose agar. Pathogen morphology and colony characteristics are similar to other *F. oxysporum* fungi. For more information see the section on Fusarium wilt of pea in this chapter. At least five distinct races have been documented. This pathogen can survive in the soil for many years.

Disease cycle

The fungus infects bean roots and travels up the vascular tissue and into the stem and above ground tissues. Disease development is favored by soil temperatures around 20° C.

Control

Use resistant cultivars if such are available. Rotate crops away from bean, though such rotations do not eliminate soil inoculum.

References

Alves-Santos, F. M., *et al.* 2002. Pathogenicity and race characterization of *Fusarium oxysporum* f. sp. *phaseoli* isolates from Spain and Greece. *Plant Pathology* 51:605–611.

Salgado, M. O. and Schwartz, H. F. 1993. Physiological specialization and effects of inoculum concentration of *Fusarium oxysporum* f. sp. *phaseoli* on common beans. *Plant Disease* 77:492–496.

Woo, S. L., *et al.* 1996. Characterization of *Fusarium oxysporum* f. sp. *phaseoli* by pathogenic races, VCGs, RFLPs, and RAPD. *Phytopathology* 86:966–973.

Glomerella lindemuthiana
(anamorph = *Colletotrichum lindemuthianum*)

ANTHRACNOSE

Introduction and significance

This is an important disease of common bean (*Phaseolus vulgaris*) and also occurs on scarlet runner bean (*P. coccineus*), other *Phaseolus* species (including *P. aureus*, *P. lunatus*, *P. mungo*, *P. radiatus*), broad bean (*Vicia faba*), and cowpea (*Vigna unguiculata*). It occurs in all production areas except hot, dry regions where furrow irrigation is used.

Symptoms and diagnostic features

Leaf symptoms are angular red-brown spots and red-brown sections of leaf veins (**330**). Stems and petioles have brown, sunken, elliptical or circular lesions with darker red-brown margins (**331**). Similar lesions form on the hypocotyl when spores wash down from the cotyledons or leaves. These lesions are typically 1–2 cm long and cause stem collapse when they girdle the stem. As disease progresses, dark brown fruiting bodies develop in the lesions; lesions later ooze pink to orange spore exudates. Pods also develop the sunken, brown, circular lesions and fruiting bodies (**332, 333**). Seed infection occurs where pod lesions extend through the pod wall. Affected seeds have yellow or brown lesions and can be shriveled, under developed, and result in poor germination. Severely affected plants show considerable loss of foliage and pod distortion.

330

330 Leaf vein discoloration from anthracnose of bean.

Causal agent

Anthracnose is caused by the ascomycete fungus *Glomerella lindemuthiana*. However, the disease is more commonly associated with the asexual stage *Colletotrichum lindemuthianum*. *Colletotrichum lindemuthianum* produces cup-shaped fruiting bodies called acervuli. Each acervulus consists of a cluster of conidiophores, long dark setae, and masses of pink conidia. The conidia are aseptate, cylindrical with rounded ends, and measure 11–20 x 2.5–5 µm. This is a complex pathogen and consists of many races, genotypes, and anastomosis groups.

Disease cycle

Anthracnose is primarily a seedborne disease, and survival on crop residues is limited. In temperate regions the fungus survives in the soil on residues for only a few months during cold winter conditions. However, if residues remain dry and undecomposed, the fungus can remain viable for as long as 5 years. In tropical and subtropical areas it is common to have crop-free periods lasting only a few months; in such cases, the anthracnose fungus can easily persist and infect subsequent bean plantings. In humid conditions, the fungus sporulates on infected seedlings and other tissues. Spores are then splash-dispersed to surrounding plants.

Control

Use resistant cultivars, as these are important for the management of anthracnose. Breeding resistant cultivars has been very successful, but there are many races and new ones continue to appear; therefore, the appropriate resistance must be deployed. The dominant ARE gene has been important in European cultivars, but in South and Central America, races have emerged that can overcome it. Use seed that does not have significant levels of the pathogen. Fungicides have been used successfully as seed treatments and as foliar sprays, but they may not be required on resistant cultivars. Ensure that there is a rotation of at least 2 years between susceptible bean crops.

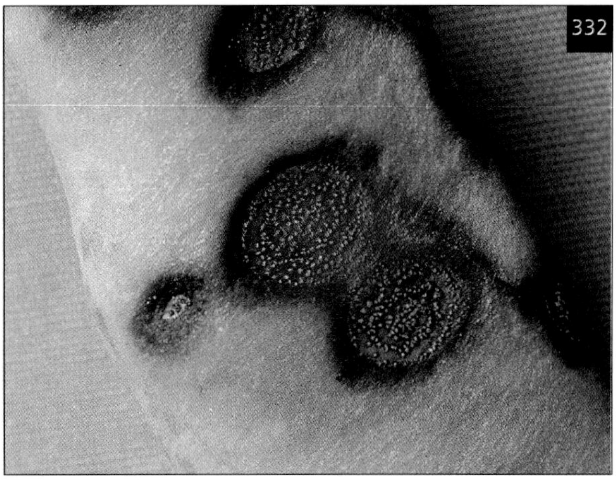

332 Pod lesions of anthracnose on bean.

References

Ansari, K. I., Palacios, N., Araya, C., Langin, T., Egan, D., and Doohan, F. M. 2004. Pathogenic and genetic variability among *Colletotrichum lindemuthianum* isolates of different geographic origins. *Plant Pathology* 53:635–642.

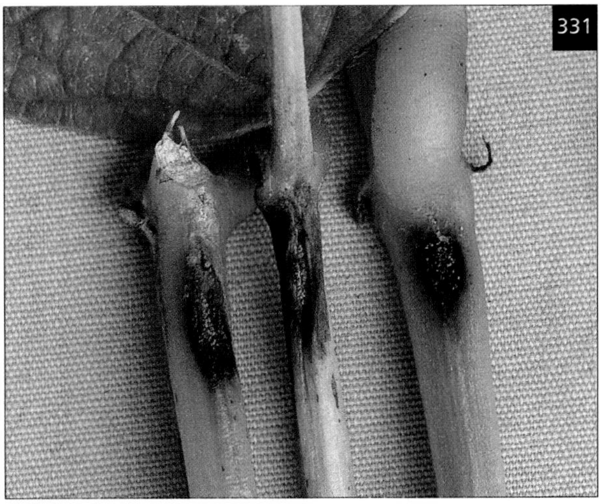

331 Stem lesions of anthracnose on bean.

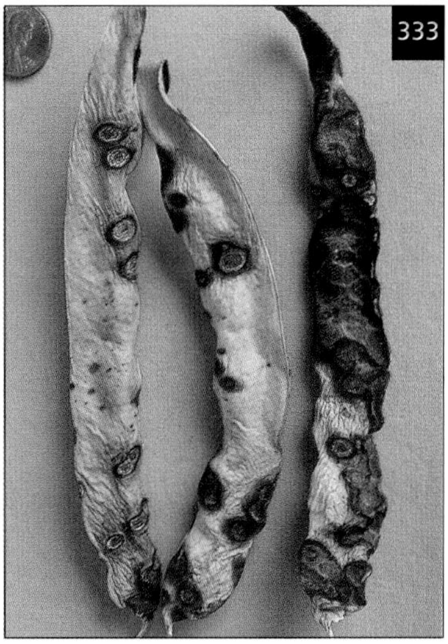

333 Dried bean pods showing anthracnose lesions.

Dillard, H. R. and Cobb, A. C. 1993. Survival of *Colletotrichum lindemuthianum* in bean debris in New York state. *Plant Disease* 77:1233–1238.

Drijfhout, E. and Davis, J. H. C. 1989. Selection of a new set of homogeneously reacting bean (*Phaseolus vulgaris*) differentials to differentiate races of *Colletotrichum lindemuthianum*. *Plant Pathology* 38:391–396.

Mahuku, G. S. and Riascos, J. J. 2004. Virulence and molecular diversity within *Colletotrichum lindemuthianum* isolates from Andean and Mesoamerican bean varieties and regions. *European Journal of Plant Pathology* 110 (3):253–263.

Menezes, J. R. and Dianese, J. C. 1988. Race characterization of Brazilian isolates of *Colletotrichum lindemuthianum* and detection of resistance to anthracnose in *Phaseolus vulgaris*. *Phytopathology* 78:650–655.

Rodríguez–Guerra, R., Ramírez-Rueda, M-T., Martínez de la Vega, O., and Simpson, J. 2003. Variation in genotype, pathotype and anastomosis groups of *Colletotrichum lindemuthianum* isolates from Mexico. *Plant Pathology* 52:228–235.

Schwartz, H. F., Pastor-Corrales, M. A., and Singh, S. P. 1982. New sources of resistance to anthracnose and angular leafspot of beans. *Euphytica* 31:741–754.

Stonehouse, J. 1994. Assessment of Andean bean diseases using visual keys. *Plant Pathology* 43:519–527.

Tu, J. C. 1981. Anthracnose on white bean in Southern Ontario: spread of the disease from an infection locus. *Plant Disease* 65:477–480.

Tu, J. C. 1982. Effect of temperature on incidence and severity of anthracnose on white bean. *Plant Disease* 66:781–783.

Tu, J. C. 1983. Epidemiology of anthracnose caused by *Colletotrichum lindemuthianum* on white bean in southern Ontario: Survival of the pathogen. *Plant Disease* 67:402–404.

Sclerotinia minor, S. sclerotiorum, S. trifoliorum

WHITE MOLD

Introduction and significance

White mold or Sclerotinia rot is an important disease on many legume crops, including bean, pea, and broad bean. Serious attacks can occur on common bean grown in temperate areas and on scarlet runner bean. *Sclerotinia sclerotiorum* is the species most commonly found on bean, broad bean, and pea; *S. minor* also affects bean, causing occasional problems in parts of North America and Europe. Older references refer to this group of pathogens as belonging in the genus *Whetzelinia*.

Symptoms and diagnostic features

The first symptoms on bean are small, irregularly shaped, water-soaked areas on stems, leaves, or pods. These quickly develop into soft, pale brown lesions. Lesions eventually support white mycelium, white mycelial mounds that are immature sclerotia, and finally mature, hard, black sclerotia (**334, 335**). Mature sclerotia usually form after tissues are rotted and breaking down. Plants with stem infections can completely collapse and dry out. The soilborne *S. minor* only causes infection on stems and crowns in contact with soil. In these situations, stems develop a brown decay that later is covered with white mycelium and small black sclerotia (**336**).

On broad bean, both young and mature plants can be affected and rapidly developing, brown lesions occur on stems, leaves, and pods. Infection often takes place at soil level, which results in watery lesions at the base of the stem and subsequent wilting and collapse of the plants. White mycelium and black sclerotia form on soft-rotted tissue.

On pea, initial symptoms are water-soaked lesions on the foliage that can quickly be covered with white mycelium. Leaf and stem lesions are pale brown, but have a bleached appearance under dry conditions. Pod lesions are pale to dark brown in color. Once established, there is rapid enlargement of leaf, stem, and pod lesions under humid conditions. When lesions girdle the stem, plants wilt and die. White mycelium and black sclerotia develop on the surface and inside rotted tissues.

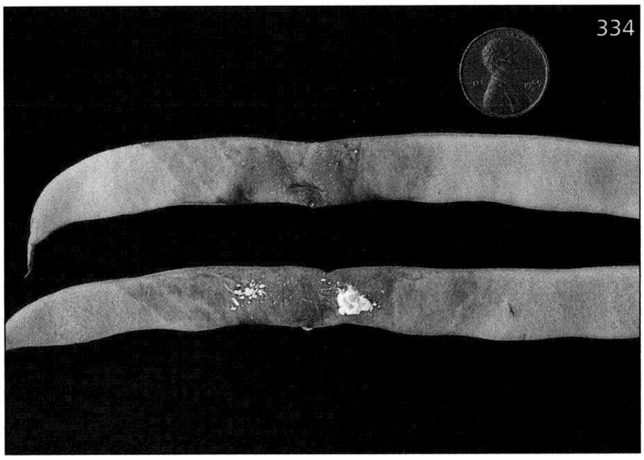

334 **Pod infections of white mold (*Sclerotinia sclerotiorum*) on bean.**

337 **The fruiting bodies, or apothecia, emerging from a sclerotium of *Sclerotinia sclerotiorum*.**

335 **Bean pods and stems infected with *Sclerotinia sclerotiorum*.**

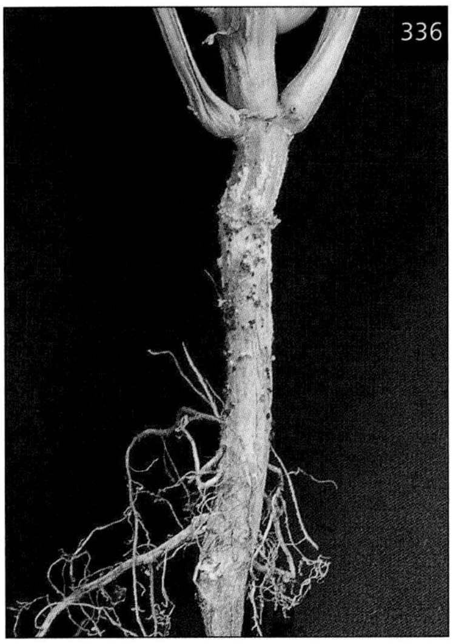

336 **Hypocotyl infection of white mold (*Sclerotinia minor*) on bean.**

Causal agents

White mold on legumes is caused by three species of the ascomycete fungus *Sclerotinia*: *S. sclerotiorum*, *S. trifoliorum*, and *S. minor*. *Sclerotinia minor* is readily distinguished from the other two species because sclerotia are significantly smaller, usually up to 2–4 mm in diameter, and are more numerous on plant tissue and in culture. In addition, *S. minor* for the most part does not produce apothecia in nature, while both S. *sclerotiorum* and *S. trifoliorum* produce these sexual fruiting bodies (337). *Sclerotinia minor* has a relatively narrow host range and in addition to occasional infections on legumes can infect a few other plants like lettuce, pepper, radicchio, tomato, basil, and cauliflower.

In contrast to *S. minor*, *S. sclerotiorum* has an extremely broad host range and has significantly larger (5–10 mm long) sclerotia. Sclerotia germinate carpogenically by forming small, tan apothecia that are cup-shaped and stalked. Optimum conditions for carpogenic germination are soil temperatures of 15° C and soil water potentials between –0.03 and –0.07 MPa. Mature apothecia contain cylindrical to clavate asci. Each ascus contains eight ascospores that are single-celled, hyaline, elliptical, and measure 9–13 x 4–5 µm. All eight ascospores are fairly uniform in size.

Sclerotinia trifoliorum forms sclerotia and apothecia that appear very similar to those of *S. sclerotiorum*. However, the eight ascospores in each ascus sometimes occur in two distinct sizes (dimorphic). In culture the mycelial growth of *S. trifoliorum* on potato dextrose agar at 26° C is more sparse and slower that of *S. sclerotiorum*. *Sclerotinia trifoliorum* has a restricted host range and is generally limited to legume plants.

Disease cycle

Sclerotinia sclerotiorum and *S. trifoliorum* survive in the soil as sclerotia which can remain viable from a few months to perhaps 10 years, depending on soil conditions, soil moisture, and depth of burial. Percent germination decreases with increasing time and burial depth. Sclerotia that are kept dry tend to survive for longer periods of time. Direct infection by soilborne sclerotia apparently is less important for these species. However, if sufficient soil moisture is present, shallowly buried sclerotia germinate carpogenically by forming small, tan apothecia that are cup-shaped and stalked.

Ascospores are released from apothecia and carried by winds to the host plant. Recent research suggests that ascospores can survive for several weeks on leaf surfaces and require very high humidity but not necessarily free moisture to germinate. Nutrients derived from senescent or damaged tissues are required for ascospores to germinate and initiate infection. Pea and bean petals are particularly good nutrient sources for these pathogens. Secondary spread occurs by mycelium where there is direct contact between diseased and healthy plant organs. In bean crops, moderate temperatures (20–25° C) and moist soils provide optimal conditions for the pathogen. Synchrony of sclerotial germination and ascospore production with flowering of bean flowers is an important aspect of white mold disease epidemiology.

Infective propagules of *S. minor* are sclerotia that are buried in the soil. Duration of soil survival depends on the same factors that govern *S. sclerotiorum* survival. These sclerotia germinate eruptively by producing mycelial plugs that directly infect taproots and crowns. To successfully infect hosts, the germinating sclerotia must be within 2 cm of the taproot, crown or senescing leaf. Optimum conditions for sclerotial germination are soil temperatures between 10–15° C and soil water potential of -0.033 MPa. Once infection has occurred, the pathogen causes a soft rot decay and production of additional sclerotia. When these decayed tissues are incorporated back into the field, sclerotia are buried and increase soil inoculum. Under field conditions the sclerotia of *S. minor* rarely produce apothecia with the notable exception of apothecial development in New Zealand. Recent studies indicate that if *S. minor* sclerotia form in aggregations, such clusters might have enough resources to produce apothecia.

Overall disease incidence in any particular field varies from season to season, and a field having severe white mold in one year may or may not have a similar severity the following year.

Control

Do not plant seed that is contaminated with sclerotia. Avoid planting susceptible legumes in severely infested fields or locations with a history of white mold problems. In a related strategy, practice crop rotations that use non-hosts, though crop rotations will not prevent disease due to the airborne nature of ascospores. Differences in cultivar susceptibility are associated with differences plant growth habit; varieties with open crop canopies may be less susceptible. Therefore, choose cultivars that may develop less severe white mold disease. Truly resistant cultivars, however, are not readily available. Irrigate in the morning so that plant foliage and the soil surface dry quickly. Or irrigate with subsurface drip systems so the soil surface is drier. Avoid excessive nitrogen fertilization, which encourages production of dense foliage, and ensure that potash levels are adequate. For *S. minor*, deep plowing soils to bury sclerotia below the root zone has been helpful for crops such as lettuce.

In some cases, foliar fungicides applied during flowering provide effective control, and new products are becoming available for control of *Sclerotinia*. Foliar fungicides include triazole, benzimidazole, dicarboximide, and strobilurin products. However, it is unlikely that fungicides will successfully prevent secondary spread among senescent leaves. Soil sterilization with chemical, steam or heat treatments can significantly reduce sclerotia in the soil, but such treatments are usually too expensive to use in legume crops and will not affect airborne ascospore inoculum entering from beyond the treated field. Biological control with commercial formulations of the mycoparasite *Coniothyrium minitans* has shown some promise against *S. sclerotiorum*, but not for *S. minor*.

Researchers find that incorporating the biomass from brassica cover crops or crops such as broccoli can also reduce soil inoculum and subsequent disease caused by *S. minor*. Researchers are attempting to define conditions required for ascospore infection and onset of symptoms; therefore, forecasting systems may contribute to disease control in the future.

References

Abawi, G. S. and Grogan, R. G. 1975. Source of primary inoculum and effects of temperature and moisture on infection of beans by *Whetzelinia sclerotiorum*. *Phytopathology* 65:300–309.

Abawi, G. S., Polach, F. J., and Molin, W. T. 1975. Infection of bean by ascospores of *Whetzelinia sclerotiorum*. *Phytopathology* 65:673–678.

Ferraz, L. C. L., Café Filho, A. C., Nasser, L. C. B., and Azevedo, J. 1999. Effects of soil moisture, organic matter and grass mulching on carpogenic germination of sclerotia and infection of bean by *Sclerotinia sclerotiorum*. *Plant Pathology* 48:77–82.

Fuller, P. A., Steadman, J. R., and Coyne, D. P. 1984. Enhancement of white mold avoidance and yield in dry bean by canopy elevation. *HortScience* 19:78–79.

Gerlagh, M., Goossen-van de Geijn, H. M., Hoogland, A. E., and Vereijken, P. F. G. 2003. Quantitative aspects of infection of *Sclerotinia sclerotiorum* sclerotia by *Coniothyrium minitans* – timing of application, concentration and quality of conidial suspension of the mycoparasite. *European Journal of Plant Pathology* 109:489–502.

Hannusch, D. J. and Boland, G. J. 1996. Influence of air temperature and relative humidity on biological control of white mold of bean (*Sclerotinia sclerotiorum*). *Phytopathology* 86:156–162.

Hao, J. J., Subbarao, K. V., and Koike, S. T. 2003. Effects of broccoli rotation on lettuce drop caused by *Sclerotinia minor* and on the population density of sclerotia in soil. *Plant Disease* 87:159–166.

Kerr, E. D., Steadman, J. R., and Nelson, L. A .1978. Estimation of white mold disease reduction on yield and yield components of dry edible beans. *Crop Science* 18:275–279.

Kohn, L. M. 1979. A monographic revision of the genus *Sclerotinia*. *Mycotaxon* 9:365–444.

Nasser, L. C. B. and Hall, R. 1997. Practice and precept in cultural management of bean diseases. *Canadian Journal of Plant Pathology* 18:176–185.

Phillips, A. J. L. 1994. Influence of fluctuating temperatures and interrupted periods of plant surface wetness on infection of bean leaves by ascospores of *Sclerotinia sclerotiorum*. *Annals of Applied Biology* 124, 413–427.

Saindon, G., Huang, H. C., and Kozub, G. C. 1995. White mold avoidance and agronomic attributes of upright common beans grown at multiple planting densities in narrow rows. *Journal of the American Society for Horticultural Science* 120:843–847.

Tu, J. C. 1989. Management of white mold of beans in Ontario. *Plant Disease* 73:281–285.

Willetts, H. J. and Wong, J. A–L. 1980. The biology of *Sclerotinia sclerotiorum*, *S. trifoliorum*, and *S. minor* with emphasis on specific nomenclature. *The Botanical Review* 46:101–165.

Uromyces appendiculatus

RUST

Introduction and significance

Most beans are susceptible to rust disease, which occurs worldwide. There are at least 35 recognized pathotypes of this pathogen, which makes it a challenge to develop resistant varieties. Rust can cause severe damage to foliage and pods and result in significant crop loss. Outbreaks of rust are sporadic and associated with extended periods of warm temperatures and high humidity. Rust is increasing in importance in the UK on both common and scarlet runner beans.

Symptoms and diagnostic features

Symptoms develop on leaves and pods, but do not usually affect the stems. Initial symptoms appear about 5 days after infection on the undersides of leaves and appear as small, raised white spots. The spot develops into a pustule (sorus), or uredinium, measuring 1–2 mm or larger in diameter, and erupts through the leaf epidermis to release red-brown powdery urediniospores (**338**).

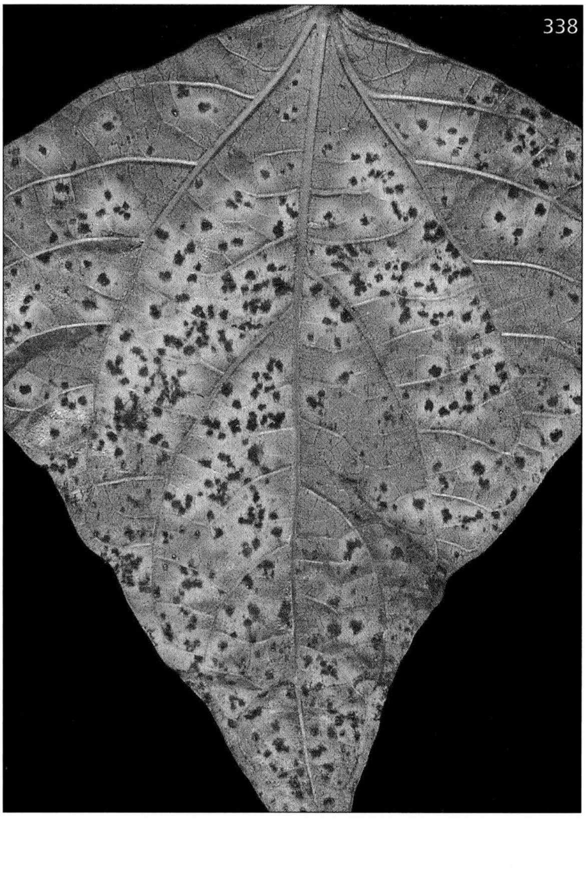

338 Rust of bean.

Uredinia may be surrounded by yellow tissue (**339**) or additional uredinia. Extensive rust development causes the leaf to turn chlorotic, dry, and decline. Black teliospores of the second spore stage are later formed in these same pustules.

Causal agent

Rust is caused by the basidiomycete fungus *Uromyces appendiculatus*. This rust is autoecious, meaning that it completes its life cycle on one host. Urediniospores are light to gold-brown, echinulate, thin-walled, unicellular, measure 18–24 x 20–37 µm, and have two pores. Teliospores are globoid to ellipsoid, unicellular, measure 20–28 x 25–35 µm, with a hemispherical hyaline papilla over the pore, and thick cell walls that are 3–4 µm. Both urediniospores and teliospores can occur in the same sorus. Urediniospores are important in both the early spread and the wider dispersal of rust, while teliospores enable the pathogen to survive between infecting crops.

Disease cycle

Optimum conditions for germination are temperatures of 17–23° C and leaf wetness duration of 6–8 hours. The fungus has low survival and infectivity rates if conditions are dry, which accounts for the seasonal variability in this disease, especially in dry regions. Temperature, light intensity, and age of host influence teliospore formation. Races may differ in their ability to form teliospores.

Control

Plant resistant cultivars, though the emergence of new races may overcome the resistance genes and create the need for new varieties. Bury or otherwise destroy crop residues, which may reduce teliospore inoculum. Avoid over-fertilizing the crop, as high nitrogen levels will increase bean susceptibility. Interestingly, increasing potash levels may lower disease severity. Apply fungicides if available; sulfur and triazoles may provide some protection.

339

339 Climbing French bean infected with rust.

References

Bassanezi, R. B., Amorim, L., Bergamin,Filho, A., Hau, B., and Berger, R. D. 2001. Accounting for photosynthetic efficiency of bean leaves with rust, angular leaf spot and anthracnose to assess crop damage. *Plant Pathology* 50:443–452.

Stavely, J. R. 1984. Pathogenic specialization in *Uromyces phaseoli* in the United States and rust resistance in beans. *Plant Disease* 68:95–99.

Stavely, J. R., Steadman, J. R., and McMillan, R. T. 1989. New pathogenic variation in *Uromyces appendiculatus* in North America. *Plant Disease* 73:428–432.

Tomkins, F. D., Canary, D. J., Mullins, C. A., and Hilty, J. W. 1983. Effect of liquid volume, spray pressure and nozzle arrangement on coverage of plant foliage and control of snap bean rust with chlorothalonil. *Plant Disease* 67:952–953.

Bean common mosaic virus

BEAN COMMON MOSAIC

Introduction and significance

Bean common mosaic virus (BCMV) occurs worldwide and is primarily seedborne, with transmission rates sometimes exceeding 80%. On susceptible cultivars, seed yield may be reduced by more than 50%.

Symptoms and diagnostic features

Symptoms consist of leaf chlorosis and mottling, downward cupping of leaves, and some reduction in leaflet size (**340, 341**). Seedborne infection can result in young plants having a leaf mosaic of light and dark green areas, with paler sections near the leaf margin. The first true leaf shows dark green blistering. Aphid-borne infection on older plants produces crinkled, chlorotic leaves with a stiff appearance, but does not cause the mottling or downward rolling of the leaf margin associated with seedborne infection. The rolling symptom is useful in distinguishing BCMV from *Bean yellow mosaic virus*. Symptoms tend to become less pronounced as the season progresses. Severe stunting and black root symptoms are produced at temperatures of 20° C or greater with virulent strains, and at higher temperatures (30° C) with less severe strains. Local vein necrosis occurs as a hypersensitive reaction in some resistant cultivars and this provides field resistance.

340 Bean seedlings infected with *Bean common mosaic virus*.

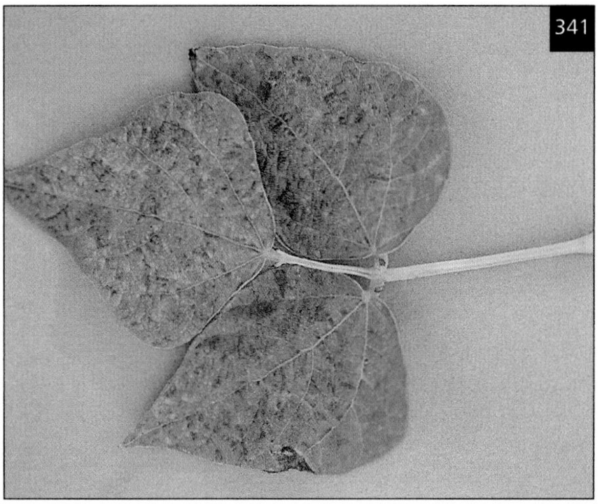

341 Foliar symptoms of bean infected with *Bean common mosaic virus*.

Causal agent and disease cycle

BCMV is a potyvirus that is transmitted in a nonpersistent manner by several aphid species, including the pea aphid (*Acyrosiphum pisi*), *Aphis fabae*, and *Myzus persicae*. Particle shape is filamentous (flexuous rods). BCMV is rarely found in broad bean.

Control

Follow general suggestions for managing virus diseases (see Part 1).

References

Spence, N. J. and Walkey, D. G. A. 1995. Variation for pathogenicity among isolates of *bean common mosaic virus* in Africa and a reinterpretation of genetic relationships between cultivars of *Phaseolus vulgaris* and pathotypes of BCMV. *Plant Pathology* 44:527–546.

Van Rheenen, H. A. and Muigai, S. G. S. 1984. Control of *bean common mosaic virus* by deployment of dominant gene I. *Netherlands Journal of Plant Pathology* 90:165–194.

Bean yellow mosaic virus

BEAN YELLOW MOSAIC

Introduction and significance

Bean yellow mosaic virus (BYMV) is a common potyvirus that affects many cultivated and weed legumes, Iridaceae plants such as gladiolus, and globe artichoke (*Cynara scolymus*).

Symptoms and diagnostic features

On bean, BYMV reduces pod numbers and overall yield may be lowered by 40%. BYMV occurs in most countries where legumes are grown. On common bean, infected leaves develop small yellow spots, and the plant eventually develops a general chlorosis (**342**). Plants become brittle during the early stages of infection. In contrast to BCMV, symptoms tend to become more severe as the season progresses. Affected plants have a stunted, bushy appearance because there is shortening of the internodes and proliferation of lateral shoots. BYMV is the most common cause of mosaic symptoms in broad bean (**343**), where symptoms depend on whether 'typical' or 'pea mosaic' strains are involved. Pea mosaic strains usually produce a distinct green and yellow mosaic, while typical strains result in a mild green mosaic, green vein banding, or a general chlorosis. Vein clearing occurs in the younger leaves 7–10 days after infection. Early infections may cause some stunting of plant growth and slight narrowing and crinkling of the leaflets. Some severe strains cause local lesions, and in the USA there is a top-rolling strain that causes mosaic and upward rolling of the leaves.

Causal agent and disease cycle

BYMV is a potyvirus and is transmitted by many aphid species, including *Acyrosiphum pisi*, *Aphis fabae*, *Macrosiphum euphorbiae*, and *Myzus persicae*. Particle shape is filamentous (flexuous rods). Seed transmission is uncommon. Clover (especially *Trifolium pratense*), vetch, and gladiolus are major sources of inoculum, so bean should not be grown near these plants.

Control

Follow general suggestions for managing virus diseases (see Part 1).

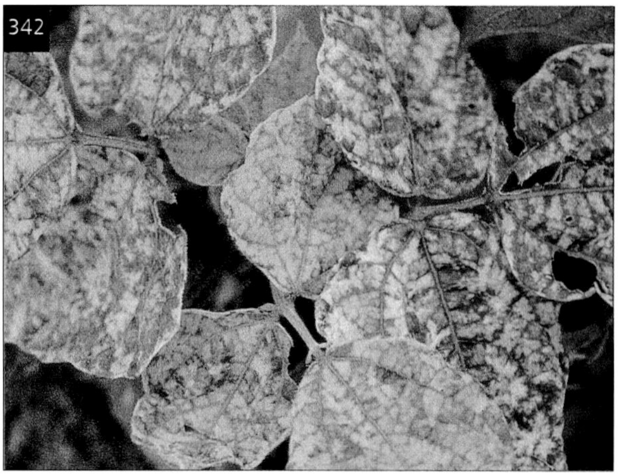

342 Leaf symptoms of *Bean yellow mosaic virus* of bean.

343 Foliar symptoms on broad bean caused by *Bean yellow mosaic virus.*

References

Bos, L., Kowalska, C. Z., and Maat, D. Z. 1974. The identification of bean mosaic, pea yellow mosaic and pea necrosis strains of *bean yellow mosaic virus*. *Netherlands Journal of Plant Pathology* 80:173–191.

Tu, J. C. 1989. Role of temperature and inoculum concentration in the development of tip necrosis and seedling death of beans infected with *bean yellow mosaic virus*. *Plant Disease* 73:405–407

PISUM SATIVUM (PEA)

Pseudomonas syringae pv. *pisi*

BACTERIAL BLIGHT

Introduction and significance

This is an important seedborne disease of pea, which is widely distributed in pea-growing regions throughout the world. Bacterial blight can cause extensive crop loss if environmental conditions are favorable. Significant losses were recorded in overwintered peas in the UK.

Symptoms and diagnostic features

Initial symptoms are water-soaked, angular spots that become darker and develop necrotic centers. Older lesions are papery in texture, with a pale center and darker brown margin (**344**). The angular shape of older lesions and a water-soaked margin are retained. Because the pathogen is seedborne, disease may be initiated when emerging seedlings are infected by bacteria from the seed coat. In these situations, the lowest stipules become the initial infection sites. The pathogen spreads from stipule lesions to the adjacent stem and other foliage. Diseased flowers may die and small, developing pods can have circular, sunken lesions. Where the pod is diseased, underlying seed may be killed or develop dark spots near the seed hilum. Severe infection can cause stand loss at the seedling stage or result in significant loss of yield and quality if foliage and pods are damaged. Bacterial blight symptoms can be confused with spots caused by *Mycosphaerella pinodes*.

Causal agent

Bacterial blight is caused by *Pseudomonas syringae* pv. *pisi*. This pathogen is a Gram-negative rod that has one to five polar flagella. The pathogen can be isolated on standard microbiological media and produces cream-colored colonies typical of most pseudomonads. When cultured on Kings medium B, this organism produces a diffusible pigment that fluoresces blue under ultraviolet light. Strains of this pathogen are host specific to pea and related legume plants. This pathogen is seedborne. There are several races of this pathogen that infect *Lathyrus, Vicia*, red clover (*Trifolium arvense*), and soybean (*Glycine max*).

344 Bacterial blight on pea leaves.

Disease cycle

Transmission from infested seed to the emerging seedling is favored if soils and weather conditions are wet. Bacteria enter through stomata or wounds. Bacterial blight can spread very rapidly from infected seedlings to nearby plants if there is rainy weather and crop injury from driving rain, high winds, hail, wind-blown soil particles, and frosts. Infection also takes place at sites of pest damage, such as the leaf notches caused by the weevil *Sitona lineatus*. The pathogen is spread between fields on machinery or clothing that has passed through the diseased crop, and in drainage or irrigation water. Seed may be contaminated due to contact with infested dust during harvesting.

Actual seed infection occurs close to harvest when dew or rain allows bacteria to penetrate the pod and colonize the seed coat. The pathogen can remain viable on the seed for at least 3 years. The ability to survive in the soil appears to be very limited, and in most situations the pathogen does not survive through the winter unless there are intact crop residues or diseased volunteer plants.

Control

Use seed that does not have significant levels of the pathogen. However, even a very low seed infestation level (0.01%) is able to cause significant problems if environmental conditions favor disease development. Avoid walking through or driving equipment into fields having obvious symptoms, as clothing and equipment can become contaminated with the bacteria. Sanitize harvest and seedcleaning machinery that is used to process pea seed. Seed treatments with dry heat, hot water, or sodium hypochlorite are reported to reduce seedborne infection. There are both race-specific and nonspecific resistance sources for pea.

References

Elvira-Recuenco, M., Bevan, J. R., Taylor, J. D. 2003. Differential responses to pea bacterial blight in stems, leaves and pods under glasshouse and field conditions. *European Journal of Plant Pathology* 109:555–564.

Roberts, S. J. 1992. Effect of soil moisture on the transmission of pea bacterial blight (*Pseudomonas syringae* pv. *pisi*) from seed to seedling. *Plant Pathology* 41:136–140.

Roberts, S. J. 1997. Effect of weather conditions on local spread and infection by pea bacterial blight (*Pseudomonas syringae* pv. *pisi*). *European Journal of Plant Pathology* 103:711–719.

Roberts, S. J., Reeves, J. C., Biddle, A. J., Taylor, J. D., Higgins, P. 1991. Prevalence of pea bacterial blight in UK seed stocks, 1986–1990. *Aspects of Applied Biology* 27:327–332.

Taylor, J. D. 1986. Bacterial blight of compounding pea. *Proceedings of the 1986 British Crop Protection Conference – Pests and Diseases* 2:733–736.

Taylor, J. D., Bevan, J. R., Crute, I. R., and Reader, S. L. 1989. Genetic relationship between races of *Pseudomonas syringae* pv. *pisi* and cultivars of *Pisum sativum*. *Plant Pathology* 38:364–375.

Aphanomyces euteiches

APHANOMYCES ROOT ROT, COMMON ROOT ROT

Introduction and significance

This disease is important in North America, Australia, New Zealand, Japan, and northern Europe. Sometimes extensive crop loss can occur in badly infested fields. In the Great Lakes area and northeastern states of the USA, Aphanomyces root rot may be the most important disease of pea.

Symptoms and diagnostic features

Aphanomyces can infect pea at any age, though the most severe problems occur as seedlings emerge. Root cortex tissue turns yellow brown 7 to 14 days after infection and gradually darkens. The cortex softens and eventually rots away, leaving only thin strands of vascular tissue (**345**). Similar decay occurs on the lower parts of stems. Secondary decay organisms contribute to the darkening and rot of affected tissues. Severely affected plants are stunted and may produce few pods.

Causal agent

The cause of Aphanomyces root rot is the oomycete *Aphanomyces euteiches*. The sexual stage develops when antheridia fuse with the homothallic oogonium and an oospore is produced. Oospores are hyaline to yellow, measure 25–35 µm in diameter, form in rotted tissues, and can survive for over 10 years in soil. The asexual stage consists of long, filamentous zoosporangia that may be up to 3–4 mm long. Swimming primary zoospores are released from the tips of the zoosporangia. These primary zoospores measure 8–11 µm, usually encyst, and later germinate and release secondary zoospores. Secondary zoospores produce hyphae that invade host tissues.

Aphanomyces root rot can be misdiagnosed because root rot symptoms resemble those caused by other soil pathogens, and nonpathogenic soil fungi are often isolated from *Aphanomyces*-infected roots. Microscopic examination of the root cortex should reveal the oospores of *A. euteiches* with their characteristic smooth outer wall and wavy (sinuous) inner wall. These features distinguish *A. euteiches* oospores from oospores of *Pythium* spp. or the larger (8–30 µm in diameter) resting sporangia of *Olpidium brassicae*. Isolate *A. euteiches* by placing symptomatic tissue from

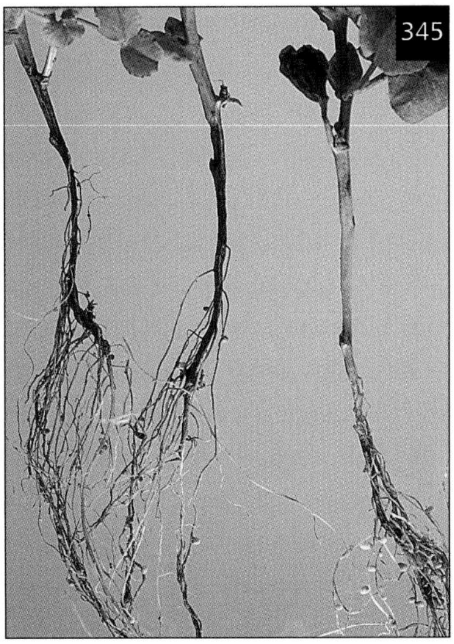

345 Aphanomyces root rot. Affected plants on the left.

346 Field of peas showing severe disease caused by Aphanomyces root rot.

young lesions onto water agar or a semi-selective media such as corn meal agar amended with antibiotics (vancomycin) and fungicides.

Aphanomyces euteiches affects alfalfa, bean, pea, clover, and leguminous weeds. Even cereals such as barley and oats can be hosts. Physiological races have been proposed based on variation in host ranges and virulence. One study characterized 11 such virulence types. Isolates also differ in host preferences; for example, isolates obtained from pea plants in France were preferentially pathogenic on pea.

Disease cycle

Oospores germinate to produce hyphae and zoosporangia. Infection by zoospores requires moist, cool soils; optimum temperature for infection is 16° C. Rain and irrigation saturate the soil and allow for rapid disease spread (**346**). Damaging attacks occur when cool wet springs are followed by warm dry summers, which create water stress for the diseased plants. Diseased plants release additional oospore inoculum when they are disked back into the soil. Mature oospores form in diseased tissues within 2 weeks of infection.

Control

Fungicide seed treatments and resistant cultivars have not provided good control in heavily infested fields. Avoid planting pea and other susceptible crops in fields having a history of this problem. Prepare fields so that drainage is enhanced. Irrigate properly so that excess water is not applied. Long rotations with non-host crops are required to reduce soilborne inoculum.

References

Holub, E. B., Grau, C. R., and Parke, J. L. 1991. Evaluation of the forma specialis concept in *Aphanomyces euteiches*. *Mycological Research* 95:147–157.

Kraft, J. M. and Boge, W. L. 1996. Identification of characteristics associated with resistance to root rot caused by *Aphanomyces euteiches* in pea. *Plant Disease* 80:1383–1386.

Muehlchen, A. M., Rand, R. E., and Parke, J. L. 1990. Evaluation of crucifer green manures for controlling Aphanomyces root rot of peas. *Plant Disease* 74:651–654.

Malvick, D. K., Percich, J. A., Pfleger, F. L., Givens, J., and Williams, J. L. 1994. Evaluation of methods for estimating inoculum potential of *Aphanomyces euteiches* in soil. *Plant Disease* 78:361–365.

Pfender, W. F. and Hagedorn, D. J. 1983. Disease progress and yield loss in Aphanomyces root rot of peas. *Phytopathology* 73:1109–1113.

Wicker, E. and Rouxel, F. 2001. Specific behaviour of French *Aphanomyces euteiches Drechs.* populations for virulence and aggressiveness on pea, related to isolates from Europe, America and New Zealand. *European Journal of Plant Pathology* 107:919–929.

Wicker, E., Hullé, M., and Rouxel, F. 2001. Pathogenic characteristics of isolates of *Aphanomyces euteiches* from pea in France. *Plant Pathology* 50:433–442.

Aphanomyces euteiches, Fusarium solani f. sp. *pisi, Phoma medicaginis* var. *pinodella, Pythium* spp., *Rhizoctonia solani, Thielaviopsis basicola*

FOOT ROT COMPLEX

Introduction and significance

The complex of root pathogens affecting pea is similar to that described for bean. Foot rot disease is extremely common, and symptoms of this problem can be found in most crops. Severe yield loss occurs where soilborne inoculum is high.

347 Poor establishment and weak growth of pea seedlings affected by the foot rot complex of *Pythium* spp., *Fusarium solani* f. sp. *pisi*, and *Phoma medicaginis* var. *pinodella*.

Symptoms and diagnostic features

In the field, pea stands are reduced due to seed death and poor plant emergence. Plants that do emerge but are infected grow poorly, are stunted, and result in fields having patches of uneven plants (**347**). Individual plants show yellow foliage, premature ripening, and possible wilting (**348**). In advanced stages of the disease, foliage can turn gray-green due to water stress and plants can collapse (**349**). Roots develop brown or black discoloration, and crown and lower stem tissues have brown to black sunken lesions (**350**). For plants infected with *Fusarium solani* f. sp. *pisi*, the stem and crown lesions tend to be red-brown. Vining pea is usually more susceptible to foot rot that other types of pea. Yield losses can be high in varieties grown for dry pea products.

One foot rot pathogen, *Phoma medicaginis* var. *pinodella*, can cause foliar symptoms as well. Lesions initially appear as small, irregularly shaped flecks that are dark brown in color. Lesions on leaves and pods enlarge, generally are round to oval in shape, and contain concentric rings of alternating shades of brown. Lesions on stems may become purple in color, and can enlarge and girdle the stem (**351**). Foliar symptoms caused by *P. medicaginis* var. *pinodella* are difficult to differentiate from those caused by *Mycosphaerella pinodes* (see the Ascochyta blight section on pea in this chapter, page 278).

348 Garbanzo beans (chickpeas) infected with black root rot.

349 Collapsing sugar pea plants affected by Fusarium root rot.

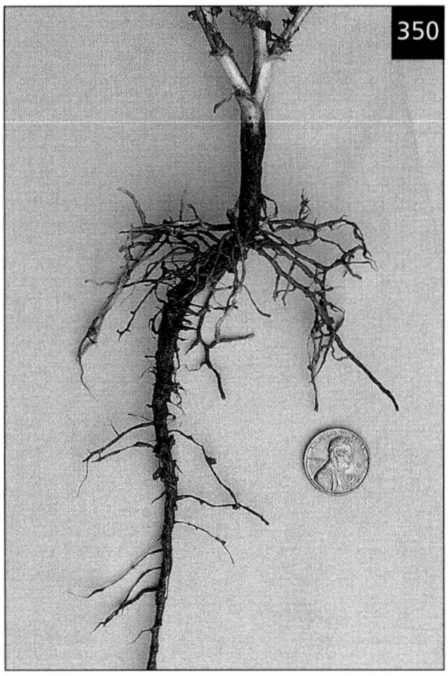
350 Garbanzo bean root infected with black root rot.

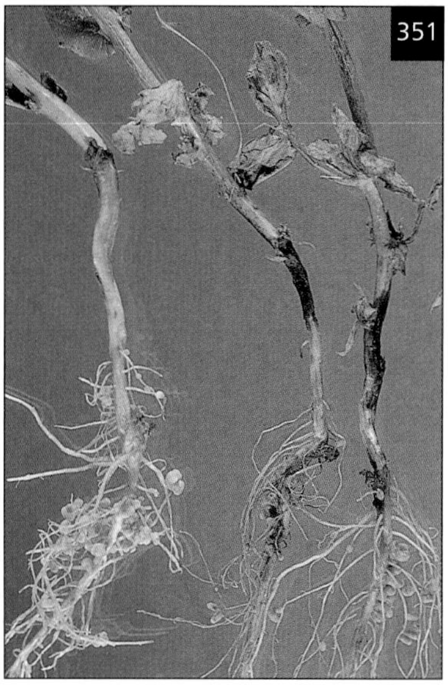
351 Lower stem and root of pea affected by ***Phoma medicaginis*** var. ***pinodella*** and ***Fusarium*** spp.

Causal agents

Foot rot is caused by one or more of the following complex of soilborne fungi: *Fusarium solani* f. sp. *pisi, Phoma medicaginis* var. *pinodella, Thielaviopsis basicola, Aphanomyces euteiches, Pythium* species., *Rhizoctonia solani.* These fungi are usually readily isolated by culturing affected tissue on agar media. *Fusarium solani* f. sp. *pisi* and *P. medicaginis* var. *pinodella* are particularly important. *Fusarium solani* f. sp. *pisi* persists in soil for long periods of time due to formation of chlamydospores. *Phoma medicaginis* var. *pinodella* can be seedborne and is capable of killing seedlings. This fungus also survives by means of chlamydospores and affects other legumes such as clover (*Trifolium* spp.). This pathogen is considered a part of the *Ascochyta* disease complex that affects peas grown in most production areas. *Phoma medicaginis* var. *pinodella* was previously named *Ascochyta pinodella.*

Different isolates, or anastomosis groups (AGs), of *R. solani* infect legumes. For *A. euteiches*, the pathogen survives in the soil via resilient oospores. Different populations appear to exist for this organism, as *A. euteiches* isolates from pea tend to infect pea but not vetch, and vetch isolates infect vetch but not pea.

Disease cycle

See the bean section on foot rot complex for more details (page 256).

Control

See the bean section on foot rot complex for more details.

References

Biddle, A.J. 1984. A prediction test for pea foot rot and the effects of previous legumes. *Proceedings of the British Crop Protection Conference – Pests & Diseases* 3:773–777.

Bowden, R. L., Wiese, M. V., Crock, J. E., and Auld, D. L. 1985. Root rot of chickpeas and lentils caused by *Thielaviopsis basicola. Plant Disease* 69:1089–1091.

Kraft, J. M. 1986. Seed electrolyte loss and resistance to Fusarium root rot of peas. *Plant Disease* 70:743–745.

Kraft, J. M. 1996. Fusarium root rot of peas. *Proceedings of the Brighton Crop Protection Conference – Pests & Diseases* 2:503–509.

Kraft, J. M. and Wilkins, D. E. 1989. The effects of pathogen numbers and tillage on root disease severity, root length, and seed yields in green peas. *Plant Disease* 73:884–887.

Levenfors, J. P., Wikström, M., Persson, L., and Gerhardson, B. 2003. Pathogenicity of *Aphanomyces* spp. from different leguminous crops in Sweden. *European Journal of Plant Pathology* 109:535–543.

Miller, D. E., Burke, D. W., and Kraft, J. M. 1980. Predisposition of bean roots to attack by the pea pathogen, *Fusarium solani* f. sp. *pisi*, due to temporary oxygen stress. *Phytopathology* 70:1221–1224.

Miller, D. E. and Burke, D. W. 1974. Influence of soil bulk density and water potential on Fusarium root rot of beans. *Phytopathology* 64:526–529.

Miller, D. E. and Burke, D. W. 1975. Effect of soil aeration on Fusarium root rot of beans. *Phytopathology* 65:519–523.

Ascochyta pisi

LEAF AND POD SPOT

Introduction and significance

In the late 1920s, researchers described three pea diseases caused by different *Ascochyta* fungi. Presently the pathogen names comprising this former *Ascochyta* complex have been changed. Leaf and pod spot, considered here, is caused by *A. pisi*. Ascochyta blight (page 278) is caused by *Mycosphaerella pinodes*; the asexual stage is *A. pinodes*. The foot rot pathogen *A. pinodella* is now classified as *Phoma medicaginis* var. *pinodella*. These three pathogens can occur together on pea crops, making it difficult to differentiate the various diseases.

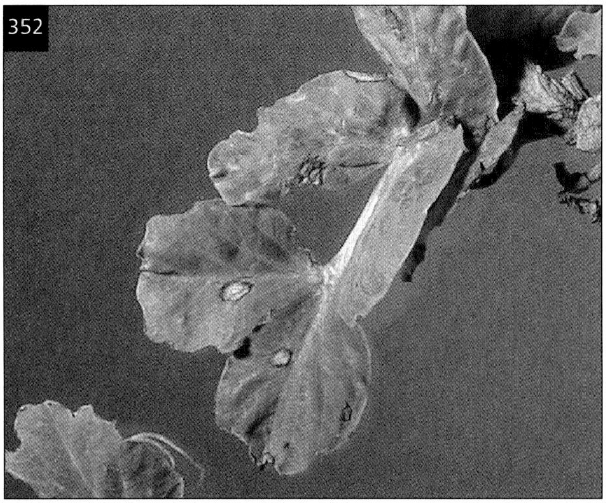

352 Leaf lesion caused by *Aschochyta pisi*.

Symptoms and diagnostic features

Foliar symptoms caused by *A. pisi* are distinct from those caused by the other two pathogens. Leaf and pod spot lesions are sunken, tan to light brown, and surrounded by a darker brown to red-brown border (352, 353). Lesions usually contain numerous fruiting bodies (pycnidia) of the fungus. On leaves and pods, the lesions tend to be round, while stem lesions are elongated. *Ascochyta pisi* does not cause a foot rot disease. Leaf and pod spots of Ascochyta blight tend to be darker brown to purple-brown in color.

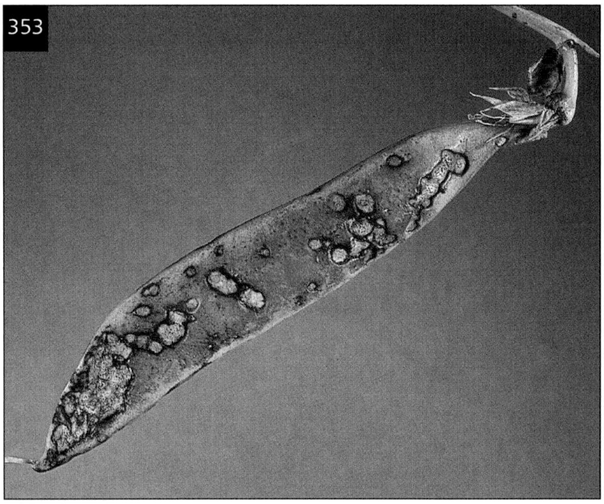

353 Diseased pea pod affected by *Ascochyta pisi*.

Causal agent

Leaf and pod spot is caused by *A. pisi*. The pathogen forms light brown pycnidia within leaf and pod lesions. Pycnidia produce conidia that are hyaline, mostly two-celled, and measure 10–16 x 3–5 µm. Conidia are exuded out of pycnidia in orange-red masses. *Ascochyta pisi* has no known sexual stage, so perithecia are not present in leaf and pod lesions or on pea crop residues. Various races of this fungus have been documented. Other *Ascochyta* species infect different legumes such as lentil and chickpea.

Disease cycle

In contrast to *M. pinodes*, *A. pisi* does not overwinter or survive in the soil. This pathogen is seedborne and primary inoculum comes from conidia and other fungal structures that are on the outside of the seedcoat. Secondary spread is by splash dispersal of conidia that germinate and penetrate host tissue via stomata or the cuticle.

Control

Use seed that does not have significant levels of the pathogen. The current practices of growing pea seed crops in arid regions and treating pea seed with fungicides has greatly reduced the incidence and importance of this disease.

References

Biddle, A. J. 1994. Seed treatment usage on peas and beans in the UK. In: *Seed Treatment: Progress and Prospects*.T. J. Martin (ed.). BCPC Monograph No. 57: 143–149. BCPC Publications, Farnham, UK.

Bowen, J. K., Peart, J., Lewis, B. G., Cooper, C., Matthews, P. 1996. Development of monoclonal antibodies against fungi of the '*Ascochyta*' complex. *Plant Pathology* 45:393–406.

Darby, P., Lewis, B. G., and Matthews, P. 1986. Diversity of virulence within *Ascochyta pisi* and resistance in the genus *Pisum*. *Plant Pathology* 35:214–223.

Kaiser, W. J. and Hannan, R. M. 1988. Seed transmission of *Ascochyta rabiei* in chickpea and its control by seed-treatment fungicides. *Seed Science and Technology* 16:625–637.

Kaiser, W. J. and Hannan, R. M. 1986. Incidence of seedborne *Ascochyta lentis* in lentil germ plasm. *Phytopathology* 76:355–360.

Khan, M. S. A., Ramsey, M. D., Corbière, R., Infantino, A., Porta-Puglia, A., Bouznad, Z., and Scott, E. S. 1999. Ascochyta blight of chickpea in Australia: identification, pathogenicity and mating type. *Plant Pathology* 48:230–234.

Wallen, V. R. 1957. The identification and distribution of physiologic races of *Ascochyta pisi* in Canada. *Canadian Journal of Plant Science* 37:337–341.

Wang, H., Hwang, S. F., Chang, K. F., Turnbull, G. D., and Howard, R. J. 2000. Characterization of *Ascochyta* isolates and susceptibility of pea cultivars to the *Ascochyta* disease complex in Alberta. *Plant Pathology* 49:540–545.

Botrytis cinerea (teleomorph = Botryotinia fuckeliana)

GRAY MOLD

Introduction and significance

Gray mold is common in pea when there are wet or humid conditions during flowering. While there is some yield loss due to damage to foliage, the greatest economic damage is from infected and diseased pods.

Symptoms and diagnostic features

Bleached spots often develop where fallen petals stick to leaves, stems, or pods. Such spots can later turn into gray to brown lesions. A more general infection of the foliage occurs later in the season when the lower leaves senesce. Senescing leaves are susceptible to infection and develop brown lesions that support extensive gray sporulation when conditions are humid. Once established in leaves or stipules, the pathogen can then spread to stems, which wilt and die when lesions girdle the stem. *Botrytis* commonly grows as a secondary invader in downy mildew lesions, particularly on tendrils, making diagnosis of the primary downy mildew problem difficult. Infected pods have brown to red-brown, circular to irregular lesions that can exhibit gray sporulation in the center of the lesions (**354**). On occasion, black sclerotia develop in the pod lesions. In storage, Botrytis can continue to rot the pods but forms white, not gray, mycelial growth. Botrytis is able to penetrate through the pod wall to produce chalky seeds in dry-harvested pea.

Causal agent

The cause of gray mold is the ascomycete fungus *Botryotinia fuckeliana*. However, this perfect stage is much less commonly seen than the anamorph, *Botrytis cinerea*. *Botrytis cinerea* is characterized by its rapidly growing gray to white mycelium, gray-brown conidiophores, and clusters of gray conidia. Conidia are obovoid to ellipsoid and measure 8–14 x 6–9 µm. In culture, some isolates sporulate very poorly if not placed under ambient or incubation light.

Disease cycle

Botrytis is a common saprophyte that grows on plant debris and has many plant hosts. Abundant conidia are produced under optimal conditions of 16–21° C temperatures and 100% relative humidity. Problems are therefore usually associated with wet weather during and after flowering. Conidia, however, do not readily infect healthy foliage and pod tissues. The presence of a food source enables conidia to germinate, colonize the nutrient source, then spread onto and infect pods and leaves. Therefore, petals and other floral parts provide a nutrient supply for the conidia and facilitate infection.

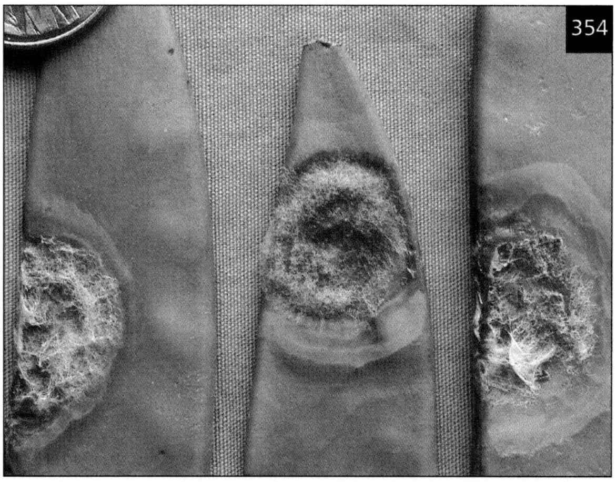

354 Gray mold on pea pods.

FUNGAL DISEASES

Control

Select sites that are well exposed to sun and winds, thereby allowing foliage to dry. Avoid over fertilizing so that excessive, succulent foliage is not grown. Apply fungicides in a timely manner during early flowering and pod formation stages if weather conditions are likely to favor the pathogen. A single spray is often used for vining pea, but two sprays may be required for dry harvested crops. Fungicide resistance in *Botrytis* has limited the effectiveness of dicarboximide fungicides and possibly the newer strobilurin products. There are no resistant pea varieties.

References

Ford, R. E. and Haglund, W. A. 1963. *Botrytis cinerea* blight of peas associated with senescent blossoms in northwestern Washington. *Plant Disease Reporter* 47:483–485.

Erysiphe pisi

POWDERY MILDEW

Introduction and significance

Powdery mildew has a worldwide distribution on pea and the disease is capable of significantly reducing yields. Powdery mildew is one of the more common foliar problems of pea.

Symptoms and diagnostic features

Early symptoms are discrete, slightly discolored areas on the upper surfaces of the oldest leaves with diffuse, gray to white radiating fungal colonies. Mycelial growth increases and colonies merge so that leaves and eventually pods become covered in powdery white growth (355). Advanced symptoms consist of yellow or purple discoloration of leaf tissues beneath powdery mildew colonies, early senescence of severely infected leaves, and blight-like symptoms due to the drying out of the foliage. Severe powdery mildew reduces seed quality, may discolor seeds, and impairs flavor of the harvested product.

Causal agent

Powdery mildew is caused by the fungus *Erysiphe pisi*; this obligate pathogen was previously known as *E. polygoni*. Conidia typically measure 20 x 30 µm and are produced singly or in chains of only two spores. The perfect stage forms on mature lesions and appears as small, black, spherical fruiting bodies (cleistothecia). Cleistothecia measure 180 µm in diameter, have 10 to 30 unbranched appendages attached to the surface, and contain three to ten asci. Each ascus holds two to eight hyaline ascospores measuring 9–14 x 10–25 µm. *Erysiphe pisi* occurs in specialized forms that infect lucerne or alfalfa (*Medicago sativa*), vetch (*Vicia* spp.), chickpea (*Cicer arietinum*), pigeon pea (*Cajanus cajan*), and lentils (*Lens* spp.). Isolates from pea infect only pea.

Disease cycle

Disease is favored by warm, dry days and cooler nights, which lead to dew formation. The fungus overwinters on infected plant debris and may be seedborne. Conidia are dispersed in the air, and spore release usually peaks about midday. Conidia can germinate within an hour and penetrate the leaf even at low humidity. The disease cycle can be completed in less than 7 days under optimal conditions. Disease severity usually is higher later in the summer season.

Control

Plant resistant cultivars or cultivars that are less severely affected by powdery mildew. Early planting and irrigating with overhead sprinklers can lower disease severity. Apply sulfur and other fungicides when early stages of disease are observed.

355

355 Sporulation of powdery mildew on pea.

References

Cook, R. T. A. and Fox, R. T. V. 1992. *Erysiphe pisi* var. *pisi* on faba beans and other legumes in Britain. *Plant Pathology* 41:506–512.

Viljanen-Rollinson, S. L. H., Frampton, C. M. A., Gaunt, R .E., Falloon, R. E., and McNeil, D. L. 1998. Spatial and temporal spread of powdery mildew (*Erysiphe pisi*) on peas (*Pisum sativum*) varying in quantitative resistance. *Plant Pathology* 47:148–156.

Fusarium oxysporum f. sp. *pisi*

FUSARIUM WILT

Introduction and significance

Fusarium wilt has been recognized in North America since the 1920s, and also is present in Europe and Australia. There are two distinct forms of the disease, called 'wilt' and 'near wilt' phases. The disease is of limited importance in the UK because of the use of race 1 resistant cultivars. Several other races are recognized and these are capable of causing economic damage in other production areas.

Symptoms and diagnostic features

Early symptoms include a downward curling of the leaves and stipules, stunted plant stature, and lack of plant vigor. Plants later show progressive leaf yellowing starting from the base of the plant and moving to the shoot tip. The roots show no external symptoms. However, the vascular system in the main roots and lower stems will show a yellow, orange or light tan streaking or discoloration.

Causal agent

Fusarium wilt is caused by the fungus *Fusarium oxysporum* f. sp. *pisi*. From infected vascular tissue in stems, the pathogen can be isolated onto standard agar media such as potato dextrose agar. Isolates show considerable variation in mycelial growth, pigmentation, and spore production in culture. Pathogen morphology and colony characteristics are similar to other *F. oxysporum* fungi. The fungus forms one- or two- celled, oval to kidney-shaped microconidia on monophialides, and four- to six-celled, fusiform, curved macroconidia. Microconidia measure 6–15 x 2.5–4.0 µm, and macroconidia range from 27–60 x 3.5–5.5 µm. Macroconidia are usually produced in cushion-shaped structures called sporodochia. Spherical chlamydospores are also formed and measure 10–11 µm in diameter. This pathogen is both seedborne and soilborne and can survive for more than 10 years. *Fusarium oxysporum* f. sp. *pisi* has distinct races; differential host inoculations are required to identify these.

Another legume, chickpea or garbanzo bean (*Cicer arietinum*), is susceptible to its own Fusarium wilt pathogen, *F. oxysporum* f. sp. *ciceris*. This pathogen is found worldwide and has at least eight distinct races.

Disease cycle

The fungus infects pea roots and travels up the vascular tissue and into the stem and above-ground tissues. Disease progresses rapidly when soil temperatures are above 20° C. Under such conditions, wilt symptoms can be severe and plants may die at or before pod formation. In the UK, Fusarium wilt is also known as St. John's disease because symptoms are most apparent around 24 June. 'Near wilt' refers to slower symptom development that occurs after flowering and is attributed to race 2.

Control

Use resistant cultivars. Resistance to race 1 has been stable and widely used for over 40 years. Resistance to other races is also available. Use seed that does not have significant levels of the pathogen. Rotate crops away from pea, though such rotations do not eliminate soil inoculum. In some regions early planting may enable peas to mature before soil temperatures are optimal for wilt development.

References

Haglund, W. A. and Kraft, J. M. 1979. *Fusarium oxysporum* f. sp. *pisi*, race 6: Occurrence and distribution. *Phytopathology* 69:818–820.

Hubbeling, N. 1974. Testing for resistance to wilt and near-wilt of peas, caused by race 1 and race 2 of *Fusarium oxysporum* f. sp. *pisi*. *Mededelingen van de Faculteit Landbouwwetenschappen Rijksuniversiteit Gent* 39:991–1000.

Jimenez-Gasco, M. d. M., Perez-Artes, E., Jimenez-Diaz, R. M. 2001. Identification of pathogenic races 0, 1B/C, 5, and 6 of *Fusarium oxysporum* f. sp. *ciceris* with random amplified polymorphic DNA (RAPD). *European Journal of Plant Pathology* 107:237–248.

Roberts, D. D. and Kraft, J. M. 1973. Enumeration of *Fusarium oxysporum* f. sp. *pisi* race 5 propagules from soil. *Phytopathology* 63:765–768.

Mycosphaerella pinodes (=Ascochyta pinodes)

ASCOCHYTA BLIGHT

Introduction and significance

Three diseases of pea, each caused by different *Ascochyta* fungi, were discovered in the late 1920s. The pathogen names comprising the original *Ascochyta* complex have recently been changed. Ascochyta blight, described here, is caused by *Mycosphaerella pinodes* (the asexual stage is *A. pinodes*). *Ascochyta pisi* (page 274) causes leaf and pod spot. The foot rot pathogen *A. pinodella* is now classified as *Phoma medicaginis* var. *pinodella* (page 272). Because these three pathogens may occur together on pea crops, it can be difficult to differentiate the various diseases.

Ascochyta blight occurs widely on pea crops grown in temperate regions, but is also known to cause damage in subtropical regions. Losses are due to foliage damage and reduced productivity, and to direct infections on pods, making them unmarketable. Ascochyta blight is considered the most serious of these three diseases.

Symptoms and diagnostic features

Mycosphaerella pinodes causes lesions on leaves, stems, and pods (**356, 357, 358**). Lesions initially appear as small, pinpoint, irregularly shaped flecks that are dark brown in color. Lesions on leaves and pods enlarge, generally are round to oval in shape, and contain concentric rings of alternating shades of brown. Lesions on stems may become purple in color, and can enlarge and girdle the stem. If conditions favor disease development, the lesions can coalesce and cause significant blighting of pea foliage (**359**). Foliar symptoms caused by *M. pinodes* are difficult to differentiate from those caused by *P. medicaginis* var. *pinodella*. However, the leaf and pod spot pathogen (*A. pisi*) causes spots that are tan to light brown, with darker brown to red-brown borders (see page 274).

Mycosphaerella pinodes can cause seedling damping-off and lesions on hypocotyls and roots if seed were infested. Therefore, the planting of infested seed can result in poor stands. However, the foot rot pathogen (*P. medicaginis* var. *pinodella*) is known to cause much more severe symptoms on below ground parts of pea.

Causal agent

Ascochyta blight is caused by the ascomycete *M. pinodes*. On senescent tissues, this fungus produces dark brown, spherical perithecia that measure 90–180 µm in diameter and contain bitunicate asci. Each ascus holds eight ascospores that measure 12–18 x 4–8 µm and are two-celled. *Mycosphaerella pinodes* has an asexual stage known as *Ascochyta pinodes*; *A. pinodes* produces spherical pycnidia that have hyaline, ellipsoid, two-celled conidia measuring 8–18 x 3–5 µm. Conidia are exuded out of pycnidia in yellowish (buff-colored) masses. *Phoma medicaginis* var. *pinodella* also produces spherical pycnidia and hyaline conidia that resemble those of *A. pinodes*. However, conidia of *P. medicaginis* var. *pinodella* are usually single-celled and slightly smaller than those of *M. pinodes*. Different races of *M. pinodes* have recently been identified.

Disease cycle

Mycosphaerella pinodes produces survival structures (sclerotia, chlamydospores) that enable this fungus to overwinter on pea crop residues and survive in the soil. Primary inoculum is by airborne ascospores that are released from perithecia present on pea crop residues. Infection occurs within 1 to 2 days at 15–25° C, and after 4 days at 5° C. A minimum of 2 hours of leaf wetness is need for germination. When environmental conditions favor disease development, symptoms can appear 2 to 4 days after ascospore infection. Severe disease is associated with repeated pea cropping and frequent wet weather after the lower leaves have started to senesce. Crops become more susceptible to disease with senescence, which has been attributed to the reduction of the phytoalexin pisatin in senescent tissues. Disease progresses from lower foliage to the top leaves by water-splashed conidia and airborne ascospores. *Mycosphaerella pinodes* is a seedborne pathogen and can infect other hosts such as *Lathyrus*, *Phaseolus*, and *Vicia*.

Control

Because the Ascochyta blight pathogen can persist in the soil, avoid planting pea in fields known to be infested with *M. pinodes*. Use seed that does not have significant levels of the pathogen. Appropriate seed treatments can also contribute to the management of seedborne inoculum. Grow seed crops in dry areas to enhance the production of clean seed. Avoid irrigating

356 **Early symptoms of Ascochyta blight of pea caused by *Mycosphaerella pinodes*.**

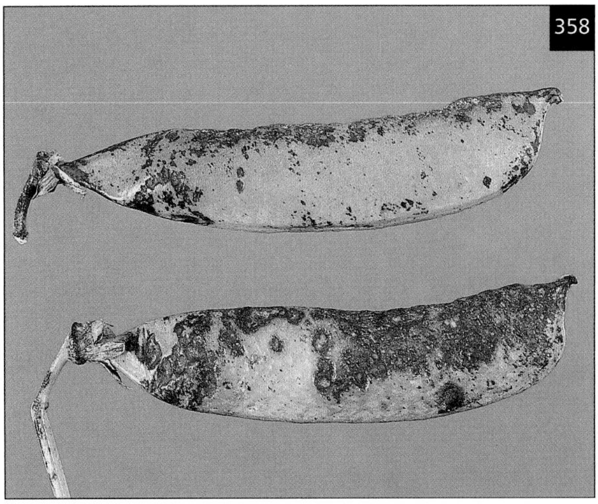

358 **Pea pods affected by *Mycosphaerella pinodes*.**

357 **Aschochyta blight symptoms caused by *Mycosphaerella pinodes* on senescent pea foliage.**

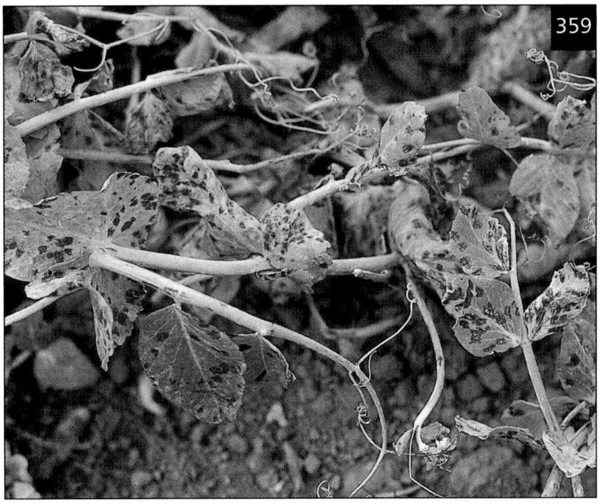

359 **Heavily diseased pea foliage affected by *Mycosphaerella pinodes*.**

with overhead sprinklers. If sprinklers are used, time irrigations so that rapid drying of the foliage takes place. Apply foliar fungicides during the flowering period when foliar symptoms first occur on the lower senescent leaves and if wet weather is anticipated. Disease forecast systems are being developed.

References

Béasse, C., Ney, B., and Tivoli, B. 1999. Effects of pod infection by *Mycosphaerella pinodes* on yield components of pea (*Pisum sativum*). *Annals of Applied Biology* 135:359–367.

Béasse, C., Ney, B., and Tivoli, B. 2000. A simple model of pea (*Pisum sativum*) growth affected by *Mycosphaerella pinodes*. *Plant Pathology* 49:187–200.

Biddle, A. J. 1994. Seed treatment usage on peas and beans in the UK. In: *Seed Treatment: Progress and Prospects*.T. J. Martin (ed.) BCPC Monograph No. 57:143–149. BCPC Publications, Farnham, UK.

Clulow, S .A., Lewis, B. G., and Matthews, P. 1992. Expression of resistance to *Mycosphaerella pinodes* in *Pisum sativum*. *Plant Pathology* 41:362–369.

Kraft, J. M., Dunne, B., Goulden, D., and Armstrong, S. 1998. A search for resistance in peas to *Mycosphaerella pinodes*. *Plant Disease* 82:251–253.

Moussart, A., Tivoli, B., Lemarchand, E., Deneufbourg, F., Roi, S., and Sicard, G. 1998. Role of seed infection by the Ascochyta blight pathogen of dried pea (*Mycosphaerella pinodes*) in seedling emergence, early disease development and transmission of the disease to aerial plant parts. *European Journal of Plant Pathology* 104:93–102.

Onfroy, C., Tivoli, B., Corbière, R., and Bouznad, Z .1999. Cultural, molecular and pathogenic variability of *Mycosphaerella pinodes* and *Phoma medicaginis* var. *pinodella* isolates from dried pea (*Pisum sativum*) in France. *Plant Pathology* 48:218–229.

Roger, C. and Tivoli, B. 1996. Spatio-temporal development of pycnidia and perithecia and dissemination of spores of *Mycosphaerella pinodes* on pea (*Pisum sativum*). *Plant Pathology* 45:518–528.

Roger, C., Tivoli, B., and Huber, L. 1999. Effects of temperature and moisture on disease and fruit body development of *Mycosphaerella pinodes* on pea (*Pisum sativum*). *Plant Pathology* 48:1–9.

Roger, C., Tivoli, B., and Huber, L. 1999. Effects of interrupted wet periods and different temperatures on the development of Ascochyta blight caused by *Mycosphaerella pinodes* on pea (*Pisum sativum*) seedlings. *Plant Pathology* 48:10–18.

Thomas, J. E., Kenyon, D. M., Biddle, A. J., and Ward, R. L. 2000. Forecasting and control of leaf and pod spot (*Mycosphaerella pinodes*) on field pea. Proceedings of the BCPC Conference – Pests & Diseases 3:871–876.

Wang, H., Hwang, S. F., Chang, K. F., Turnbull, G. D., and Howard, R. J. 2000. Characterization of Ascochyta isolates and susceptibility of pea cultivars to the Ascochyta disease complex in Alberta. *Plant Pathology* 49:540–545.

Wroth, J. M. 1998. Variation in pathogenicity among and within *Mycosphaerella pinodes* populations collected from field pea in Australia. *Canadian Journal of Botany* 76, 1955–1966.

Xue, A. G., Warkentin, T. D., and Kenaschuk, E. O. 1997. Effects of timing of inoculation with *Mycosphaerella pinodes* on yield and seed infection of field pea. *Canadian Journal of Plant Science* 78:685–689.

360 Pea seedling infected with downy mildew.

Peronospora viciae

DOWNY MILDEW

Introduction and significance

Downy mildew can be a major problem in many pea-growing areas and is particularly a concern in northern Europe and irrigated seed growing areas of the western USA and Canada. The pathogen is adaptable, and new races continue to emerge and grow on previously resistant cultivars. This is the same pathogen as the downy mildew that affects broad bean, though there may be physiological differences which limit cross infection between pea and broad bean.

Symptoms and diagnostic features

Seedlings infected at emergence are stunted and may die (**360**). Local, non-systemic leaf lesions typically begin as small, yellow, vein delimited blotches on the upper surface of leaves (**361**). The undersides of these lesions support sporulation that is gray to purple (**362**). As leaf lesions age they become brown and dry. Systemically infected shoots are usually lighter in color than healthy foliage, have a silvery appearance, and also develop gray to purple sporulation (**363**). Pods can be infected and typically show large, yellow blotches on the pod surface and white mycelial growth within the pod (**364**). Pod lesions later turn brown and in some cases become slightly distorted. Tendrils, particularly in semi-leafless cultivars, can be severely affected and become bleached and later colonized by *Botrytis*.

Causal agent

Downy mildew on pea and broad bean is caused by the oomycete *Peronospora viciae*. In some reports the pathogen is designated as a pathogen that is specific to pea: *Peronospora viciae* f. sp. *pisi*. The pathogen produces clusters of dichotomously branched sporangiophores that emerge through leaf stomata. Oval to elliptic sporangia are borne on pointed branch tips and measure 11–22 x 13–39 µm. The sexual structures, or oospores, are produced in local foliar lesions and in pods and can appear as early as 3 weeks after infection. Oospores are spherical, yellow to light brown, have reticulate ridges, and measure 25–37 µm in diameter. Isolates may be homothallic.

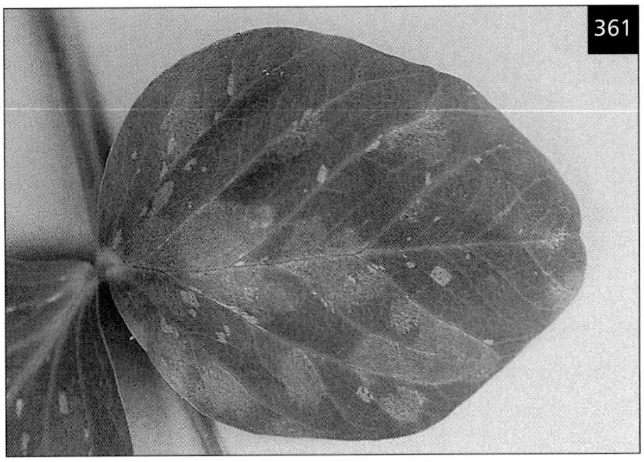

361 Upper surface of pea leaf with downy mildew.

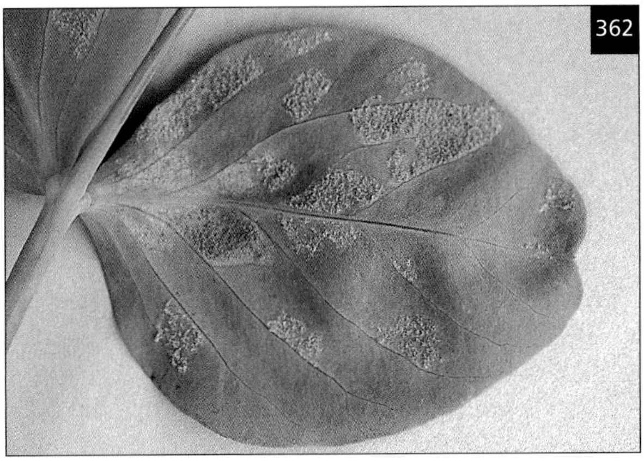

362 Underside of pea leaf with downy mildew sporulation.

363 Pea shoots with systemic infection of downy mildew.

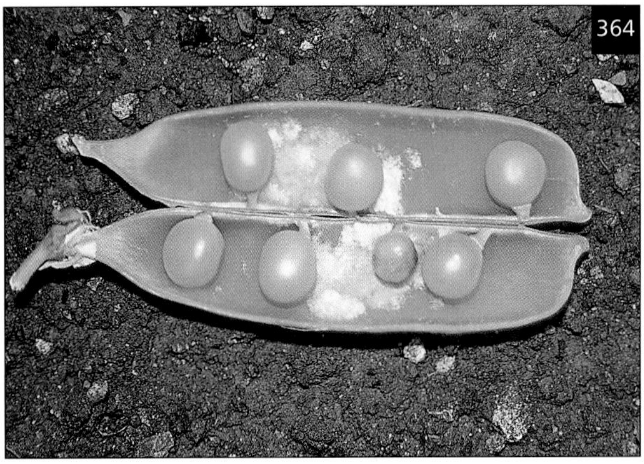

364 Internal pea pod tissues showing downy mildew sporulation.

Disease cycle

Though *P. viciae* is seedborne, infected seeds usually fail to germinate. Oospores in the soil are therefore the most important source of initial inoculum and mostly form at 10–15° C. In naturally infested soils, oospore numbers can range from 2 to 21 spores per gram of soil. Germinating seedlings infected by oospores develop systemic symptoms and support abundant sporangia that spread to other plants. Disease development is most rapid when temperatures are below 10° C and leaves are wet. Sporangial production decreases rapidly if humidity is below 95%. Wind and rain disperse sporangia, which penetrate the foliage by means of germ tubes if there is at least 4 hours of leaf wetness. Leaves become less susceptible to infection as they age. Maximum infection requires 6 hours of leaf wetness at 8–20° C.

Control

Practice good crop rotation with non-hosts to avoid buildup of soilborne inoculum. Plant resistant cultivars. However, when new races of the pathogen emerge, such resistance is overcome and new cultivars will need to be developed. Use seed treatments to protect the seed and emerging seedlings from infection during plant germination. Captan, thiram, and systemic phenylamide fungicides have provided some control as seed treatments, and newer fungicides are being used in Europe. Foliar fungicides have provided only limited control of downy mildew.

References

Dickinson, C. H. and Singh, H. 1982. Colonization and sporulation of *Peronospora viciae* on cultivars of *Pisum sativum*. *Plant Pathology* 31:333–341.

Miller, M. W. and de Whalley, C. V. 1981. The use of metalaxyl seed treatment to control pea downy mildew. *Proceedings 1981 British Crop Protection Conference – Insecticides and Fungicides* 1:341–348.

Pegg, G. F. and Mense, M. J. 1970. The biology of *Peronospora viciae* on pea: laboratory experiments on the effect of temperature, relative humidity and light on the production, germination and infectivity of sporangia. *Annals of Applied Biology* 66:417–428.

Stegmark, R. 1994. Downy mildew on peas (*Peronospora viciae* f. sp. *pisi*). *Agronomie* 14: 641–647.

Taylor, P. N., Lewis, B. G., and Mathews, P. 1990. Factors affecting systemic infection of *Pisum sativum* by *Peronospora viciae*. *Mycological Research* 94:179–181.

Van der Gaag, D. J. and Frinking, H. D. 1997. Extraction of oospores of *Peronospora viciae* from soil. *Plant Pathology* 46:675–679.

Van der Gaag, D. J. and Frinking, H. D. 1997. Survival characteristics of oospore populations of *Peronospora viciae* f. sp. *pisi* in soil. *Plant Pathology* 46: 978–988.

Pythium spp.

PYTHIUM ROOT ROT, DAMPING-OFF

Introduction and significance

Damping-off and root rot in pea are commonly caused by *Pythium* species, though several other soilborne pathogens cause similar symptoms (see the pea section on foot rot complex in this chapter). If soil conditions favor *Pythium* activity, significant stand loss can result.

Symptoms and diagnostic features

With pre-emergence damping-off, pea seed and newly germinated seedlings are attacked and rotted prior to the above-ground emergence of the seedling. Symptoms of post-emergence damping-off and root rot consist of stunted plants, yellowed lower leaves, general poor growth, wilting, and eventual collapse and death of plants. Roots of infected plants appear water-soaked or light brown in color. In severe cases, nearly all roots may be girdled or rotted off.

Causal agents

Damping-off and Pythium root rot are caused by the oomycetes *Pythium aphanidermatum*, *P. debaryanum*, and *P. ultimum*. All these species survive in the soil as saprophytes and are favored by wet soil conditions. With the exception of *P. ultimum*, these pathogens usually produce zoospores that swim to and infect susceptible tissues. Sexual structures, antheridia, oogonia, and oospores, are produced by all species. *Pythium debaryanum* and *P. ultimum* are favored when temperatures are below 20° C, while *P. aphanidermatum* is favored by temperatures above 20° C.

Disease cycle

Pythium species survive in soil as mycelium, sporangia, or oospores. The infective agent is the swimming zoospore. The pathogen is stimulated to produce zoospores during seed imbibition and germination, and seedling emergence, when seed and seedlings leak nutrients into the soil. Infection is favored by high soil moisture. In northern Europe, problems are associated with cold wet conditions that delay germination, while in parts of the USA the most severe attacks occur when temperatures are 18–24° C (above the optimum for germination of pea seeds). This latter condition is known as Pythium wilt and often leads to death of plants.

Control

Rotate away from susceptible pea and legume crops, though crop rotation is of limited value because of the soilborne nature of *Pythium*. Plant seed that has been treated with fungicides such as thiram, captan, or drazoxolon. Avoid using cultivars that are particularly sensitive to this pathogen; for example, cultivars with wrinkled seeds are generally more severely affected than those with round or dimpled seeds. Select sites and manage fields so that soils drain well, soil compaction is reduced, and low areas in the field are avoided.

References

Kraft, J. M. 1974. The influence of seedling exudates on the resistance of peas to Fusarium and Pythium root rot. *Phytopathology* 64:190–193.

Matthews, S. 1971. A study of seed lots of peas (*Pisum sativum* L.) differing in predisposition to pre-emergence mortality in soil. *Annals of Applied Biology* 68:177–183.

Short, G. E. and Lacy, M. L. 1976. Carbohydrate exudation from pea seeds: Effect of cultivar, seed age, seed color, and temperature. *Phytopathology* 66:182–187.

Sclerotinia minor, S. sclerotiorum, S. trifoliorum

WHITE MOLD, SCLEROTINIA ROT

Introduction and significance

Sclerotinia has a wide host range and pea is included. Severe outbreaks occur sporadically in production areas worldwide. For dry-harvested pea, the black sclerotia of the pathogen can be found as a contaminant in harvested seed.

Symptoms and diagnostic features

Symptoms of white mold on pea are very similar to white mold symptoms found on other legumes. See the phaseolus bean section on white mold (page 262).

Causal agents

White mold on pea is caused by *Sclerotinia sclerotiorum* and *S. trifoliorum*. See also the bean section on white mold in this chapter.

Disease cycle

See the bean section on white mold in this chapter.

Control

For a description of control options, see the bean section on white mold in this chapter. Fungicides applied at flowering can provide good control.

References

Huang, H. C. and Kokko, E. G. 1992. Pod rot of dry peas due to infection by ascospores of *Sclerotinia sclerotiorum*. *Plant Disease* 76:597–600.

Bean leaf roll virus

BEAN LEAF ROLL

Introduction and significance

Bean leaf roll virus (BLRV) is common in Europe, the Middle East, and the USA on both pea and broad bean.

Symptoms and diagnostic features

Early infection of pea causes severe stunting, chlorosis, and death of plants prior to flowering. If infected later in the growth stage, pea shows chlorosis of apical shoots and new leaves, upward rolling of leaves, and occasional brown leaf spots. Early infection of BLRV on broad bean causes stunting and reduction in size of developing leaves. BLRV also causes thickening of the leaf, leaf rolling, and interveinal yellowing (**365**, **366**). The number of pods on affected plants is reduced.

Causal agent and disease cycle

BLRV is a luteovirus with isometric particles. The pathogen is transmitted in a persistent manner by pea (*Acyrthosiphon pisum*) and other (*Aphis craccivora, Myzus persicae*) aphids. BLRV is also known as *Pea leaf roll virus*. In addition to pea and broad bean, BLRV affects weed legumes, lucerne (alfalfa), and red and white clovers.

Control

Follow general suggestions for managing virus diseases (see Part 1).

365 **Rolled leaflets caused by *Bean leaf roll virus*.**

366 **Apical yellowing of broad bean caused by *Bean leaf roll virus*.**

Pea early browning virus

PEA EARLY BROWNING

Introduction and significance

Pea early browning virus (PEBV) is thus far only found in Europe.

Symptoms and diagnostic features

Early symptoms on seedlings are chlorosis and death of the shoot tips, resulting in formation of secondary shoots (**367**). At this stage, the stem usually has a distinctive orange-brown color (**368**). Older plants develop purple-brown lesions on leaves, petioles, stems, and pods (**369, 370**). Many of the lesions become pale brown or bleached and can be confused with scorch symptoms caused by herbicide or fertilizer applications.

Causal agent and disease cycle

PEBV belongs to the tobravirus group, has rod-shaped particles, and is transmitted by *Paratrichodorus* and *Trichodorus* nematodes. It is a seedborne virus. PEBV is found particularly in the Netherlands and UK where light sandy soils favor the nematode vector.

Control

Follow general suggestions in Part 1. Plant seed that is free of the virus. Do not plant susceptible pea crops in fields that have a history of this disease.

References

Hampton, R., Waterworth, H., Goodman, R. M., and Lee, R. 1982. Importance of seedborne viruses in crop germplasm. *Plant Disease* 66:977–978.

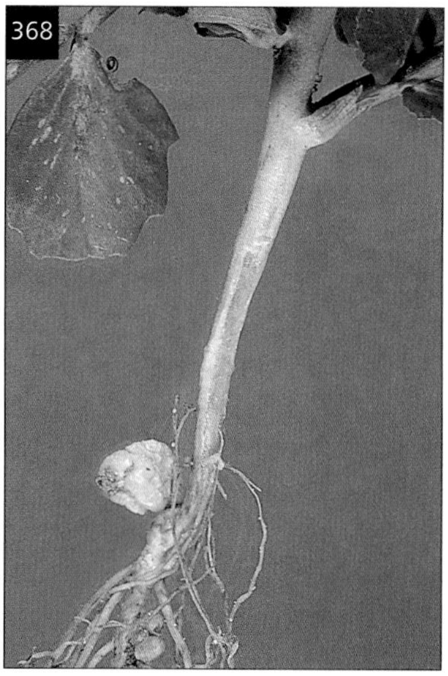

368 Orange discoloration inside pea stem infected with *Pea early browning virus.*

369 Pea infected with *Pea early browning virus*, showing necrotic leaf symptoms.

367 Pea seedlings infected with *Pea early browning virus.*

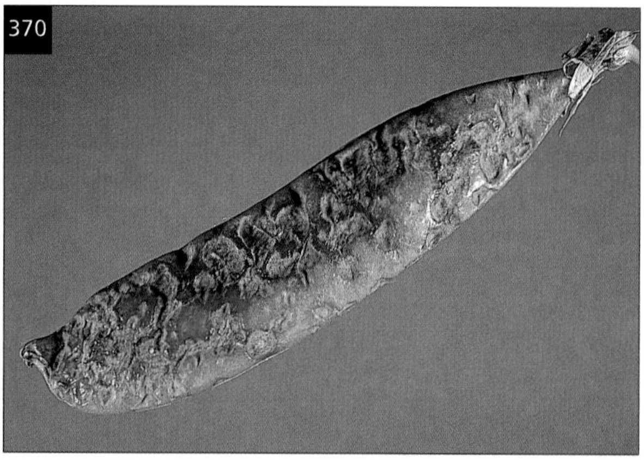

370 Pea pod with necrotic symptoms caused by *Pea early browning virus.*

Pea enation mosaic virus

PEA ENATION MOSAIC

Introduction and significance

Pea enation mosaic virus (PEMV) is a common virus and was first reported in the USA in 1935. It is widespread in north temperate areas and can cause losses up to 50%. PEMV affects many of the major genera of legumes worldwide, including pea, broad bean, chickpea, and lentil.

Symptoms and diagnostic features

Although early infections cause severe distortion of foliage and even death of plants, more typically there is vein clearing and chlorotic or transluscent flecking of the leaves (**371**). Translucent areas may be produced along the veins, together with leaf and pod distortion and formation of enations. Enations are small tissue proliferations or growths that usually form on the undersides of leaves and on pods. Severe infection causes leaf deformities and plant stunting.

Causal agent and disease cycle

PEMV is the only member of the enamovirus group and consists of two isometric particles measuring 25 nm and 28 nm. PEMV is transmitted in a persistent manner by several aphid species and especially by the pea aphid (*Acyrothosiphon pisum*). Some strains are readily transmitted mechanically.

Control

Follow general suggestions for managing virus diseases (see Part 1).

References

Fargette, D., Jenniskens, M. J., and Peters, D. 1982. Acquisition and transmission of pea enation mosaic virus by the individual pea aphid. *Phytopathology* 72:1386–1390.

371 **Pea shoot tips infected with *Pea enation mosaic virus*, showing yellowing and mosaic symptoms.**

Pea seedborne mosaic virus

PEA SEEDBORNE MOSAIC

Introduction and significance

Pea seedborne mosaic virus (PSbMV) was first reported in the USA in 1968 but has now been disseminated worldwide due to its seedborne nature. Up to 90% seed infection has been reported.

Symptoms and diagnostic features

Pea plants are stunted and can be distorted. Leaves roll downwards and develop chlorosis, mosaic, and other general virus-like foliar symptoms. Affected plants may fail to set pods or produce distorted pods or seed with split seed coats. The disease delays plant maturity. Seed infection occurs when plants are infected before flowering. Broad bean may show more severe rolling and yellowing of the upper leaves.

Causal agent

PSbMV is a potyvirus that is seedborne in pea, lentil (*Lens culinaris*), and vetch. The particle shape is a flexuous rod. Chickpea is also a host but does not carry the virus in its seed. There are several strains of the virus and many have a restricted host range. For example, a lentil strain is unable to infect pea that carries the mo gene for resistance to *Bean yellow mosaic virus*.

Disease cycle

PSbMV is transmitted in a nonpersistent manner by aphids, particularly the pea aphid (*Acyrthosiphon pisum*), *Myzus persicae*, and *Macrosiphum euphorbiae*. The main source of infection is usually seed, and the elimination of infected seed is the most important control measure; even a low level of seed infection can lead to widespread disease.

Control

Follow general suggestions for managing virus diseases (see Part 1). International regulations are used to prevent spread of this virus in seed. Resistance genes have been identified for this virus.

References

Alconero, R., Provvidenti, R., and Gonsalves, D. 1986. Three pea seedborne mosaic virus pathotypes from pea and lentil germ plasm. *Plant Disease* 70:783–786.

Hampton, R. O., Kraft, J. M., and Muehlbauer, F. J. 1993. Minimizing the threat of seedborne pathogens in crop germ plasm: elimination of pea seedborne mosaic virus from the USDA-ARS germ plasm collection of *Pisum sativum*. *Plant Disease* 77:220–224.

Hampton, R., Waterworth, H., Goodman, R. M., and Lee, R. 1982. Importance of seedborne viruses in crop germ plasm. *Plant Disease* 66:977–978.

Kohen, P. D., Doughterty, W. G., and Hampton, R. O. 1992. Detection of pea seedborne potyvirus by sequence specific enzymatic amplification. *Journal of Virological Methods* 37:253–258.

Masmoudi, K., Suhas, M., Khetarpal, R. K., and Maury, Y. 1994. Specific serological detection of the transmissible virus in pea seed infected by pea seedborne mosaic virus. *Phytopathology* 84:756–760.

Pea streak virus

PEA STREAK

Introduction and significance

Pea streak virus (PeSV) is known mainly from the USA, though it has been identified in clover in Canada and Germany.

Symptoms and diagnostic features

Early infection kills the plants of most pea cultivars, while later infection results in purple or brown streaks on stems and petioles. Necrotic spots can develop on leaves. Pods show sunken, brown lesions, or may not develop fully if the plant is infected at an early stage.

Causal agent and disease cycle

Researchers differ on whether or not PeSV is a carlavirus; it also shares characteristics with potexviruses. Transmission is in a nonpersistent manner by the pea aphid (*Acyrthosiphon pisum*). Problems often occur close to alfalfa, which is a significant reservoir of the pathogen.

Control

Follow general suggestions in Part 1.

References

Hampton, R. O. and Webster, K. A. 1983. Pea streak and alfalfa mosaic viruses in alfalfa: reservoir of viruses infectious to *Pisum* peas. *Plant Disease* 67:308–310.

VICIA FABA (BROAD BEAN)

Aphanomyces euteiches, Fusarium spp., *Pythium* spp., *Phytophthora megasperma, Rhizoctonia solani*

FUSARIUM AND OTHER ROOT ROTS

Introduction and significance

There are a number of root rot diseases caused by various pathogens on broad (or faba) bean. Five different *Fusarium* species are particularly important. One of them, *F. solani* f. sp. *fabae*, has been very damaging in China, Japan, and Sudan, with losses of up to 40%. Significant root rot problems are also caused by species of *Rhizoctonia, Pythium*, and *Phytophthora*. Three other pathogens, *Thielaviopsis basicola, Macrophomina phaseoli*, and *Helicobasidium purpureum*, cause only occasional problems.

373 Lower stem and root lesions and fungal growth from Fusarium root rot of broad bean.

372 Wilting broad bean foliage caused by Fusarium root rot.

Symptoms and diagnostic features

The pathogens cause various plant problems. Shortly after planting, these soil inhabitants cause seed decay and damping-off of emergent seedlings. Roots of both young and mature plants develop rots that turn the root water-soaked, then black. Mature plants show the effects of below-ground infection by wilting and collapsing of foliage (372). If *Fusarium* species are responsible, then masses of white mycelium and pink or orange spore clusters can be seen on the stem base and roots (373). *Fusarium* infection may be associated with insect damage if there is splitting of broad bean stem tissues, which allows midge larvae (*Resseliella* species) to become established.

Causal agents

Fusarium species affecting broad bean include *F. oxysporum, F. avenaceum, F. culmorum, F. graminearum*, and *F. solani*. Some *Fusarium* pathogens exhibit host preferences or limitations and are designated with the forma specialis (f. sp.) notation. An example of this is *F. solani* f. sp. *fabae*. The various *Fusarium* spp. are identified by morphological features, characteristics when grown in culture, and host ranges.

A second group of root rot pathogens are in the oomycete category: *Pythium* spp., *Phytophthora megasperma*, and *Aphanomyces euteiches*. *Phytophthora megasperma* causes severe root rot and wilt in spring planted broad bean in the UK, but isolates from broad bean were not pathogenic to pea, lupin, clover, sugar beet, or brassicas. *Phytophthora erythroseptica* var. *pisi* from pea is reported to infect broad bean. Isolates of *A. euteiches* show considerable variation in their host range and pathogenicity, and there may be limited cross infection between various legumes. For more details, see the pea section on Aphanomyces root rot in this chapter (page 270).

Rhizoctonia solani is another root pathogen of broad bean. This fungus is a common pathogen of many other crops and survives for long periods in the soil. The fungus has a basidiomycete perfect stage (*Thanatephorus cucumeris*), though the role of this perfect stage in disease development is not clear.

Disease cycle

All pathogens considered here are soil inhabitants. Therefore, initial inoculum is always from propagules in field soils. These diseases are usually more severe where the field drains poorly, soil is compacted, seed beds are not properly prepared, and excess water is applied during irrigation.

Control

Practice crop rotation so that legume crops are not planted too frequently. A break of at least 4 years between legume plantings is helpful. However, note that crop rotations will not eradicate soilborne pathogens from fields. Manage the field soil to minimize soil compaction and enhance good drainage. Irrigate properly so that plant stress is minimized but overwatering is avoided. Seed treatments may provide partial control of seed rot and damping-off phases of root rot diseases.

References

Salt, G.A. 1983. Root diseases of *Vicia faba* L. In: *The Faba Bean (Vicia faba L.): A Basis for Improvement*. (Ed. By P.D. Hebblethwaite) pp.393–419. Butterworths, London.

***Botrytis fabae*,**
***B. cinerea* (teleomorph = *Botryotinia fuckeliana*)**

CHOCOLATE SPOT

Introduction and significance

This disease occurs in most production areas throughout the world. Disease severity varies from year to year depending on environmental conditions. Severe attacks may significantly reduce yields.

Symptoms and diagnostic features

The first symptoms are small, circular, red to brown leaf spots measuring up to 5 mm in diameter. These spots are visible on both sides of the leaf surface (**374**). These smaller, discrete lesions make up the 'non-aggressive' phase of chocolate spot disease. The 'aggressive' phase develops if there is suitable humid or wet weather and occurs when the small spots coalesce into large, irregularly shaped gray to black leaf blotches that can result in death of leaflets and premature defoliation (**375**, **376**). Aggressive lesions may have concentric lines within the diseased area, causing them to resemble leaf symptoms from other diseases. Stems develop elongated red, dark brown, or black lesions. Occasionally, severe disease can kill overwintered plants before they reach flowering stage. Similar lesions also appear on the flowers and pods. Diseased pods are usually unmarketable.

Causal agents

Chocolate spot is caused by the fungus *Botrytis fabae*. Overnight incubation of leaves with aggressive lesions will sometimes cause the fungus to sporulate; otherwise, standard isolations will be necessary to confirm this pathogen. *Botrytis fabae* produces ellipsoidal to ovoid conidia that measure 14–29 x 11–20 µm. In culture the pathogen forms profuse, small (1–3 mm), spherical to oblong, black sclerotia. In addition to *B. fabae*, *B. cinerea* is thought to contribute to chocolate-spot disease by either initiating and forming its own leaf spots, or by co-infecting spots along with *B. fabae*. *Botyrtis cinerea* conidia are significantly smaller (6–18 x 4–11µm) than conidia of *B. fabae*. *Botryotinia fuckeliana*, the teleomorph stage of *B. cinerea*, is rarely observed on this host.

374 **Foliar symptoms of chocolate spot of broad bean.**

375 **Broad bean with aggressive and non-aggressive symptoms on leaflets.**

Disease cycle

While *B. fabae* can be seedborne, this aspect is probably of little importance. Disease is initiated when sclerotia on infected debris produce conidia. Conidia move via wind currents and land on susceptible broad bean tissues. Overwintered and volunteer broad bean plants that support established infections allow the pathogen to survive and infect new crops in the spring. Frost and other damage to the foliage during the winter or early spring also allow early build up of inoculum. Optimum temperatures for disease development are 15–20° C, but activity occurs over the range of 4–30° C. Infection does not require the presence of free water and occurs when relative humidity is between 85–100%. The disease can develop rapidly, with leaf infection able to take place in less than 12 hours and new lesions appearing in less than 48 hours. Under optimum conditions, leaves can collapse within 4 days. The pathogen is favored by high humidity, with little lesion development below 70% relative humidity. After initial infection, conidia are produced when the relative humidity is high (80–90%); these spores move in the air and further spread the disease.

Control

Implement cultural practices to enhance broad bean growth; stressed plants suffering from nutritional deficiencies, poor drainage, and other factors are more susceptible. Avoid low-lying or sheltered sites. To improve drying of plant foliage and decrease humidity, select sites that have good exposure to wind and sun, increase

376 **Broad bean field affected with chocolate spot.**

row spacing between plants, and orient crop rows parallel with prevailing winds. For overwintered crops, adjust the planting date so that vigorous shoot growth prior to winter is reduced. For the UK, planting should be from late October onwards. Fungicides are only partially successful in controlling severe chocolate spot. Foliar sprays are usually initiated at early to mid-flowering when non-aggressive spots start to increase on the lower leaves. Triazole or strobilurin fungicides are often applied with chlorothalonil products.

References

Creighton, N. F., Bainbridge, A., and Fitt, B. D. L. 1985. Epidemiology and control of chocolate spot (*Botrytis fabae*) on winter field beans (*Vicia faba*). *Crop Protection* 4:235–243.

Gladders, P., Ellerton, D. R., and Bowerman, P. 1991. Optimising the control of chocolate spot. *Aspects of Applied Biology* 27, *Production and Protection of Legumes*, pp. 105–110.

Harrison, J.G. 1980. Effects of environmental factors on growth of lesions on field bean leaves infested by *Botrytis fabae*. *Annals of Applied Biology* 95:53–61.

Harrison, J.G. 1984. Effect of humidity on infection of field bean leaves by *Botrytis fabae* and on germination of conidia. *Transactions of the British Mycological Society* 82:245–248.

Harrison, J.G. 1984. The biology of *Botrytis* spp. on Vicia beans and chocolate spot disease – a review. *Plant Pathology* 37:168–201.

Koike, S. T. 1998. Severe outbreak of chocolate spot of fava bean, caused by *Botrytis fabae*, in California. *Plant Disease* 82:831.

Lane, A. and Gladders, P. 2000. Pests and Diseases of Oilseeds, Brassica seed crops and Field beans. Chapter 3, *BCPC Pest and Disease Management Handbook* (ed. D V Alford) pp. 52–83. Blackwell Science, Oxford.

Didymella fabae (anamorph = *Ascochyta fabae*)

LEAF AND POD SPOT

Introduction and significance

This important seedborne disease is present in all the major production areas. Seed tests and pathogen thresholds are applied to seed stocks, and *Ascochyta* is a major cause of rejection for seed lots. Distinct strains may differentially infect vetch (*Vicia sativa*), soybean (*Glycine max*), and bean.

Symptoms and diagnostic features

Symptoms can occur first on the initial leaves of seedlings growing from infected seed. Symptoms are slightly sunken, gray to black leaf spots that measure 1–2 cm in diameter and have distinct dark brown or black margins (**377**). Leaf spots usually contain small but visible dark brown fruiting bodies called pycnidia. Pycnidia are conspicuous in the gray-colored lesions and may be produced in concentric rings, while these structures are more difficult to see in darker, black leaf spots. Therefore, the black leaf spots may resemble chocolate spot lesions. On stems and pods, lesions are usually sunken and penetrate deeply (**378**). Stem lesions can cause wilting of the foliage and stem collapse. Infected pods may drop from the plant; seed inside diseased pods become dark brown and stained.

377 Close-up of leaf and pod spot lesions of broad bean.

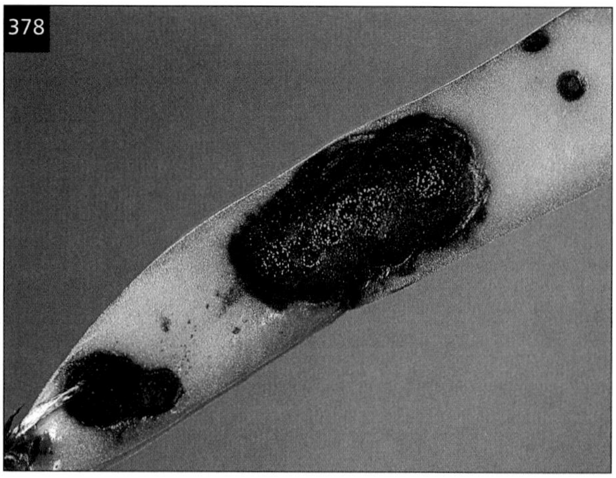

378 Broad bean pods with pink spore ooze from the leaf and pod spot pathogen.

Causal agent

Leaf and pod spot is caused by *Didymella fabae*. This is an ascomycete fungus and produces subglobose, dark brown pseudothecia and ascospores that are hyaline, two-celled, and measure 15–18 x 10–14 µm. The upper cell is wider than the lower cell and there is a constriction at the septum. In diseased tissues, pycnidia of the *Ascochyta fabae* imperfect stage are dark brown, spherical, and produce asexual conidia. Conidia are guttulate, straight or slightly curved, measure 16–24 x 3.5–6 µm, and have from one to three septa. The pathogen is seedborne.

Disease cycle

The fungus can remain viable in seed for up to 3 years, but survival on soil-incorporated crop residues is only a few months. Perithecia release ascospores that can function as primary inoculum and be blown long distances by winds. However, secondary spread via conidia from pycnidia depends on splashing water and will therefore only travel as far as the rain and sprinklers disperse the conidia. The disease cycle takes 12 to 18 days and is favored by wet and cool conditions.

Control

Use seed that does not have significant levels of the pathogen. For broad bean, thresholds for infested seed are 0 positive seed per 600 seeds tested. Fungicide seed treatments are usually only partially effective. Foliar fungicides applied for controlling other diseases may have some efficacy against *Ascochyta*. There are few products specifically registered for leaf and pod spot. Use resistant cultivars if available.

References

Bond, D. A. and Pope, M. 1980. *Ascochyta fabae* on winter beans (*Vicia faba*): pathogen spread and variation in host resistance. *Plant Pathology* 29:59–65.

Hewett, P. D. 1973. The field behaviour of seed-borne *Ascochta fabae* and disease control in field beans. *Annals of Applied Biology* 74:287–295.

Gaunt, R. E. 1983. Shoot diseases caused by fungal pathogens. In: *The Faba Bean (Vicia faba L.): A Basis for Improvement*. (Ed. By P.D. Hebblethwaite) pp.463–492. Butterworths, London.

Jellis, G. J. and Punithalingam, E. 1991. Discovery of *Didymella fabae* sp. nov., teleomorph of *Ascochyta fabae*, on faba bean straw. *Plant Pathology* 40:150–157.

Madeira, A. D., Fyrett, K. P., Rossall, S., and Clarke, J. A. 1993. Interactions between *Ascochyta fabae* and *Botrytis fabae*. *Mycological Research* 97:1217–1222.

Peronospora viciae

DOWNY MILDEW

Introduction and significance

Damaging attacks of downy mildew occur regularly on broad bean. This is the same downy mildew that affects pea, though there may be physiological differences that limit cross infection between broad bean and pea.

Symptoms and diagnostic features

Foliar symptoms consist of various sized blotches and lesions that first are chlorotic, later turn red-brown, and can reach up to 2–3 cm in diameter (**379, 380**). The pathogen produces purple-brown sporulation on the undersides of lesions. Systemic symptoms are sometimes found and appear as a general chlorosis in

379 Downy mildew lesion on the upper leaf surface of broad bean.

380 Underside of downy mildew lesion of broad bean.

the upper leaves, accompanied by profuse sporulation on the leaf underside. Old downy mildew lesions become dry, brown to black in color, and can be colonized by *Botrytis* fungi.

Causal agent

Downy mildew is caused by the oomycete *Peronospora viciae*. For a description of this pathogen see the pea section on downy mildew in this chapter (page 280).

Disease cycle

Primary inoculum is mainly soilborne oospores, and secondary spread is by airborne sporangia. This disease is favored by cool, moist conditions. For more details, see the pea section on downy mildew in this chapter.

Control

Use resistant cultivars. Rotate broad bean and other legume crops with non-hosts so that soil inoculum does not build up to high levels; crop rotations, however, will not prevent disease because of the far ranging aerial sporangia. Some seed treatments can help prevent seedling infection from soilborne oospores. Apply foliar fungicides to reduce secondary spread.

References

Biddle, A. J., Thomas, J., Kenyon, D., Hardwick, N. V., and Taylor, M. C. 2003. The effect of downy mildew (*Peronospora viciae*) on the yield of spring sown field beans (*Vicia faba*) and its control. *Proceedings of the BCPC International Congress – Crop Science & Technology 2003* 2:947–952.

Glasscock, H. H. 1963. Downy mildew of broad bean. *Plant Pathology* 12:91–92.

Van der Gaag, D. J., Frinking, H. D., and Geerrds, C. F. 1993. Production of oospores by *Peronospora viciae* f. sp. *fabae*. *Netherlands Journal of Plant Pathology* 99, Supplement 3:83–91.

Sclerotinia minor, S. sclerotiorum, S. trifoliorum

WHITE MOLD, SCLEROTINIA ROT

381 Broad bean, like field bean pictured here, can be severely affected by *Sclerotinia*.

382 Young broad bean plants dying from *Sclerotinia trifoliorum* infections.

Introduction and significance

Sclerotinia diseases are important on many legume crops, including broad bean. Crops in coastal regions of North Africa, Asia, North America, and Western Europe can experience significant problems with *Sclerotinia* pathogens. Three species of *Sclerotinia* occur on broad bean. *Sclerotinia sclerotiorum* is the most commonly found species on this crop. In the UK, *S. trifoliorum* is the important species on overwintered crops; severe disease is often associated with previous red clover (*Trifolium pratense*) plantings. *Sclerotinia minor* affects broad bean but is less of a problem. Interestingly, in California broad bean appears to be immune to *S. minor*.

Symptoms and diagnostic features
Symptoms of white mold on broad bean, caused by all three species of *Sclerotinia*, are very similar to white mold symptoms found on other legumes (**381**). For symptom descriptions, see the phaseolus bean section on white mold in this chapter (page 262).

Causal agents
White mold on broad bean can be caused by any of the three *Sclerotinia* species: *S. sclerotiorum*, *S. trifoliorum*, (**382**), and *S. minor*. For descriptions of these pathogens see the bean section on white mold in this chapter.

Disease cycle
For a description of the disease cycle, see the bean section on white mold in this chapter (page 262).

Control
For a description of control options, see the bean section on white mold in this chapter.

References

Jellis, G. J., Davies, J. M. L., and Scott, E. S. 1984. *Sclerotinia* on oilseed rape: implications for crop rotation. *Proceedings of the British Crop Protection Conference – Pests & Diseases 1984* 2:709–716.

Koike, S. T., Smith, R. F., Jackson, L. E., Wyland, L. J., Inman, J. I., and Chaney, W. E. 1996. Phacelia, lana woollypod vetch, and Austrian winter pea: three new cover crop hosts of *Sclerotinia minor* in California. *Plant Disease* 80:1409–1412.

Jellis, G. J., Smith, D. B., and Scott, E. S. 1990. Identification of *Sclerotinia* spp. on *Vicia faba*. *Mycological Research* 94:407–409.

Willets, H. J. and Wong, J. A-L. 1980. The biology of *Sclerotinia sclerotiorum*, *S. trifoliorum*, and *S. minor* with emphasis on specific nomenclature. *Botanical Review* 46: 101–165.

Williams, G. H. and Western, J. H. 1965. The biology of *Sclerotinia trifoliorum* Erikss. and other species of sclerotium-forming fungi. I Apothecium formation from sclerotia. *Annals of Applied Biology* 56:253–260.

Williams, G. H. and Western, J. H. 1965. The biology of *Sclerotinia trifoliorum* Erikss. and other species of sclerotium-forming fungi. II The survival of sclerotia in soil. *Annals of Applied Biology* 56:261–268.

Uromyces vicia-fabae

RUST

Introduction and significance
This common and widely occurring rust disease affects broad bean, pea, lentil, and other plants in the Fabaceae. There is some specialization of races, which vary in their host ranges. Rust is capable of reducing yields and is particularly severe in North Africa, the Mediterranean region, and the Middle East. More typically, however, rust epidemics develop after flowering, especially on autumn-planted crops, and its effects on yield and quality are limited.

Symptoms and diagnostic features
Initial rust symptoms are small, chlorotic leaf spots that first develop on leaves lower in the plant canopy. Typical orange-brown, raised pustules, called uredinia, develop from these yellow spots (**383**). The pustules are usually bordered with a chlorotic halo and later break open, releasing the dusty orange-brown spores. As rust develops, the number of spots and pustules increases and leads to drying and death of entire leaves. A general epidemic causes an overall bronzed appearance and extensive defoliation. Rust also has an aecia stage that

383 Rust uredinia on broad bean.

384 Rust aecia on upper leaf surfaces of broad bean.

385 Rust aecia on lower leaf surfaces of broad bean.

386 Telial stage of rust of broad bean.

causes yellow spots on leaves (**384, 385**). The conspicuous dark brown teliospore stage forms on plant tissues in late summer when crops are maturing (**386**).

Causal agent

Rust of broad bean is caused by the fungus *Uromyces vicia-fabae*. This pathogen is a macrocyclic, autoecious rust that produces sub-globoid to ovoid, pale brown urediniospores. Urediniospores are echinulate, have three to four pores, and measure 18–28 x 18–22 µm. Teliospores, a second spore type, are similar in shape to the urediniospores but are smooth walled and darker brown. Teliospores are also larger, measuring 24–35 x 18–25 µm, and are attached to brown stalks (pedicels) that can be up to 100 µm long. A third spore type, aecidiospores, are yellow, polygonoid–globoid in shape, and measure 18–36 x 16–24 µm. Aecidiospores are reported to overwinter in Mediterranean climates but are unable to survive severe winters. In the UK, aecidia on *V. faba* have been reported only occasionally in late autumn or at the end of winter. The previous name for this pathogen was *Uromyces fabae*.

Disease cycle

The type of initial inoculum appears to depend on the nature of the winter season in various regions. If winters are relatively mild, then urediniospores from volunteer plants are the initial inoculum. In regions having more severe winters, teliospores from crop residues germinate and result in aecidia on the spring planted broad beans. Teliospores may remain viable for up to 2 years. There are reports that seeds may harbor this pathogen. High humidity and warm temperatures favor development of this disease.

Control

Destroy volunteer plants and bury crop residues so that they decompose. Apply fungicides prior to or at an early stage of rust development. Effective products are dithiocarbamates, strobilurins, and chlorothalonil, which are protectants, and triazoles and morpholines, which have some eradicative activity. Sources of rust resistance have been identified but may not be completely satisfactory. Plant resistant cultivars as they become available.

References

Conner, R. L. and Bernier, C. C. 1982. Host range of *Uromyces viciae-fabae*. *Phytopathology* 72:687–689.

Conner, R. L. and Bernier, C. C. 1982. Race identification of *Uromyces viciae-fabae*. *Plant Pathology* 4:157–160.

Murray, D. C. and Walters, D. R. 1992. Increased photosynthesis and resistance to rust infection in upper, uninfected leaves of rusted broad beans (*Vicia faba* L.). *New Phytologist* 120:235–242.

Rashid, K. Y. and Bernier, C. C. 1986. Selection for slow rusting in faba bean to *Uromyces viciae–fabae*. *Crop Protection* 5:218–224.

Rubiales, D. and Sillero, J. C. 2003. Uromyces viciae-fabae haustorium formation in susceptible and resistant faba bean lines. European Journal of Plant Pathology 109:71–73.

Sache, I., and Zadoks, J. C. 1995. Life-table analysis of faba bean rust. *European Journal of Plant Pathology* 101:431-439.

Sillero, J. C., Moreno, M. T., and Rubiales, D. 2000. Characterization of new sources of resistance to *Uromyces viciae-fabae* in a germplasm collection of *Vicia faba*. *Plant Pathology* 49: 389–395.

Williams, P. F. 1978. Growth of broad beans infected by *Uromyces viciae-fabae*. *Annals of Applied Biology* 24: 329–334.

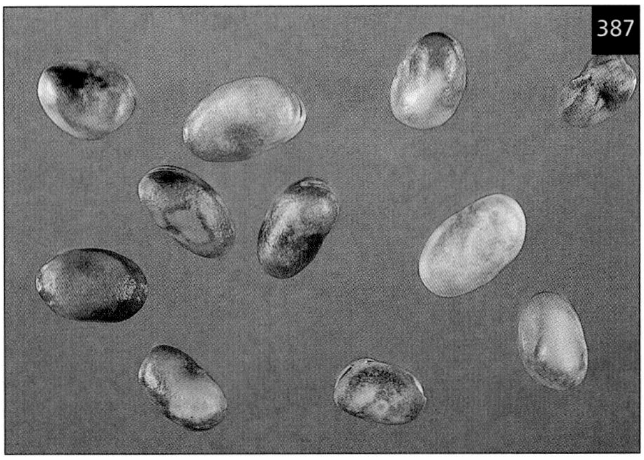

387 Broad bean seed coat showing necrosis due to *Broad bean stain virus*.

Broad bean stain virus, Broad bean true mosaic virus

BROAD BEAN STAIN, BROAD BEAN TRUE MOSAIC

Introduction and significance

Some viruses of interest that infect broad bean are the following: *Bean yellow mosaic virus, Bean leaf roll virus, Broad bean stain virus, Broad bean true mosaic virus*. In this chapter, *Bean yellow mosaic virus* is discussed in the virus section for bean, and *Bean leaf roll virus* is discussed in the virus section for pea. *Broad bean stain virus* (BBSV), in nature, infects only broad bean and occurs in Europe, the Mediterranean area, and Australia.

Symptoms and diagnostic features

BBSV causes severe mosaic symptoms on leaves, though symptom expression can vary between successive leaves. Plants are stunted and produce deformed pods. BBSV induces a brown necrosis or stain around the periphery of infected seed (**387**) in cultivars such as Aquadulce; however, other cultivars such as Triple White show no such effect. Yield loss may be as high as 30 to 40%.

Causal agents and disease cycle

BBSV has isometric particles that measure 28 nm in diameter and belongs to the comovirus group. Infected seed is an important source of BBSV as transmission rates to seedlings may approach 40%. BBSV can be detected in seed by immuno-scanning electron microscopy (ISEM). In the field, BBSV is vectored by weevils, particularly *Apion vorax* and *Sitona lineatus*.

Broad bean true mosaic virus (BBTMV) is also a comovirus that has particle shape, symptoms, biology, epidemiology, and vectors very similar to BBSV. Serological tests, however, show that BBTMV and BBSV are distinct from each other. BBTMV is commonly seedborne and affected seedlings are stunted and produce few pods. There is some evidence of cultivar resistance for BBTMV.

Control

Follow general suggestions for managing virus diseases (see Part 1).

Lactuca sativa Lettuce

Lettuce (*Lactuca sativa*) is in the Asteraceae (aster family) and is the only commercially grown *Lactuca* species. Lettuce is perhaps the most popular fresh salad vegetable in the world and is almost exclusively used as a fresh product. (Some lettuce is used in stir-fried dishes in Asian cuisine, and lettuce is cooked in a few dishes in western countries.) Lettuce most likely originated in the Mediterranean area. It has a long history of use for human consumption and today is grown throughout the world. Several lettuce types are grown commercially. Head-forming types are crisphead or iceberg, and butterhead or bibb lettuce. Loose, non-head-forming types are romaine or cos, and leaf or loose-leaved (mostly red and green leaf) cultivars.

Aster yellows phytoplasma

ASTER YELLOWS

Introduction and significance

Aster yellows is periodically important on lettuce crops throughout the world. Disease incidence is usually low, so this problem is considered a minor disease.

Symptoms and diagnostic features

Symptoms and disease severity of aster yellows on lettuce can be highly variable, depending on the strain of the phytoplasma, age of the plant when infected, and other factors. Plants are generally stunted and yellowed (**388**). Leaves can be malformed in various ways, and often remain small and thickened (**389**). When the plant is sliced lengthwise, the arrangement of the stems sometimes shows a twisted, spiral configuration (**390**). Characteristic, diagnostic pink to orange-tan latex oozes to leaf and petiole surfaces and results in oblong to circular deposits in the interior parts of lettuce heads (**391**). Plants infected early in their development will not form heads.

388 Chlorotic and stunted lettuce infected by aster yellows.

389 Deformed lettuce infected by aster yellows.

390 Lettuce infected with aster yellows can develop a twisting of the petioles.

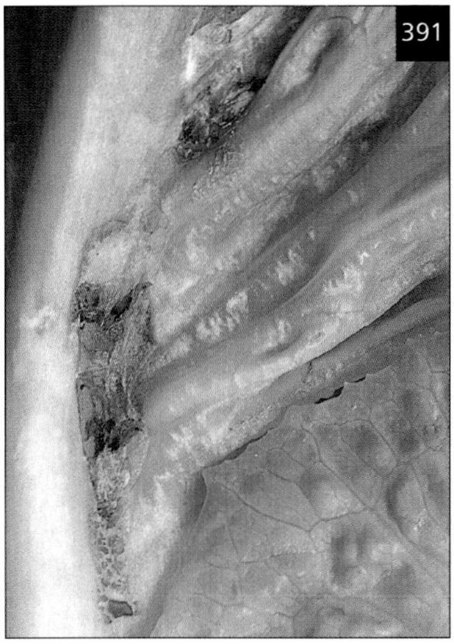

391 Aster yellows causes pink latex deposits to form in lettuce.

Causal agent

Aster yellows disease is caused by the aster yellows phytoplasma. Phytoplasmas, like typical bacteria, are prokaryotes but are placed in a distinct category called mollicutes. Mollicutes are single-celled organisms that lack a cell wall, appear in various shapes (called pleomorphism), and have very small genomes. Phytoplasmas inhabit the phloem tissue of their host plants. This pathogen affects a very wide host range of cultivated and wild plants.

Disease cycle

The aster yellows phytoplasma is vectored by adult leafhoppers, especially the aster leafhopper (*Macrosteles quadrilineatus*). The phytoplasma overwinters in perennial or biennial host plants and in the body of the leafhopper vector. The distribution of aster yellows follows the pattern of leafhopper migration from foothills, pastures, and weedy areas. The aster yellows phytoplasma has an extremely broad host range and can infect hundreds of different plant species. Phytoplasmas have not yet been cultured. Nucleic acid preparations or serological methods are necessary to confirm the presence of this pathogen. Sieve elements from plant vascular tissue may contain particulate material generated by the pathogen that can be stained with Azure A and viewed with a light microscope.

Control

Do not plant lettuce or other sensitive crops in fields or areas having a history of the disease. Such locations are frequented by the vector or have in the vicinity a natural reservoir of the pathogen; therefore, lettuce planted here will be exposed to the problem.

References

Deely, J., Stevens, W. A., and Fox, R. T. V. 1979. Use of Dienes' stain to detect plant diseases induced by mycoplasmalike organisms. *Phytopathology* 69:1169–1171.

Deng, S. and Hiruki, C. 1991. Genetic relatedness between two nonculturable mycoplasmalike organisms revealed by nucleic acid hybridization and polymerase chain reaction. *Phytopathology* 81:1475–1479.

Lee, I.-M., Davis. R. E., Chen, T.-A., Chiykowski, L. N., Fletcher, J., Hiruki, C., and Schaff, D. A. 1992. A genotype-based system for identification and classification of mycoplasmalike organisms (MLOs) in the aster yellows MLO strain cluster. *Phytopathology* 82:977–986.

Lee, I.-M., Davis. R. E., and Hsu, H.-T. 1993. Differentiation of strains in the aster yellows mycoplasma-like organism strain cluster by serological assay with monoclonal antibodies. *Plant Disease* 77:815–817.

Severin, H. H. P. and Frazier, N. W. 1945. California aster yellows on vegetable and seed crops. *Hilgardia* 16:573–596.

Zhang, J., Hogenhout, S. A., Nault, L. R., Hoy, C. W., and Miller, S. A. 2004. Molecular and symptom analyses of phytoplasma strains from lettuce reveal a diverse population. *Phytopathology* 94:842–849.

Zhou, X., Hoy, C. W., Miller, S. A., and Nault, L. R. 2002. Spacially explicit simulation of aster yellows epidemics and control on lettuce. *Ecological Modeling* 151:293–307.

Pseudomonas cichorii

VARNISH SPOT

Introduction and significance

Varnish spot is a disease that is associated with contaminated water sources, and therefore tends to be limited to certain production areas. The disease is of periodic importance in California and also occurs in most lettuce producing areas elsewhere in the world.

Symptoms and diagnostic features

The disease affects only the inner leaves of lettuce varieties that form an enclosed head; hence this disease will not usually be found on romaine or leaf lettuce varieties that have an open architecture. To see the symptoms, remove the outer wrapper leaves of the headed plant. Varnish spot appears as dark brown, shiny, firm, necrotic lesions on these inner leaves (**392**, **393**). The disease initially results in small lesions (1–3 mm diameter) that can later expand and coalesce into extensive necrotic sections encompassing entire leaves. Lesion borders are usually not delimited by veins. A notable feature is that lesions caused by this pathogen remain intact; the lesions are not soft, mushy, or broken down. This firmness is in contrast to secondary bacterial soft rots and slime of lettuce in which tissues are disintegrated (**394**). The disease is particularly challenging because the symptoms cannot be observed without cutting into the lettuce heads. Therefore, even if the disease is detected in a few field samples, the likely incidence of the disease remains unknown. Growers and harvesters must speculate on the advisability of harvesting such fields.

Causal agent

Varnish spot is caused by the bacterium *Pseudomonas cichorii*. *Pseudomonas cichorii* is a fluorescent, Gram-negative rod in the Group I pseudomonad group. The pathogen can be isolated on standard microbiological media. Colonies are cream to light yellow in color, smooth, and produce a fluorescent pigment on King's medium B. While *P. cichorii* is known to infect other vegetable crops, little information is available on whether lettuce strains can infect other plants or vice versa. On lettuce, *P. cichorii* has also been reported to cause leaf spot and stem rot diseases, which are distinct from varnish spot.

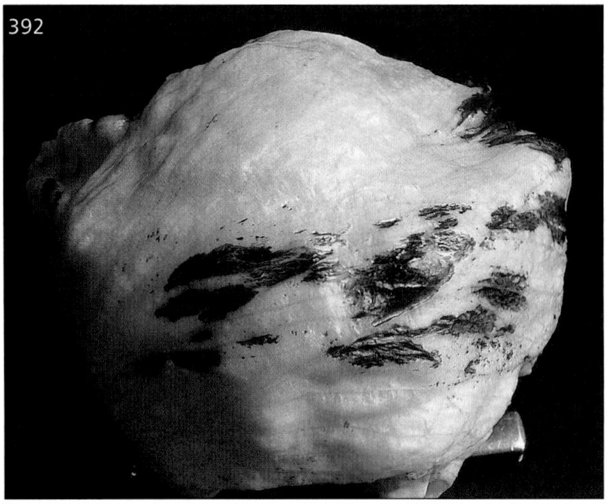

392 Lesions of varnish spot on lettuce.

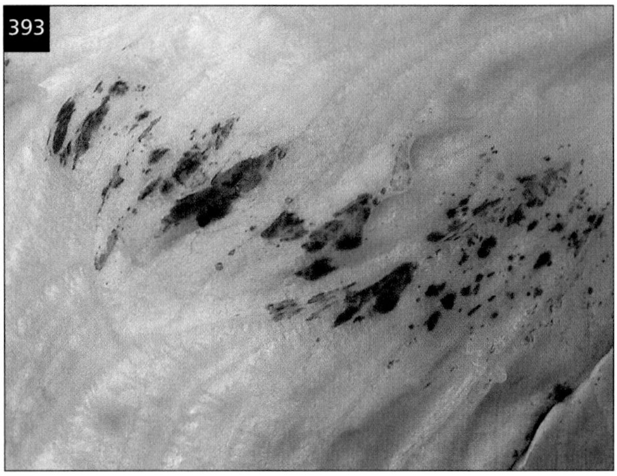

393 Close-up of lesions of varnish spot on lettuce.

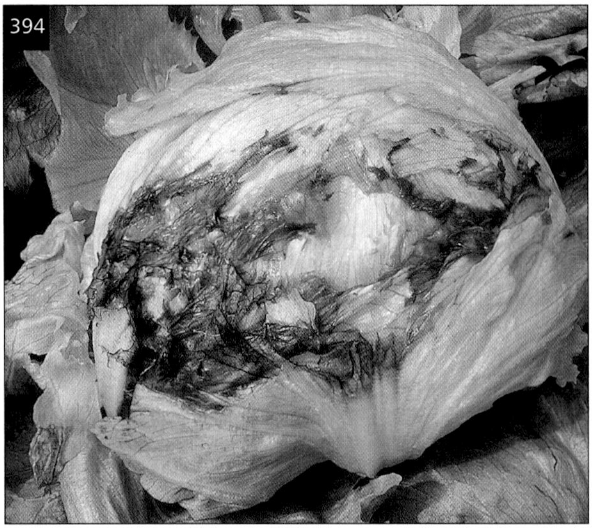

394 In contrast to varnish spot, bacterial slime infections are soft and watery.

Disease cycle

Varnish spot occurs when water contaminated with *P. cichorii* is used to sprinkle irrigate head-forming lettuce crops that are at the rosette stage of development. There is some evidence that *P. cichorii* also survives for brief periods in the soil and could be splashed up onto plants via sprinkler irrigation or rain. As infested lettuce heads begin to grow and become enclosed, the bacteria are trapped within the head, infect the leaves, and initiate disease.

Control

Alter the irrigation system so that overhead sprinklers are not used. If sprinklers must be used, do not plant cultivars that form a totally enclosed head. Do not plant consecutive lettuce crops because infested crop residues also harbor the pathogen. Rotate with non-host crops so that soilborne inoculum levels are reduced. Treat the irrigation water with chemicals such as copper compounds, which perhaps provide some control.

References

Dhanvantari, B. N. 1990. Occurrence of bacterial stem rot caused by *Pseudomonas cichorii* in greenhouse-grown lettuce in Ontario. *Plant Disease* 74:394.

Grogan, R. G., Misaghi, I. J., Kimble, K. A., Greathead, A. S., Ririe, D., and Bardin, R. 1977. Varnish spot, destructive disease of lettuce in California caused by *Pseudomonas cichorii. Phytopathology* 67:957–960.

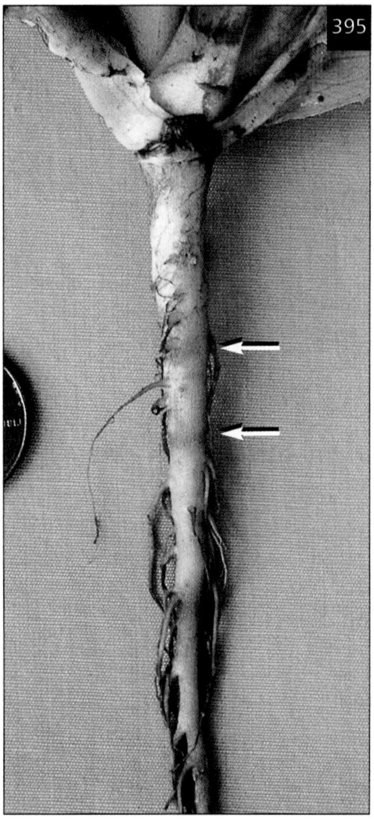

395 Initial root lesions of corky root of lettuce.

Rhizomonas suberifaciens

CORKY ROOT

Introduction and significance

In some lettuce producing areas, corky root is a limiting factor in lettuce production. In parts of California the disease is widespread and affects large areas of the crop.

Symptoms and diagnostic features

Early symptoms consist of yellow bands and patches on taproots of lettuce seedlings (**395**). These yellow areas gradually expand, taking on a green-brown color and developing cracks and rough areas on the root surface. As disease severity increases, the entire taproot may become brown, severely cracked, brittle, and nonfunctional; the feeder root system can also be reduced and damaged (**396**). As the disease continues to progress, the taproot may be reduced to a short (3–6 cm long)

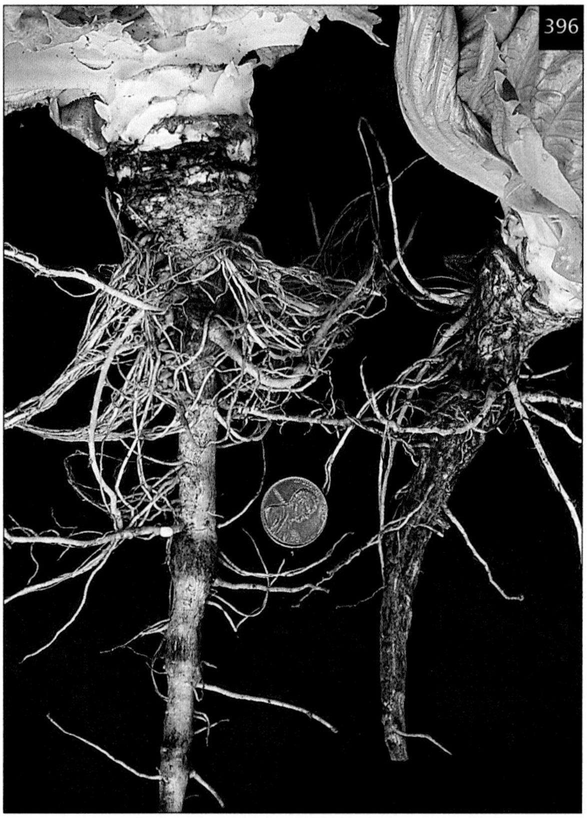

396 Severe root lesions of corky root of lettuce.

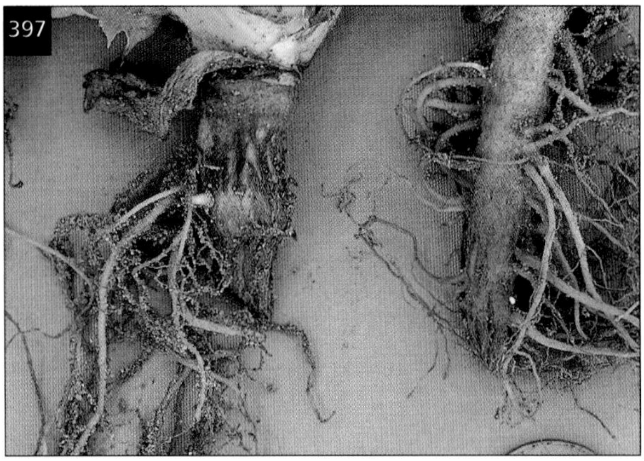

397 Stubbed-off taproot of lettuce affected by corky root.

stump and the plant forced to survive via adventitious roots that develop from this stump (**397**). The disease may cause some internal discoloration and hollowing of the root pith, though root symptoms are mostly on the exterior root surfaces. When the root is severely diseased, above-ground symptoms consist of wilting during warm weather, stunting of plants, and general poor, uneven, and delayed growth.

Causal agent

Corky root disease is caused by the bacterium *Rhizomonas suberifaciens*. *Rhizomonas suberifaciens* is a soilborne, Gram-negative bacterium. The bacterium is an oligotrophic organism that grows only on media with low carbon content and has a very slow growth rate. For these reasons, *R. suberifaciens* is difficult to isolate and grow in culture. Attempt isolations by using the semi-selective S-medium. Colonies are cream to off-white in color, wrinkled, and after one week have attained a colony diameter of only 1 mm (incubated at 28° C). *Rhizomonas suberifaciens* causes disease on lettuce, endive, and sowthistle (*Sonchus oleraceus*) and prickly lettuce (*Lactuca serriola*) weeds. The bacterium can also live as a rhizosphere organism, and it has been recovered from the roots of barley and other non-hosts.

Disease cycle

As a soil inhabitant, the pathogen survives in the soil for a long period of time. Corky root disease is typically more of a problem when soil temperatures are relatively warmer, with severity steadily increasing as temperatures rise from 10 to 31° C. High rates of nitrogen fertilizers, especially nitrates, also enhance disease severity.

Control

Plant corky root resistant cultivars, though some strains have been isolated that can overcome such resistance. Avoid planting consecutive lettuce crops and rotate with non-host plants. Reduce the amount of nitrogen fertilizers used on the lettuce crop; slow-release fertilizers may be useful in this regard. Use drip irrigation as this can lessen corky root severity. For fields with heavy disease pressure, use lettuce transplants instead of direct seeding. Corky root is much less severe on the fibrous root systems of transplants compared with the central taproot system of direct seeded plants.

References

Koike, S. T. and Schulbach, K. F. 1994. Evaluation of lettuce cultivars for resistance to corky root disease. *Biological and Cultural Tests* 9:29.

Mou, B. and Bull, C. 2004. Screening lettuce germplasm for new sources of resistance to corky root. *Journal of the American Society for Horticultural Science* 129:712–716.

O'Brien, R. D. and van Bruggen, A. H. C. 1991. Populations of *Rhizomonas suberifaciens* on roots of host and nonhost plants. *Phytopathology* 81:1034–1038.

van Bruggen, A. H. C., Brown, P. R., and Jochimsen, K. N. 1990. Host range of *Rhizomonas suberifaciens*, the causal agent of corky root of lettuce. *Plant Disease* 74:581–584.

van Bruggen, A. H. C., Brown, P. R., Shennan, C., and Greathead, A. S. 1990. The effect of cover crops and fertilization with ammonium nitrate on corky root of lettuce. *Plant Disease* 74:584–589.

van Bruggen, A. H. C., Grogan, R. G., Bogdanoff, C. P., and Waters, C. M. 1988. Corky root of lettuce in California caused by a Gram negative bacterium. *Phytopathology* 78:1139–1145.

van Bruggen, A. H. C., Jochimsen, K. N., and Brown, P. R. 1990. *Rhizomonas suberifaciens* gen. nov., sp. nov., the causal agent of corky root of lettuce. *International Journal of Systematic Bacteriology* 40:175–188.

Xanthomonas campestris pv. *vitians*

BACTERIAL LEAF SPOT

Introduction and significance

Bacterial leaf spot was previously considered a minor disease of lettuce. However, in recent years the disease has become more important and can at times cause significant economic damage. The disease is prevalent in both the USA and in Europe.

Symptoms and diagnostic features

Early symptoms of bacterial leaf spot are small (2–5 mm), water-soaked leaf spots on the older leaves of the plant. These lesions are typically bordered by leaf veins and are angular in shape. Lesions quickly turn black – this is a diagnostic character of the disease (**398**). If disease is severe, numerous lesions may coalesce, resulting in the collapse of the leaf. Older lesions dry up and become papery in texture, but retain the black color. Lesions rarely occur on newly developing leaves. If diseased heads are packed in cartons, secondary decay organisms can colonize the lesions and result in postharvest problems. Bacterial leaf spot occurs on the leaves of both leaf and head lettuce varieties as well as on flower bracts of lettuce seed crops (**399**).

Causal agent

Bacterial leaf spot is caused by *Xanthomonas campestris* pv. *vitians*. The pathogen can be isolated on standard microbiological media and produces yellow, mucoid, slow-growing colonies typical of most xanthomonads. However, *X. campestris* pv. *vitians* weakly hydrolyzes starch, so starch-based semi-selective media such as SX and MXP media are not diagnostic for isolating and identifying this pathogen. Tween medium is useful because this bacterium forms characteristic white calcium salt crystals when growing on it. Early research indicated that *X. campestris* pv. *vitians* was a pathogen of both lettuce and ornamental aroid plants. Based on fatty acid analyses and reactions to monoclonal antibodies, it is now known that this lettuce pathogen is distinct from *X. campestris* strains isolated from aroids. *Xanthomonas campestris* pv. *vitians* is mostly limited to lettuce hosts. Recent research indicates that *X. campestris* pv. *vitians* is composed of homogeneous strains, and that genetically distinct strains are not yet found.

Disease cycle

The pathogen is highly dependent on wet, cool conditions for infection and disease development. Splashing water from overhead irrigation and rain disperses the pathogen in the field. The pathogen can be seedborne. If infested seed is used to grow lettuce transplants in a greenhouse, the pathogen may become established on these plants because of favorable conditions in the greenhouse. The bacterium can survive for several months in the soil and be splashed onto subsequent lettuce crops. The bacterium has also been found growing epiphytically on weed plants, but the significance of this factor in disease development is not

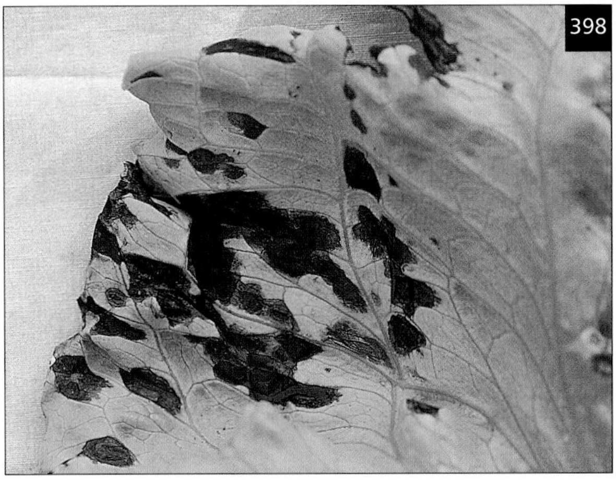

398 Black spots of bacterial leaf spot on lettuce.

399 Lettuce flower bracts infected with bacterial leaf spot.

known. Research indicates that *X. campestris* pv. *vitians* can move systemically in lettuce vascular tissue.

Control

Research in California indicates that commercial lettuce seed is infrequently infested with this pathogen. In addition, infestation thresholds have not been established for seedborne inoculum of this disease. However, if seedborne inoculum becomes important, use seed that does not have significant levels of the pathogen. Examine lettuce transplants and remove symptomatic plants. Reduce or eliminate overhead sprinkler irrigation in the production field. Some resistant cultivars are available. Do not plant consecutive lettuce crops, and use crop rotations having non-host plants so that soilborne inoculum levels can decline. Copper based fungicides appear to provide only marginal control.

References

Barak, J. D. and Gilbertson, R. L. 2003. Genetic diversity of *Xanthomonas campestris* pv. *vitians*, the causal agent of bacterial leafspot of lettuce. *Phytopathology* 93:596–603.

Barak, J. D., Koike, S. T., and Gilbertson, R. L. 2001. Role of crop debris and weeds in the epidemiology of bacterial leaf spot of lettuce in California. *Plant Disease* 85:169–178.

Barak, J. D., Koike, S. T., and Gilbertson, R. L. 2002. Movement of *Xanthomonas campestris* pv. *vitians* in the stems of lettuce and seed contamination. *Plant Pathology* 51:506–512.

Carisse, O., Ouimet, A., Toussaint, V., and Philion, V. 2000. Evaluation of the effect of seed treatments, bactericides, and cultivars on bacterial leaf spot of lettuce caused by *Xanthomonas campestris* pv. *vitians*. *Plant Disease* 84:295–299.

McGuire, R. G., Jones, J. B., and Sasser, M. 1986. Tween media for semiselective isolation of *Xanthomonas campestris* pv. *vesicatoria* from soil and plant material. *Plant Disease* 70:887–891.

Pernezny, K., Raid, R. N., Stall, R. E., Hodge, N. C., and Collins, J. 1995. An outbreak of bacterial spot of lettuce in Florida caused by *Xanthomonas campestris* pv. *vitians*. *Plant Disease* 79:359–360.

Sahin, F. and Miller, S. A. 1997. Identification of the bacterial leaf spot pathogen of lettuce, *Xanthomonas campestris* pv. *vitians*, in Ohio, and assessment of cultivar resistance and seed treatment. *Plant Disease* 81:1443–1446.

Bremia lactucae

DOWNY MILDEW

Introduction and significance

Downy mildew is probably the most important foliar disease of lettuce worldwide and can attack all lettuce types. Losses are experienced when severe downy mildew causes lowered yield and quality and when diseased leaves need to be trimmed from the harvested produce.

Symptoms and diagnostic features

Downy mildew results in light green to yellow angular spots on the upper surfaces of leaves (**400**). White fluffy growth of the pathogen develops primarily on the undersides of these spots (**401**). With time, the lesions turn brown and dry up. Older leaves are usually attacked first; such leaves that have lesions and are in contact with the soil can become soft and rotted due to secondary decay organisms, such as *Botrytis cinerea*. On occasion the pathogen can cause systemic infections that result in dark discoloration and streaking of internal vascular and pith tissues. If downy mildew infects the cotyledons of young seedlings, the plants can die. Greenhouse-grown lettuce transplants can also be infected. Cultivated lettuce is the main host, though *B. lactucae* infects wild *Lactuca* species, artichoke, and other plants in the Asteraceae.

Causal agent

Downy mildew is caused by the obligate parasite *Bremia lactucae*, which belongs in the oomycete group of organisms. *Bremia lactucae* conidiophores emerge from leaf stomates, branch dichotomously, and have distinctively flared tips on which sterigmata are formed. Conidia are hyaline, ovoid, measure 12–30 x 11–28 μm, and have apical papilla. *Bremia lactucae* is heterothallic and requires B1 and B2 mating types to be present for the production of the sexual oospore. Oospores are spherical, thick walled, and measure 20–31 μm. However, in some lettuce production areas oospores are not observed even if both mating types have been found.

Bremia lactucae is a complex pathogen and consists of multiple races (pathotypes). Pathotypes are identified by testing their virulence on differential sets of lettuce cultivars having various resistance genes. The worldwide nomenclature of these pathotypes is confusing because different countries use their own numbering system (e.g. NL1, NL2, etc. in the Netherlands; IL1, IL2, etc. in Israel; CA V, CA VI, etc. in California). Standardization is improving and BL (for *B. lactucae*) numbers are now widely used in Europe.

Lettuce is subject to a second downy mildew disease caused by *Plasmopara lactucae-radicis*. This pathogen is apparently restricted to the roots of lettuce and causes tan to brown necrotic lesions. Profuse sporulation can be observed on infected root surfaces, and oospores develop in the root cortex. This downy mildew disease has only been detected on lettuce grown in hydroponic systems.

Disease cycle

Humid, cool conditions are required for *B. lactucae* to sporulate and to infect lettuce. Initial inoculum consists of conidia from surrounding plants (weeds, other lettuce fields) and, in some regions, oospores. The conidia are produced on lesions at night, and are then released into the air in the early morning if relative humidity is high. The spores are dispersed by winds but are short-lived. Infection takes place in only 3 or 4 hours if free moisture is on the leaves or near saturation conditions are present and temperatures are optimal (10–22° C).

Control

Plant resistant cultivars. However, because there are a number of different pathotypes and also diverse isolates of uncharacterized genetics, resistant cultivars are continually challenged and overcome by *B. lactucae* isolates. Apply fungicides prior to the development of the disease. Isolates from several countries have developed insensitivity to the phenylamide fungicide metalaxyl, and in California *B. lactucae* isolates were found to be resistant to fosetyl-Al. Resistance management strategies dictate that different fungicides be used in rotation to slow down the development of insensitivity. Research has been conducted on disease prediction models for *B. lactucae*. Such systems potentially can help farmers reduce the number of fungicide applications made to lettuce; however, a consistently reliable and commercially available model is not yet available. Culturally, use irrigation systems that reduce leaf wetness and humidity, such as drip irrigation. Greenhouse-grown transplants should especially be protected so that downy mildew is not brought to the field on these plants.

401 Sporulation of downy mildew on lettuce leaf underside.

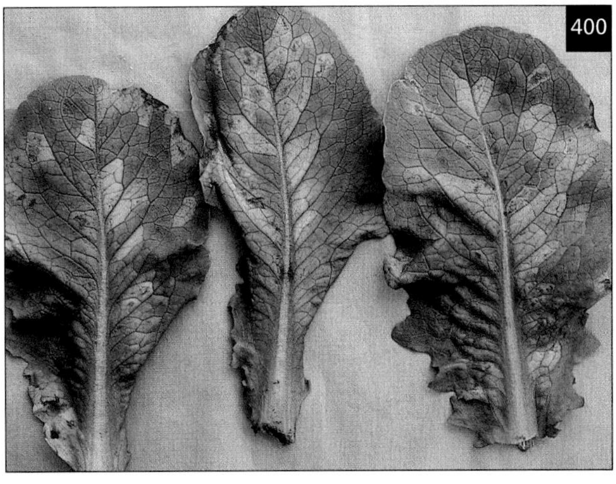

400 Chlorotic and angular lesions of downy mildew on lettuce.

References

Brown, S., Koike, S. T., Ochoa, O. E., Laemmlen, F., and Michelmore, R. W. 2003. Insensitivity to the fungicide fosetyl-Aluminium in California isolates of the lettuce downy mildew pathogen, *Bremia lactucae*. *Plant Disease* 87:502–508.

Cobelli, L., Collina, M., and Brunelli, A. 1998. Occurrence in Italy and characteristics of lettuce downy mildew (*Bremia lactucae*) resistant to phenylamide fungicides. *European Journal of Plant Pathology* 104:449–455.

Crute, I. R. 1987. Occurrence, characteristics, distribution, genetics, and control of a metalaxyl-resistant pathotype of *Bremia lactucae* in the United Kingdom. *Plant Disease* 71:763–767.

Davies, J. M.Ll. 1994. Integrated control of downy mildew in crisp lettuce. *Proceedings of the Brighton Crop Protection Conference - Pests and Diseases* 2: 817–822.

Garibaldi, A., Minuto, A., Gilardi, G., and Gullino, M. L. 2003. First report of *Bremia lactucae* causing downy mildew on *Helichrysum bracteatum* in Italy. *Plant Disease* 87:315.

Ilott, T. W., Hulbert, S. H., and Michelmore. R. W. 1989. Genetic analysis of the gene-for-gene interaction between lettuce (*Lactuca sativa*) and *Bremia lactucae*. *Phytopathology* 79:888–897.

Lebeda, A. and Reinink, K. 1991. Variation in the early development of *Bremia lactucae* on lettuce cultivars with different levels of field resistance. *Plant Pathology* 40:232–237.

Lebeda, A. and Schwinn, F. J. 1994. The downy mildews – an overview of recent research progress. *Journal of Plant Disease and Protection* 101:225–254.

Lebeda, A. and Zinkernagel, V. 2003. Evolution and distribution of virulence in the German population of *Bremia lactucae*. *Plant Pathology* 52:41–51.

Norwood, J. M. and Crute, I. R. 1985. Further characterization of field resistance in lettuce to *Bremia lactucae* (downy mildew). *Plant Pathology* 34:481–486.

O'Neill, T. M., Gladders, P., and Ann, D. M. 1997. Prospects for integrated control of lettuce diseases. *Proceedings of the BCPC/ANPP Conference, University of Canterbury, Kent, UK: Crop Protection and Food Quality: Meeting Customer Needs*, pp. 485–490.

Scherm, H., Koike, S. T., Laemmlen, F. F., and van Bruggen, A. H. C. 1995. Field evaluation of fungicide spray advisories against lettuce downy mildew (*Bremia lactucae*) based on measured or forecast morning leaf wetness. *Plant Disease* 79:511–516.

Scherm, H. and van Bruggen, A. H. C. 1994. Weather variables associated with infection of lettuce by downy mildew (*Bremia lactucae*) in coastal California. *Phytopathology* 84:860–865.

Schettini, T. M., Legg, E. J., and Michelmore, R. W. 1991. Insensitivity to metalaxyl in California populations of *Bremia lactucae* and resistance in California lettuce cultivars to downy mildew. *Phytopathology* 81:64–70.

Stanghellini, M. E., Adaskaveg, J. E., and Rasmussen, S. L. 1990. Pathogenesis of *Plasmopara lactucae-radicis*, a systemic root pathogen of cultivated lettuce. *Plant Disease* 74:173–178.

Stanghellini, M. E., and Gilbertson, R. L. 1988. *Plasmopara lactucae-radicis*, a new species on roots of hydroponically grown lettuce. *Mycotaxon* 31:395–400.

Su, H., van Bruggen, A. H. C., Subbarao, K. V., and Sherm, H. 2004. Sporulation of *Bremia lactucae* affected by temperature, relative humidity, and wind in controlled conditions. *Phytopathology* 94:396–401.

***Botrytis cinerea* (teleomorph = *Botryotinia fuckeliana*)**

GRAY MOLD

Introduction and significance

Gray mold is a common but usually minor lettuce disease. However, if environmental conditions are favorable for the pathogen, gray mold can cause significant crop loss in greenhouse-grown lettuce as well as in the field. Lettuce that is transplanted into the field can at times be severely affected.

Symptoms and diagnostic features

The initial symptom of gray mold is a water-soaked, brown-gray to brown-orange, mushy rot that occurs on the oldest leaves and at the base of damaged or senescent leaves and stems. Injured tissues that are wet or in contact with the soil are especially susceptible. Such leaves need to be trimmed off the plant at harvest. The pathogen uses this compromised tissue as a food base and later progresses into the healthy parts of the lettuce crown, causing a similar decay of the main stem and attached leaves (**402**). The characteristic gray fuzzy growth of the fungus can usually be readily seen on diseased areas. Black sclerotia may form on these diseased tissues, although some isolates produce few or no sclerotia. In advanced stages of the disease, a severely infected plant is girdled at the crown, wilts, and collapses, resulting in symptoms that closely resemble lettuce drop or Phoma basal rot. Young seedlings and transplants can also die from gray mold (**403**). In California, early spring plantings of transplanted romaine cultivars are particularly susceptible and gray mold can result in significant stand reduction.

Gray mold can also develop on upper lettuce leaves if such tissues are damaged. For example, if young, inner leaf tips of romaine cultivars have tipburn symptoms, *Botrytis cinerea* conidia can land on such tissue, colonize the necrotic tissue, and proceed to rot the rest of the leaf.

Causal agent

The causal agent of gray mold is *Botrytis cinerea*. The perfect stage, *Botryotinia fuckeliana*, apparently is not found on lettuce. Conidiophores of *B. cinerea* are long (1–2 mm), become gray-brown with maturity, and branch irregularly near the apex. Conidia are clustered at the branch tips and are single-celled, pale brown, ellipsoid to obovoid, and measure 6–18 x 4–11 µm.

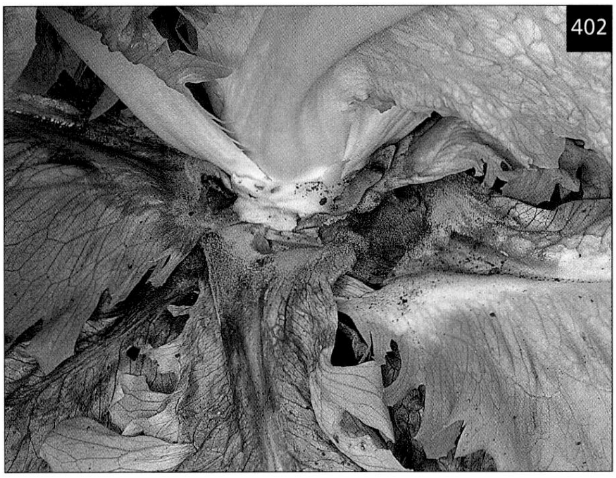

402 Crown decay of lettuce caused by gray mold.

403 Lettuce transplants infected with gray mold.

The pathogen can be isolated on standard microbiological media. Some isolates sporulate poorly in culture unless incubated under lights (12 h light/12 h dark). If formed, sclerotia are black, oblong or dome-shaped, and measure 4–10 mm. The fungus grows best at 18–23° C but is inhibited at warm temperatures above 32° C. On host tissue the fungus produces characteristically profuse sporulation that is dense, velvety, and grayish brown in color.

Disease cycle

Botrytis cinerea survives in and around fields as a saprophyte on crop debris, as a pathogen on numerous crops and weed plants, and as sclerotia in the soil. Conidia develop from these sources and become windborne. When conidia land on senescent or damaged lettuce tissue, they will germinate if free moisture is available and rapidly colonize this food base. Once established, the pathogen will grow into adjacent healthy stems and leaves, resulting in disease symptoms and the production of additional conidia. Cool temperatures, free moisture, and high humidity favor the development of the disease. Lettuce tissues are predisposed to infection by frost or heat damage, physiological problems such as tipburn, or the activity of other pathogens such as *Bremia lactucae, Phoma exigua, Rhizoctonia solani,* and *Sclerotinia* species.

Control

Because *B. cinerea* initiates infection on damaged tissues, minimize damage to lettuce that is caused by cultural practices, environmental extremes, or other pathogens and pests. Reduce leaf wetness by avoiding or reducing sprinkler irrigation. Schedule crop residue incorporation and soil preparation so that excessive plant residues at planting are minimized. Plant transplants in a timely manner so that the plants are not too large and overly mature; old transplants are subject to additional leaf breakage and damage during planting, and hence are more susceptible to gray mold infection. Because romaine transplants are especially prone to gray mold disease, plant this lettuce type by using direct seeding when possible. To reduce overall humidity, adequately ventilate or heat greenhouses. Apply fungicides to protect plants from gray mold. However, *B. cinerea* strains resistant to dicarboximide fungicides are already widespread in some geographic areas. Use diverse fungicide products with different modes of action to reduce the risk of pathogen insensitivity.

References

Delon, R., Kiffer, E., and Mangenot, F. 1977. Ultrastructural study of host-parasite interactions: II. Decay of lettuce caused by *Botrytis cinerea* and phyllosphere bacteria. *Canadian Journal of Botany* 55:2463–2470.

O'Neill, T. M., Gladders, P., and Ann, D. M. 1997. Prospects for integrated control of lettuce diseases. *Proceedings of the BCPC/ANPP Conference, University of Canterbury, Kent, UK: Crop Protection and Food Quality: Meeting Customer Needs*, pp. 485–490.

Wang, Z.-N., Coley-Smith, J. R., and Wareing, P. W. 1986. Dicarboximide resistance in *Botrytis cinerea* in protected lettuce. *Plant Pathology* 35:427–433.

Wareing, P. W., Wang, Z.-N., Coley-Smith, J. R., and Jeves, T. M. 1986. Fungal pathogens in rotted basal leaves of lettuce in Humberside and Lancashire with particular reference to *Rhizoctonia solani*. *Plant Pathology* 35:390–395.

Fusarium oxysporum f. sp. *lactucae*

FUSARIUM WILT

Introduction and significance

The first report indicating that lettuce was susceptible to Fusarium wilt was made in 1960 in Asia (Japan). The disease was later found in the USA in the early 1990s (California) and late 1990s (Arizona), and most recently was reported in Europe (Italy) in 2002. This disease is becoming a serious concern in certain areas in the southwest USA.

Symptoms and diagnostic features

Fusarium wilt causes infected seedlings to wilt and possibly die. Vascular stem and taproot tissues of affected seedlings are red or brown. In older plants, leaves turn yellow, wilt, and become necrotic. Internally, the stem vascular system is red-brown to dark brown and a red-brown discoloration develops in the cortex of the crown and taproot (**404**). The taproot may develop a hollow cavity. Plants are usually stunted and may fail to form heads (**405**). Fusarium wilt symptoms resemble those caused by ammonium toxicity and Verticillium wilt.

404 Vascular discoloration of Fusarium wilt of lettuce.

405 Stunted, declining lettuce plants with Fusarium wilt.

Causal agent

The cause of Fusarium wilt of lettuce is *Fusarium oxysporum* f. sp. *lactucae*. The pathogen morphology and colony characteristics are similar to other *F. oxysporum* fungi. The fungus forms one- or two-celled, oval to kidney shaped microconidia on monophialides, and three- to five-celled, fusiform, curved macroconidia. Macroconidia are usually produced in cushion-shaped structures called sporodochia. Chlamydospores are also formed. The pathogen is usually readily isolated from symptomatic vascular tissue. Semi-selective media like Komada's medium can help isolate the pathogen if secondary rot organisms are present in the sample. Inoculation experiments indicate that the fungus appears to be host specific to lettuce, and that lettuce is not susceptible to other forma speciales of the *F. oxysporum* group. Currently, 3 races (1, 2, and 3) of this pathogen are known to exist.

Disease cycle

Like other Fusarium wilt pathogens, *F. oxysporum* f. sp. *lactucae* is a soil inhabitant that can survive in the soil for indefinite periods of time due to the production of resilient chlamydospores. The disease tends to be more severe on lettuce planted in the warmer months of the season. Seedborne inoculum of *F. oxysporum* f. sp. *lactucae* has been hypothesized, but has not yet been documented.

Control

Avoid planting lettuce in fields known to be infested, or plant lettuce in these locations only in the spring or early summer when symptoms are likely to be less severe. Plant tolerant or less susceptible cultivars if available. Romaine lettuce appears to be more tolerant than other types. Minimize the movement of contaminated, infested soil to clean fields.

References

Fujinaga, M., Ogiso, H., Tsuchiya, N., and Saito, H. 2001. Physiological specialization of *Fusarium oxysporum* f. sp. *lactucae*, a causal organism of Fusarium root rot of crisp head lettuce in Japan. *Journal of General Plant Pathology* 67:205–206.

Fujinaga, M., Ogiso, H., Tsuchiya, N., Saito, H., Yamanaka, S., Nozue, M., and Kojima, M. 2003. Race 3, a new race of *Fusarium oxysporum* f. sp. *lactucae* determined by a differential system with commercial cultivars. *Journal of General Plant Pathology* 69:23–28.

Garibaldi, G., Gilardi, G., and Gullino, M. 2002. First report of *Fusarium oxysporum* on lettuce in Europe. *Plant Disease* 86:1052.

Hubbard, J. C. and Gerik, J. S. 1993. A new wilt disease of lettuce incited by *Fusarium oxysporum* f. sp. *lactucum forma specialis nov*. *Plant Disease* 77:750–754.

Komada, H. 1975. Development of a selective medium for quantitative isolation of *Fusarium oxysporum* from natural soil. *Review of Plant Protection Research*. 8:114–124.

Matheron, M. E. and Koike, S. T. 2003. First report of Fusarium wilt of lettuce caused by *Fusarium oxysporum* f. sp. *lactucae* in Arizona. *Plant Disease* 87:1265.

Matuo, T. and Motohashi, S. 1967. On *Fusarium oxysporum* f. sp. *lactucae n. f.* causing root rot of lettuce. *Transactions of the Mycological Society of Japan* 8:13–15.

Pasquali, M., Dematheis, F., Gilardi, G., Gullino, M. L., and Garibaldi, A. 2005. Vegetative compatibility groups of *Fusarium oxysporum* f. sp. *lactucae* from lettuce. *Plant Disease* 89:237–240.

Yamauchi, N., Horiuchi, S., and Satou, M. 2001. Pathogenicity groups in *Fusarium oxysporum* f. sp. *lactucae* on horticultural types of lettuce cultivars. *Journal of General Plant Pathology* 67:288–290.

Golovinomyces cichoracearum
(= Erysiphe cichoracearum)

POWDERY MILDEW

Introduction and significance

Powdery mildew is generally considered a minor disease of lettuce and is primarily reported in North America and Europe. However, in some regions and under certain conditions, powdery mildew can cause significant reductions in crop quality.

Symptoms and diagnostic features

This powdery mildew fungus grows ectophytically and appears as a white, powdery growth on both upper and lower sides of lettuce leaves (**406**). Older leaves are always infected first and most severely. Such infections may cause chlorosis and deformity and buckling of the leaves. In advanced stages, leaves may begin to dry out and turn brown. On occasion, the sexual phase will occur and small, brown cleistothecia may be seen on the leaves. All lettuce types are susceptible.

Causal agent

The causal agent, *Golovinomyces cichoracearum* (previously named *Erysiphe cichoracearum*), produces epiphytic mycelium that grows superficially on host surfaces. Conidiophores are borne on this surface mycelium. Conidia are produced in long chains, are hyaline, ellipsoid to barrel-shaped, and measure 25–45 x 14–25 µm. Globose cleistothecia have numerous hypha-like appendages and contain up to 10–25 asci. Asci contain two ascospores that are ovoid to ellipsoid.

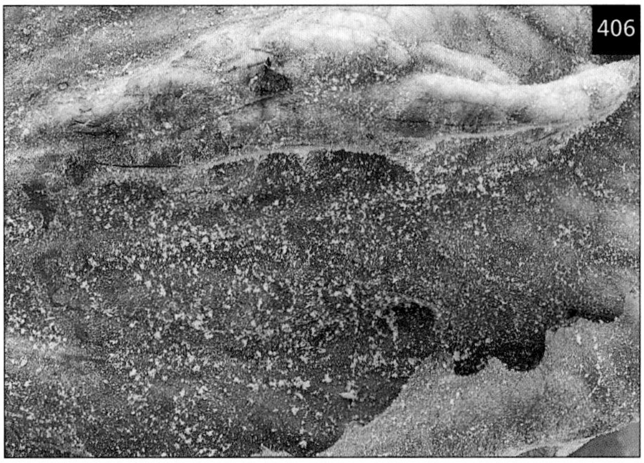

406 Powdery mildew sporulation on lettuce leaf.

Golovinomyces cichoracearum is listed in the literature as infecting over 150 plant species. However, there is evidence that distinct strains, or physiological races, exist within *G. cichoracearum* populations. Under experimental conditions, powdery mildew from lettuce could infect the following plants: vegetables – artichoke, cantaloupe, chicory, cucumber, endive; ornamentals – calendula, dahlia, delphinium, florist's cineraria, sunflower, and zinnia; weeds – prickly or wild lettuce (*Lactuca serriola*) and milk thistle (*Silybum marianum*). *G. cichoracearum* isolates from calendula, zinnia, and wild lettuce, however, failed to infect lettuce.

Disease cycle

Initial inoculum can be either airborne conidia from diseased host plants or ascospores emerging from cleistothecia. Optimum temperatures for fungal growth and development are 18–20° C. Maximum spore germination requires high humidity, but free moisture may actually inhibit germination.

Control

Plant resistant cultivars or apply fungicides such as sulfur compounds.

References

Crute, I. R. and Burns, I. G. 1983. New or unusual records: Powdery mildew of lettuce (*Lactuca sativa*). *Plant Pathology* 32:455–457.

Lebada, A., Mieslerova, B., Dolezalova, I., and Kristkova, E. 2002. Occurrence of powdery mildew on *Lactuca viminea* subsp. *chondrilliflora* in south France. *Mycotaxon* 84:83–87.

Schnathorst, W. C. 1959. Spread and life cycle of the lettuce powdery mildew fungus. *Phytopathology* 49:464–468.

Schnathorst, W. C. 1960. Effects of temperature and moisture stress on the lettuce powdery mildew fungus. *Phytopathology* 50:304–308.

Schnathorst, W. C., Grogan, R. G., and Bardin, R. 1958. Distribution, host range, and origin of lettuce powdery mildew. *Phytopathology* 48:538–543.

Turini, T. A. and Koike, S. T. 2002. Comparison of fungicides for control of powdery mildew on iceberg lettuce, 2001. *Fungicide and Nematicide Reports*. Vol. 57:V042.

Microdochium panattonianum

ANTHRACNOSE, RING SPOT

Introduction and significance

Anthracnose disease is also called shot hole, ring spot, or rust and is found in most lettuce producing areas of the world. If weather conditions favor disease development, this problem can cause widespread crop loss.

Symptoms and diagnostic features

Initial symptoms are small (2–3 mm), water-soaked spots occurring on outer leaves. Spots enlarge, turn yellow, and are usually angular in shape. Under cool, moist conditions, white to pink spore masses of the fungus will be visible in the centers of the tan-colored lesions (**407**). If disease is severe, the lesions will coalesce and cause significant dieback of the leaf, and in some cases will result in significant stunting of the plant (**408**). As spots age, the affected tissue will dry up and become papery in texture. Eventually the centers of these spots can fall out, resulting in a shot hole appearance and hence the use of that name. Anthracnose lesions are often clustered along the midribs of lower leaves.

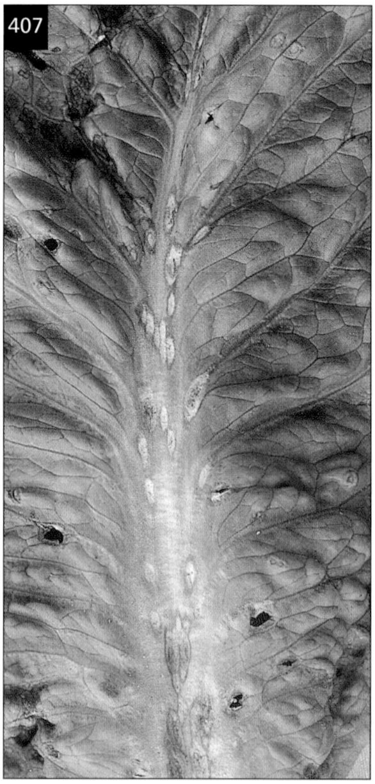

407 Leaf spots of anthracnose on lettuce.

Causal agent

The pathogen is *Microdochium panattonianum*. Older literature lists the pathogen as *Marssonina panattoniana*. The pathogen can be isolated on standard microbiological media, though confirmation of anthracnose might be better accomplished by microscopic examination of leaf spot tissue that has been collected from the field. The fungus is very slow growing and on agar media will form small, raised, slimy appearing colonies of white to pink mycelium after several days (10 mm colony diameter, 7 days, 24° C). Though sometimes difficult to observe, conidiophores are solitary or grouped in sporodochia. Conidia are hyaline, smooth, fusiform, slightly curved, with one septum. The lower cell is slightly smaller and tapers towards the base. Dimensions of the conidia are 3–4 x 13–16 µm. Microsclerotia consist of groups of multicellular, thick-walled cells that measure 60–100 µm and may be observed in infected tissue. *M. panattonianum* infects lettuce and other *Lactuca* species and has also been reported on chicory and endive. At least five different races of the pathogen have been identified.

Disease cycle

The fungus can survive for up to 4 years as microsclerotia in soil. The anthracnose pathogen requires cool, wet conditions for infection and symptom development and hence is associated with rainy weather. Splashing water moves microsclerotia and conidia from soil onto leaves, resulting in infection. Optimum temperatures for disease development are approximately 18–20° C and symptoms can appear 4 to 8 days after infection. While all lettuce types are susceptible, romaine cultivars are particularly sensitive.

Control

Avoid planting lettuce in fields having a history of the disease. Rotate with non-host plants to help reduce soil inoculum levels, though such rotations will not eliminate the pathogen unless hosts are not planted for over 4 years. Use irrigation systems (furrow or drip irrigation) that reduce or eliminate leaf wetting. Plant resistant cultivars if they are available. Apply protectant fungicides, such as strobilurins, which are effective for controlling this disease.

408 Severe stunting of lettuce affected with anthracnose.

References

Couch, H. B. and Grogan, R. G. 1955. Etiology of lettuce anthracnose and host range of the pathogen. *Phytopathology* 45:375–380.

Galea, V. J., Price, T. V., and Sutton, B. C. 1986. Taxonomy and biology of lettuce anthracnose fungus. *Transactions of the British Mycological Society* 86:619–628.

Moline, H. E. and Pollack, F. G. 1976. Conidiogenesis of *Marssonina panattoniana* and its potential as a serious postharvest pathogen of lettuce. *Phytopathology* 66:669–674.

Ochoa, O., Delp, B., and Michelmore, R. W. 1987. Resistance in *Lactuca* spp. to *Microdochium panattonianum* (lettuce anthracnose). *Euphytica* 36:609–614.

Parman, T. and Price, T. V. 1991. Production of microsclerotia by *Microdochium panattonianum*. *Australasian Plant Pathology* 20:41–46.

Patterson, C. L. and Grogan, R. G. 1985. Source and survival of primary inoculum of *Marssonina panattoniana*, causal agent of lettuce anthracnose. *Phytopathology* 75:13–19.

Patterson, C. L. and Grogan, R. G. 1991. Role of microsclerotia as primary inoculum of *Microdochium panattonianum*, incitant of lettuce anthracnose. *Plant Disease* 75:134–138.

Wicks, T. J., Hall, B., and Pezzaniti, P. 1994. Fungicidal control of anthracnose (*Microdochium panattonianum*) on lettuce. *Australian Journal of Experimental Agriculture* 34: 277–283.

Phoma exigua

PHOMA BASAL ROT

Introduction and significance

Phoma basal rot was reported as a problem in greenhouse lettuce production in the United Kingdom in the early 1990s. More recently, this disease has caused significant losses on romaine grown in the USA (California).

Symptoms and diagnostic features

Above ground symptoms are similar to those caused by lettuce drop and gray mold diseases and consist of wilting and yellowing of lower leaves, one-sided growth of plants, overall stunting, and eventual plant collapse (**409**). However, in contrast to these other crown rots, there is no visible fungal growth at the crowns of infected plants. Rather, Phoma basal rot causes distinct, dark brown to black, sunken cavities to develop at the crown and upper taproot tissues (**410, 411, 412**). Lesions are notably dry and firm. Lesions can extend deep into the crowns and roots, resulting in extensive weakening of the plant structure; such plants can easily be broken off at the ground level. Secondary fungi, especially *Botrytis cinerea*, can colonize these stem lesions. In greenhouse grown lettuce in Europe this disease also results in circular, dark gray to black leaf spots that can expand up to 3 cm in diameter. With time the inner leaf spot tissue dries out, cracks, and falls out, resulting in a shot hole effect. The fruiting bodies of the pathogen can be observed in leaf spot tissues.

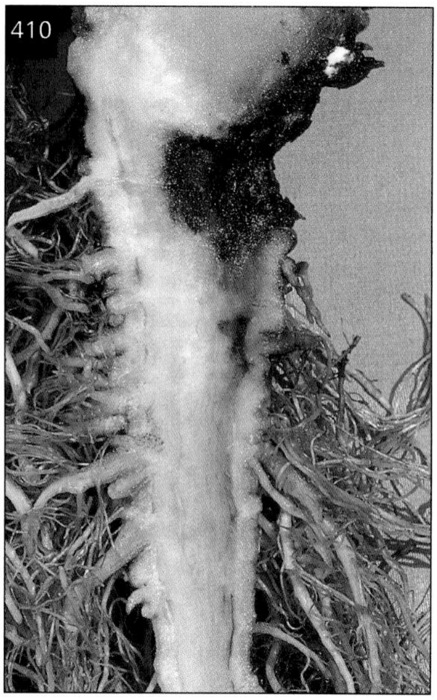

410 Transverse section of lettuce showing sunken crown cavity of Phoma basal rot.

411 Black crown cavity of Phoma basal rot of lettuce.

409 Collapsed lettuce plants infected with Phoma basal rot.

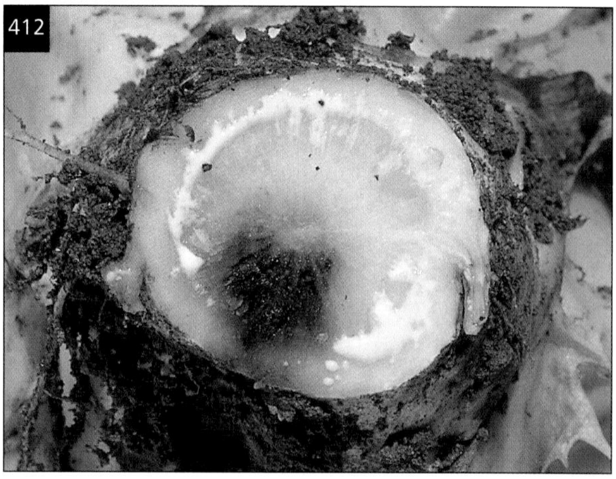

412 Cross section showing lettuce crown infection from Phoma basal rot.

Causal agent

The causal agent of Phoma basal rot is *Phoma exigua*, a soilborne fungus. The fungus produces dark, ostiolate pycnidia that bear hyaline, straight or slightly curved, ellipsoid or cylindrical conidia that measure 5–10 x 2–4 µm. Conidia are mostly single-celled but can become one septate. Pycnidia can be found in leaf spots of this disease, but rarely in the crown rot phase. Most lettuce types show some susceptibility, though romaine cultivars appear to be the most sensitive to damage.

Disease cycle

The epidemiology of this disease has not yet been documented. For greenhouse grown lettuce in the UK, the disease can be more common during the winter months.

Control

Apply fungicides to the base of young lettuce plants. Avoid keeping soils overly wet, as the disease is sometimes associated with wet conditions. If possible, plant susceptible romaine in fields that do not have a history of the disease.

References

Koike, S. T. 2001. Investigation of a new crown rot disease of lettuce. *California Lettuce Research Board Annual Report*.

Koike, S. T., Subbarao, K. V., Verkley, G. J. M., O'Neill, T., and Fogle, D. 2003. Phoma basal rot of lettuce caused by *P. exigua* in California. *Phytopathology* 93:S47.

O'Neill, T. M. and McPherson, G. M. 1991. Rots, spots, and blotches—new disease problems affecting protected lettuce. *Grower*, June 6 issue, pp. 11–18.

Rhizoctonia solani
(teleomorph = *Thanatephorus cucumis*)

BOTTOM ROT

Introduction and significance

Bottom rot is generally a minor disease problem. The pathogen is found worldwide and can cause some limited damage to lettuce. In some regions, greenhouse-grown lettuce may develop significant bottom rot.

Symptoms and diagnostic features

Bottom rot typically infects lettuce plants as the heads begin to form, though the responsible pathogen can at times cause damping-off of seedlings. Brown, sunken lesions form on leaf midribs that are in contact with the soil (**413**, **414**). If conditions are favorable, the lesions

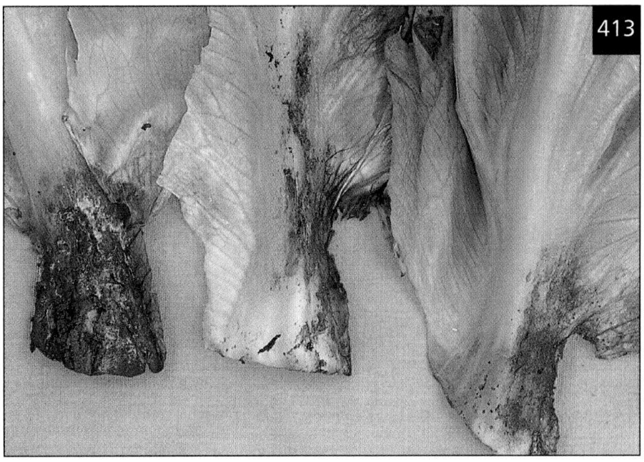

413 Infected petiole bases of lettuce with bottom rot disease.

414 Decayed crown of lettuce affected by bottom rot.

can enlarge rapidly and rot the lower leaves. Tan to brown mycelium and amber-colored drops of liquid may be evident in the rotted areas. The fungus can sometimes penetrate and infect leaves inside the head, resulting in a rot of the entire plant. Symptoms are most evident on plants that are near or at maturity. The activity of secondary decay organisms can result in a very soft, slimy decay of the base of infected plants and contribute to collapse of the head.

Causal agent

Bottom rot is caused by the basidiomycete fungus *Rhizoctonia solani*, which is a soilborne fungus with a very broad host range. *Rhizoctonia solani* has no asexual fruiting structures or spores, but produces characteristically coarse, brown, approximately right-angle branching hyphae. The hyphae are distinctly constricted at branch points, and cross walls with dolipore septa are deposited just after the branching. Hyphal cells are multi-nucleate. Small, tan to brown loosely aggregated clumps of mycelia function as sclerotia. The teleomorph *Thanatephorus cucumis* is not commonly observed on symptomatic lettuce, but when it occurs it is a whitish, thin, delicate, flat hymenial layer present on plant surfaces near the soil surface. Basidia and basidiospores (measuring 7–13 x 4–7 µm) are produced in this layer. The role of the perfect stage in disease development is not known. *Rhizoctonia solani* isolates are divided up into a series of distinct groups, called anastomosis groups (AGs), which have differing virulence levels and in some cases preferred host ranges. AG 4 isolates have been associated with bottom rot of lettuce.

Disease cycle

Rhizoctonia solani is a soil inhabitant that survives as sclerotia or as mycelium in and on crop residue and soil organic matter. Disease develops when lower portions of developing lettuce leaves come in contact with infested soil. Warm (25–27° C), moist soil conditions are conducive to infection and disease development. This disease can also occur under cool, moist conditions in over-wintered greenhouse crops where lettuce has been regularly grown.

Control

Bottom rot is controlled in some cases by the use of fungicides; apply chemicals to the base of the plant. Rotate crops with non-hosts to help reduce soil inoculum. Do not plant back into fields having undecomposed plant residue from a previous crop. Plant in well-draining soil and avoid over irrigation so that the bed tops can be kept dry. Select appropriate cultivars, as lettuce varieties that have upright growth habits experience less disease severity and incidence.

References

Grosch, R., Schneider, J. H. M., and Kofoet, A. 2004. Characterisation of *Rhizoctonia solani* anastomosis groups causing bottom rot in field-grown lettuce in Germany. *European Journal of Plant Pathology* 110:53–62.

Herr, L. J. 1992. Characteristics of *Rhizoctonia* isolates associated with bottom rot of lettuce in organic soils in Ohio. *Phytopathology* 82:1046–1050.

Herr, L. J. 1993. Host sources, virulence, and overwinter survival of *Rhizoctonia solani* anastomosis groups isolated from field lettuce with bottom rot symptoms. *Crop Protection* 12:521–526.

Koike, S. T. and Martin, F. N. 2005. Evaluation of fungicides for controlling bottom rot of iceberg lettuce, 2001 and 2002. *Fungicide and Nematicide Tests* (online). Report 60:V150. DOI: 10.1094/FN60. American Phytopathological Society, St. Paul, MN.

Kuramae, E. E., Buzeto, A. L., Ciampi, M. B., and Souza, N. L. 2003. Identification of *Rhizoctonia solani* AG 1-IB in lettuce, AG 4 HG-I in tomato and melon, and AG 4 HG-III in broccoli and spinach, in Brazil. *European Journal of Plant Pathology* 109:391–395.

Mahr, S. E. R., Stevenson, W. R., and Sequeira, L. 1986. Control of bottom rot of head lettuce with iprodione. *Plant Disease* 70:506–509.

Pieczarka, D. J. and Lorbeer, J. W. 1975. Microorganisms associated with bottom rot of lettuce grown on organic soil in New York state. *Phytopathology* 65:16–21.

Wareing, P. W., Wang, Z.-N., Coley-Smith, J. R., and Jeves, T. M. 1986. Fungal pathogens in rotted basal leaves of lettuce in Humberside and Lancashire with particular reference to *Rhizoctonia solani*. *Plant Pathology* 35:390–395.

Sclerotinia minor, S. sclerotiorum

LETTUCE DROP

Introduction and significance

Lettuce drop is an economically important problem in North America, Europe, Australia, and New Zealand, and the pathogens are found throughout the world. In some lettuce producing regions, such as California, this disease is generally considered the most important soilborne problem of the crop. If severe, 75% or more of a particular planting may be lost to this disease. In Europe, lettuce drop has increased in importance since the introduction of crisphead types which have longer growing periods than butterhead cultivars. On crops other than lettuce, this disease is called white mold.

Symptoms and diagnostic features

Lettuce drop disease is caused by two species of *Sclerotinia*. *Sclerotinia minor* only infects lettuce stems and leaves that are in contact with the soil. Once infection takes place, the fungus causes a brown, soft, watery decay of the lower leaves and crown (**415**). The outer leaves then wilt and the entire plant can collapse, usually when lettuce is near maturity (**416**). White mycelium and small (3–5 mm), black, irregularly shaped sclerotia form on the decayed crown, on upper taproot tissues, and on leaves attached to decaying crowns (**417**). The overall above-ground symptoms resemble those of gray mold and Phoma basal rot.

Sclerotinia sclerotiorum can also infect lower leaves and stems, causing symptoms similar to those of *S. minor* (**418**). In addition, *S. sclerotiorum* produces

416 Collapsed lettuce plant infected by *Sclerotinia minor*.

417 Sclerotia and mycelium of *Sclerotinia minor* on lettuce.

415 Rotted lettuce crown infected by *Sclerotinia minor*.

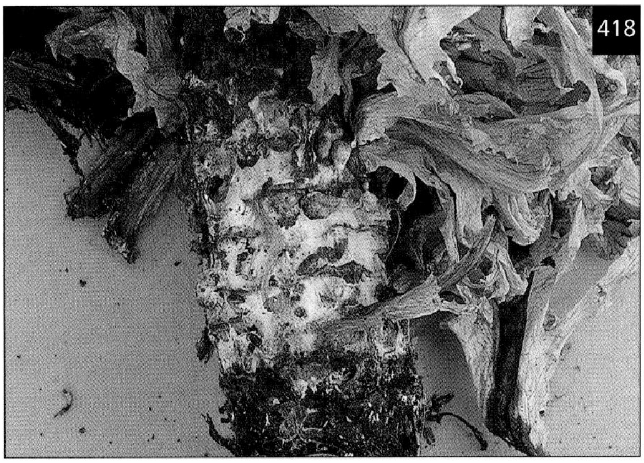

418 Rotted lettuce crown caused by *Sclerotinia sclerotiorum*.

aerial spores that can infect the upper parts of plants. Dispersed by winds, these spores usually infect damaged or senescing tissue. Both crown and upper leaf infections likewise result in a brown, soft, watery rot that is accompanied by white mycelium and black, oblong or dome-shaped sclerotia that are 5–10 mm long. Plant collapse also tends to occur near harvest.

Causal agents

White mold is caused by two species of *Sclerotinia*, *S. minor* and *S. sclerotiorum*. The two pathogens are distinguished primarily by the size of sclerotia. *Sclerotinia minor* sclerotia typically are significantly smaller than those of *S. sclerotiorum* (**419**). In addition, *S. sclerotiorum* produces apothecia (**420**), while *S. minor* does not generally do so in the field. For detailed descriptions of these pathogens, see the bean white mold section in the chapter on legume diseases (page 262).

Disease cycle

For detailed descriptions of the disease cycles, see the bean white mold section in the chapter on legume diseases. Recent studies on lettuce indicated that ascospores of *S. sclerotiorum* may be able to survive several weeks on leaves. Such survival means that lettuce can become infected some time after the arrival of the inoculum.

Control

For an integrated approach to *Sclerotinia* control, see the bean white mold section in the chapter on legume diseases. Because *S. minor* completes its life cycle in the soil, control this lettuce drop pathogen by applying fungicides to lettuce after plants have been thinned, rotating to non-host crops, and deep plowing soils to bury sclerotia below the root zone. Note that the planting of some cover crops can significantly increase *S. minor* inoculum. Crop residues from broccoli plantings show suppressive effects against *S. minor*.

Control of lettuce drop caused by *S. sclerotiorum* is more difficult because inoculum usually consists of aerial ascospores that enter lettuce fields at unpredictable intervals. Apply fungicides at rosette stage to provide some protection. Biological control with commercial formulations of the mycoparasite *Coniothyrium minitans* has shown some promise against *S. sclerotiorum*, but not for *S. minor.*

419

419 Comparison of sclerotia from the two *Sclerotinia* species that infect lettuce: *S. sclerotiorum* (left); *S. minor* (right).

420

420 Apothecia of *Sclerotinia sclerotiorum* forming from one sclerotium.

References

Bell, A. A., Liu, L., Reidy, B., Davis, R. M., and Subbarao, K. V. 1998. Mechanisms of subsurface drip-mediated suppression of lettuce drop caused by *Sclerotinia minor*. *Phytopathology* 88:252–259.

Clarkson, J. P., Phelps, K., Whipps, J. M., Young, C. S., Smith, J. A., and Watling, M. 2004. Forecasting Sclerotinia disease on lettuce: towards developing a prediction model for carpogenic germination of sclerotia. *Phytopathology* 94:268–279.

Clarkson, J. P., Staveley, J., Phelps, K., Young, C. S., and Whipps, J. M. 2003. Ascospore release and survival in *Sclerotinia sclerotiorum*. *Mycological Research* 107:213–222.

Dillard, H. R., and Grogan, R. G. 1985. Relationship between sclerotial spatial pattern and density of *Sclerotinia minor* and the incidence of lettuce drop. *Phytopathology* 75:90–94.

Grube, R., and Ryder, E. 2004. Identification of lettuce (*Lactuca sativa*) germplasm with genetic resistance to drop caused by *Sclerotinia minor*. *Journal of the American Society for Horticultural Science* 129:70–76.

Hao, J. J., Subbarao, K. V., and Duniway, J. M. 2003. Germination of *Sclerotinia minor* and *S. sclerotiorum* sclerotia under various soil moisture and temperature combinations. *Phytopathology* 93:443–450.

Hao, J. J., Subbarao, K. V., and Koike, S. T. 2003. Effects of broccoli rotation on lettuce drop caused by *Sclerotinia minor* and on the population density of sclerotia in soil. *Plant Disease* 87:159–166.

Hawthorne, B. T. 1975. Observations on the development of apothecia of *Sclerotinia minor* Jagg. in the field. *New Zealand Journal of Agricultural Research* 19:383–386.

Hubbard, J. C., Subbarao, K. V., and Koike, S. T. 1997. Development and significance of dicarboximide resistance in *Sclerotinia minor* isolates from commercial lettuce fields in California. *Plant Disease* 81:148-153.

Imolehin, E. D., and Grogan, R. G. 1980. Factors affecting survival of sclerotia, and effects of inoculum density, relative position, and distance of sclerotia from the host on infection of lettuce by *Sclerotinia minor*. *Phytopathology* 70:1162–1167.

Imolehin, E. D., Grogan, R. G., and Duniway, J. M. 1980. Effect of temperature and moisture tension on growth, sclerotial production, germination, and infection by *Sclerotinia minor*. *Phytopathology* 70:1153–1157.

Jones, E. E., and Whipps, J. M. 2002. Effect of inoculum rates and sources of *Coniothyrium minitans* on control of *Sclerotinia sclerotiorum* disease in glasshouse lettuce. *European Journal of Plant Pathology* 108:527–538.

Koike, S. T., Smith, R. F., Jackson, L. E., Wyland, L. J., Inman, J. I., and Chaney, W. E. 1996. Phacelia, lana woollypod vetch, and Austrian winter pea: three new cover crop hosts of *Sclerotinia minor* in California. *Plant Disease* 80:1409–1412.

Meltzer, M. S., Smith, E. A., and Boand, G. J. 1997. Index of plant hosts of *Sclerotinia minor*. *Canadian Journal of Plant Pathology* 19:272–280.

Patterson, C. L., and Grogan, R. G. 1985. Differences in epidemiology and control of lettuce drop caused by *Sclerotinia minor* and *Sclerotinia sclerotiorum*. *Plant Disease* 69:766–770.

Subbarao, K. V. 1998. Progress towards integrated management of lettuce drop. *Plant Disease* 82:1068–1078.

Subbarao, K. V., Hubbard, J. C., and Schulbach, K. F. 1997. Comparison of lettuce diseases and yield under subsurface drip and furrow irrigation. *Phytopathology* 87:877–883.

Subbarao, K. V., Koike, S. T., and Hubbard, J. C. 1996. Effects of deep plowing on the distribution and density of *Sclerotinia minor* sclerotia and lettuce drop incidence. *Plant Disease* 80:28–33.

Wu, B. M., and Subbarao, K. V. 2003. Effects of irrigation and tillage on temporal and spatial dynamics of *Sclerotinia minor* sclerotia and lettuce drop incidence. *Phytopathology* 93:1572–1580.

Young, C. S., Clarkson, J. P., Smith, J. A., Watling, M., Phelps, K., and Whipps, J. M. 2004. Environmental conditions influencing *Sclerotinia sclerotiorum* infection and disease development in lettuce. *Plant Pathology* 53:387–397.

Verticillium dahliae

VERTICILLIUM WILT

Introduction and significance

This disease is not yet widespread in the world. It has been reported from Greece and the USA (California). In some lettuce producing areas in California the pathogen can cause significant damage.

Symptoms and diagnostic features

Initial symptoms appear at rosette stage when the lower leaves wilt. Those leaves closest to the lettuce head can yellow, die, and remain closely appressed to the head. Discolored streaks occur in the vascular tissues of the taproot and crown, and can be a combination of green, brown, or black (**421**). As disease develops, the plant's

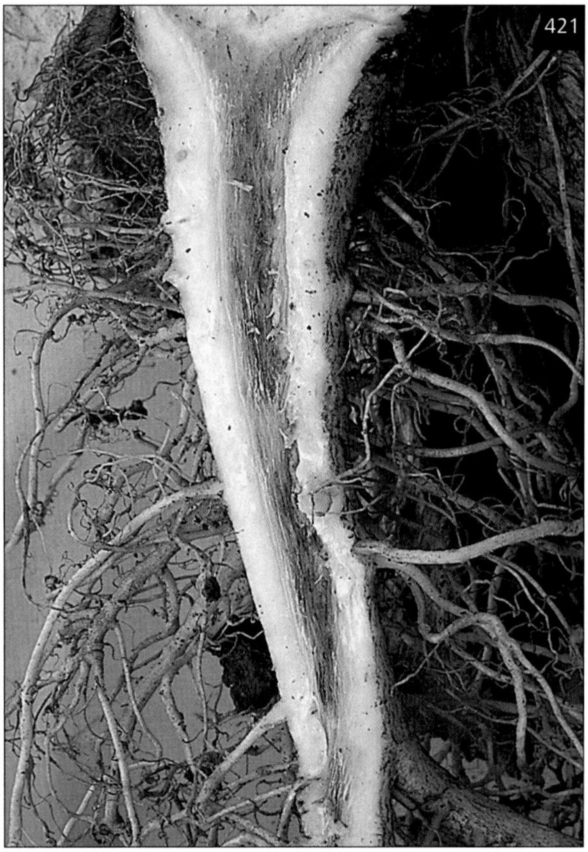

421 Vascular discoloration of Verticillium wilt of lettuce.

422 Collapsed plants of Verticillium wilt of lettuce.

outer whorl of leaves turns yellow, wilts, and dies (**422**). Both wilting and vascular discoloration symptoms may appear that are similar to those caused by Fusarium wilt or ammonium toxicity.

Causal agent

The causal pathogen is *Verticillium dahliae*. It can be isolated on standard microbiological media, though semi-selective media such as NP-10 can also be used. On general purpose media, the pathogen forms the characteristic hyaline, verticillate conidiophores bearing 3–4 phialides at each node, and hyaline, single-celled, ellipsoidal conidia that measure 2–8 x 1–3 µm. Older cultures form dark brown to black torulose microsclerotia that consist of groups of swollen cells formed by repeated budding. Microsclerotia size varies greatly and is in the range of 15–100 µm diameter. Microsclerotia enable the pathogen to survive in the soil for extended periods of time (up to 8 to 10 years). In infested lettuce fields in California, inoculum levels are extremely high and can reach levels as high as 400–600 microsclerotia/gram of soil. An important discovery is that this lettuce pathogen can be seedborne in lettuce. Therefore, infested lettuce seed could be a means of spreading the pathogen to previously uninfested fields.

Disease cycle

Researchers are still investigating the relationship of lettuce *V. dahliae* isolates to isolates from other plants. For example, *V. dahliae* from lettuce infects artichoke, cotton, eggplant, pepper (chile type), potato, strawberry (variable response), tomato, and watermelon. However, the lettuce isolates could not readily infect alfalfa, cabbage, cauliflower, or pepper (bell type) which are hosts of *V. dahliae*. Another potentially important aspect to lettuce Verticillium wilt is the confirmation that several weeds found in and around lettuce fields are hosts to *V. dahliae*: shepherd's purse (*Capsella bursa-pastoris*), hairy nightshade (*Solanum sarrachoides*), sowthistle (*Sonchus oleraceus*), and groundsel (*Senecio vulgaris*). Some of these isolates from weed hosts are more strongly virulent on lettuce than lettuce isolates.

Control

Plant resistant or tolerant cultivars as they become available. Pre-plant treatment of soil with effective fumigants will give short-term control, though the use of such materials is not usually economically feasible for lettuce production. Rotate crops so that lettuce is not planted in fields having a history of the problem. Minimize spread of infested soil to other, uninfested areas. Be aware that infested lettuce seed and possibly weeds and weed seeds are sources of the pathogen.

References

Ligoxigakis, E. K. and Vakalounakis, D. J. 2000. Hosts of *Verticillium dahliae* race 2 in Greece. In: Tjamos, E. C., Rowe, R. C., Heale, J. B., and Fravel, D. R. *Advances in Verticillium: Research and Disease Management*. American Phytopathological Society Press.

Sorensen, L. H., Schneider, A. T., and Davis., J. R. 1991. Influence of sodium polygalacturonate sources and improved recovery of *Verticillium* species from soil. (Abstract) *Phytopathology* 81:1347.

Subbarao, K. V. 2000. Biology and epidemiology of Verticillium wilt of lettuce. *California Lettuce Research Board Annual Report* pp. 96–103.

Vallad, G. E., Bhat, R. G., Koike, S. T., Ryder, E. J., and Subbarao, K. V. 2005. Weedborne reservoirs and seed transmission of *Verticillium dahliae* in lettuce. *Plant Disease* 89:317–324.

Beet western yellows virus

BEET WESTERN YELLOWS

Introduction and significance

Beet western yellows is an occasional virus problem of lettuce in North America and Europe. Diagnosis of this disease may be complicated because symptoms caused by this virus can closely resemble those of nutrient deficiencies

Symptoms and diagnostic features

Symptoms usually do not develop until plants reach rosette stage or later. At this point in crop development, the leaves develop chlorotic blotches in interveinal tissue. This yellowing continues until the oldest, lower leaves are bright yellow to sometimes almost white in color, with the main leaf veins remaining green (**423**, **424**). Yellowed leaves often have a thick, brittle texture. Yellowing can progress until the wrapper leaves adjacent to the head also turn yellow, and head color may be unacceptably light green. In most lettuce varieties, significant stunting or reduction in plant size does not occur. Overall symptoms of this yellows disease may resemble iron chlorosis.

Causal agent and disease cycle

The pathogen is the *Beet western yellows virus* (BWYV), which is a polerovirus. BWYV has isometric particles with diameters of 25 nm and single-stranded, linear RNA genomes. BWYV may be found in numerous crop and weed plants, and the host list is extensive, including over 150 documented plant species. Some isolates or strains of this virus have different abilities to infect certain plants; thus, not all strains of BWYV may be able to infect all known hosts, greatly complicating the etiology of this disease for lettuce. BWYV is vectored by several aphid vectors, especially the green peach aphid (*Myzus persicae*), in a persistent manner.

Control

Follow general suggestions for managing virus diseases (see Part 1). Plant tolerant cultivars. Remove weed hosts and keep aphid populations low.

423 Leaf chlorosis of beet western yellows of lettuce.

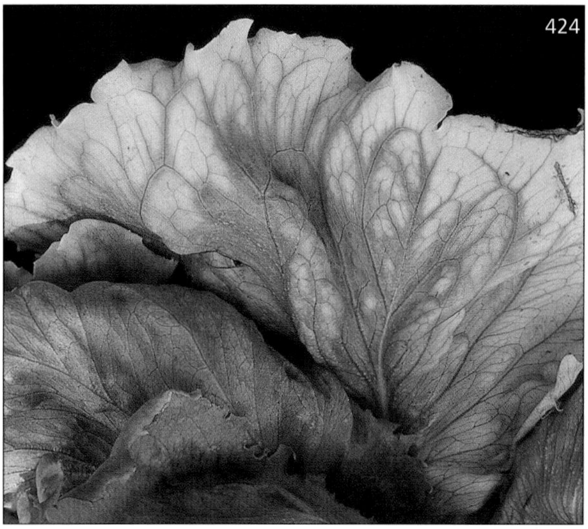

424 Close-up of leaf chlorosis of beet western yellows of lettuce, showing characteristic green veins.

References

Falk, B. W. and Guzman, V. I. 1981. A virus as the causal agent of spring yellows of lettuce and escarole. *Proceedings of the Florida State Horticultural Society* 94:149–152.

Pink, D. A. C., Walkey, D. G. A., and McClement, S. J. 1991. Genetics of resistance to beet western yellows virus in lettuce. *Plant Pathology* 40:542–545.

Walkey, D. G. A. and Pink, D. A. C. 1990. Studies on resistance to beet western yellows virus in lettuce (*Lactuca sativa*) and the occurrence of field sources of the virus. *Plant Pathology* 39:141–155.

Zink, F. W. and Duffus, J. E. 1972. Association of beet western yellows and lettuce mosaic viruses with internal rib necrosis of lettuce. *Phytopathology* 62:1141–1144.

Lettuce mosaic virus

LETTUCE MOSAIC

Introduction and significance

Lettuce mosaic is one of the most important virus diseases of lettuce. This disease occurs in all lettuce-producing areas in the world. Integrated disease management programs help limit the impact of this virus. However, outbreaks still occur and can possibly cause crop loss.

Symptoms and diagnostic features

Symptoms can vary greatly depending on the particular type or cultivar of lettuce infected, age of infected plant, virus strain, and environmental conditions. Plants that are infected at a young stage are stunted, deformed, and (in some varieties) show a mosaic or mottling pattern (**425, 426**). Such plants rarely grow to full size. Plants that are infected later in the growth cycle will show a different set of symptoms. These plants may reach full size, but the older outer leaves will be yellow, twisted, or otherwise deformed. On head lettuce the wrapper leaves will often curve back away from the head (**427**). Developing heads may be deformed. In some cases brown, necrotic flecks occur on the wrapper leaves and leaf margins may be more serrated than normal.

Causal agent

Lettuce mosaic is caused by *Lettuce mosaic virus* (LMV). LMV is a potyvirus and consists of filamentous rods measuring approximately 750 x 15 nm and single-stranded RNA genomes. Throughout the world, LMV isolates are divided into four distinct groups, or pathotypes (I, II, III, IV), based upon the ability of the isolate to infect various differential cultivars having mo and g resistance genes. LMV is vectored in a nonpersistent manner by many different aphids and can also be sap transmitted. LMV is the only economically important lettuce virus that is seedborne. In addition to lettuce, the virus can infect numerous other plants. However, in commercial settings it appears that LMV primarily infects plants in the Asteraceae. Ornamentals such as *Gazania* species and weeds like bristly oxtongue (*Picris echioides*) can be important reservoir hosts in California.

Disease cycle

Initial inoculum of the virus comes from infected seed or diseased lettuce and alternate hosts. Aphids carry the virus from the diseased hosts to uninfected, new lettuce. If aphid activity is extensive, the virus can be rapidly spread throughout a field and result in a high incidence of lettuce mosaic disease. An important aphid vector in California is the green peach aphid (*Myzus persicae*). However, the lettuce aphid (*Nasonovia ribis-nigri*) does not vector the virus.

425 Stunting and chlorosis of lettuce caused by lettuce mosaic.

426 Mosaic symptom on leaf lettuce caused by lettuce mosaic.

427 Lettuce leaf curling symptom of lettuce mosaic.

Control

Use seed that does not have significant levels of the pathogen. Infection thresholds differ according to the epidemiological features of a particular area. Researchers found that a zero in 30,000 seed infection threshold is required for LMV control in California's Salinas Valley. However, in parts of Europe the seed infection threshold is zero in 2,000 seed. Control weeds in and around the lettuce production areas, because weeds can be a significant reservoir of the virus and the source from which aphids obtain the virus. Plow down, in a timely manner, old lettuce plantings because infected lettuce plants, like weeds, are a source of virus. A 2-week lettuce-free period in California helps prevent continuous, year-to-year buildup of LMV and helps reduce the amount of virus that would 'bridge' over from one season to the next. Plant resistant lettuce cultivars if available. While controlling aphids does not prevent the transmission of LMV, managing these vectors is helpful in slowing LMV spread.

References

Dinant, S. and Lot, H. 1992. Lettuce mosaic virus: a review. *Plant Pathology* 41:528–542.

Falk, B. W and Purcifull, D. E. 1983. Development and application of ELISA test to index lettuce seeds for lettuce mosaic virus in Florida. *Plant Disease* 67:413–416.

Grogan, R. G. 1980. Control of lettuce mosaic with virus-free seed. *Plant Disease* 64:446–449.

Grogan, R. G. 1983. Lettuce mosaic virus control by the use of virus-indexed seed. *Seed Science and Technology* 11:1043–1049.

Krause-Sakate, R., Le Gall, O., Fakhfakh, H., Peypelut, M., Marrakchi, M., Varveri, C., Pavan, M. A., Souche, S. Lot, H., Zerbini, F. M., and Candresse, T. 2002. Molecular and biological characterization of *Lettuce mosaic virus* (LMV) isolates reveals a distinct and widespread type of resistance-breaking isolate: LMV-Most. *Phytopathology* 92:563–572.

Nebreda, M., Moreno, A., Perez, N., Palacios, I., Seco-Fernandez, V., and Fereres, A. 2004. Activity of aphids associated with lettuce and broccoli in Spain and their efficiency as vectors of Lettuce mosaic virus. *Virus Research* 100:83–88.

Pink, D. A. C., Kostova, D., and Walkey, D. G. A. 1992. Differentiation of pathotypes of lettuce mosaic virus. *Plant Pathology* 41:5–12.

Van Vuurde, J. W. L. and Maat, D. Z. 1983. Routine application of ELISA for the detection of lettuce mosaic virus in lettuce seeds. *Seed Science and Technology* 11:505–513.

Zerbini, F. M., Koike, S. T., and Gilbertson, R. L. 1995. Biological and molecular characterization of lettuce mosaic potyvirus isolates from the Salinas Valley of California. *Phytopathology* 85:746–752.

Zerbini, F. M., Koike, S. T., and Gilbertson, R. L. 1997. *Gazania* spp.: A new host of lettuce mosaic potyvirus, and a potential inoculum source for recent lettuce mosaic outbreaks in the Salinas Valley of California. *Plant Disease* 81:641–646.

Lettuce necrotic stunt virus

LETTUCE DIEBACK

Introduction and significance

Romaine cultivars consistently show the most pronounced and serious symptoms of lettuce dieback, though several leaf lettuce and butterhead cultivars are also susceptible. This is a new disease that has been documented only since the late 1990s.

Symptoms and diagnostic features

Infected lettuce can be severely stunted with mature, diseased plants failing to develop past the eight to ten leaf stage. Extensive chlorosis is present on the outermost leaves (**428**). The younger, inner leaves often remain dark green in color, but become rough and leathery in texture. The chlorotic outer leaves usually develop necrotic spotting that can turn into extensive areas of brown, dead tissue (**429**). Affected redleaf lettuce takes on an orange color (**430**). The roots of severely affected plants can be necrotic and rotted, although it is not clear if such symptoms are caused directly by the pathogen or by secondary decay factors. Currently used iceberg cultivars are immune to this problem.

Causal agent

The causal agent is *Lettuce necrotic stunt virus* (LNSV). LNSV is a tombusvirus and consists of isometric particles measuring up to 30 nm in diameter and double-stranded RNA genomes. Strains of this lettuce pathogen form a new, highly significant cluster within the group and have 3'-terminal genomic nucleotide sequences that are distinct from previously described *Tomato bushy stunt virus* (TBSV) sequences. Based on this sequence information, this lettuce pathogen appears to be a new tombusvirus that is closely related to TBSV. Presently, diagnosis directly from lettuce tissue is not reliable, even using sensitive methods such as Western blot analysis, reverse transcriptase-PCR, and immunocapture-PCR. Instead, symptomatic lettuce tissues are used to mechanically inoculate indicator plants (*Chenopodium quinoa, Nicotiana benthamiana, N. clevelandii*); serological or molecular methods are then employed to test local lesions or plant sap from the indicator species.

428 **Lettuce infected with lettuce dieback showing brown necrotic lesions.**

429 **Stunted and chlorotic lettuce infected with lettuce dieback disease.**

430 **Red leaf lettuce cultivar infected with lettuce dieback disease.**

Disease cycle

LNSV has no known invertebrate or fungal vector. Rather, the virus resides in soil and water and is spread in river water, irrigation runoff, flood waters, and infested soil and mud. Research is needed to determine the exact host range of LNSV, but preliminary information indicates that the LNSV and TBSV host ranges are probably very similar.

Control

Plant resistant romaine and leaf lettuce cultivars as they become available. Recently released romaine cultivars have good resistance. Iceberg lettuce can be planted in infested fields, as these cultivars do not develop disease. Preliminary experiments indicated that the disease was not controlled by the application of soil fumigants. If susceptible romaine must be used, select fields that do not have a history of the problem. Avoid spreading infested soil and mud to clean fields.

References

Gerik, J. S., Duffus, J. E., Perry, R., Stenger, D. C., and Van Maren, A. F. 1990. Etiology of tomato plant decline in the California desert. *Phytopathology* 80:1352–1356.

Grube, R. C. and Ryder, E. J. 2003. Romaine lettuce breeding lines with resistance to lettuce dieback caused by tombusviruses. *HortScience* 38:627–628.

Liu, H.-Y., Sears, J. L., Obermeier, C., Wisler, G. C., Ryder, E. J., Duffus, J. E., and Koike, S. T. 1999. First report of tomato bushy stunt virus isolated from lettuce. *Plant Disease* 83:301.

Obermeier, C., Sears, J. L., Liu, H. Y., Schlueter, K. O., Ryder, E. J., Duffus, J. E., Koike, S. T., and Wisler, G. C. 2001. Differentiation of tombusvirus strains that cause diseases of lettuce and tomato in the southwestern U.S. *Phytopathology* 91:797–806.

Tomlinson, J. A. and Faithfull, E. M. 1984. Studies on the occurrence of *tomato bushy stunt virus* in English rivers. *Annals of Applied Biology* 104:485–495.

Mirafiori lettuce virus

LETTUCE BIG VEIN

Introduction and significance

Big vein disease is generally a minor disease, though severe symptoms can reduce the quality of the commodity. This disease occurs in the USA, Brazil, Europe, and Japan.

Symptoms and diagnostic features

Lettuce big vein disease causes veins in leaves to become greatly enlarged and clear (**431**). Such deformities are easily seen if the leaf is examined with a light source behind it. The enlarged veins cause the rest of the leaf to be ruffled, buckled, and malformed (**432**). If infected early in the growth cycle, lettuce plants can be stunted.

431 Enlarged veins of big vein of lettuce.

432 Lettuce plant on left is infected with big vein. Plant on the right is healthy.

Severely affected plants can be so deformed as to be unmarketable, and head lettuce varieties may fail to form a head or be delayed in development.

Causal agent

The causal agent of lettuce big vein disease eluded identification for many years. Researchers believed that the pathogen was a virus, and had confirmed that this agent could be graft-transmitted and vectored by a soil microorganism, *Olpidium brassicae*. Only recently has the pathogen, *Mirafiori lettuce virus* (MiLV), been identified. MiLV is an ophiovirus with particles that are highly kinked filaments. These particles are difficult to see with an electron microscope because they have low contrast and greatly vary in size and shape. Confirming the etiology of lettuce big vein disease was complicated by the presence of other non-pathogenic viruses in the *O. brassicae* vector. One such virus, *Lettuce big vein virus*, is now known to be non-pathogenic to lettuce. The genome of MiLV consists of four RNA segments.

Disease cycle

MiLV is a soilborne virus and is introduced into lettuce plants by a soil chytrid fungus, *Olpidium brassicae*. This vector survives in the soil for long periods of time and produces swimming zoospores that move about in soil water and attach themselves to lettuce roots. When zoospores infect and colonize the epidermal cells of the roots, they themselves do not cause disease but transmit MiLV into the plant. The lettuce big vein disease does not occur in nature if *O. brassicae* is absent; however, not all *O. brassicae* isolates contain MiLV. The disease is spread by the movement of infested soil and water that contain viruliferous *O. brassicae*. Seedlings can be infected within 8 days of planting, and symptoms can appear 18 days after infection. Disease symptoms develop most extensively if air temperatures are relatively cool (below 16-20° C). Temperatures above 22° C will generally prevent symptom development. *Olpidium brassicae* can persist in the soil as resilient resting sporangia, thereby ensuring that the virus will also be present for many years, even in the absence of its lettuce host.

Control

Use resistant cultivars. In badly infested fields with histories of severe disease, delay planting lettuce until the warmer time of the year. Crop rotations with non-hosts will not prevent the disease, but will help reduce disease severity for subsequent lettuce crops. When producing lettuce transplants, do not allow the plants and trays to come into contact with infested soil.

References

Campbell, R. N. 1965. Weeds as reservoir hosts of the *lettuce big vein virus*. *Canadian Journal of Botany* 43:1141–1149.

Campbell, R. N. and Grogan, R. G. 1963. Big vein of lettuce and its transmission by *Olpidium brassicae*. *Phytopathology* 53:252–259.

Colariccio, A., Chaves, A. L. R., Eiras, M., Chagas, C. M., Lenzi, R., and Roggero, P. 2003. Presence of lettuce big-vein disease and associated viruses in a subtropical area of Brazil. *Plant Pathology* 52:792.

Lot, H., Campbell, R. N., Souche, S., Milne, R. G., and Roggero, P. 2002. Transmission by *Olpidium brassicae* of *Mirafiori lettuce virus* and *lettuce big vein virus*, and their roles in lettuce big vein etiology. *Phytopathology* 92:288–293.

Navarro, J. A., Botella, F., Maruhenda, A., Sastre, A., Sanchez-Pina, M. A., and Pallas, V. 2004. Comparative infection progress analysis of *lettuce big vein virus* and *Mirafiori lettuce virus* in lettuce crops by developed molecular diagnosis techniques. *Phytopathology* 94:470–477.

Roggero, P., Lot, H., Souche, S., Lenzi, R., and Milne, R. G. 2003. Occurrence of *mirafiori lettuce virus* and *lettuce big-vein virus* in relation to development of big-vein symptoms in lettuce crops. *European Journal of Plant Pathology* 109:261–267.

van der Wilk, F., Dullemans, A. M., Verbeek, M., and van den Heuvel, J. F. J. M. 2002. Nucleotide sequence and genomic organization of an ophiovirus associated with lettuce big vein disease. *Journal of General Virology* 83:2869–2877.

White, J. G. 1980. Control of lettuce big-vein disease by soil sterilisation. *Plant Pathology* 29:124–130.

Tomato spotted wilt virus

TOMATO SPOTTED WILT

Introduction and significance

Tomato spotted wilt is an important, though sporadic, problem that can cause significant losses in some lettuce areas, such as the USA (Hawaii) and Tasmania. The virus is found worldwide on many different hosts.

Symptoms and diagnostic features

Symptoms tend to develop on one side of a plant and include yellowing of leaves, brown to dark brown necrotic spots on leaves and petioles, and a twisting of the plant foliage (**433, 434**). Margins of leaves can wilt. Young foliage may especially show many brown necrotic spots. Plants infected early may die.

Causal agent

Tomato spotted wilt disease is caused by the *Tomato spotted wilt virus* (TSWV). TSWV is a tospovirus and has isometric particles, measuring approximately 80–110 nm, which are surrounded with membranes. TSWV has a genome consisting of three linear single-stranded RNAs and is vectored by several species of thrips; at least 8 species are found to be vectors. The western flower thrips (*Frankliniella occidentalis*) is probably the most important vector, though *Thrips tabaci* has been implicated as the main vector in Tasmania. TSWV has one of the most extensive host ranges of any known plant virus and can infect over 900 different cultivated and weedy plant species.

Disease cycle

Only thrips in their larval stage can acquire the virus via feeding on infected plants. The larvae can transmit TSWV after a 4–12 day incubation period. Viruliferous larvae then pupate and the emerging adults move away and spread the virus. Thrips transmit the virus in the circulative propagative manner. Weed hosts are an important reservoir of the virus and source of the thrips.

Control

Controlling TSWV is difficult. Use non-host plants in crop rotations to help limit inoculum. Control weeds adjacent to and near lettuce fields. Insecticide applications to manage thrips will not prevent the disease from occurring but might help limit major tomato spotted wilt outbreaks.

433 Early symptoms of tomato spotted wilt of lettuce.

434 Advanced symptoms of tomato spotted wilt of lettuce.

References

Cho, J. J., Mau, R. F. L., German, T. L., Hartmann, R. W., Yudin, L. S., Gonsalves, D., and Provvidenti, R. 1989. A multidisciplinary approach to management of *tomato spotted wilt virus* in Hawaii. *Plant Disease* 73:375–383.

Cho, J. J., Mitchell, W. C., Mau, R. F. L., and Sakimura, K. 1987. Epidemiology of *tomato spotted wilt virus* disease on crisphead lettuce in Hawaii. *Plant Disease* 71:505–508.

Groves, R. L., Walgenbach, J. F., Moyer, J. W., and Kennedy, G. G. 2002. The role of weed hosts and tobacco thrips, *Frankliniella fusca*, in the epidemiology of *tomato spotted wilt virus*. *Plant Disease* 86:573–582.

Mumford, R. A., Barker, I., and Wood, K. R. 1996. The biology of tospoviruses. *Annals of Applied Biology* 128:159–183

Wilson, C. R. 1998. Incidence of weed reservoirs and vectors of *tomato spotted wilt tospovirus* on southern Tasmania. *Plant Pathology* 47:171–176.

Yudin, L. S., Tabashnik, B. E., Cho, J.J., and Mitchell, W. C. 1990. Disease-prediction and economic models for managing *tomato spotted wilt virus* disease in lettuce. *Plant Disease* 74:211–216.

Turnip mosaic virus

TURNIP MOSAIC

Introduction and significance

Turnip mosaic is an occasional virus problem that affects lettuce.

Symptoms and diagnostic features

Early symptoms include numerous, small, light green to yellow, circular and irregular spots on leaves (**435**). Later symptoms include curvature of the midrib and distortion of the leaf blade. Infected young lettuce may remain stunted and small with deformed, ruffled foliage (**436**).

Causal agent and disease cycle

Turnip mosaic is caused by *Turnip mosaic virus* (TuMV). TuMV is a potyvirus and has filamentous particles measuring approximately 720 nm in length and is vectored by several aphids in a nonpersistent manner. TuMV has a very wide host range.

Control

Plant resistant cultivars and avoid susceptible ones. For example, crisphead cultivars with downy mildew resistance derived from P. I. 91532 are very susceptible to TuMV. Remove weed hosts. Keep aphid populations under control, though the direct effect of such steps on disease incidence is questionable.

References

Duffus, J. E. and Zink, F. W. 1969. A diagnostic host reaction for the identification of *turnip mosaic virus*. *Plant Disease Reporter* 53:916–917.

Zink, F. W. and Duffus, J. E. 1970. Linkage of *turnip mosaic virus* susceptibility and downy mildew, *Bremia lactucae*, resistance in lettuce. *Journal of the American Society for Horticultural Science* 95:420–422.

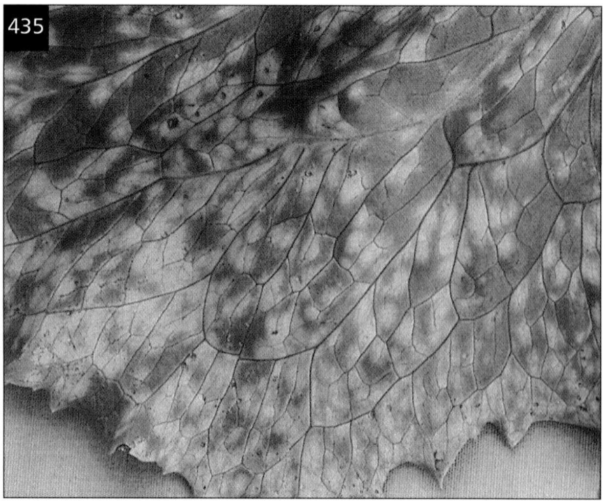

435 Eye spot symptom of turnip mosaic of lettuce.

436 Stunting and deformity of turnip mosaic of lettuce.

AMMONIUM TOXICITY

Introduction and significance

If conditions favor development of this abiotic disorder, significant loss of plant stands can occur. In general, this disorder is of low to moderate concern.

Symptoms and diagnostic features

Lettuce roots are sensitive to high levels of ammonium in the soil. The central core of the root first turns yellow to light brown, then becomes dark brown to red in color, though typically a brick red discoloration is seen (437, 438). In severe cases, the root xylem collapses and a central cavity forms throughout the length of the root and even into the lower crown of the plant. Lateral roots may be short, with blackened tips. In some situations the external surface of the root turns yellow or light brown and develops cracks, which mimic corky root symptoms. Above ground symptoms indicate that plants have damaged root systems and consist of poor growth, stunting, yellowing of older foliage, and wilting. Plants damaged by excess ammonium have symptoms in the vascular tissue that resemble those caused by Fusarium and Verticillium wilts. In general, the hollow root cavity symptom is more characteristic of ammonium toxicity, while Verticillium wilt tends to have black, not red, vascular discoloration.

Causal factor

This problem results from the buildup of ammonium in the root zone of lettuce plants. Ammonium toxicity occurs when soils are cool, compacted, and waterlogged. Sealing of the soil surface from sprinkler irrigation or rain can also contribute to this problem. Such conditions result in slow nitrification rates in the soil. Low soil pH and excessive fertilization also contribute to this problem. Ammonium toxicity tends to occur most often in direct seeded lettuce.

Control

Few management steps are practical for correcting ammonium toxicity problems. To help prevent ammonium toxicity development, prepare soils so that they are not compacted or sealed.

References

Marlatt, R. B. 1967. Nonpathogenic diseases of lettuce: their identification and control. *Bulletin 721*. University of Florida.

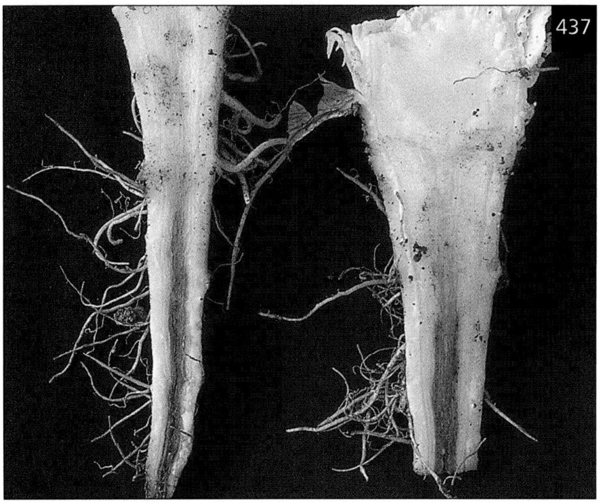

437

437 Discolored lettuce vascular tissue caused by ammonium toxicity.

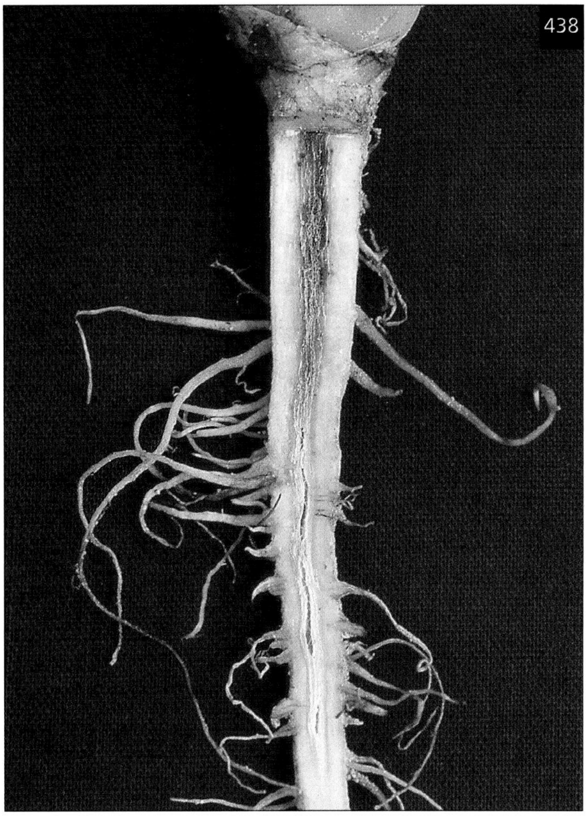

438

438 Discolored lettuce vascular tissue caused by ammonium toxicity.

TIPBURN

Introduction and significance

Tipburn is a physiological disorder of lettuce. If environmental conditions favor development of this abiotic disorder, significant quality loss can take place.

Symptoms and diagnostic features

All lettuce types can develop tipburn, though lettuces that form an entirely enclosed head, such as iceberg and butterhead, develop this disorder more frequently and severely. Symptoms mostly occur on the margins of developing leaf tips and consist of light to dark brown speckling, lesions, and necrosis (**439**, **440**). In severe cases tipburn can result in extensive damage. Secondary decay can result in soft bacterial rot or 'slime.' The gray mold pathogen, *Botrytis cinerea*, can colonize the necrotic areas, progress into adjacent healthy leaf tissue, and cause brown, soft decay. Symptomatic leaves are usually found within the inner whorls of open-head cultivars and underneath the enclosing wrapper leaves of closed-head types. The disease causes significant concerns for harvesters because enclosed head varieties do not show external symptoms, and tipburn can only be detected after the lettuce is cut open.

439 **Tipburn symptoms on butterhead lettuce.**

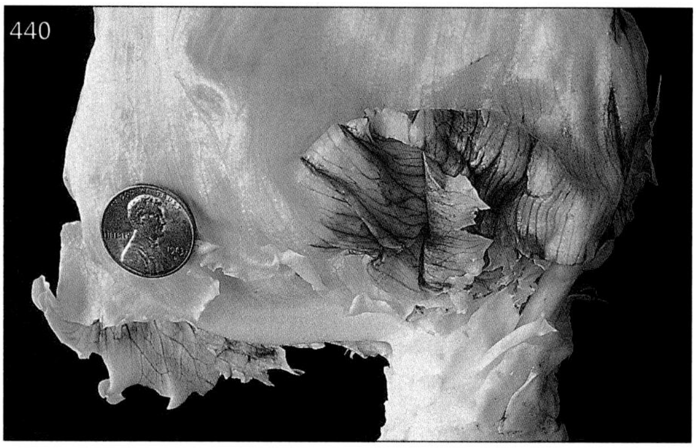

440 **Tipburn symptoms on iceberg lettuce.**

Causal factor

Tipburn develops when the plant cannot provide itself with sufficient calcium. Conditions that interfere with calcium uptake or that favor rapid plant growth may trigger tipburn. These include warm temperatures combined with high fertilization rates and dry root-zone soil. The disorder is similar to blackheart of celery, and to tipburn of spinach and other leafy vegetables.

Control

Plant resistant cultivars. Avoid applying excessive amounts of fertilizer, and irrigate with drip irrigation systems. Foliar calcium supplements may have some benefit on open-head lettuce cultivars.

References

Ashkar, S. A. and Ries, S. K. 1971. Lettuce tipburn as related to nutrient imbalance and nitrogen composition. *Journal of the American Society for Horticultural Science* 96:448–452.

Barta, D. J. and Tibbetts, T. W. 1991. Calcium localization in lettuce leaves with and without tipburn: comparison of controlled-environment and field-grown plants. *Journal of the American Society for Horticultural Science* 116:870–875.

Barta, D. J. and Tibbitts, T. W. 2000. Calcium localization and tipburn development in lettuce leaves during early enlargement. *Journal of the American Society for Horticultural Science* 125:294–298.

Collier, G. F. and Tibbitts, T. W. 1984. Effects of relative humidity and root temperature on calcium concentration and tipburn development in lettuce. *Journal of the American Society for Horticultural Science* 109:128–131.

Frantz, J. M., Ritchie, G., Cometti, N. N., Robinson, J., and Bugbee, B. 2004. Exploring the limits of crop productivity: beyond the limits of tipburn in lettuce. *Journal of the American Society for Horticultural Science* 129:331–338.

Misaghi, I. J. and Grogan, R. G. 1978. Effect of temperature on tipburn development in head lettuce. *Phytopathology* 68:1738–1743.

Misaghi, I. J. and Grogan, R. G. 1978. Physiological basis for tipburn development in head lettuce. *Phytopathology* 68:1744–1753.

Misaghi, I. J., Grogan, R. G., and Westerlund, F. V. 1981. A laboratory method to evaluate lettuce cultivars for tipburn tolerance. *Plant Disease* 65:342–344.

Misaghi, I. J., Matyac, C. A., and Grogan, R. G. 1981. Soil and foliar applications of $CaCl_2$ and $Ca(NO_3)_2$ to control tipburn of head lettuce. *Plant Disease* 65:821–822.

Solanum lycopersicum Tomato

TOMATO (*Solanum lycopersicum*) is one of the most widely produced plants in the Solanaceae (nightshade family). Fruit is valued for its high nutrition, and the crop is grown throughout the world. Tomato originated in South America. Field-grown tomatoes are widely grown in warm temperate and tropical climates, and glasshouse tomatoes are produced in many additional regions. There is little commercial, outdoor production in northern Europe, although tomatoes are grown outdoors in gardens. A wide diversity of fruit types is available, including large red beefsteak, small cherry, yellow pear, and others. Tomatoes are grown for fresh market produce as well as a range of processed products such as tomato juice, canned peeled tomatoes, tomato purée and sauce, ketchup, soups, and sun-dried tomatoes.

Clavibacter michiganensis subsp. *michiganensis*

BACTERIAL CANKER

Introduction and significance

Bacterial canker was first noted in 1909 and today is still an important tomato disease found throughout the world. Periodic outbreaks can cause significant damage to tomato crops in numerous geographic regions.

Symptoms and diagnostic features

Initial symptoms are the result of primary, systemic infections. These symptoms usually affect the lower foliage first and consist of curling of leaves, wilting of leaves and branches, chlorosis of leaves, and brown necrosis and shriveling of leaf tissue. These leaf symptoms sometimes develop on only one side of the leaf, with the other leaf side appearing normal. Internal vascular tissue begins to turn light yellow to tan in color. As disease develops the vascular tissue turns darker brown to red-brown (**441**; also **451**, page 333). The central pith tissue of the stem becomes discolored, turning yellow and sometimes brown. The pith tissue also becomes damaged, resulting in a mealy texture, separation of the pith from the adjacent vascular tissue, and hollowing of the stem center. In advanced stages the pathogen becomes systemic in plant phloem tissues and causes the overall plant to grow poorly, wilt, and possibly die. Foliage throughout the canopy will continue to wilt, turn yellow then brown, and finally collapse. Stems can split and result in the formation of open breaks, or cankers. Stems break off easily at branch nodes.

In addition to the symptoms caused by primary infections, secondary infections and their resulting symptoms occur when bacteria are splashed onto the surfaces of foliage, stems, and fruit. Leaves and stems can develop white to cream-colored, raised, blister-like spots. Such spots on young stems may turn tan in color.

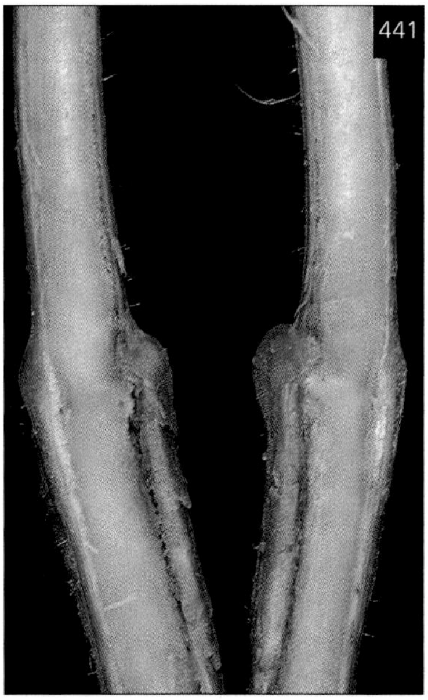

441 Vascular streaking caused by bacterial canker of tomato.

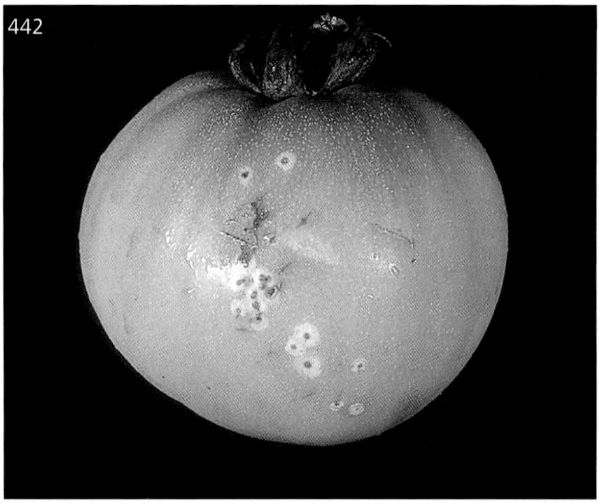

442 Fruit lesions caused by bacterial canker of tomato.

Secondary fruit symptoms are very characteristic and are a useful diagnostic feature. Spots occur on green fruit and are at first small, round, and white or yellow. Such spots enlarge slightly (up to 3–4 mm in diameter) and develop raised, brown centers that remain encircled by the white to cream-colored spot (**442**; also **451**, page 333). They are commonly called 'bird's eye spots.'

Causal agent

The pathogen is *Clavibacter michiganensis* subsp. *michiganensis*. This organism is an aerobic, non-spore-forming, Gram-positive bacterium. This pathogen can be isolated on standard microbiological media and produces yellow colonies on nutrient agar; colonies reach 3–4 mm in diameter after 5 days incubation. Specialized media such as SCM medium are very useful for recovering this pathogen. Bacterial cells can be pleomorphic in culture, but are rod shaped when isolated from plants. Optimum growth occurs at 24–27° C. The pathogen is not a true soilborne pathogen, but it can survive in soil for long periods in dried tomato plant residues. In older references the pathogen is listed as *Corynebacterium michiganense*. *Clavibacter michiganensis* subsp. *michiganensis* is most important as a pathogen on tomato. However, the pathogen can infect other solanaceous plants such as pepper, aubergine (eggplant), *Nicotiana glutinosa*, and the weeds black nightshade (*Solanum nigrum*), perennial nightshade (*S. douglasii*), and cutleaf nightshade (*S. triflorum*).

Disease cycle

Primary inoculum can come from many sources and includes infested plant debris in soil, infected volunteer tomato plants, infected weed hosts, contaminated wood stakes, diseased transplants, and infested seed. Depending on weather conditions, infested tomato leaves and stems that are not buried below ground allow the pathogen to survive for long periods of time; such crop residues can be sources of initial inoculum for new tomato plantings.

Infested seed is a particularly important inoculum source because bacteria can spread and disease can develop when plants are grown under greenhouse conditions. A 1% seed transmission rate is sufficient to give 100% disease. The practices of clipping or mowing transplants and using overhead sprinkler irrigation can significantly spread the pathogen, though young, diseased transplants may show few or no symptoms of bacterial canker until later in their growth. During transplanting, the diseased or contaminated plants result in bacterial spread to equipment, workers' hands, and previously clean transplants.

Once bacterial canker is established on plants in the field, the pathogen can be spread plant-to-plant via splashing water (rain, sprinkler irrigation), contaminated tools and implements, and workers' hands. Bacteria enter the tomato host via stomates, small wounds, and broken trichomes. After entering through these openings, the pathogen can become systemic and will colonize and move through the xylem tissue. Disease development is favored by warm conditions (24–32° C) and factors that create succulent growth of the tomato.

Control

Monitor tomato seed fields regularly so that bacterial canker problems can be detected at an early stage and managed appropriately. Obtain and plant high-quality seed that does not have detectable, economically important levels of *C. michiganensis* subsp. *michiganensis*. Use a hot water seed treatment or treat seed with hydrochloric acid, calcium hypochlorite, or other recommended materials. Hot water treatments can reduce seed viability and germination percentages. Seed health testing and certification programs help regulate the availability and cleanliness of such seed. Such seed tests usually involve the washing of a 10,000 seed sample and subsequent plating of the liquid onto semi-selective medium. Discard heavily infested seed. Use resistant cultivars if such are available.

Though young transplants may not exhibit bacterial canker symptoms, inspect transplants and remove suspect plants and surrounding transplant trays. Sanitize benches that hold transplants, transplant trays, and equipment that comes into contact with plants. Roguing of affected plants may be worthwhile if disease incidence is low. Lower the water pressure in irrigation equipment to reduce damage to leaves. Consider applying preventative fungicides (copper-based materials) for protecting transplants. Realize that the practice of mowing transplants to regulate transplant height can readily spread the pathogen. Instead, implement growing practices that involve air movement or differential temperature management.

Minimize damage to plants during transplanting, transplant only when foliage is dry, and periodically sanitize transplanting equipment. Avoid using overhead sprinkler irrigation in the field. In the field, use new wooden stakes or stakes that have been steamed or treated with chlorine. With an appropriate disinfectant, periodically and regularly sanitize tools such as clippers and pruning shears. Do not allow equipment or workers to pass through fields when foliage is wet. Copper sprays may provide some disease control. Once the tomato crop is finished, incorporate the crop residues to enhance plant decomposition and the dissipation of bacteria. Rotate to a non-host crop before returning to tomato and do not allow volunteer tomato or weed hosts to survive.

References

Chang, R. J., Ries, S. M., and Pataky, J. K. 1992. Effects of temperature, plant age, inoculum concentration, and cultivar on the incubation period and severity of bacterial canker of tomato. *Plant Disease* 76:1150–1155.

Chang, R. J., Ries, S. M., and Pataky, J. K. 1991. Dissemination of *Clavibacter michiganensis* subsp. *michiganensis* by practices used to produce tomato transplants. *Phytopathology* 81:1276–1281.

Chang, R. J., Ries, S. M., and Pataky, J. K. 1992. Local sources of *Clavibacter michiganensis* ssp. *michiganensis* in the development of bacterial canker on tomatoes. *Phytopathology* 82:553–560.

Fatmi, M. and Schaad, N. W. 1988. Semiselective agar medium for isolation of *Clavibacter michiganense* subsp. *michiganense* from tomato seed. *Phytopathology* 78:121–126.

Fatmi, M., Schaad, N. W., and Bolkan, H. A. 1991. Seed treatments for eradicating *Clavibacter michiganensis* subsp. *michiganensis* from naturally infected tomato seeds. *Plant Disease* 75:383–385.

Forster, R. L. and Echandi, E. 1973. Relation of age of plants, temperature, and inoculation concentration to bacterial canker development in resistant and susceptible *Lycopersicon* spp. *Phytopathology* 63:773–777.

Gitaitis, R., McCarter, S., and Jones, J. 1992. Disease control of tomato transplants produced in Georgia and Florida. *Plant Disease* 76:651–656.

Gitaitis, R. D., Beaver, R. W., and Voloudakis, A. E. 1991. Detection of *Clavibacter michiganensis* subsp. *michiganensis* in symptomless tomato transplants. *Plant Disease* 75:834–838.

Gleason, M. L., Gitaitis, R. D., and Ricker, M. D. 1993. Recent progress in understanding and controlling bacterial canker of tomato in Eastern North America. *Plant Disease* 77:1069–1076.

Gleason, M. L., Braun, E. J., Carlton, W. M., and Peterson, R. H. 1991. Survival and dissemination of *Clavibacter michiganensis* subsp. *michiganensis* in tomatoes. *Phytopathology* 81:1519–1523.

Hausbeck, M. K., Bell, J., Medina-Mora, C., Podolsky, R., and Fulbright, D. W. 2000. Effect of bactericides on population sizes and spread of *Clavibacter michiganensis* subsp. *michiganensis* on tomatoes in the greenhouse and on disease development and crop yield in the field. *Phytopathology* 90:38–44.

Thompson, E., Leary, J. V., and Chun, W. W. C. 1989. Specific detection of *Clavibacter michiganensis* subsp. *michiganensis* by homologous DNA probe. *Phytopathology* 79:311–314.

Fatmia, M., and Schaad, N. W. 2002. Survival of *Clavibacter michiganensis* ssp. *michiganensis* in infected tomato stems under natural field conditions in California, Ohio and Morocco. *Plant Pathology* 51:149–154.

Pseudomonas syringae pv. *tomato*

BACTERIAL SPECK

Introduction and significance

Bacterial speck is a relatively minor disease of tomato that can cause economic damage if environmental conditions favor severe infections. The disease occurs worldwide.

Symptoms and diagnostic features

Foliar symptoms are dark brown to black spots on leaves and stems (**443**). Leaf spots can be circular to angular in shape, and individual spots are generally smaller than 5 mm in diameter. Spots are visible from both top and bottom sides of the infected leaf and are often surrounded by a yellow halo. Leaf spots can coalesce and result in the necrosis of large areas of the leaf. Stems and petioles also develop dark brown to black spots that are irregular in shape but tend to be slightly elongated along the axis of the stem (**444, 445**). In severe cases, bacterial speck can result in stunted growth, delayed crop maturity, and reduced yields. Symptoms on green fruit consist of small (less than 3 mm in diameter), slightly raised, superficial black lesions or specks (**446** also **451**, page 333). Such specks usually are not surrounded by halos.

Causal agent

Bacterial speck is caused by the aerobic, Gram-negative bacterium *Pseudomonas syringae* pv. *tomato*. The pathogen can be isolated on standard microbiological media and produces cream-colored colonies typical of most pseudomonads. When cultured on Kings medium B, it produces a diffusible pigment that fluoresces blue under ultraviolet light. Strains of this pathogen can be host specific to tomato, while other strains can infect both tomato and pepper. This pathogen is seedborne, which results in infection of tomato seedlings. Two distinct races (races 0 and 1) have been documented.

Disease cycle

Primary inoculum can come from infested seed, soilborne plant debris, or reservoir plant hosts. Infested seed is a particularly important inoculum source because bacteria can spread and disease can develop when plants are grown under greenhouse conditions. The practice of using overhead sprinkler irrigation can significantly spread the pathogen. Once diseased transplants are in the field, the pathogen can spread plant-to-plant via splashing water (rain, sprinkler irrigation), contaminated tools and implements, and workers' hands. Disease development is favored by high humidity (greater than 80% relative humidity), free moisture, and cool temperatures (18–24° C). Seedlings are particularly susceptible. The pathogen is not a soilborne organism, but it can survive in the soil on infested plant residues for up to 30 weeks. The pathogen can overwinter on volunteer tomato plants; weeds can also support pathogen populations on root and leaf surfaces.

443 Leaf spots caused by bacterial speck of tomato.

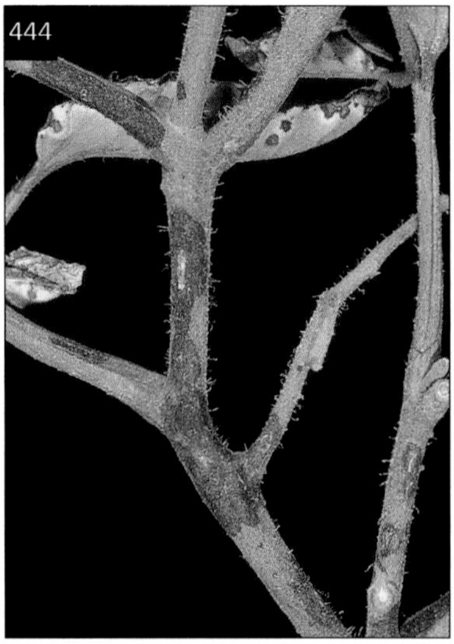

444 Stem lesions on field-grown tomato caused by bacterial speck.

445 Stem lesions on tomato transplants caused by bacterial speck.

446 Fruit lesions caused by bacterial speck of tomato.

Control

Regularly monitor tomato seed fields so that early outbreaks can be controlled. Obtain and plant high-quality seed that does not have detectable, economically important levels of *P. syringae* pv. *tomato*. Use a hot water seed treatment or treat seed with hydrochloric acid, calcium hypochlorite, or other recommended materials. Hot water treatments can reduce seed viability and germination percentages. Seed health testing and certification programs help regulate the availability and cleanliness of such seed. Such seed tests usually involve the washing of a 10,000 seed sample and subsequent plating of the liquid onto semi-selective medium. Discard heavily infested seed. Use resistant cultivars if such become available.

Inspect transplants and remove symptomatic plants and surrounding transplant trays. Sanitize benches that hold transplants, transplant trays, and equipment that comes into contact with plants. Consider applying preventative spray applications (copper-based materials combined with maneb fungicides, newer chemistry such as acibenzolar-S-methyl) for protecting transplants. Avoid using overhead sprinkler irrigation in the field. With an appropriate disinfectant, periodically and regularly sanitize tools such as clippers and pruning shears. Do not allow equipment or workers to pass through fields when foliage is wet. Copper+mancozeb sprays provide some control. Once the tomato crop is finished, incorporate the crop residues to enhance plant decomposition and the dissipation of bacteria. Rotate to a non-host crop before returning to tomato, and do not allow volunteer tomato or weed hosts to survive.

References

Bashan, Y. 1986. Field dispersal of *Pseudomonas syringae* pv. *tomato*, *Xanthomonas campestris* pv. *vesicatoria*, and *Alternaria macrospora* by animals, people, birds, insects, mites, agricultural tools, aircraft, soil particles, and water sources. *Canadian Journal of Botany* 64:276–281.

Bashan, Y., Okon, Y., and Henis, Y. 1982. Long-term survival of *Pseudomonas syringae* pv. *tomato* and *Xanthomonas campestris* pv. *vesicatoria* in tomato and pepper seeds. *Phytopathology* 72:1143–1144.

Cuppels, D. A. and Elmhirst, J. 1999. Disease development and changes in the natural *Pseudomonas syringae* pv. *tomato* population on field tomato plants. *Plant Disease* 83:759–764.

Gitaitis, R., McCarter, S., and Jones, J. 1992. Disease control of tomato transplants produced in Georgia and Florida. *Plant Disease* 76:651–656.

Goode, M. J. and Sasser, M. 1980. Prevention – the key to controlling bacterial spot and bacterial speck of tomato. *Plant Disease* 64:831–834.

McCarter, S. M., Jones, J. B., Gitaitis, R. D., and Smitley, D. R. 1983. Survival of *Pseudomonas syringae* pv. *tomato* in association with tomato seed, soil, host tissue, and epiphytic weed hosts in Georgia. *Phytopathology* 73:1393–1398.

Smitley, D. R. and McCarter, S. M. 1982. Spread of *Pseudomonas syringae* pv. *tomato* and role of epiphytic populations and environmental conditions in disease development. *Plant Disease* 66:713–717.

Voloudakis, A. E., Gitaitis, R. D., Westbrook, J. K., Phatak, S. C., and McCarter, S. M. 1991. Epiphytic survival of *Pseudomonas syringae* pv. *syringae* and *Pseudomonas syringae* pv. *tomato* on tomato transplants in Southern Georgia. *Plant Disease* 75:672–675.

Xanthomonas campestris pv. *vesicatoria*

BACTERIAL SPOT

Introduction and significance

While bacterial spot occurs throughout the world, this disease is most serious in tropical and sub-tropical tomato growing areas, such as the southeastern USA and Brazil, where both high humidity and rainfall are present.

Symptoms and diagnostic features

Initial foliar symptoms consist of circular to irregularly shaped, water-soaked spots on leaves (447). These spots later turn dark brown and usually remain smaller than 5 mm in diameter. As disease progresses, the spots may coalesce and result in leaves having significant necrotic areas (448). Severely infected seedlings can become defoliated. While infected leaves may show some chlorosis, the individual spots usually are not surrounded by yellow halos. Dark streaks may develop on petioles and stems. In severe cases, bacterial spot can result in stunted growth, a scorched appearance to the foliage, defoliation, and reduced yields. Early symptoms on green fruit consist of small (less than 3 mm in diameter), raised blisters (449, 450, 451). These diseased spots enlarge into brown, rough scabs that measure 5–8 mm in diameter.

Causal agent

Bacterial spot is caused by the aerobic, Gram-negative bacterium *Xanthomonas campestris* pv. *vesicatoria*. The pathogen can be isolated on standard microbiological media and produces yellow, mucoid, slow-growing colonies typical of most xanthomonads. However, *X. campestris* pv. *vesicatoria* only weakly hydrolyzes starch, so starch-based semi-selective media such as SX and MXP media are not diagnostic for isolating and identifying this pathogen. Tween medium is useful, because this bacterium forms characteristic white calcium salt crystals when growing on it. This pathogen is seedborne, which results in infection of tomato seedlings.

Xanthomonad pathogens from tomato and pepper hosts are a complex group of organisms. At least three tomato strains of this pathogen are host specific to tomato and are designated as T1, T2, and T3. Other strains are host specific to pepper. Finally, other strains

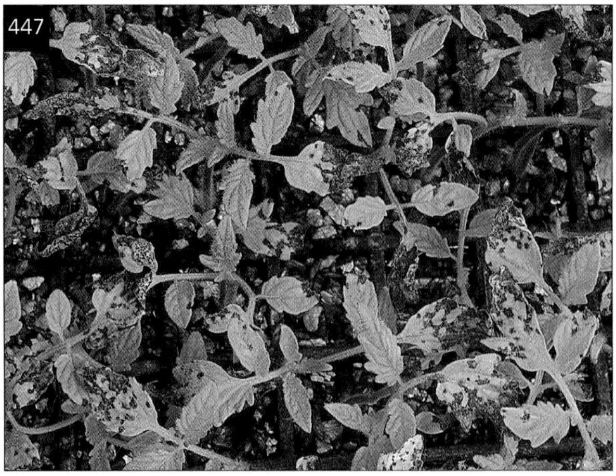

447 Leaf spots due to bacterial spot on tomato transplants.

448 Leaf spots of bacterial spot on mature tomato leaves.

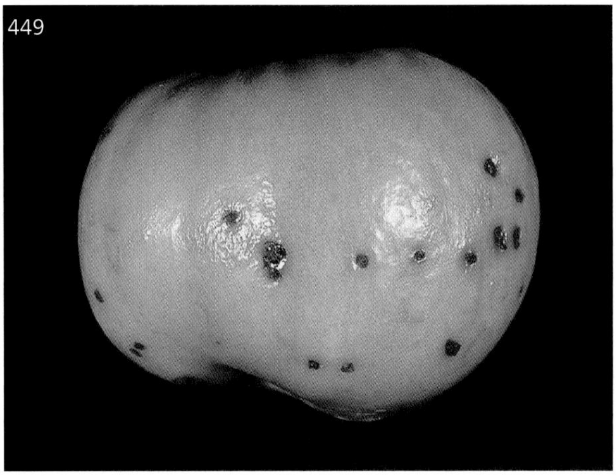

449 Fruit lesions of bacterial spot of tomato.

are pathogenic on both of these hosts; these pepper–tomato strains are designated as PT strains. Researchers further find that xanthomonad pathogens from tomato and pepper can be divided into various groups (A, B, C, and D) based on genetic and biochemical parameters. The assignment of these groups to various species and pathovars is still being debated.

Disease cycle

Primary inoculum can come from infested seed, soilborne plant debris, or reservoir plant hosts. Infested seed is a particularly important inoculum source because bacteria can spread and disease can develop when plants are grown under greenhouse conditions. The practice of using overhead sprinkler irrigation can significantly spread the pathogen. Once diseased transplants are in the field, the pathogen can be spread plant-to-plant via splashing water (rain, sprinkler irrigation), contaminated tools and implements, and workers' hands. Disease development is favored by high humidity and warm temperatures in the 24–30° C range. The pathogen is not a soilborne organism, but it can survive in the soil on infested plant residues. The pathogen can also overwinter on volunteer tomato plants and on weeds such as black nightshade (*Solanum nigrum*) and ground cherry (*Physalis minima*).

Control

Use seed that does not have detectable, economically important levels of the pathogen. Regularly monitor tomato seed fields so that early outbreaks can be controlled. Use a hot water seed treatment or treat seed with hydrochloric acid, calcium hypochlorite, or other recommended materials. Hot water treatments can reduce seed viability and germination percentages. Seed health testing and certification programs help regulate the availability and cleanliness of such seed. Such seed tests usually involve the washing of a 10,000 seed sample and subsequent plating of the liquid onto semi-selective medium. Discard heavily infested seed.

Inspect transplants and remove symptomatic plants and surrounding transplant trays. Sanitize benches that hold transplants, transplant trays, and equipment that comes into contact with plants. Consider applying preventative fungicide applications (copper-based materials combined with maneb fungicides) for protecting transplants. Avoid using overhead sprinkler irrigation in the field. With an appropriate disinfectant, periodically and regularly sanitize tools such as clippers and pruning shears. Do not allow equipment or workers to pass through fields when foliage is wet. Copper+mancozeb sprays provide some disease control. Once the tomato crop is finished, incorporate the crop residues to enhance plant decomposition and the dissipation of bacteria. Rotate to a non-host crop before returning to tomato and do not allow volunteer tomato or weed hosts to survive.

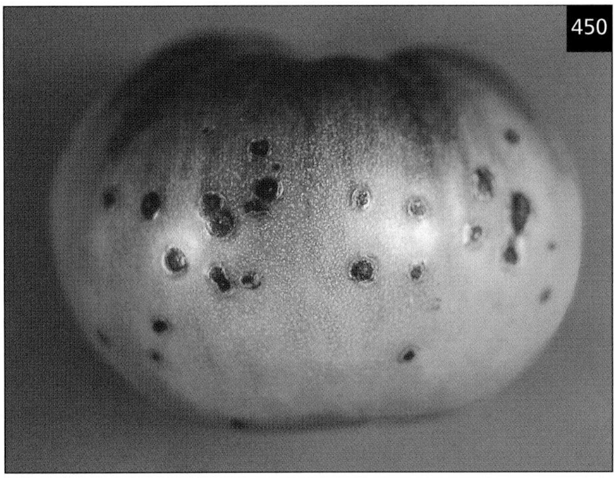

450 Bacterial spot lesions on tomato fruit.

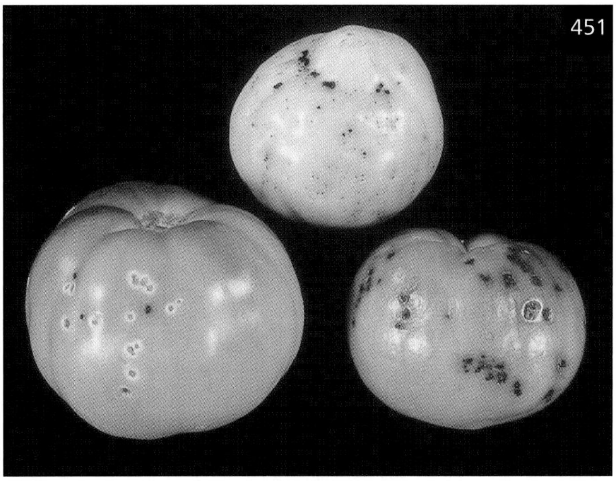

451 Comparison of bacterial spot (right), speck (top), and canker (left) symptoms on tomato fruit.

References

Bashan, Y., Okon, Y., and Henis, Y. 1982. Long-term survival of *Pseudomonas syringae* pv. *tomato* and *Xanthomonas campestris* pv. *vesicatoria* in tomato and pepper seeds. *Phytopathology* 72:1143–1144.

Bouzar, H., Jones, J. B., Stall, R. E., Hodge, N. C., Minsavage, G. V., Benedict, A. A., and Alvarez, A. M. 1994. Physiological, chemical, serological, and pathogenic analyses of a worldwide collection of *Xanthomonas campestris* pv. *vesicatoria* strains. *Phytopathology* 84:663–671.

Gitaitis, R., McCarter, S., and Jones, J. 1992. Disease control of tomato transplants produced in Georgia and Florida. *Plant Disease* 76:651–656.

Goode, M. J. and Sasser, M. 1980. Prevention – the key to controlling bacterial spot and bacterial speck of tomato. *Plant Disease* 64:831–834.

Jones, J. B., Bouzar, H., Somodi, G. C., Stall, R. E., Pernezny, K., El-Morsy, G., and Scott, J. W. 1998. Evidence for the preemptive nature of tomato race 3 of *Xanthomonas campestris* pv. *vesciatoria* in Florida. *Phytopathology* 88:33–38.

Jones, J. B., Pohronezny, K. L., Stall, R. E., and Jones, J. P. 1986. Survival of *Xanthomonas campestris* pv. *vesicatoria* in Florida on tomato crop residue, weeds, seeds, and volunteer tomato plants. *Phytopathology* 76:430–434.

Jones, J. B., Stall, R. E., and Bouzar, H. 1998. Diversity among Xanthomonads pathogenic on pepper and tomato. *Annual Review of Phytopathology* 36:41–58.

Jones, J. B., Stall, R. E., Scott, J. W., Somodi, G. C., Bouzar, H., and Hodge, N. C. 1995. A third tomato race of *Xanthomonas campestris* pv. *vesicatoria*. *Plant Disease* 79:395–398.

McGuire, R. G., Jones, J. B., and Sasser, M. 1986. Tween media for semiselective isolation of *Xanthomonas campestris* pv. *vesicatoria* from soil and plant material. *Plant Disease* 70:887–891.

McGuire, R. G., Jones, J. B., Stanley, C. D., and Csizinszky, A. A. 1991. Epiphytic populations of *Xanthomonas campestris* pv. *vesicatoria* and bacterial spot of tomato as influenced by nitrogen and potassium fertilizer. *Phytopathology* 81:656–660.

Obradovic, A., Mavridis, A., Rudolph, K., Janse, J. D., Arsenijevic, M., Jones, J. B., Minsavage, G. V., Wang, J.-F. 2004. Characterization and PCR-based typing of *Xanthomonas campestris* pv. *vesicatoria* from peppers and tomatoes in Serbia. *European Journal of Plant Pathology* 110:285–292.

Scott, J. W. and Jones, J. B. 1986. Sources of resistance to bacterial spot in tomato. *HortScience* 21:304–306.

Sijam, K., Chang, C. J., and Gitaitis, R. D. 1992. A medium for differentiating tomato and pepper strains of *Xanthomonas campestris* pv. *vesicatoria*. *Canadian Journal of Plant Pathology* 14:182–184.

Stahl, R. E., Beaulieu, C., Egel, D., Hodge, N. C., Leite, R. P., Minsavage, G. V., Bouzar, H., Jones, J. B., Alvarez, A. M., and Benedict, A. A. 1994. Two genetically diverse groups of strains are included in *Xanthomonas campestris* pv. *vesicatoria*. *International Journal of Systematic Bacteriology* 44:47–53.

Pseudomonas corrugata

TOMATO PITH NECROSIS

Introduction and significance

Tomato pith necrosis has been reported in the USA, UK, and other countries. This disease is widespread in parts of South America. Pith necrosis affects older plants and symptoms usually do not show until fruit begin to develop.

Symptoms and diagnostic feature

Early symptoms include wilting of young foliage and chlorosis and wilting of older leaves. Affected leaves can curl up and turn brown on their edges. Dark brown to black lesions develop on the surfaces of lower stems. Internally, such stems have pith tissue that contains cavities, is darkly discolored, and can become hollow (**452, 453**). Adventitious roots may grow profusely from these symptomatic stems. Vascular tissue may be brown. In advanced stages the internal pith symptom may progress high up into the plant, and severely affected plants may collapse and die. Overall, tomato pith necrosis symptoms may resemble those of bacterial canker.

Causal agents

Tomato pith necrosis is caused by *Pseudomonas corrugata*. The pathogen is an aerobic, Gram-negative bacterium that can be isolated on standard microbiological media and produces buff-colored, raised, wrinkled colonies, with or without green centers, on nutrient dextrose agar. On King's medium B, this bacterium produces a yellow-green, nonfluorescent, diffusible pigment. Though considered mostly a pathogen of tomato, *P. corrugata* can cause small brown lesions on alfalfa.

Recent discoveries show that pith necrosis can also be caused by other species such as *P. viridiflava*, *P. fluorescens* biotype I, and *Erwinia carotovora* pathogens. This indicates that the cause or etiology of pith necrosis may be complex. *Pseudomonas corrugata* is reported to cause pith necrosis on pepper, as well.

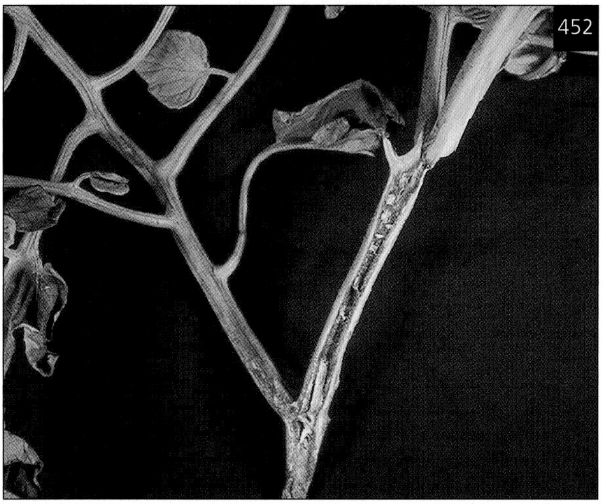

452 Discolored central pith tissue of stem affected by *Pseudomonas corrugata*.

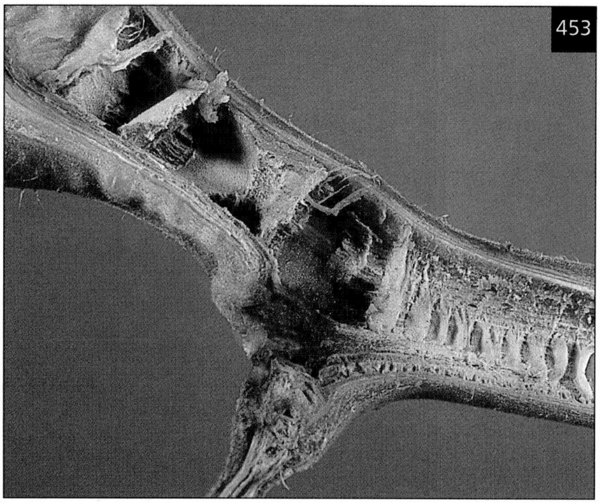

453 Pith necrosis of tomato showing laddering of central pith in stems.

Disease cycle

Tomato pith necrosis appears to develop when there are low night temperatures, high nitrogen levels, and high humidity. Disease is expressed when the first fruit set is near the mature green stage. Inoculum is possibly seedborne, but disease epidemiology is poorly understood.

Control

Few control recommendations have been established. Avoid using excessively high nitrogen fertilizer rates. Do not use overhead sprinkler irrigation in the field. With an appropriate disinfectant, periodically and regularly sanitize tools such as clippers and pruning shears. Do not allow equipment or workers to pass through fields when foliage is wet. Once the tomato crop is finished, incorporate the crop residues to enhance plant decomposition and the dissipation of bacteria. Rotate to a non-host crop before returning to tomato and do not allow volunteer tomato or weed hosts to survive.

References

Alippi, A. M., Dal Bo, E., Ronco, L. B., López, M. V., López, A. C., and Aguilar, O. M. 2003. *Pseudomonas* populations causing pith necrosis of tomato and pepper in Argentina are highly diverse. *Plant Pathology* 52:287–302.

Jones, J. B., Jones, J. P., Stall, R. E., and Miller, J. W. 1983. Occurrence of stem necrosis on field-grown tomatoes incited by *Pseudomonas corrugata* in Florida. *Plant Disease* 67:425–426.

Lai, M., Opgenorth, D. C., and White, J. B. 1983. Occurrence of *Pseudomonas corrugata* on tomato in California. *Plant Disease* 67:110–112.

Lukezic, F. L. 1979. *Pseudomonas corrugata*, a pathogen of tomato, isolated from symptomless alfalfa roots. *Phytopathology* 69:27–31.

Scarlett, C. M., Fletcher, J. T., Roberts, P., and Lelliott, R. A. 1978. Tomato pith necrosis caused by *Pseudomonas corrugata* n. sp. *Annals of Applied Biology* 88:105–114.

Alternaria alternata

BLACK MOLD

Introduction and significance

Black mold disease affects only ripe tomato fruit. Significant fruit losses can occur if environmental conditions favor disease development.

Symptoms and diagnostic features

On ripe tomato fruit the initial symptoms consist of irregularly shaped flecks and stains on the cuticle. The affected areas are small and tan to brown. With suitable environmental conditions, the small infected areas expand into large, sunken, circular to oval-shaped lesions that can extend deep into the fruit (**454**). The black, velvety growth of the pathogen covers the surface of the lesions (**455**). Lesions break down and the fruit will rot. The large infected areas can also support the growth of other decay fungi such as *Stemphylium* species, which are considered by some researchers to be weak tomato fruit pathogens, and *Cladosporium* and *Aspergillus* species, which are saprobes. Postharvest spread of the disease can occur if infected fruit are stored for long periods.

454 Fruit lesions caused by black mold of tomato in the field.

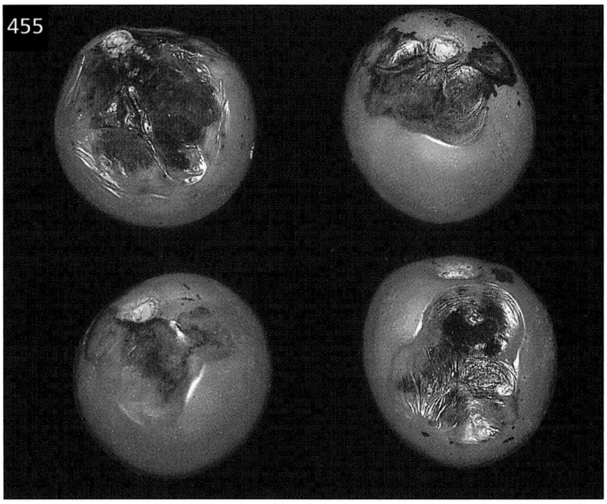

455 Fruit lesions caused by black mold of tomato.

Causal agent

Black mold is caused by the fungus *Alternaria alternata*. The pathogen can be isolated on standard microbiological media. On potato dextrose agar, colonies are usually dark green to black and have limited aerial mycelium. Conidia are obclavate, ovoid, or ellipsoidal in shape and often have a short, conical beak that does not exceed one-third of the length of the spore. Conidia are pale to gold brown, smooth or verruculose, measure 20–63 x 9–18 µm, and have three to five cross septa and occasional longitudinal septa. Conidia are produced in long chains that sometimes branch. In contrast to *A. alternata* f. sp. *lycopersici*, the causal agent of Alternaria stem canker, the black mold *A. alternata* is not host specific to tomato. This pathogen is an active saprobe and can colonize damaged tissues of other plants and readily grows on dead organic matter.

Disease cycle

Alternaria alternata is commonly found on decaying plant material in and around fields, including senescent and dead tomato leaves of the current crop. Conidia are blown onto ripe fruit via winds and splashing water. If fruit are damaged or wet from dew, rains, and irrigations, conidia germinate and infect the fruit. Fruit damaged by sunburn or blossom end rot (a calcium deficiency) are particularly susceptible to colonization by *A. alternata*. Disease development is most rapid at 24–28° C.

Control

Harvest fruit in a timely manner so that ripe fruit are not left in the field longer than necessary. Do not irrigate with overhead sprinklers. Apply protectant fungicides 4 to 6 weeks before harvest.

References

Bartz, J. A. 1971. Studies on the causal agent of black fungal lesions on stored tomato fruit. *Proceedings of the Florida State Horticultural Society* 84:117–119.

Davis, R. M., Miyao, E. M., Mullen, R. J., Valencia, J., May, D. M., and Gwynne, B. J. 1997. Benefits of applications of chlorothalonil for the control of black mold of tomato. *Plant Disease* 81:601–603.

Morris, P. F., Connolly, M. S., and St. Clair, D. A. 2000. Genetic diversity of *Alternaria alternata* isolated from tomato in California assessed using RAPDs. *Mycological Research* 104:286–292.

Pearson, R. C. and Hall, D. H. 1975. Factors affecting the occurrence and severity of blackmold of ripe tomato fruit caused by *Alternaria alternata*. *Phytopathology* 65:1352–1359.

Simmons, E. G. 1999. *Alternaria* themes and variations (236–243): host specific toxin producers. *Mycotaxon* 70:325–369.

Simmons, E. G. 2000. *Alternaria* themes and variations (244–286) species on Solanaceae. *Mycotaxon* 75:1–115.

456 Stem cankers on tomato caused by *Alternaria alternata* f. sp. *lycopersici*.

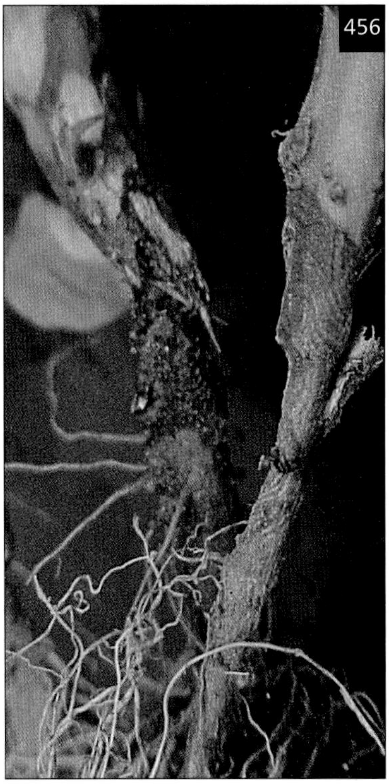

Alternaria alternata f. sp. *lycopersici*

ALTERNARIA STEM CANKER

Introduction and significance

Alternaria stem canker is primarily a problem on tomatoes grown in coastal regions in California. The disease is of limited distribution.

Symptoms and diagnostic features

Large, irregularly shaped, dark brown to almost black cankers form on stems. Cankers characteristically contain light and dark concentric zonation within their borders. Cankers continue to enlarge as the plant develops, resulting in girdled stems (**456**) and the death of the stems and even the entire plant. Vascular and pith tissue beneath the cankers can show brown streaking and become dry and cracked. Infected leaves develop irregularly shaped, dark brown to black areas that are primarily interveinal. Affected plants may be stunted. Fruit symptoms always begin on green, unripe fruit and appear as brown, circular to oval, sunken lesions that also have concentric rings in them (**457**). This green fruit disease is distinct from blackmold disease which is caused by a different *Alternaria* pathogen that affects only ripe fruit.

Causal agent

The causal agent is *Alternaria alternata* f. sp. *lycopersici*. The pathogen can be isolated on standard microbiological media. On potato dextrose agar, colonies are initially whitish and cottony. With age, colonies darken and become gray to almost black and show appressed growth. Conidia are light olive-brown to dark brown and are dictyospores having three to five cross septa, occasional longitudinal septa, and short-beaked apical

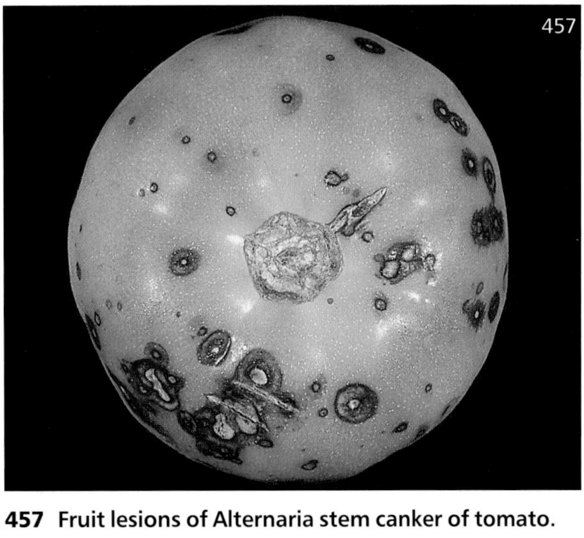

457 Fruit lesions of Alternaria stem canker of tomato.

cells on the terminal ends. Conidia are produced in chains of three to five and measure 18–50 x 7–18 µm. The fungus is pathogenic on tomato only and produces a host-specific toxin (AAL-toxin, analogs TA and TB) that moves systemically from stem cankers to leaves, causing the interveinal necrotic lesions.

Disease cycle

The pathogen can survive in the soil for up to a year on infested tomato residue. Infection takes place when plants come in contact with infested residue or when conidia are spread, via winds, to host tomato plants. Cankers also occur around wounds such as pruning cuts on pole tomato. Spore germination and infection requires free moisture on the host tissue. Optimum temperature for disease development is 25° C.

Control

Plant resistant cultivars, as this will virtually eliminate serious disease outbreaks. Resistance to this pathogen is achieved by a single gene (Asc) with two alleles, which confers complete protection against this pathogen. If using susceptible cultivars, avoid irrigating with overhead sprinklers. Rotate crops so that tomato does not immediately follow a previous tomato crop. Use effective fungicides if available.

References

Akamatsu, H., Itoh, Y., Kodama, M., Otani, H., and Kohmoto, K. 1997. AAL-toxin-deficient mutants of *Alternaria alternata* tomato pathotype by restriction enzyme-mediated integration. *Phytopathology* 87:967–972.

Clouse, S. D. and Gilchrist, D. G. 1987. Interaction of the asc locus in F8 paired lines of tomato with *Alternaria alternata* f. sp. *lycopersici* and AAL-toxin. *Phytopathology* 77:80–82.

Fuson, G. B. and Pratt, D. 1988. Effects of the host-selective toxins of *Alternaria alternata* f. sp. *lycopersici* on suspension-cultured tomato cells. *Phytopathology* 78:1641–1648.

Gilchrist, D. G. and Grogan, R. G. 1976. Production and nature of a host-specific toxin from *Alternaria alternata* f. sp. *lycopersici*. *Phytopathology* 66:165–171.

Grogan, R. G., Kimble, K. A., and Misaghi, I. 1975. A stem canker disease of tomato caused by *Alternaria alternata* f. sp. *lycopersici*. *Phytopathology* 65:880–886.

Alternaria solani (= Alternaria tomatophila)

EARLY BLIGHT

Introduction and significance

Early blight occurs on tomato throughout the world and can be an important disease.

Symptoms and diagnostic features

The disease begins as small, brown to black, circular lesions on mature leaves (**458**). Such lesions may be surrounded by chlorotic tissue. As disease develops, the spots enlarge and can be 8–10 mm or larger in diameter. At this later stage the leaf spots contain characteristic concentric rings. With severe infections, plants can become defoliated and the exposed fruit are subject to sunburn damage. Infections on stems initially consist of small, brown, sunken lesions. Stem lesions expand and become elongated or oval lesions with concentric rings. Stem lesions can eventually girdle the stem and result in stem or plant death. Seedlings can show stem lesions at the soil level when they are less than 3 weeks old. Fruit infections, at either the green or ripe stage, consist of sunken, dark brown to black, circular spots that also contain concentric rings. Early blight green fruit and stem symptoms may resemble fruit and stem infections caused by the Alternaria stem canker pathogen.

458

458 Leaf spots of early blight of tomato.

Causal agent

Early blight is caused by the fungus *Alternaria solani*. The pathogen can be isolated on standard microbiological media. Cultures appear gray-brown in color and usually produce a yellow or red diffusable pigment. Sporulation in culture is enhanced by exposing colonies to light. Conidia are olivaceous brown to dark brown and are obclavate to obpyriform, having seven to eight cross septa, and occasional longitudinal septa. Conidia usually have long-beaked apical cells that can be the same as or exceed the length of the spore body. Conidia are produced singly or in chains of two and measure 150–300 x 15–19 µm. *Alternaria solani* produces resilient structures called chlamydospores, which enable the pathogen to survive in soil for a period of time. This pathogen is seedborne in tomato and can result in diseased seedlings or transplants. Affected seedlings may develop collar rot in which the base of the plant is girdled due to a stem canker. Plants with collar rot can become stunted, wilted, and dead. This pathogen also affects other solanaceous host such as potato and aubergine (eggplant).

Reports indicate that the early blight *Alternaria* pathogens on tomato and potato, historically both designated as *A. solani*, may be distinct species. Morphologically and genetically the early blight isolates from tomato are distinct from those from potato. Tomato isolates are more virulent on tomato, and the potato isolates are more aggressive on potato. The tomato isolates apparently do not produce spores very well in culture, while the potato isolates sporulate extensively. The tomato early blight pathogen is proposed to be *A. tomatophila*, while the potato pathogen retains the *A. solani* name.

Disease cycle

Initial inoculum of *A. solani* is present on infested tomato crop debris and on tomato seed. In addition, the pathogen can overwinter on volunteer tomato plants and on other solanaceous plants such as aubergine (eggplant), potato, and the weeds black nightshade (*Solanum nigrum*) and horsenettle (*S. carolinense*). Conidia are blown onto plants via winds and splashing water. Wet, mild conditions (with a temperature range of 24–29° C) favor disease development. In Europe, early blight is most important when summer temperatures are high.

Control

Select and plant resistant cultivars as available. Use seed that does not have significant levels of the pathogen. Treat infested seed lots with hot water or fungicides. Inspect and discard infected, symptomatic transplants. Multiple applications of fungicides are usually required to control this disease. Some forecasting programs, such as TOMCAST, are used to assist with fungicide timing for early blight control. Practice crop rotation so that the selected field did not have tomato in the previous year. Remove volunteer tomato and host weed species.

References

Brammall, R. A. 1993. Effect of foliar fungicide treatment on early blight and yield of fresh market tomato in Ontario. *Plant Disease* 77:484–488.

Frazer, J. T. and Zitter, T. A. 2003. Two species of *Alternaria* cause early blight of potato (*Solanum tuberosum*) and tomato (*Lycopersicon esculentum*). *Phytopathology* 93:S27.

Gleason, M. L., MacNab, A. A., Pitblado, R. E., Ricker, M. D., East, D. A., and Latin, R. X. 1995. Disease early warning systems for processing tomatoes in eastern North America: are we there yet? *Plant Disease* 79:113–121.

Gleason, M. L., Parker, S. K., Pitblado, R. E., Latin, R. X., Speranzini, D., Hazzard, R. V., Maletta, M. J., Cowgill, W. P., and Biederstedt, D. L. 1997. Validation of a commercial system for remote estimation of wetness duration. *Plant Disease* 81:825–829.

Miller, D. J., Coffman, C. B., Teasdale, J. R., Everts, K. L., Abdul-Baki, A. A., Lydon, J., and Anderson, J. D. 2002. Foliar disease in fresh-market tomato grown in differing bed strategies and fungicide spray programs. *Plant Disease* 86:955–959.

Patterson, C. L. 1991. Importance of chlamydospores as primary inoculum for *Alternaria solani*, incitant of collar rot and early blight of tomato. *Plant Disease* 75:274–278.

Pennypacker, S. P., Madden, L., and MacNab, A. A. 1983. Validation of an early blight forcasting system for tomatoes. *Plant Disease* 67:287–289.

Simmons, E. G. 2000. *Alternaria* themes and variations (244–286): species on Solanaceae. *Mycotaxon* 75:1–115.

Vakalounakis, D. J. 1991. Control of early blight of greenhouse tomato, caused by *Alternaria solani*, by inhibiting sporulation with ultraviolet-absorbing vinyl film. *Plant Disease* 75:795–797.

Weir, T. L., Huff, D. R., Christ, B. J., and Romaine, C. P. 1998. RAPD-PCR analysis of genetic variation among isolates of *Alternaria solani* and *Alternaria alternata* from potato and tomato. *Mycologia* 90:813–821.

Athelia rolfsii (anamorph = *Sclerotium rolfsii*)

SOUTHERN BLIGHT

Introduction and significance

Southern blight, or Sclerotium stem rot, occurs on a large number of vegetable and ornamental plants especially in warmer regions of the world. In the UK this tomato disease is not established in the field though southern blight has been recorded on ornamentals and imported vegetables. This disease is found in many tomato growing regions of the USA.

Symptoms and diagnostic features

On tomato, the early symptoms consist of a water-soaked lesion on crown and lower stem tissue that is in contact with the soil. These infection sites turn light to dark brown and can rapidly girdle the entire crown. Crown tissue is often cracked or split. Above-ground symptoms consist of wilting and a quick collapse of all foliage (**459**). If soil moisture conditions are suitable, the pathogen will form a thick, white mycelial mat or layer on crown, lower stem, and even on the surrounding soil around the crown (**460, 461**). Small (1–2 mm in diameter), spherical, tan to light brown sclerotia form profusely on and in this white growth. Sclerotia are characterized by having an outer, differentiated, pigmented rind. Fruit in contact with infested soil can also become infected, developing sunken, water-soaked lesions that later support the typical white mycelial growth and sclerotia.

Causal agent

Southern blight is caused by *Sclerotium rolfsii,* which is an imperfect fungus in the mycelia sterilia category and produces no asexual spores (older references list the fungus as *Corticium rolfsii*). *Sclerotium rolfsii* has a broad host range and forms a basidiomycete perfect stage (*Athelia rolfsii*), though it is unknown whether this stage is involved in the disease.

Disease cycle

Because of its resilient sclerotia, the pathogen can survive in the soil and in crop debris for many years. The fungus is favored by high temperatures above 30° C. In the soil, sclerotia near host tissues will germinate, form infective mycelium, and directly penetrate root and crown tissues of tomato.

459 Wilting of tomato plants caused by southern blight disease.

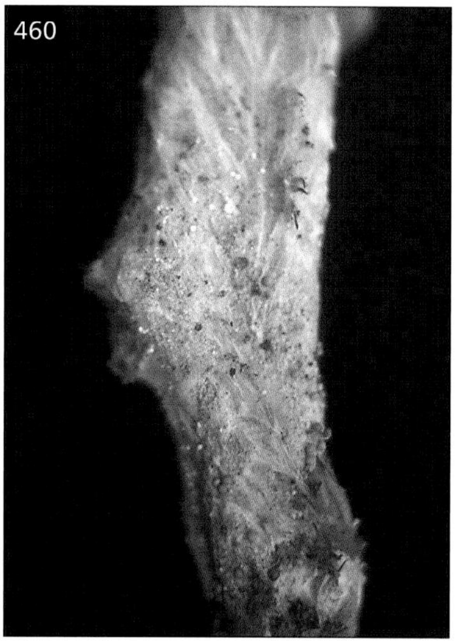

460 Tomato stem colonized by the southern blight pathogen.

461 White mycelial fan of southern blight of tomato.

Control

Rotate with non-host plants so that soil inoculum levels are reduced. Deep plowing of fields prior to planting, which inverts the soil profile, may help reduce inoculum levels. Pre-plant treatment of soil with effective fumigants will give short-term control; however, such treatments may be too expensive to implement for processing tomato crops.

References

Bulluck, L. R. and Ristaino, J. B. 2002. Effect of synthetic and organic soil fertility amendments on southern blight, soil microbial communities, and yield of processing tomatoes. *Phytopathology* 92:181–189.

Hasan, A. and Khan, M. N. 1985. The effect of *Rhizoctonia solani, Sclerotium rolfsii*, and *Verticillium dahliae* on the resistance of tomato to *Meloidogyne incognita. Nematologia Mediterranea.* 13:133–136.

McCarter, S. M., Jaworski, C. A., Johnson, A. W., and Williamson, R. E. 1976. Efficacy of soil fumigants and methods of application for controlling southern blight of tomatoes grown for transplants. *Phytopathology* 66:910–913.

Punja, Z. K. 1985. The biology, ecology, and control of *Sclerotium rolfsii. Annual Revue of Phytopathology* 23:97–127.

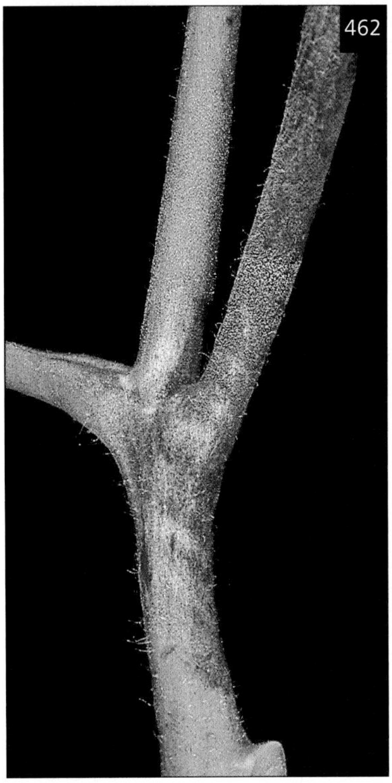

462 Stem lesion caused by gray mold on tomato.

Botrytis cinerea (teleomorph = *Botryotinia fuckeliana*)

GRAY MOLD

Introduction and significance

Gray mold is a commonly found disease of tomato and is manifested in several ways. Gray mold can be particularly damaging in greenhouse environments, due to the elevated humidity in such structures.

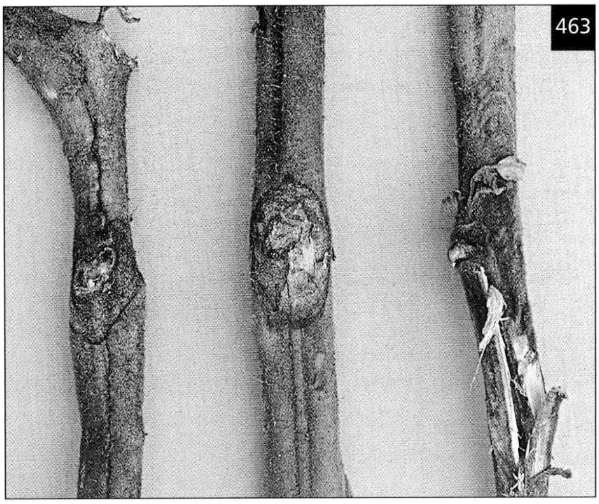

463 Stem lesion caused by gray mold of tomato, showing initiation at wounds.

Symptoms and diagnostic features

Petioles and stems become infected and develop tan to darker brown lesions (**462**). The developing lesions can eventually girdle the entire petiole or stem and show concentric rings due to the sporulation of the pathogen and coloration of the lesion. Leaves may also have brown lesions and sporulation. The gray mold pathogen often infects petioles, stems, and leaves that are damaged or senescing (**463**). Senescing petals are also subject to gray mold infection; if such petals are in contact with the developing fruit, the fungus can grow from the petal and into the fruit tissue.

Infections on green or ripe fruit result in a soft, decayed, circular rot that can eventually envelop the entire fruit. The fungus usually sporulates on the fruit calyx or in the center of the fruit lesion where the epidermis has split. A distinct fruit infection, called 'ghost spots', occurs when the pathogen invades the fruit but

464 **Numerous 'ghost spots' on a ripening tomato.**

465 **Gray sporulation of *Botrytis cinerea* on infected tomato stem.**

then dies prior to causing decay. The resulting symptom is a white to yellow ring that can range from 3–10 mm in diameter (**464**). Ghost spots do not result in fruit decay but cause the fruit to lose market quality. Tomato transplants that are damaged during the planting process can develop gray mold stem infections and die.

On greenhouse tomatoes, stem infections are the most prominent problem. Cracked or wounded stems and cut stubs remaining after fruit are harvested are primary locations for gray mold infections. Gray mold colonies that start on such wounded tissues usually progress into the main stem and cause girdling.

Causal agent

Gray mold is caused by asexual fungus *Botrytis cinerea*. The sexual stage, *Botryotinia fuckeliana*, is rarely found on the crop. Conidiophores of *B. cinerea* are long (1–2 mm), become gray-brown with maturity, and branch irregularly near the apex. Conidia are clustered at the branch tips and are single-celled, pale brown, ellipsoid to obovoid, and measure 6–18 x 4–11 μm. The pathogen can be isolated on standard microbiological media. Some isolates sporulate poorly in culture unless incubated under lights (12 h light/12 h dark). If formed, sclerotia are black, oblong or dome-shaped, and measure 4–10 mm. The fungus grows best at 18–23° C but is inhibited at warm temperatures above 32° C. On host tissue the fungus produces characteristically profuse sporulation that is dense, velvety, and grayish brown in color (**465**).

Disease cycle

Botrytis cinerea survives in and around fields as a saprophyte on crop debris, as a pathogen on numerous crops and weed plants, and as sclerotia in the soil. Conidia develop from these sources and become windborne. When conidia land on senescent or damaged tomato tissue, they will germinate if free moisture is available and rapidly colonize this food base. Once established, the pathogen will grow into adjacent healthy stems and leaves, resulting in disease symptoms and the production of additional conidia. Cool temperatures, free moisture, and high humidity favor the development of the disease. Tomato tissues that are damaged from other diseases can become colonized by *B. cinerea* that acts as a secondary decay organism.

Control

Reduce plant wetness by avoiding or reducing sprinkler irrigation. Adequately ventilate greenhouses, or heat them to reduce overall humidity. Fungicides may be useful in protecting fruit from gray mold. It is important to use a diversity of fungicides with different modes of action because *B. cinerea* commonly develops resistance to such materials. For example, resistance to benzimidazole and dicarboximide materials is prevalent and often limits their effectiveness. Dichlofluanid is useful for control of the ghost spot symptom as it is one of the few fungicides that prevents spore germination.

References

Chastagner, G. A., Ogawa, J. M., and Manji, B. T. 1978. Dispersal of conidia of *Botrytis cinerea* in tomato fields. *Phytopathology* 68:1172–1176.

Chastagner, G. A. and Ogawa, J. M. 1979. A fungicide-wax treatment to protect and suppress *Botrytis cinerea* on fresh market tomatoes. *Phytopathology* 69:59–63.

Ferrer, J. B. and Owen, J. H. 1959. *Botrytis cinerea*, the cause of ghost-spot disease of tomato. *Phytopathology* 49:411–417.

Miller, M. W. and Jeves, T. M. 1979. The persistence of benomyl tolerance in *Botrytis cinerea* in glasshouse tomato crops. *Plant Pathology* 28:119–122.

Morgan, W. M. 1984. The effect of night temperatures and glasshouse ventilation on the incidence of *Botrytis cinerea* in a late planted tomato crop. *Crop Protection* 3:243–251.

O'Neill, T. M., Shtienberg, D., and Elad, Y. 1997. Effect of some host and microclimate factors on infection of tomato stems by *Botrytis cinerea*. *Plant Disease* 81:36–40.

Sasaki, T., Honda, Y., Umekawa, M., and Nemoto, M. 1985. Control of certain diseases of greenhouse vegetables with ultraviolet-absorbing vinyl film. *Plant Disease* 69:530–533.

Shtienberg, D., Elad, Y., Niv, A., Nitzani, Y., and Kirshner, B. 1998. Significance of leaf infection by *Botrytis cinerea* in stem rotting of tomatoes grown in non-heated greenhouses. *European Journal of Plant pathology* 104:753–763.

Vakalounakis, D. J. 1992. Control of fungal diseases of greenhouse tomato under long-wave infrared-absorbing plastic film. *Plant Disease* 76:43–46.

Colletotrichum coccodes, C. gloeosporioides, C. dematium

ANTHRACNOSE

Introduction and significance

Anthracnose disease occurs in various parts of the world where tomato is grown. The disease is usually of minor importance unless environmental conditions favor development of the pathogen.

Symptoms and diagnostic features

The disease primarily affects the fruit. Young, green fruit may be infected, but disease symptoms are not expressed until fruit begin to ripen. Ripe fruit initially show small, circular, depressed lesions (**466**). Lesions can then become quite large (12–15 mm in diameter), sunken, and contain concentric rings. Lesion centers are usually tan, but become black as fungal structures (microsclerotia and acervuli) form in the tissues (**467**). If humid, wet weather occurs, the fruiting bodies in the lesions will release pink-colored spore masses. Harvested fruit infected with anthracnose will not ship or store well, and are very susceptible to secondary fruit decay organisms.

Vegetative parts of the tomato plant are also susceptible to anthracnose. Leaves develop small, circular, tan to brown spots that often are ringed with yellow halos. Roots initially show brown lesions and later rot. As root cortex tissue breaks down, the black microsclerotia of the pathogen form profusely, giving this phase of the disease the name black dot root rot. Black dot root rot is part of the brown root rot disease complex that occurs on greenhouse-grown tomato in Europe.

Causal agents

Anthracnose is caused by several species of the fungus *Colletotrichum*: *C. coccodes*, *C. gloeosporioides*, and *C. dematium*. *C. coccodes* is the species most frequently associated with the fruit disease and appears to be the only causal agent of black dot root rot. The minute (about 0.3 mm in diameter), cup-shaped acervuli fruiting bodies are usually present in fruit lesions. Acervuli release single-celled, hyaline conidia that are cylindrical with obtuse ends. Conidia measure 16–24 x 2–5 µm. Long, brown, septate setae are usually present in the acervuli. The pathogen forms small (0.2–0.4 mm), irregularly shaped survival structures called microsclerotia. *Colletotrichum coccodes* has a broad host range and can infect a number of other plants such as cucurbits, legumes, potato, and weeds.

466 Fruit lesions caused by anthracnose.

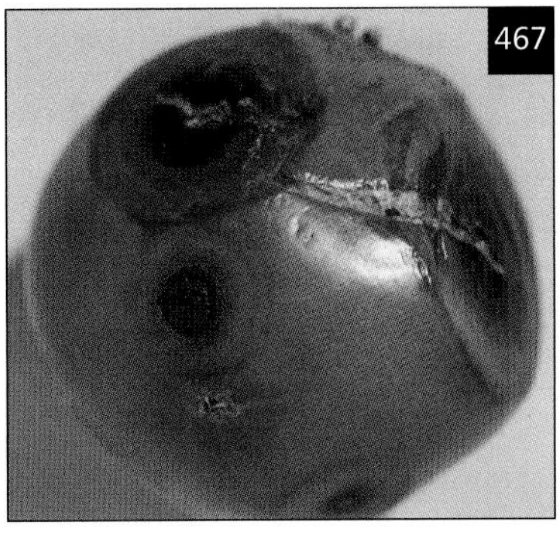

467 Dark fungal structures inside fruit lesions caused by anthracnose.

Disease cycle

The fungus survives in soil in the form of microsclerotia or as acervuli and microsclerotia on dried plant residue. The fungus can be seedborne. The pathogen is splashed from the soil onto tomato foliage and fruit and initiates infections. In addition, fruit that are in contact with the soil become infected by soilborne inoculum. Ripe fruit are particularly susceptible to infection. The root phase of anthracnose disease is often found in infested greenhouse situations due to high concentrations of inoculum and favorable conditions for disease development. Optimum temperatures for disease development are 20–24° C. Wet, humid weather favors the development of acervuli and conidia; conidia are spread by splashing water.

Control

Rotate crops so that non-hosts are grown at least every other year. Many weeds can support the pathogen, so practice good weed control. Stake plants or use mulch materials to reduce the number of fruit in contact with soil. Avoid sprinkler irrigation which spreads the conidia. Apply fungicides as necessary and use disease forecasting programs such as TOMCAST to schedule applications. Harvest fruit in a timely manner so that they are not overly ripe. Researchers are attempting to develop resistant cultivars.

References

Batson, W. E. and Roy, K. W. 1982. Species of *Colletotrichum* and *Glomerella* pathogenic to tomato fruit. *Plant Disease* 66:1153–1155.

Blakeman, J. P. and Hornby, D. 1966. The persistence of *Colletotrichum coccodes* and *Mycosphaerella ligulicola* in soil with special reference to sclerotia and conidia. *Transactions of the British Mycological Society* 49:227–240.

Byrne, J. M., Hausbeck, M. K., and Latin, R. X. 1997. Efficacy and economics of management strategies to control anthracnose fruit rot in processing tomatoes in the midwest. *Plant Disease* 81:1167–1172.

Dillard, H. R. 1989. Effect of temperature, wetness duration, and inoculum density on infection and lesion development of *Colletotrichum coccodes* on tomato fruit. *Phytopathology* 79:1063–1066.

Farley, J. D. 1976. Survival of *Colletotrichum coccodes* in soil. *Phytopathology* 66:640–641.

Gleason, M. L., MacNab, A. A., Pitblado, R. E., Ricker, M. D., East, D. A., and Latin, R. X. 1995. Disease early warning systems for processing tomatoes in eastern North America: are we there yet? *Plant Disease* 79:113–121.

Last, F. T. and Ebben, M. H. 1966. The epidemiology of tomato brown root rot. *Annals of Applied Biology* 57:95–112.

Lees, A.K. and Hilton, A.J. 2003. Black dot (*Colletotrichum coccodes*) an increasingly important disease of potato. *Plant Pathology* 52:3–12.

Miller, D. J., Coffman, C. B., Teasdale, J. R., Everts, K. L., Abdul-Baki, A. A., Lydon, J., and Anderson, J. D. 2002. Foliar disease in fresh-market tomato grown in differing bed strategies and fungicide spray programs. *Plant Disease* 86:955–959.

Raid, R. N. and Pennypacker, S. P. 1987. Weeds as hosts for *Colletotrichum coccodes*. *Plant Disease* 71:643–646.

Schneider, R. W., Grogan, R. G., and Kimble, K. A. 1978. Colletotrichum root rot of greenhouse tomatoes in California. *Plant Disease Reporter* 62:969–ß971.

Fusarium oxysporum* f. sp. *lycopersici

FUSARIUM WILT

Introduction and significance

Fusarium wilt was first characterized in 1895 and continues to be present worldwide. If resistant cultivars are not used, the disease can still be a serious production problem.

Symptoms and diagnostic features

If young seedlings are infected, plants can be stunted and exhibit poor growth. However, the more familiar symptoms of Fusarium wilt occur on older plants. Initial symptoms on such plants are chlorosis of lower leaves followed by wilting of that foliage. Characteristically, the yellow to yellow-gold discoloration and wilting symptoms often first occur on only one side of the plant. As disease progresses, the entire plant will turn chlorotic, wilt, and then collapse and dry up (**468**). The vascular tissue is discolored and turns brown (**469**) with the discoloration extending into the upper stems. This extensive browing is a helpful feature in distinguishing this disease from Fusarium crown and root rot, in which the vascular browning is found only in the lower stem. Fusarium wilt vascular browning tends to be darker than the vascular discoloration caused by Verticillium wilt, though this distinction is not always clear. Disease symptoms can be accentuated if the infected plant is bearing a heavy load of fruit or is stressed by some other factor. The overall symptoms are similar to those caused by Verticillium wilt; hence disease confirmation will require laboratory analysis.

468 **Fusarium wilt causing severe yellowing of tomato.**

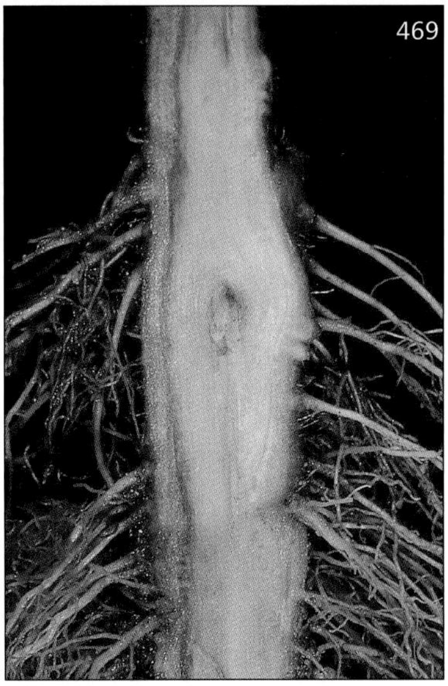

469 **Vascular discoloration caused by Fusarium wilt of tomato.**

Causal agent

Fusarium wilt is caused by the fungus *Fusarium oxysporum* f. sp. *lycopersici*. The pathogen morphology and colony characteristics are similar to other *F. oxysporum* fungi. The fungus forms one- or two-celled, oval- to kidney-shaped microconidia on monophialides, and four- to six-celled, fusiform, curved macroconidia. Microconidia measure 5–12 x 2–4 µm, while macroconidia range from 25–45 x 3–5 µm (four-celled) to 35–60 x 3–5 µm (six-celled). Macroconidia are usually produced in cushion-shaped structures called sporodochia. Chlamydospores are also formed. The pathogen is usually readily isolated from symptomatic vascular tissue. Semi-selective media like Komada's medium can help isolate the pathogen if secondary rot organisms are present. This pathogen is apparently host specific to tomato and can be seedborne. Three races (races 1, 2, and 3) have been documented. A Petri-plate technique can be used to differentiate this vascular wilt *Fusarium* from the crown and root rot *Fusarium* (see Sanchez, *et al.*).

Disease cycle

Like other Fusarium wilt pathogens, *F. oxysporum* f. sp. *lycopersici* is a soil inhabitant that can survive in the soil for indefinite periods of time due to the production of overwintering chlamydospores. The fungus is favored by warm temperatures, and optimum wilt development takes place at 28° C. Researchers find that Fusarium wilt may be more severe if plants are grown under certain nutrient conditions. For example, low nitrogen, low phosphorus, high potassium, and ammoniacal forms of nitrogen may enhance disease.

Control

Use resistant cultivars. Use nitrate-based fertilizers instead of ammoniacal ones. Do not plant tomato in fields having high populations of root knot nematode (*Meloidogyne* species) as this nematode can cause the plant's Fusarium wilt resistance to be overcome. Practice good field sanitation so that infested soil and mud are not spread to uninfested fields.

References

Bao, J. R. and Lazarovitis, G. 2001. Differential colonization of tomato roots by nonpathogenic and pathogenic *Fusarium oxysporum* strains may influence Fusarium wilt control. *Phytopathology* 91:449–456.

Borrero, C., Trillas, M. I., Ordovas, J. Tello, J. C., and Aviles, M. 2004. Predictive factors for the suppression of Fusarium wilt of tomato in plant growth media. *Phytopathology* 94:1094–1101.

Boyer, A. and Charest, P. M. 1989. Use of lectins for differentiating between *Fusarium oxysporum* f. sp. *radicis-lycopersici* and *Fusarium oxysporum* f. sp. *lycopersici* in pure culture. *Canadian Journal of Plant Pathology* 11:14–21.

Cai, G., Gale, L. R., Schneider, R. W., Kistler, H. C., Davis, R. M., Elias, K. S., and Miyao, E. M. 2003. Origin of race 3 of *Fusarium oxysporum* f. sp. *lycopersici* at a single site in California. *Phytopathology* 93:1014–1022.

Davis, R. M., Kimble, K. A., and Farrar, J. J. 1988. A third race of *Fusarium oxysporum* f. sp. *lycopersici* identified in California. *Plant Disease* 72:453.

Elias, K. S. and Schneider, R. W. 1991. Vegetative compatibility groups in *Fusarium oxysporum* f. sp. *lycopersici*. *Phytopathology* 81:159–162.

Erb, W. A. and Rowe, R. C. 1992. Screening tomato seedlings for multiple disease resistance. *Journal of the American Society for Horticultural Science* 117:622–627.

Katan, T., Shlevin, E., and Katan, J. 1997. Sporulation of *Fusarium oxysporum* f. sp. *lycopersici* on stem surfaces of tomato plants and aerial dissemination of inoculum. *Phytopathology* 87:712–719.

Mai, W. F. and Abawi, G. S. 1987. Interactions among root knot nematodes and Fusarium wilt fungi on host plants. *Annual Review of Phytopathology* 25:317–338.

Marlatt, M. L., Correll, J. C., Kaufmann, P., and Cooper, P. E. 1996. Two genetically distinct populations of *Fusarium oxysporum* f. sp. *lycopersici* race 3 in the United States. *Plant Disease* 80:1336–1342.

Mes, J. J., Weststeijn, E. A., Herlaar, F., Lambalk, J. J. M., Wijbrandi, J., Haring, M. A., and Cornelissen, B. J. C. 1999. Biological and molecular characterization of *Fusarium oxysporum* f. sp. *lycopersici* divides race 1 isolates into separate virulence groups. *Phytopathology* 89:156–160.

Sanchez, L. E., Endo, R. M., and Leary, J. V. 1975. Rapid technique for identifying clones of *Fusarium oxysporum* f. sp. *lycopersici* causing crown and root rot of tomato. *Phytopathology* 65:726–727.

Scott, J. W., Agrama, H. A., and Jones, J. P. 2004. RFLP-based analysis of recombination among resistance genes to Fusarium wilt races 1, 2, and 3 in tomato. *Journal of the American Society for Horticultural Science* 129:394–400.

Weststeijn, G. 1973. Soil sterilization and glasshouse disinfection to control *Fusarium oxysporum* f. sp. *lycopersici* in tomatoes in the Netherlands. *Netherlands Journal of Plant Pathology* 79:36–40

Fusarium oxysporum f. sp. *radicis-lycopersici*

FUSARIUM CROWN & ROOT ROT

Introduction and significance

Fusarium crown and root rot is found in many parts of the world on both field and greenhouse grown tomatoes. The disease can be particularly severe in greenhouse production environments.

Symptoms and diagnostic features

The initial symptom is chlorosis of the lower leaves that often is initiated along the margins of the leaves. Such leaves later become necrotic and then wither. In many cases, successively younger leaves develop chlorosis and necrosis until only the upper part of the plant has healthy, functional foliage. Infected plants can be stunted and not productive. In other cases plants decline more rapidly and collapse completely. A tan to brown discoloration develops in the vascular tissue of the root and extends into the adjacent tissues of the lower stem, as well (**470**). However, such internal stem discoloration remains in the lower stem and does not extend beyond 10–30 cm above the soil line. This limited, lower discoloration is a helpful feature in distinguishing this disease from Fusarium wilt, in which the vascular browning can extend far into the upper stems. Examination of the outside surface of plant crowns and lower stems may reveal the presence of large, irregular, brown, necrotic cankers (**471**). On occasion orange spore deposits may form on these cankers.

470 Internal discoloration caused by *Fusarium oxysporum* f. sp. *radicis-lycopersici*.

Causal agent

Fusarium crown and root rot is caused by the fungus *Fusarium oxysporum* f. sp. *radicis-lycopersici*. The pathogen morphology and colony characteristics are similar to other *F. oxysporum* fungi. The fungus forms one- or two-celled, oval to kidney shaped microconidia on monophialides, and four- to six-celled, fusiform, curved macroconidia. Microconidia measure 5–12 x 2–4 μm, while macroconidia range from 25–45 x 3–5 μm (four-celled) to 35–60 x 3–5 μm (six-celled). Macroconidia are usually produced in cushion-shaped structures called sporodochia and appear orange-colored in culture or on infected stem cankers. Chlamydospores are also formed. The pathogen is usually readily isolated from symptomatic vascular tissue. Semi-selective media like Komada's medium can help isolate the pathogen if secondary rot organisms are present.

A Petri-plate technique can be used to differentiate this crown and root rot *Fusarium* from the vascular wilt *Fusarium* (see Sanchez, *et al.*). Under experimental conditions, researchers found that this pathogen can also infect plants such as bean, beet, clover, cucumber, aubergine (eggplant), pepper, and others.

Disease cycle

Like Fusarium wilt pathogens, *F. oxysporum* f. sp. *radicis-lycopersici* is a soil inhabitant that can survive in the soil for indefinite periods of time due to the production of overwintering chlamydospores. In addition to soil inoculum, the pathogen can also grow saprophytically and produce conidia on decaying organic matter. In greenhouses, microconidia can reach tomato plants by becoming airborne or by being transported by fungus gnats. Optimum disease development takes place at temperatures between 20–22° C. Following initial infection at the base of the stem, there appears to be an incubation period of several days before there is secondary spread through the vascular system of susceptible cultivars.

Control

In greenhouses, steam the soil and then apply fungicides prior to transplanting tomato. For infested outdoor fields, no control measures have been developed. In such cases, use crop rotations that do not include host plants. Some resistant cultivars are being developed for this disease.

471 Tomato stem lesions caused by Fusarium crown and root rot

References

Boyer, A. and Charest, P. M. 1989. Use of lectins for differentiating between *Fusarium oxysporum* f. sp. *radicis-lycopersici* and *Fusarium oxysporum* f. sp. *lycopersici* in pure culture. *Canadian Journal of Plant Pathology* 11:14–21.

Chellemi, D. O., Olson, S. M., and Mitchell, D. J. 1994. Effects of soil solarization and fumigation on survival of soilborne pathogens of tomato in northern Florida. *Plant Disease* 78:1167–1172.

Gillespie, D. R. and Menzies, J. G. 1993. Fungus gnats vector *Fusarium oxysporum* f. sp. *radicis-lycopersici. Annals of Applied Biology* 123:539–544.

Hartman, J. R. and Fletcher, J. T. 1991. Fusarium crown and root rot of tomatoes in the U. K. *Plant Pathology* 40:85–92.

Jarvis, W. R. and Thorpe, H. J. 1980. Effects of nitrate and ammonium nitrogen on severity of Fusarium foot and root rot and on yield of greenhouse tomatoes. *Plant Disease* 64:309–310.

Krikun, J., Nachmias, A., Cohn, R., and Lahkim-Tsror, L. 1982. The occurrence of Fusarium crown and root rot of tomato in Israel. *Phytoparasitica* 10:113–115.

Menzies, J. G., Koch, C., and Seywerd, F. 1990. Additions to the host range of *Fusarium oxysporum* f. sp. *radicis-lycopersici. Plant Disease* 74:569–572.

Mihuta-Grimm, L., Erb, W. A., and Rowe, R. C. 1990. Fusarium crown and root rot of tomato in greenhouse rock wool systems: sources of inoculum and disease management with benomyl. *Plant Disease* 74:996–1002.

Rekah, Y., Shtienberg, D., and Katan, J. 1999. Spatial distribution and temporal development of Fusarium crown and root rot of tomato and pathogen distribution in field soil. *Phytopathology* 89:831–839.

Rekah, Y., Shtienberg, D., and Katan, J. 2001. Population Dynamics of *Fusarium oxysporum* f. sp. *radicis-lycopersici* in relation to the onset of Fusarium crown and root rot of tomato. *European Journal of Plant pathology* 107:367–375.

Rodriquez-Molina, M. C., Medina, I., Torres-Vila, L. M. and Cuartero, J. 2003. Vascular colonisation patterns in susceptible and resistant tomato cultivars inoculated with *Fusarium oxysporum* f. sp. *radicis-lycopersici* races 0 and 1. *Plant Pathology* 52:199–203.

Rosewich, U. L., Pettway, R. E., Katan, T., and Kistler, H. C. 1999. Population genetic analysis corroborates dispersal of *Fusarium oxysporum* f. sp. *radicis-lycopersici* from Florida to Europe. *Phytopathology* 89:623–630.

Rowe, R. C. 1980. Comparative pathogenicity and host ranges of *Fusarium oxysporum* isolates causing crown and root rot of greenhouse and field-grown tomatoes in North America and Japan. *Phytopathology* 70:1143–1148.

Rowe, R. C., Farley, J. D., and Coplin, D. L. 1977. Airborne spore dispersal and recolonization of steamed soil by *Fusarium oxysporum* in tomato greenhouses. *Phytopathology* 67:1513–1517.

Rowe, R. C. and Farley, J. D. 1981. Strategies for controlling Fusarium crown and root rot in greenhouse tomatoes. *Plant Disease* 65:107–112.

Sanchez, L. E., Endo, R. M., and Leary, J. V. 1975. Rapid technique for identifying clones of *Fusarium oxysporum* f. sp. *lycopersici* causing crown and root rot of tomato. *Phytopathology* 65:726–727.

Leveillula taurica (anamorph = *Oidiopsis taurica*), *Oidium neolycopersici*, *O. lycopersici*

POWDERY MILDEW

Introduction and significance

Powdery mildew disease can significantly limit tomato production in various parts of the world. Severe disease can cause early plant senescence and reduced yields.

Symptoms and diagnostic features

There are two different types of powdery mildew fungi that infect tomato. *Leveillula taurica* (anamorph = *Oidiopsis taurica*) initially causes light green, irregularly shaped leaf spots. Spots can have diffuse margins, but often tend to appear angular and vein-delimited. As the spots age, the tissue becomes chlorotic and then necrotic (472, 473). Careful examination of the undersides of these leaves reveals the white powdery growth of this pathogen (474). *Leveillula taurica* usually infects the older leaves; the younger leaves escape infection until they mature. In some regions, severe disease can result in extensive shriveling of foliage, reduced growth, and significant yield loss when fruit are exposed to the sun and become sunburned.

The other powdery mildew on tomato is caused by one or more species of *Oidium*. This powdery mildew disease results in white colonies on upper and lower surfaces of leaves and on stems and petioles (475). On occasion the colonies may take on a grayish white color. The infected, underlying tissue may initially turn purple but later becomes chlorotic and necrotic. Severe infections cause leaves to be twisted and deformed.

Causal agents

For *L. taurica*, the asexual (anamorph) stage is named *Oidiopsis taurica*. Conidiophores develop only from endophytic mycelium and emerge through stomata in the lower leaf epidermis. Conidiophores can be branched and carry one, or sometimes two, conidia. Conidia are hyaline, single-celled, and dimorphic. Primary (terminal) conidia are lanceolate with distinct apical points. Secondary conidia are ellipsoid–cylindric and lack the apical point. Conidial dimensions for both types vary according to the host plant but generally are 50–70 x 16–24 µm. For both conidial types, length-to-width ratios are greater than 3.

Globose cleistothecia of the *Leveillula* teleomorph have numerous hypha-like appendages, contain up to 20 asci, but are rarely observed. Asci contain two ascospores that are cylindrical to pyriform and measure 24–40 x 12–22 µm. *Leveillula taurica* appears to have a broad host range of numerous crops and weeds, and has been reported to infect plants in over 50 plant families. However, there may be distinct subpopulations that have more restricted host ranges. This pathogen is found on tomato throughout the world and is often associated with drier climates.

In contrast to *L. taurica*, *Oidium* species produce epiphytic, conidiophore-bearing mycelium that grows superficially on host surfaces. *Oidium neolycopersici* produces conidia singly; conidia measure 22–46 x 10–20 µm and are ellipsoid–ovoid in shape. This species is widespread and can be found on tomato in many regions of the world. *Oidium lycopersici* produces conidia in chains of three to five spores; conidia

472 Yellow lesions caused by *Leveillula taurica*.

473 Yellow lesions caused by *Leveillula taurica*.

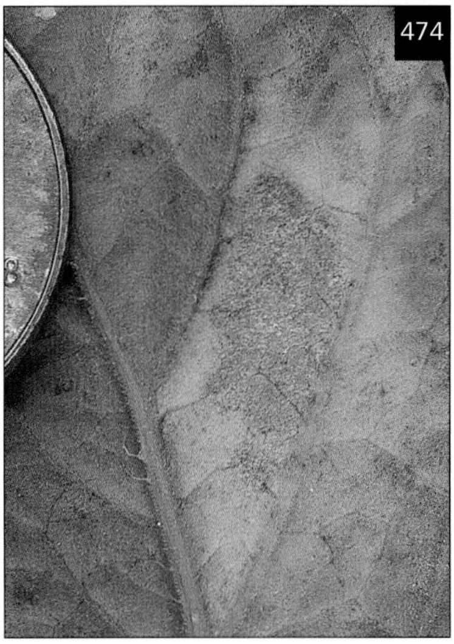

474 Sporulation of powdery mildew of tomato caused by *Leveillula taurica*.

475 Sporulation of powdery mildew of tomato caused by *Oidium* spp.

measure 25–40 x 12–18 μm and are elliptical to doliform in shape. This species has only been confirmed in Australia. The sexual cleistothecia stage has yet to be observed for either of these *Oidium* species.

Disease cycle

Powdery mildew species are obligate pathogens and survive on overwintering tomato, alternate hosts, or possibly as cleistothecia from their respective perfect stages. The asexual conidia of these fungi are dispersed by winds. Powdery mildew development is favored by mild temperatures below 30° C. Interestingly, in greenhouses where humidity is extremely high (95% RH), powdery mildew caused by *O. lycopersici* may be suppressed. Many new reports document recent occurrences of *O. neolycopersici* in various regions. However, researchers found that this pathogen is present on preserved herbarium plants that are over 50 years old.

Control

Fungicides, such as sulfur, may be needed if the disease becomes severe. Irrigation systems may influence disease severity; studies indicate that furrow irrigated tomato may have more severe powdery mildew than sprinkler irrigated plantings. No other control options are available.

References

Arredondo, C. R., Davis, R. M., and Rizzo, D. M. 1996. First report of powdery mildew of tomato in California caused by an *Oidium* sp. *Plant Disease* 80:1303.

Correll, J. C., Gordon, T. R., and Elliott, V. J. 1987. Host range, specificity, and biometrical measurements of *Leveillula taurica* in California. *Plant Disease* 71:248–251.

Correll, J. C., Gordon, T. R., and Elliott, V. J. 1988. Powdery mildew of tomato: the effect of planting date and triadimefon on disease onset, progress, incidence and severity. *Phytopathology* 78:512–519.

Fletcher, J. T., Smewin, B. J., and Cook, R. T. A. 1988. Tomato powdery mildew. *Plant Pathology.* 37:594–598.

Khodaparast, S. A., Takamatsu, S., and Hedjaroude, G.-A. 2001. Phylogenetic structure of the genus *Leveillula* (Erysiphales:Erysiphaceae) inferred from the nucleotide sequences of the rDNA ITS region with special reference to the *L. taurica* species complex. *Mycological Research* 105:909–918.

Kiss, L., Cook, R. T. A., Saenz, G. S., Cunnington, J. H., Takamatsu, S., Pascoe, I., Bardin, M., Nicot, P. C., Sato, Y., and Rossman, A. Y. 2001. Identification of two powdery mildews, *Oidium neolycopersici* sp. *nov*. and *Oidium lycopersici*, infecting tomato in different parts of the world. *Mycological Research* 105:684–697.

Palti, J. 1971. Biological characteristics, distribution, and control of *Leveillula taurica*. *Phytopathologia Mediterranea* 10:139–153.

Price, T. V. 1981. Powdery mildew of tomato in Australia. *Australasian Plant Pathology* 10:38–40.

Reuveni, R. and Rotem, J. 1973. Epidemics of *Leveillula taurica* on tomatoes and peppers as affected by conditions of humidity. *Phytopathologische Zeitschrift* 76:153–157.

Vakalounakis, D. J. and Papadakis, A. 1992. Occurrence of a new powdery mildew of greenhouse tomato in Greece, caused by *Erysiphe* sp. *Plant Pathology* 41:372–373.

Whipps, J. M. and Budge, S. P. 2000. Effect of humidity on development of tomato powdery mildew (*Oidium lycopersici*) in the glasshouse. *European Journal of Plant Pathology* 106:395–397.

Whipps, J. M., Budge, S. P., and Fenlon, J. S. 1998. Characteristics and host range of tomato powdery mildew. *Plant Pathology* 47:36–48.

Phytophthora capsici, P. cryptogea, P. drechsleri, P. parasitica (= P. nicotianae var. *parasitica)*

PHYTOPHTHORA ROOT ROT

Introduction and significance

There are several soilborne *Phytophthora* species that cause diseases on tomato. Different tomato-producing regions throughout the world may have different species involved in this disease. In recent years these pathogens have increased in importance on various vegetable crops in the USA.

Symptoms and diagnostic features

Diseases caused by *Phytophthora* spp. are manifested as seed decays, seedling damping-off, root rots, and fruit rots. Symptoms of Phytophthora root rot initially consist of water-soaked root lesions that later turn dark gray to brown. The discoloration can occur on both the fine feeder and larger taproots. As lesions expand, individual roots become girdled or entirely rotted. The discoloration will affect both vascular and stele tissues of the root and can move up the main taproot and into the plant crown and lower main stem (**476**). In advanced stages, the roots will be soft and decayed. The plant crown can show surface and internal discoloration. Above-ground symptoms consist of foliage that first turns dull gray-green, then later wilts. The entire plant canopy can rapidly collapse and die (**477, 478**).

476 Darkened tomato stems caused by *Phytophthora capsici*.

477 Foliar dieback caused by *Phytophthora*.

Phytophthora fruit disease is called buckeye rot. Buckeye rot almost always occurs on fruit that are touching infested soil. Green and ripe fruit can show similar symptoms. The disease begins as small, brown spots on fruit surfaces in contact with soil. Spots grow into large, circular or irregularly oblong lesions that can cover more than half of the fruit. The lesions are characterized by concentric rings of alternating light and dark brown discoloration (**479**). Diseased fruit are initially firm, but will later become soft and rotted. The white mycelium of the pathogen can sometimes be observed when the lesion breaks open and rots. The early, firm lesion symptoms on the fruit may resemble the fruit infections caused by the late blight pathogen.

Causal agents

Phytophthora root rot is caused by several species including *P. capsici, P. cryptogea,* and *P. parasitica* (= *P. nicotianae* var. *parasitica*). Buckeye rot is caused by *P. capsici, P. drechsleri,* and *P. parasitica*. All three species are oomycetes, soil inhabitants, and can persist in soils for extended periods of time. *Phytophthora capsici* forms irregularly shaped sporangia that can be

478 Collapsing plants caused by Phytophthora root rot of tomato.

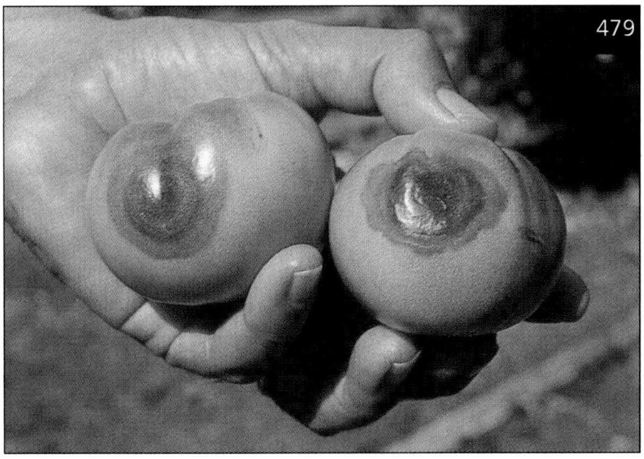

479 Buckeye rot symptoms on tomato fruit caused by *Phytophthora*.

spherical, ovoid, elongated, or have more than one apex. Sporangia are papillate, deciduous, have pedicels that are 10 or more µm in length, and measure 30–60 x 25–35 µm. *Phytophthora parasitica* sporangia vary greatly and can be ellipsoidal, ovoid, pyriform, or spherical with distinct papilla. Sporangia are not deciduous and measure 11–60 x 20–45 µm.

Disease cycle

Phytophthora species are spread by surface water and movement of infested soil. Both fruit and root diseases require wet soil conditions. Compacted, finely textured, and poorly draining soils create conditions favorable for root rot. Excess soil moisture or splashing water is required for significant fruit rot development. These *Phytophthora* species can infect pepper, cucurbits, and other hosts.

Control

Plant tomato in fields having soils that drain well. Prepare soil so that drainage is enhanced and low areas are avoided. Carefully manage irrigation so that excess soil water is reduced. Stake plants to keep fruit off the ground, or use plastic mulches on bed tops. Keep bed tops dry by using subsurface drip irrigation. Some fungicides may help manage both root and fruit infections.

References

Blaker, N. S. and Hewitt, J. D. 1987. Comparison of resistance to *Phytophthora parasitica* in tomato. *Phytopathology* 77:1113–1116.

Cafe-Filho, A. C. and Duniway, J. M. 1995. Dispersal of *Phytophthora capsici* and *Phytophthora parasitica* in furrow-irrigated rows of bell pepper, tomato, and squash. *Plant Pathology* 44:1025–1032.

Neher, D. and Duniway, J. M. 1992. Dispersal of *Phytophthora parasitica* in tomato fields by furrow irrigation. *Plant Disease* 76:582–586.

Neher, D. A., McKeen, C. D., and Duniway, J. M. 1993. Relationships among Phytophthora root rot development, *Phytophthora parasitica* populations in soil, and yield of tomatoes under commercial field conditions. *Plant Disease* 77:1106–1111.

Ristaino, J. B. and Duniway, J. M. 1989. Effect of preinoculation and postinoculation water stress on severity of Phytophthora root rot in processing tomatoes. *Plant Disease* 73:349–352.

Ristaino, J. B., Duniway, J. M., and Marois, J. J. 1988. Influence of frequency and duration of furrow irrigation on development of Phytophthora root rot. *Phytopathology* 78:1701–1706.

Ristaino, J. B., Duniway, J. M., and Marois, J. J. 1989. Phytophthora root rot and irrigation schedule influence growth and phenology of processing tomatoes. *Journal of the American Society for Horticultural Science* 114:556-561.

Swiecki, T. J. and MacDonald, J. D. 1991. Soil salinity enhances Phytophthora root rot of tomato but hinders asexual reproduction by *Phytophthora parasitica*. *Journal of the American Society for Horticultural Science* 116:471–477.

Workneh, F., van Bruggen, A. H. C., Drinkwater, L. E., and Shennan C. 1993. Variables associated with corky root and Phytophthora root rot of tomatoes in organic and conventional farms. *Phytopathology* 83:581–589.

Phytophthora infestans

LATE BLIGHT

Introduction and significance

Late blight disease of potato is one of the world's most well known plant diseases because of its historic significance and role in the Irish potato famine of the 1840s. Late blight is also an important tomato disease. Worldwide, late blight reemerged in the 1990s through early 2000s as a very serious disease concern for both potato and tomato crops, and devastating epidemics have been chronicled on various continents.

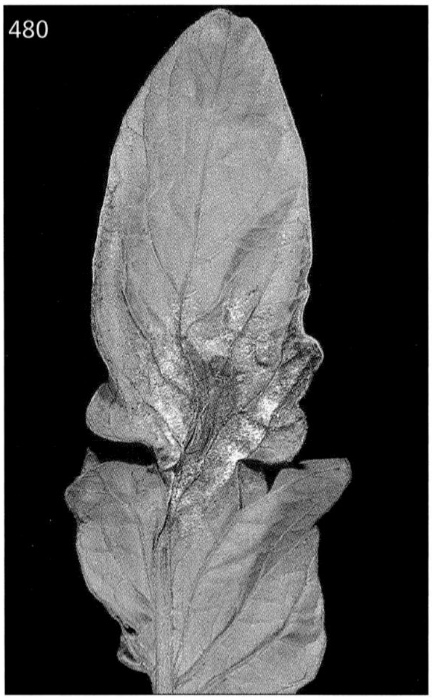

480 Late blight sporulation on tomato leaf.

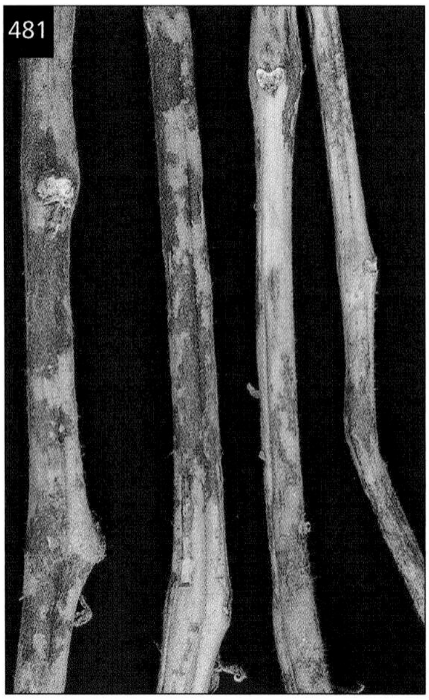

481 Stem lesions caused by late blight of tomato.

Symptoms and diagnostic features

Initial late blight symptoms on tomato foliage are irregularly shaped, pale green, water-soaked spots and areas on leaves. These areas rapidly expand into large brown to gray leaf lesions that in severe cases can kill the entire leaf. If environmental conditions are suitable, the leaf undersides can be covered with the velvet white growth of the pathogen (**480**). Stems and petioles can also develop the water-soaked and later brown to gray lesions (**481**). The pathogen can rapidly colonize tomato foliage and cause entire plants to turn brown, shrivel, and die (**482**). Infected green fruit develop circular to oval to irregularly shaped lesions that are green-brown to dark brown in color (**483**). Such lesions can become quite large, sometimes covering the entire fruit. Characteristically these green fruit lesions are very firm and do not break down and rot unless the infections are old and secondary decay organisms are present in the fruit.

Causal agent

Late blight is caused by the oomycete *Phytophthora infestans*. *Phytophthora infestans* is heterothallic and requires two distinct mating types (A1 and A2) to undergo sexual reproduction and produce the sexual spore, the oospore. The presence and importance of oospores appears to differ depending on where the crop is grown. In parts of the USA including California, even if both mating types are present in one area, it appears that sexual recombination is not common; the oospore is rarely found in host plant tissue in these areas. In Canada (British Columbia), Central Mexico, and parts of Europe (Netherlands, Poland, Russia), the oospore is more readily found and genetic evidence indicates a greater diversity is present in populations, which presupposes sexual recombination.

The fungus is sometimes difficult to maintain in culture, though cultures can be grown on V-8 juice, pea, rye B, and other media. Incubating the symptomatic tissue at low temperatures (below 20° C) and with high humidity will also encourage the fungus to grow and sporulate after only one to a few days. Mycelium is coenocytic (lacking cell cross walls) and produces ellipsoidal to ovoid, hyaline, semi-papillate sporagia that measure 21–38 x 12–23 µm. Sporangia are deciduous (=caducous), have pedicels measuring up to 3 µm long, and are distributed by winds and splashing water. Sporangia can germinate directly and infect tomato

482 Tomato plants infected with late blight.

tissues, or can produce swimming zoospores that are released and infect the host. Optimum temperature for growth of the fungus is 20° C. Tomato and potato are the main hosts, though other plants in the Solanaceae can also support the pathogen. Examples of other hosts are hairy nightshade (*S. sarrachoides*), *S. physalifolium*, and petunia (*Petunia* species).

The recent increase in late blight severity coincides with worldwide changes in the population genetics of *P. infestans*. Recent advances in molecular methods and DNA technology have enabled researchers to precisely characterize isolates and obtain some insight into reasons for this increased severity. Research reveals that *P. infestans* is a very complex organism. While all *P. infestans* isolates belong to one of two mating types, all individuals also simultaneously belong to one of a series of distinct clonal lineages that are asexual descendants from single genotypes. Prior to the 1980s, the various regions of the world were primarily populated by *P. infestans* isolates that were of the A1 mating type and overwhelmingly belonged to the US-1 clonal lineage group that is sensitive to the widely used fungicide metalaxyl.

However, the recent serious outbreaks were often caused by novel lineages. In North America, four such lines caused significant damage to crops: US-6 MR (A1 mating type; resistant to metalaxyl; highly virulent to tomato), US-7 (A2 mating type; resistant to metalaxyl; highly virulent to tomato); US-8 (A2 mating type;

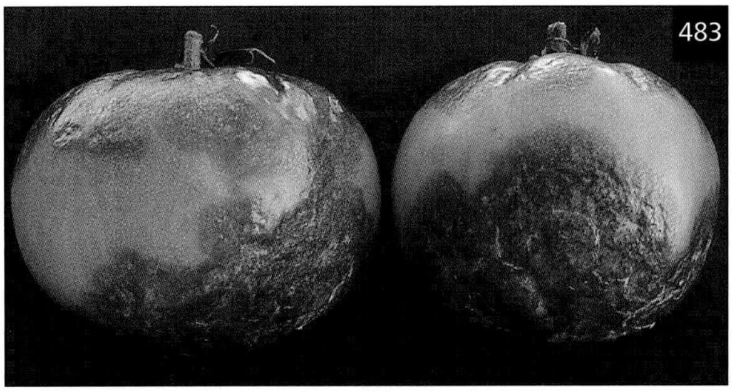

483 Fruit infections caused by late blight of tomato.

resistant to metalaxyl; highly virulent to potato), and US-11 (A1 mating type; resistant to metalaxyl; highly virulent on potato and tomato). The occurrence and spread of these new pathogen lines is most likely due to the shipment of potato seed tubers, tomato fruit, and tomato transplants between areas and continents.

These population genetics are in flux, because novel clonal lineages replace older ones after only a few seasons. Significant changes in *P. infestans* population structures have also been documented in Europe, South America, and Asia. These new aggressive isolates can reproduce more rapidly than older genotypes, thereby shortening the disease cycle to 5–7 days under favorable conditions. Additional changes in virulence and fungicide insensitivity could occur more frequently in the future because sexual reproduction will likely be more common now that both mating types are widely distributed.

One additional parameter of *P. infestans* diversity is aggressiveness of isolates. Studies indicate that only some isolates are highly virulent to tomato, while all tested isolates were highly virulent on potato. Therefore, *P. infestans* can be further divided up into tomato-aggressive and tomato-nonaggressive isolates. This factor is an important management consideration; if a tomato-nonaggressive isolate is present in a tomato field, then crop loss due to late blight is likely to be limited and fungicides may be needed less frequently. There is some evidence that the A2 mating type occurs more frequently on tomato than on potato.

Disease cycle

In contrast to most other *Phytophthora* species, *P. infestans* usually is not considered a soilborne fungus, though this aspect might change if both mating types are present and oospores are formed that can persist in the soil. The fungus overwinters in volunteer host plants, potato cull piles, and residential gardens. Potato plants and cull tubers may be an important source of inoculum for tomato fields. When high humidity and moderate temperatures are present, sporangia are produced, released into the air, and blown onto susceptible crops. Optimum growth of the fungus is at 18–22° C. A striking feature of late blight is the speed of disease development and spread. If conditions are suitable for the pathogen, entire fields can become infected after only a few days.

Control

Fungicides are key tools for managing late blight. Apply protectant fungicides prior to late blight infection. The pathogen has developed resistance to some fungicide products such as metalaxyl. Late blight forecasting systems may be helpful in deciding when fungicides should best be applied. Several forecasting systems are used for identifying late blight infection periods in potato crops. In the UK, Smith periods define when weather conditions are favorable for blight infection. A full Smith period requires relative humidity greater than 90% for a minimum of 11 hours on two consecutive days and temperatures above 10° C.

Avoid or reduce the use of sprinklers for irrigation. Destroy old tomato fields after harvest is completed. Eliminate pathogen reservoir sources such as potato tuber or tomato fruit cull piles and volunteer potato or tomato plants. Because disease can develop on transplants inside greenhouses, carefully inspect tomato transplants and remove suspect plants and trays. Some resistant cultivars are available.

References

Adler, N. E., Erselius, L. J., Chacon, M. G., Flier, W. G., Ordonez, M. E., Kroon, L. P. N. M., and Forbes, G. A. 2004. Genetic diversity of *Phytophthora infestans* sensu lato in Ecuador provides new insight into the origin of this important plant pathogen. *Phytopathology* 94:154–162.

Anderson, B., Johansson, M., Jonsson, B. 2003. First report of *Solanum physalifolium* as a host plant for *Phytophthora infestans* in Sweden. *Plant Disease* 87:1538.

Bakonyi, J., Láday, M., Dula, T., and Érsek, T. 2002. Characterisation of isolates of *Phytophthora infestans* from Hungary. *European Journal of Plant pathology* 108:139–146.

Cohen, Y., Farkash, S., Reshit, Z., and Baider, A. 1997. Oospore production of *Phytophthora infestans* in potato and tomato leaves. *Phytopathology* 87:191–196.

Cook, D. E. L., Young, V., Birch, P. R. J., Toth, R., Gourlay, F., Day, J. P., Carnegie, S. F., and Duncan, J. M. 2003. Phenotypic and genotypic diversity of *Phytophthora infestans* populations in Scotland (1995–1997). *Plant Pathology* 52:181–192.

Deahl, K. L., Pagani, M. C., Vilaro, F. L., Perez, F. M., Moravec, B., Cooke, L. R. 2003. Characteristics of *Phytophthora infestans* isolates from Uruguay. *European Journal of Plant Pathology* 109:277–281.

Deahl, K. L., Shaw, D. S., and Cooke, L. R. 2004. Natural occurrence of *Phytophthora infestans* on black nightshade (*Solanum nigrum*) in Wales. *Plant Disease* 88:771.

Fry, W. E. and Goodwin, S. B. 1997. Re-emergence of potato and tomato late blight in the United States. *Plant Disease* 81:1349–1357.

Fry, W. E., Goodwin, S. B., et al. 1993. Historical and recent migrations of *Phytophthora infestans*: chronology, pathways, and implications. *Plant Disease* 77:653–661.

Fry, W. E., Goodwin, S. B., Matuszak, J. M., Spielman, L. J., and Milgroom, M. G. 1992. Population genetics and intercontinental migrations of *Phytophthora infestans*. *Annual Review of Phytopathology* 30:107–129.

Goodwin, S. B. 1997. The population genetics of *Phytophthora*. *Phytopathology* 87:462–473.

Goodwin, S. B., Cohen, B. A., Deahl, K. L., and Fry, W. E. 1994. Migration from northern Mexico as the probable cause of recent genetic changes in populations of *Phytophthora infestans* in the United States and Canada. *Phytopathology* 84:553–558.

Goodwin, S. B. and Drenth, A. 1997. Origin of the A2 mating type of *Phytophthora infestans* outside Mexico. *Phytopathology* 87:992–999.

Goodwin, S. B., Smart, C. D., Sandrock, R. W., Deahl, K., Punja, Z. K., and Fry, W. E. 1998. Genetic change within populations of *Phytophthora infestans* in the United States and Canada during 1994 to 1996: Role of migration and recombination. *Phytopathology* 88:939–949.

Goodwin, S. B., Sujkowski, L. S., and Fry, W. E. 1996. Widespread distribution and probable origin of resistance to metalaxyl in clonal genotypes of *Phytophthora infestans* in the United States and western Canada. *Phytopathology* 86:793–800.

Knapova, G. and Gisi, U. 2002. Phenotypic and genotypic structure of *Phytophthora infestans* populations on potato and tomato in France and Switzerland. *Plant Pathology* 51:641–653.

Koh, Y. J., Goodwin, S. B., Dyer, A. T., Cohen, B. A., Ogoshi, A., Sato, N., and Fry, W. E. 1994. Migrations and displacements of *Phytophthora infestans* populations in East Asian countries. *Phytopathology* 84:922–927.

Legard, D. E., Lee, T. Y., and Fry, W. E. 1995. Pathogenic specialization in *Phytophthora infestans*: aggressiveness on tomato. *Phytopathology* 85:1356–1361.

Matuszak, J. M., Fernandez-Elquezabal, J., and Villarreal-Gonzalez, M.. 1994. Sensitivity of *Phytophthora infestans* populations to metalaxyl in Mexico: distribution and dynamics. *Plant Disease* 78:911–916.

Porter, L. D. and Johnson, D. A. 2004. Survival of *Phytophthora infestans* in surface water. *Phytopathology* 94:380–387.

Tumwine, J., Frinking, H. D., and Jeger, M. J. 2002. Integrating cultural control methods for tomato late blight (*Phytophthora infestans*) in Uganda. *Annals of Applied Biology* 141:225–236.

Phytophthora* spp., *Pythium* spp., *Rhizoctonia solani

DAMPING-OFF, FRUIT ROTS

Introduction and significance

Several soilborne pathogens cause seed, seedling, and transplant diseases of tomato. Pathogens include species of *Phytophthora*, *Pythium*, and *Rhizoctonia*. These pathogens can also infect tomato fruit and cause field and postharvest fruit rots.

Symptoms and diagnostic features

These diseases have various phases. Tomato seeds can be infected prior to germination and result in seed death. Newly germinated seedlings can be infected to such a degree that plants do not emerge above the soil (pre-emergence damping-off). Seedlings might emerge from the ground but become diseased after soil emergence (post-emergence damping-off). Finally, transplants placed in the ground can develop root rots or stem lesions from these same pathogens. Initial symptoms of post-emergence damping-off occur on seedling stems in contact with the soil. These symptoms consist of shriveled stems that have discolored, tan to dark brown lesions. With time the lower stem collapses, roots decay, and the cotyledons and leaves wilt. Such plants usually bend over. Damping-off diseases often result in death of the seedling and subsequent reduction of plant stands. However, even if plants do not succumb to these pathogens, the surviving plant may be stunted and delayed in development.

These three pathogens can also infect roots and crowns of older plants as well as fruit. *Phytophthora* fruit infections are called buckeye rot (see Phytophthora root rot section, page 359) and can be caused by several species of *Phytophthora*. *Pythium* fruit infections are called watery rot and usually infect ripe tomato fruit. Symptoms consist of irregular, water-soaked lesions on fruit that are in contact with the soil. Once infected, the fruit is rapidly colonized by the *Pythium* pathogen, the epidermis breaks, and the fruit becomes soft and watery. White mycelial growth can be observed on the rotted tissues. *Rhizoctonia* fruit rot can affect immature green fruit but is mostly a problem on ripe fruit. Dark brown, circular to oblong lesions develop on ripe fruit that are touching the soil. These lesions enlarge and break down into a mushy, soft rot. *Rhizoctonia solani* will form characteristic brown, persistent hyphae on the fruit surface.

Causal agents
Phytophthora and *Pythium* species are in the oomycete group. *Phytophthora capsici* and *P. parasitica* usually infect roots of older plants, but can cause damping-off. A number of *Pythium* species cause damping-off: *P. aphanidermatum*, *P. arrhenomanes*, *P. debaryanum*, *P. myriotylum*, *P. ultimum*. All these organisms survive in the soil as saprophytes and are favored by wet soil conditions. With the exception of *P. ultimum*, these pathogens usually produce zoospores that swim to and infect susceptible tissues. Sexual structures—antheridia, oogonia, and oospores—are produced by all species. In addition to tomato, these pathogens can infect numerous other plants.

Rhizoctonia solani is a soilborne fungus with a very broad host range. *Rhizoctonia solani* has no asexual spores, but produces characteristically coarse, brown, approximately right-angle branching hyphae. The hyphae are distinctly constricted at branch points, and cross walls with dolipore septa are deposited just after the branching. Hyphal cells are multi-nucleate. Small, tan to brown loosely aggregated clumps of mycelia function as sclerotia. This fungus can survive by infecting and thriving on a great number of plant hosts, besides tomato, and can also persist in the soil as a saprophyte. The teleomorph stage, *Thanatephorus cucumis*, is not commonly observed.

Disease cycle
Most of these soilborne pathogens survive in fields for indefinite amounts of time. Wet soil conditions and cool temperatures (15–20° C) generally favor these organisms and their ability to grow and infect hosts. However, for pathogens such as *P. aphanidermatum* and *P. myriotylum*, warmer soil conditions (32–37° C) are favorable. The seedling stage is most susceptible to infection, though these pathogens can infect the feeder roots of mature plants.

Control
Damping-off is primarily controlled by creating conditions unfavorable for the pathogens. Plant on raised beds and in soils that drain well so that overly wet soil conditions are reduced. Carefully manage irrigation and do not apply excess water. Plant seed that has been treated with fungicides, and avoid planting seed too deeply, which delays seedling emergence and increases the chance of infection. Post-plant fungicides may provide control for some of these pathogens. Avoid planting too soon into fields that still have extensive crop residue in the soil. Rotate crops, as consecutive tomato plantings will increase the populations of the soilborne pathogens. For transplants, prepare good-quality beds and do not plant too deep.

To avoid fruit rots, keep bed tops dry by carefully managing the irrigation or by using buried drip irrigation. Prepare soil to enhance drainage and avoid low areas. Stake plants or use plastic mulches on the bed tops to keep fruit off the ground. Some fungicides may help manage both root and fruit infections.

References
Cafe-Filho, A. C. and Duniway, J. M. 1995. Dispersal of *Phytophthora capsici* and *Phytophthora parasitica* in furrow-irrigated rows of bell pepper, tomato, and squash. *Plant Pathology* 44:1025–1032.

Golden, J. K. and Van Gundy, S. D. 1975. A disease complex of okra and tomato involving the nematode *Meloidogyne incognita*, and the soil-inhabiting fungus, *Rhizoctonia solani*. *Phytopathology* 65:265–273.

Pearson, R. C. and Hall, D. H. 1973. Ripe fruit rot of tomato caused by *Pythium ultimum* and *Pythium aphanidermatum*. *Plant Disease Reporter* 57:1066–1069.

Pyrenochaeta lycopersici

CORKY ROOT ROT

Introduction and significance
Corky root rot, or brown rot, was first described from Europe but now also occurs in North America. Corky root rot is important where tomato crops are grown repeatedly in the same soil.

Symptoms and diagnostic features
Initial symptoms consist of plants that show poor vigor, are stunted, and begin to wilt. Leaves may show interveinal chlorosis and later fall off the plant. The most characteristic symptoms occur on larger roots and consist of brown lesions that have a rough corky or wrinkled texture (**484, 485**). Such lesions often appear as horizontal bands across the length of the root; lesions are dry and have cracks that run lengthwise along the root. Smaller feeder roots may either show the brown, rough lesions or may be completely rotted. Internal tissues of the larger roots do not exhibit discoloration or symptoms. Affected plants rarely collapse and die, but can experience a reduction in yield.

Causal agent

Corky root is caused by the fungus *Pyrenochaeta lycopersici*. This pathogen is a slow-growing fungus that forms gray colonies on standard microbiological media but is difficult to isolate without using semi-selective media. In culture, the pathogen forms brown to black pycnidia that measure 150–300 µm in diameter. Pycnida release spores through a circular pore (ostiole) that is ringed with three to twelve light brown, septate setae. Single-celled, cylindrical to allantoid, hyaline conidia measure 4–8 x 1.5–2 µm and are borne on conidiophores within the pycnidial body. The fungus forms microsclerotia that measure 63.5 x 448 µm.

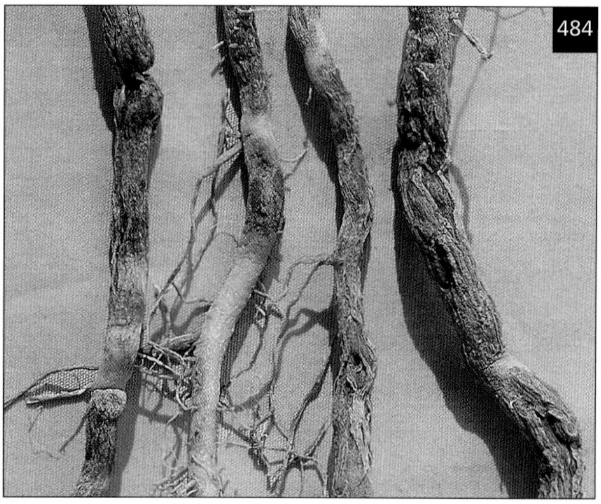

484 Root banding caused by corky root rot of tomato.

Disease cycle

Pyrenochaeta lycopersici is a soilborne organism and can survive for long periods of time as microsclerotia in soil or on old tomato roots. The fungus prefers cool conditions and optimum disease development takes place at 15–20° C, though the range is from 8–32° C. The pathogen can also infect aubergine (eggplant), melon, pepper, safflower, spinach, and squash. *Pyrenochaeta lycopersici* often co-infects tomato roots with the black dot pathogen (*Colletotrichum coccodes*).

485 Root banding caused by corky root rot of tomato.

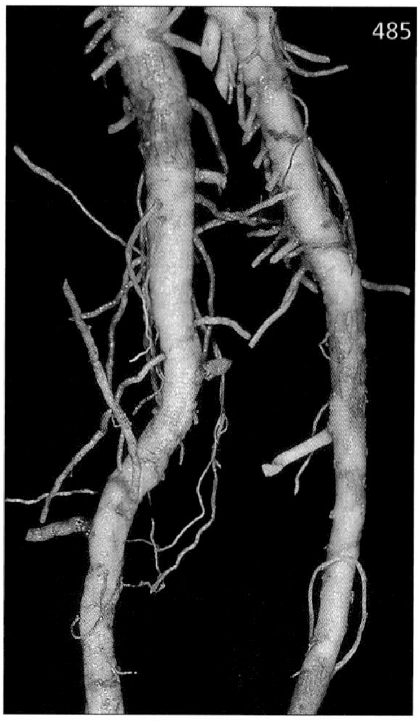

Control

Apply fumigants to field soil, and steam or fumigants to greenhouse planting areas. Rotate away from tomato to avoid buildup of inoculum. Delay planting until later in the spring when soils are warmer. Some European tomato cultivars are resistant to this pathogen. Additional control measures include grafting to resistant rootstocks and mounding soil around the stem base to allow new adventitious roots to grow.

References

Campbell, R. N., Schweers, V. H., and Hall, D.H. 1982. Corky root of tomato in California caused by *Pyrenochaeta lycopersici* and control by soil fumigation. *Plant Disease* 66:657–661.

Grove, G. G. and Campbell, R. N. 1987. Host range and survival in soil of *Pyrenochaeta lycopersici*. *Plant Disease* 71:806–809.

Hockey, A. G. and Jeves, T. M. 1984. Isolation and identification of *Pyrenochaeta lycopersici*, causal agent of tomato brown root rot. *Transactions of the British Mycological Society* 82:151–152.

Shishkoff, N. and Campbell, R. N. 1990. Survival of *Pyrenochaeta lycopersici* and the influence of temperature and cultivar resistance on the development of corky root of tomato. *Plant Disease* 74:889–894.

Workneh, F., van Bruggen, A. H. C., Drinkwater, L. E., and Shennan C. 1993. Variables associated with corky root and Phytophthora root rot of tomatoes in organic and conventional farms. *Phytopathology* 83:581–589.

Workneh, F. and van Bruggen, A. H. C. 1994. Suppression of corky root of tomatoes in soils from organic farms associated with soil microbial activity and nitrogen status of soil and tomato tissue. *Phytopathology* 84:688–694.

Sclerotinia minor, S. sclerotiorum

WHITE MOLD, SCLEROTINIA ROT

Introduction and significance

White mold, or Sclerotinia rot, is an occasional tomato problem. Both pathogens occur on tomato throughout the world. Disease caused by the *S. sclerotiorum* species is the more important *Sclerotinia* disease.

Symptoms and diagnostic features

The types of symptoms seen on tomato depend on which species of *Sclerotinia* is involved. *Sclerotinia minor* only infects tomato tissues that are in contact with the soil. *Sclerotinia minor* causes a water-soaked lesion to develop at the crown and lower stem (**486**). The lesion enlarges and can girdle the plant, resulting in the collapse of the canopy and foliage (**487**). With time, the crown and stem lesions turn light tan to off-white in color. White mycelium and small (3–5 mm), black, irregularly shaped sclerotia form around and within the decayed crown.

Sclerotinia sclerotiorum is the other species that can attack tomato. Because *S. sclerotiorum* has an airborne spore stage, infections can occur throughout the tomato canopy. These above-ground infections usually occur on damaged stems or petioles, or where a nutrient source, such as a senescent flower petal, falls onto stems or petioles. These infections are water-soaked lesions that gradually enlarge and encircle the stems. Older infections turn off-white to white-gray in color (**488**). White mycelium can be observed on infected lesions if conditions are favorable. The large, black sclerotia can grow on the outer surface or in the central cavity of stems (**489**). *S. sclerotiorum* can also infect tomato crowns and lower stems. Fruit can become infected and develop a soft, watery rot (**490**). This pathogen produces white mycelium and black, oblong or dome-shaped sclerotia. Sclerotia are significantly larger (5–10 mm long) than those of *S. minor*.

Causal agent

For detailed descriptions of *S. sclerotiorum* and *S. minor*, see the bean white mold section in the chapter on legume diseases (page 262).

Disease cycle

For detailed descriptions of the disease cycles, see the bean white mold section in the chapter on legume diseases.

Control

For an integrated approach to *Sclerotinia* control, see the bean white mold section in the chapter on legume diseases. For high value greenhouse tomato crops, soil sterilization with heat or chemical treatments may be cost-effective.

References

Abawi, G. S. and Grogan, R. G. 1979. Epidemiology of diseases caused by *Sclerotinia* species. *Phytopathology* 69:899–904.

Kohn, L. M. 1979. Delimitation of the economically important plant pathogenic *Sclerotinia* species. *Phytopathology* 69:881–886.

Lobo, M., Lopes, C. A., and Silva, W. L. C. 2000. Sclerotinia rot losses in processing tomatoes grown under centre pivot irrigation in central Brazil. *Plant Pathology* 49:51–56.

Purdy, L. H. 1979. *Sclerotinia sclerotiorum*: history, diseases and symptomatology, host range, geographic distribution, and impact. *Phytopathology* 69:875–880.

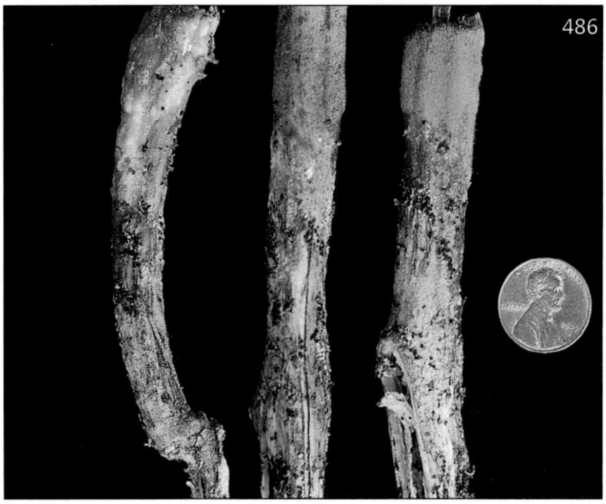

486 Tomato stems infected by *Sclerotinia minor*.

488 Bleached stems and black sclerotia from *Sclerotinia sclerotiorum* on tomato.

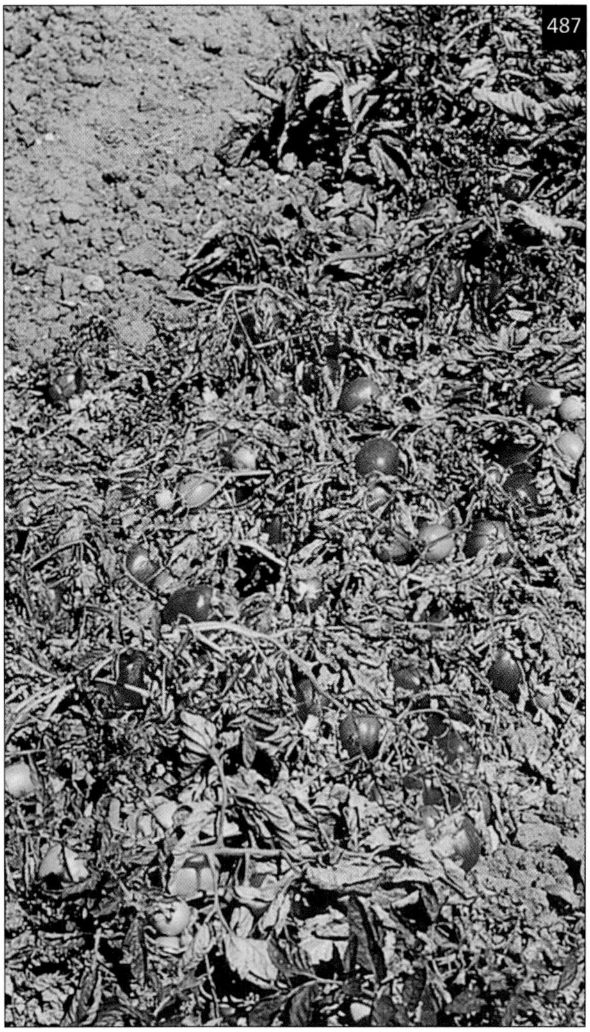

487 Collapsed tomato plants caused by *Sclerotinia minor*.

489 Black sclerotia from *Sclerotinia sclerotiorum* forming inside tomato stems.

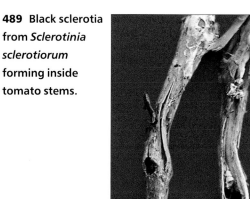

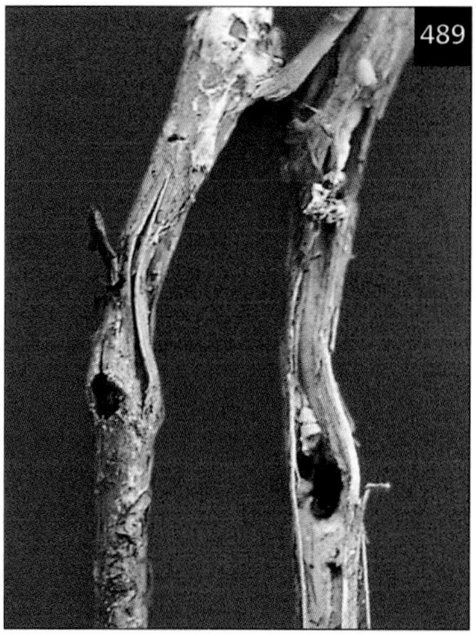

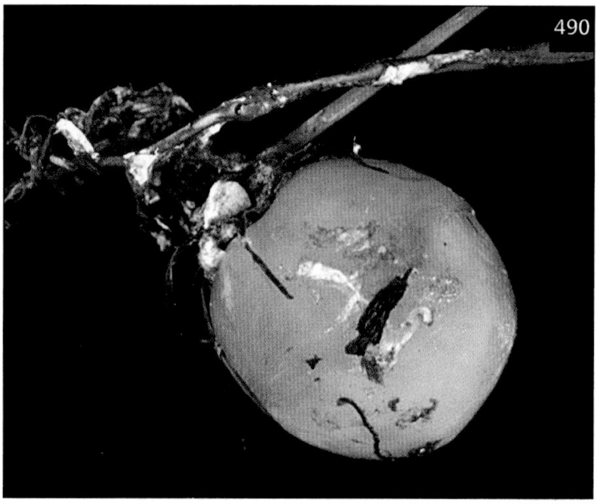

490 Tomato fruit showing white mycelium of *Sclerotinia sclerotiorum.*

Verticillium dahliae

VERTICILLIUM WILT

Introduction and significance

Verticillium wilt is a well-known disease that affects hundreds of different crops and is an important tomato production factor throughout the world. The closely related pepper and aubergine (eggplant) are also subject to this disease.

Symptoms and diagnostic features

On tomato, early symptoms consist of the chlorosis of leaf margins and tips of older, lower leaves; these yellowed areas are sometimes angular in shape and interveinal (**491**). The chlorotic sections later turn necrotic and die (**492**). Shoot tips and foliage wilt, especially during the warmer times of the day, and recover at night. Internal vascular tissue discolors to a tan to light brown color (**493**). This coloring is most evident in the main stems closer to the crown; such discoloration may not be evident in the upper, smaller stems. Verticillium wilt vascular discoloration tends to be lighter and subtler than the vascular discoloration caused by Fusarium wilt, though this distinction is not always clear. Disease symptoms can be accentuated if the infected plant is bearing a heavy load of fruit or is stressed by some other factor. Even if diseased plants do not collapse completely, plant growth and yields can be significantly reduced, sometimes by over 20%. The overall symptoms are similar to those caused by Fusarium wilt; hence disease confirmation will require laboratory analysis. On tomato, Verticillium wilt tends to develop more slowly than Fusarium wilt.

Causal agent

The causal agent is *Verticillium dahliae*. The pathogen can be isolated on standard microbiological media, though semi-selective media such as NP-10 can be useful for isolation. On general purpose media, the pathogen forms the characteristic hyaline, verticillate conidiophores bearing three to four phialides at each node, and hyaline, single-celled, ellipsoidal conidia that measure 2-8 x 1-3 µm. Older cultures form dark brown to black torulose microsclerotia that consist of groups of swollen cells formed by repeated budding. Microsclerotia size varies greatly and is in the range of 15–100 µm in diameter. Microsclerotia enable the pathogen to survive in the soil for extended periods of time (up to 8 to 10 years). *Verticillium dahliae* has an extensive host range of crops and weeds. Two distinct tomato races have been documented.

491 Angular tomato leaf lesion caused by Verticillium wilt.

492 Dieback of tomato foliage due to Verticillium wilt.

493 Vascular discoloration of tomato stems affected by Verticillium wilt.

Disease cycle

The pathogen survives in the soil as dormant microsclerotia, but can also persist as epiphytes on non-host roots. Cool to moderate weather conditions favor the pathogen, and disease is enhanced at temperatures between 20–24° C. The fungus enters host roots through wounds, and later systemically infects tomato vascular tissue.

Control

Plant resistant or tolerant cultivars. Plants with the Ve gene are resistant to tomato race 1; however, resistance has not yet been identified for tomato race 2. It seems likely that new races of *V. dahliae* will continue to emerge and overcome the currently available genetic resistance. Pre-plant treatment of soil with effective fumigants will give short-term control but will not eradicate the pathogen from fields. For greenhouse production, steaming of soil can also provide short-term control. Rotate crops so that tomato is not planted in fields having a history of the problem. Rotation with non-host crops, such as small grains and corn, can lower inoculum levels but will not eradicate the pathogen. Minimize spread of infested soil to other, uninfested areas.

References

Ashworth, L. J., Huisman, O. C., Harper, D. M., and Stromberg, L. K. 1979. Verticillium wilt disease of tomato: influence of inoculum density and root extension upon disease severity. *Phytopathology* 69:490–492.

Baergen, K. D., Hewitt, J. D., and St. Clair, D. A. 1993. Resistance of tomato genotypes to four isolates of *Verticillium dahliae* race 2. *HortScience* 28:833–836.

Bletsos, F. A., Thanassoulopoulos, C. C., and Roupakias, D. G. 1999. Water stress and Verticillium wilt severity on eggplant (*Solanum melongena* L.). *Journal of Phytopathology* 147:243–248.

Erb, W. A. and Rowe, R. C. 1992. Screening tomato seedlings for multiple disease resistance. *Journal of the American Society for Horticultural Science* 117:622–627.

Grogan, R. G., Ioannou, N., Schneider, R. W., Sall, M. A., and Kimble, K. A. 1979. Verticillium wilt on resistant tomato cultivars in California: virulence of isolates from plants and soil and relationship of inoculum density to disease incidence. *Phytopathology* 69:1176–1180.

Harrington, M. A. and Dobinson, K. F. 2000. Influences of cropping practices on *Verticillium dahliae* populations in commercial processing tomato fields in Ontario. *Phytopathology* 90:1011–1017.

Nagao, H., Shiraishi, T., Oshima, S., Koike, M., Iijima, T. 1997. Assessment of vegetative compatibility of race-2 tomato wilt isolates of *Verticillium dahliae* in Japan. *Mycoscience* 38:379–385.

Alfalfa mosaic virus

ALFALFA MOSAIC

Introduction and significance

Alfalfa mosaic virus (AMV) is present throughout the world but is a serious production concern in only certain regions.

Symptoms and diagnostic features

Tomato plants that are infected early in their development may die. Leaves will develop irregularly shaped, bright yellow patches and a bronze discoloration. Foliage of severely diseased plants will curl downward. A red to red-brown discoloration occurs in the vascular tissue of the lower main stem. Symptomatic fruit can be distorted and exhibit irregularly shaped, sunken, dark brown, necrotic spots, rings, and patches (**494**).

Causal agent

AMV is the only member of the alfamovirus group and has particles that are bacilliform and measure 30–56 x 18 nm. The particles contain three single-stranded genomic RNAs and a fourth subgenomic RNA. AMV is transmitted by a number of aphid vectors and can be seedborne in tomato.

Disease cycle

Disease is usually most severe when tomato fields are planted close to infected alfalfa plantings. AMV is also seedborne in tomato.

Control

Follow general suggestions in Part 1. Use seed that does not have significant levels of the pathogen

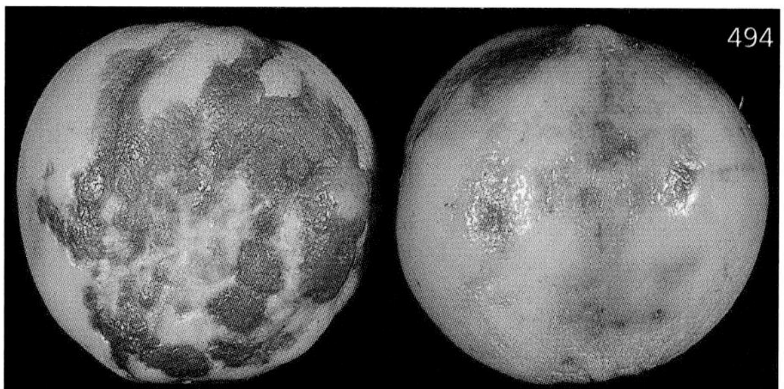

494 Fruit symptoms caused by *Alfalfa mosaic virus* on tomato.

Beet curly top virus

BEET CURLY TOP

Introduction and significance

Beet curly top virus (BCTV) occurs in North and South America, Asia, the Middle East, and the Mediterranean region. This virus is an important pathogen of many crops such as tomato, pepper, and *Chenopodium* plants.

Symptoms and diagnostic features

If infected early in development, tomato plants will die. Plants that are infected later will remain extremely stunted and become very chlorotic with a bronze or purple tinge (**495**). Leaves become thick and brittle in texture, chlorotic with purple veins, and roll upwards (**496**). Fruit ripen prematurely and are small, wrinkled, and red.

495 Stunted and chlorotic symptoms of *Beet curly top virus* on tomato.

496 Leaf curling and purple vein symptoms of *Beet curly top virus* on tomato.

Causal agent

BCTV is a geminivirus with isometric particles that measure 18–22 nm in diameter and which occur singly or in pairs. The BCTV genome is a single-stranded circular DNA. BCTV is vectored in a persistent manner by the beet leafhopper (*Circulifer tenellus*). *Circulifer opacipennis* is a vector in the Mediterranean region. In the plant, BCTV is restricted to the phloem tissue. On a molecular level, researchers have compared strains of BCTV from North America and the Middle East and found them to be similar, providing evidence that these various BCTV strains share a common origin.

Disease cycle

This virus infects many weed and crop hosts. Recent research on curly top disease as it occurs in Amaranthaceae, Fabaceae, Solanaceae, and other crops indicates that the viral agent may differ depending upon the host being considered. Beet curly top as a disease may actually be caused by one of four different curly top virus species: *Beet curly top virus* (BCTV), *Beet mild curly top virus* (BMCTV), *Beet severe curly top virus* (BSCTV), and *Spinach curly top virus* (SCTV). Research is ongoing to further determine the relationships of these various viruses.

Control

Follow general suggestions for managing virus diseases (see Part 1).

References

Briddon, R. W., Stenger, D. C., Bedford, I. D., Stanley, J., Izadpanah, K., Markham, P. G. 1998. Comparison of a *beet curly top virus* isolate originating from the old world with those from the new world. *European Journal of Plant Pathology* 104:77–84.

Martin, M. W. and Thomas, P. E. 1986. Increased value of resistance to infection if used in integrated pest management control of tomato curly top. *Phytopathology* 76:540–542.

Soto, M. J. and Gilbertson, R. L. 2003. Distribution and rate of movement of the curtovirus *Beet mild curly top virus* (family Geminiviridae) in the beet leafhopper. *Phytopathology* 93:478–484.

Thomas, P. E. and Martin, M. W. 1972. Characterization of a factor of resistance in *curly top virus*-resistant tomatoes. *Phytopathology* 62:954–958.

Wang, H., Gurusinghe, P. de A., and Falk, B. W. 1999. Systemic insecticides and plant age affect *beet curly top virus* transmission to selected host plants. *Plant Disease* 83:351–355.

Cucumber mosaic virus

CUCUMBER MOSAIC

Introduction and significance

Cucumber mosaic virus (CMV) is commonly found throughout the world and can cause disease on over 800 crop and weed hosts, including tomato and many other vegetable crops. CMV is most prevalent in temperate regions.

Symptoms and diagnostic features

CMV severely stunts tomato plant growth and early infections result in small, yellow, bushy plants. Leaves may show a mottled pattern. A most striking symptom occurs when leaf blades do not develop and the leaf takes on an elongated, filiform shape known as 'shoe-string' or 'strap-leaf' (**497**). Infected plants may not produce many fruit, or fruit that do develop are small and slow to mature.

Causal agent and disease cycle

CMV is a cucumovirus with particles that are isometric in shape (29 nm in diameter) and contain three single-stranded RNAs. Many CMV strains have been documented, and on tomato alone many different strains have been found. CMV is transmitted in a nonpersistent manner by several aphid vectors.

Control

CMV is difficult to control because of its extremely wide host range. Follow general suggestions for managing virus diseases (see Part 1). Use seed that does not have significant levels of the pathogen.

References

Fuchs, M., Provvidenti, R., Slightom, J. L., and Gonsalves, D. 1996. Evaluation of transgenic tomato plants expressing the coat protein gene of *cucumber mosaic virus* strain WL under field conditions. *Plant Disease* 80:270–275.

Gallitelli, D. 2000. The ecology of *cucumber mosaic virus* and sustainable agriculture. *Virus Research* 71:9–21.

Hellwald, K.-H., Zimmermann, C., and Buchenauer, H. 2000. RNA 2 of *cucumber mosaic virus* subgroup I strain NT-CMV is involved in the induction of severe symptoms in tomato. *European Journal of Plant Pathology* 106:95–99.

Palukaitis, P., Roossinck, M. J., Dietzgen, R. G., and Francki, R. I. B. 1992. *Cucumber mosaic virus*. *Advances in Virus Research* 41:281–348.

497 Strap-leaf symptoms of *Cucumber mosaic virus* on tomato.

Potato virus Y

POTATO VIRUS Y

Introduction and significance

Potato virus Y (PVY) is a pathogen of solanaceous plants around the world and is of major economic importance.

Symptoms and diagnostic features

Symptoms can vary greatly but generally consist of veinbanding, in which dark green bands form along leaf veins, and a downward rolling of the leaves. Mottle and mosaic patterns and some leaf distortions can also occur. As disease develops, leaves can show brown, necrotic interveinal lesions. Fruit usually do not exhibit symptoms.

Causal agent and disease cycle

PVY is a potyvirus with long (730 x 11 nm) flexuous rods that contain single-stranded RNA. It is vectored in a nonpersistent manner by several aphids, with *Myzus persicae* being particularly important. This virus is also readily transmitted mechanically.

Control

Follow general suggestions for managing virus diseases (see Part 1).

Tobacco mosaic virus, Tomato mosaic virus

TOBACCO MOSAIC, TOMATO MOSAIC

Introduction and significance

Tobacco mosaic (TMV) and *Tomato mosaic* (ToMV) *viruses* are two closely related virus pathogens. Both viruses can infect tomato and other solanaceous plants.

Symptoms and diagnostic features

TMV and ToMV both can infect tomato and cause light and dark green mottling or mosaics on foliage. Leaflets may be deformed and narrow, giving the leaf a fern-like appearance (**498**). Fruit set may be reduced, and fruit may ripen unevenly.

Causal agents and disease cycle

TMV and ToMV are viruses in the tobamovirus group. These viruses have straight rod particles that measure 300 x 18 nm and contain single-stranded, linear RNA genomes. TMV has no known vector and is readily transmitted mechanically. ToMV also lacks a known vector, is transmitted mechanically, and can be seedborne in tomato.

Control

Follow general suggestions for managing virus diseases (see Part 1). Because ToMV is seedborne in tomato, use seed that does not have significant levels of the pathogen. Seed infested with ToMV can also be treated with dry heat (70° C for 2 to 4 days) or with trisodium phosphate (10% for 15 minutes).

498 Distorted tomato leaves caused by *Tobacco mosaic virus*. Healthy leaf is in the center.

References

Broadbent, L. 1976. Epidemiology and control of *tomato mosaic virus*. *Annual Review of Phytopathology* 14:75–96.

Mayhew, D. E., Hedin, P., and Thomas, D. L. 1984. Corky ringspot: a new strain of *tomato mosaic virus* in California. *Plant Disease* 68:623–625.

Pelham, J. 1972. Strain-genotype interaction of *tobacco mosaic virus* in tomato. *Annals of Applied Biology* 71:219–228.

Tomato spotted wilt virus

TOMATO SPOTTED WILT

Introduction and significance

In some regions this virus is very common on tomato, pepper, and many other crops.

Symptoms and diagnostic features

Tomato spotted wilt virus (TSWV) causes leaves to develop irregularly shaped to circular, black, small (3-8 mm in diameter) spots (**499**). Stems and shoots may have black streaks or lesions on the epidermis (**500**). Severely affected plants may wilt or be stunted. Symptomatic fruit will develop chlorotic rings, patches, and lesions (**501, 502**).

Causal agent

TSWV is a tospovirus and has isometric particles, measuring approximately 80–110 nm, which are surrounded with membranes. TSWV has a genome consisting of three linear single-stranded RNAs and is vectored by several species of thrips; at least eight species are found to be vectors. The western flower (*Frankliniella occidentalis*) and tobacco (*F. fusca*) thrips are probably the most important vectors.

Disease cycle

TSWV has one of the most extensive host ranges of any known plant virus and can infect over 900 different cultivated and weedy plant species. Therefore initial inoculum can come from any number of landscape plants, weeds, and other plants. Thrips insects vector the virus to the tomato crops. It is well documented that thrips insects can only acquire the virus as larvae that feed on diseased plants; after acquiring the virus, the insects carry the virus for the rest of their lives.

499 **Leaf lesions due to *Tomato spotted wilt virus* on tomato.**

501 **Green fruit symptoms caused by *Tomato spotted wilt virus* on tomato.**

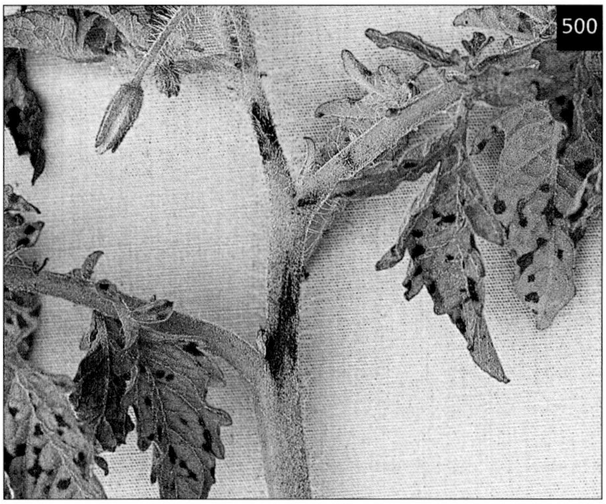

500 **Stem lesions due to *Tomato spotted wilt virus* on tomato.**

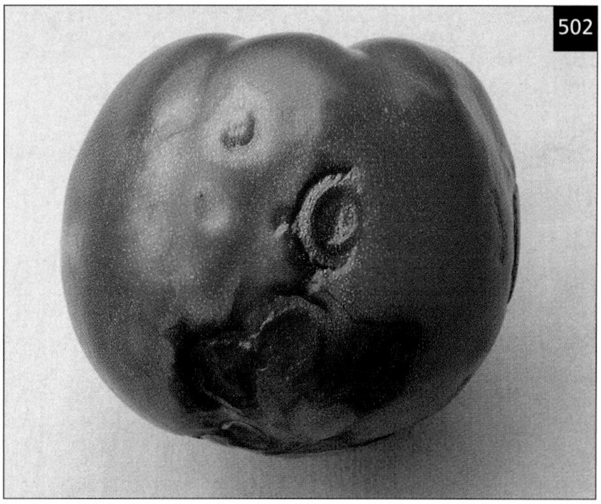

502 **Ripe fruit symptoms caused by *Tomato spotted wilt virus* on tomato.**

Control

Follow general suggestions for managing virus diseases (see Part 1). The broad host range of TSWV and the difficulty in controlling thrips makes this disease particularly difficult to manage.

References

Best, R. J. 1968. *Tomato spotted wilt virus. Advances in Virus Research* 13:66–146.

Cho, J. J., Mau, R. F. L., Gonsalves, D., and Mitchell, W. C. 1986. Reservoir weed hosts of *tomato spotted wilt virus*. *Plant Disease* 70:1014–1017.

Greenough, D. R., Black, L. L., and Bond, W. P. 1990. Aluminum-surfaced mulch: an approach to the control of *tomato spotted wilt virus* in solanaceous crops. *Plant Disease* 74:805–808.

Groves, R. L., Walgenbach, J. F., Moyer, J. W., and Kennedy, G. G. 2002. The role of weed hosts and tobacco thrips, *Frankliniella fusca*, in the epidemiology of *tomato spotted wilt virus*. *Plant Disease* 86:573–582.

Krishna Kumar, N. K., Ullman, D. E., and Cho, J. J. 1993. Evaluation of *Lycopersicon* germ plasm for *tomato spotted wilt tospovirus* resistance by mechanical and thrips transmission. *Plant Disease* 77:938–941.

Momol, M. T., Olson, S. M., Funderburk, J. E., Stavisky, J., and Marois, J. J. 2004. Integrated management of tomato spotted wilt on field grown tomatoes. *Plant Disease* 88:882–890.

Tomato yellow leaf curl virus

TOMATO YELLOW LEAF CURL

Introduction and significance

Tomato yellow leaf curl virus (TYLCV) is one of the most damaging tomato virus diseases in tropical and sub-tropical areas, where entire tomato plantings can become infected if vector populations are high. The disease has become widely distributed and has been reported in the Mediterranean region (particularly in southern Europe), the Middle East (Israel), Africa, and recently in North America (Caribbean area, southeastern USA, and Mexico).

Symptoms and diagnostic features

If infected at a young stage, tomato plants can be severely stunted and will not produce fruit (**503**). Foliage shows an upright or erect growth habit, leaves curl upwards and may be crumpled. Interveinal chlorosis is also observed in the leaves (**504**).

Causal agent and disease cycle

TYLCV is a geminivirus that consists of twinned icosahedral particles and single-stranded DNA genomes. Bean, pepper, and several weeds are also hosts. TYLCV is vectored in a persistent manner by the sweet potato whitefly (*Bemisia tabaci*). This virus is apparently not seedborne in tomato. Researchers in Europe believe two distinct species of this virus exist: *Tomato yellow leaf curl-Sardinia* (TYLCV-Sar), *Tomato yellow leaf curl-Israel* (TYLCV-Is).

503 Stunted tomato plant affected by *Tomato yellow leaf curl virus*.

504 Yellow, distorted tomato foliage caused by *Tomato yellow leaf curl virus*.

Control

Follow general suggestions for managing virus diseases (see Part 1).

References

Accotto, G. P., Navas-Castillo, J., Noris, E., Moriones, E., and Louro, D. 2000. Typing of *tomato yellow leaf curl viruses* in Europe. *European Journal of Plant Pathology* 106:179–186.

Delatte, H., Dalmon, A., Rist, D., Soustrade, I., Wuster, G., Lett, J. M., Goldbach, R. W., Peterschmitt, M., and Reynaud, B. 2003. *Tomato yellow leaf curl virus* can be acquired and transmitted by *Bemisia tabaci* (*Gennadius*) from tomato fruit. *Plant Disease* 87:1297–1300.

Lapidot, M., Friedmann, M., Lachman, O., Yehezkel, A., Nahon, S., Cohen, S., and Pilowsky, M. 1997. Comparison of resistance level to *tomato yellow leaf curl virus* among commercial cultivars and breeding lines. *Plant Disease* 81:1425–1428.

McGlashan, D., Polston, J. E., and Bois, D. 1994. *Tomato yellow leaf curl virus* in Jamaica. *Plant Disease* 78:1219.

Nakhla, M. K., Maxwell, D. P., Martinez, R. P., Carvalho, M. G., and Gilbertson, R. L. 1994. Widespread occurrence of the Eastern Mediterranean strain of *tomato yellow leaf curl geminivirus* in tomatoes in the Dominican Republic. *Plant Disease* 78:926.

Polston, J. E. and Anderson, P. K. 1997. The emergence of whitefly-transmitted geminiviruses in tomato in the western hemisphere. *Plant Disease* 81:1358–1369.

Salati, R., Nahkla, M. K., Rojas, M. R., Guzman, P., Jaquez, J., Maxwell, D. P., and Gilbertson, R. L. 2002. *Tomato yellow leaf curl virus* in the Dominican Republic: characterization of an infectious clone, virus monitoring in whiteflies, and identification of reservoir hosts. *Phytopathology* 92:487–496.

Zubiaur, Y. M., Zabalgogeazcoa, I., de Blas, C., Sanchez, F., Peralata, E. L., Romero, J., and Ponz, F. 1996. Geminivirus associated with diseased tomatoes in Cuba. *Journal of Phytopathology* 144:277–279.

BLOSSOM END ROT

Symptoms and diagnostic features

Blossom end rot is a physiological disorder of tomato fruit. Initial symptoms can occur on green fruit and consist of small, light brown flecks and lesions that are usually clustered on the blossom end of the developing fruit. As the disorder worsens, the blossom end has a circular to oblong lesion that is dark brown to black in color, firm in texture, and sunken (**505, 506, 507**). The lesion can take up most of the blossom end. Secondary decay organisms may colonize the lesion and result in a fruit rot.

Causal agent

Blossom end rot is caused by a calcium deficiency in the tissues at that end of the fruit. Calcium is not particularly mobile in plant tissues, so the distal end of the fruit can experience shortages of this nutrient. Plants that are growing rapidly, subject to water stress, and fertilized with high nitrogen rates may be more susceptible. Uneven watering practices contribute to the development of the disorder. Pepper and squash fruit can also develop similar symptoms due to calcium deficiencies.

Control

To minimize damage from this disorder, do not over fertilize with high-nitrogen fertilizers. Irrigate crops regularly and with appropriate amounts of water. Calcium fertilizer supplements may help in some cases, but these treatments are generally ineffective. Choose cultivars that tend to develop this problem less frequently than others.

References

Bangerth, F. 1979. Calcium related physiological disorders of plants. *Annual Review of Phytopathology* 17:97–122.

Banuelos, G. S., Offermann, G. P., and Seim, E. C. 1985. High relative humidity promotes blossom end rot on growing tomato fruit. *HortScience* 20:894–895.

Evans, H., and Troxler, R. 1953. Relation of Ca nutrition to the incidence of blossom end rot in tomatoes. *Proceedings of the American Society for Horticultural Science* 61:346–357.

Taylor, M. D. Locascio, S. J., and Alligood, M. R. 2004. Blossom-end rot incidence of tomato as affected by irrigation quantity, calcium source, and reduced potassium. *HortScience* 39:1110–1115.

Van Goor, B. 1968. The role of Ca and cell permeability in the disease blossom end rot of tomatoes. *Physiologia Plantarum* 21:1110–1121.

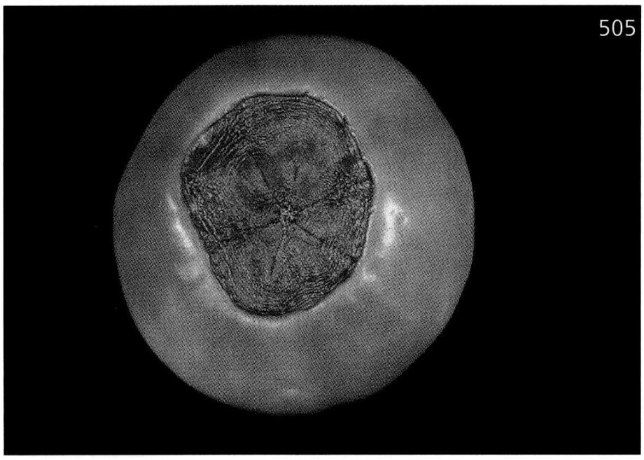

505 Fruit symptom of blossom end rot.

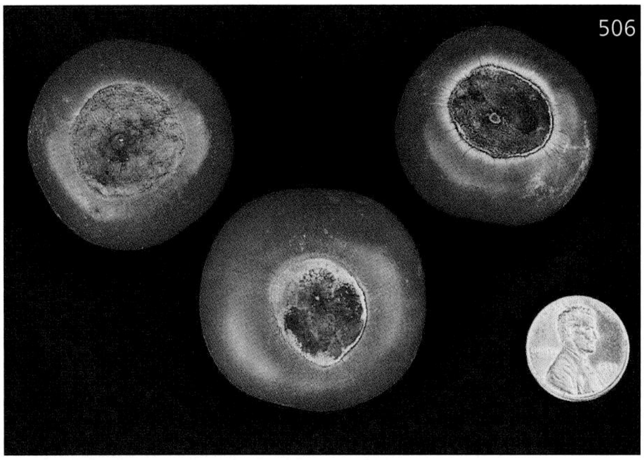

506 Fruit symptom of blossom end rot.

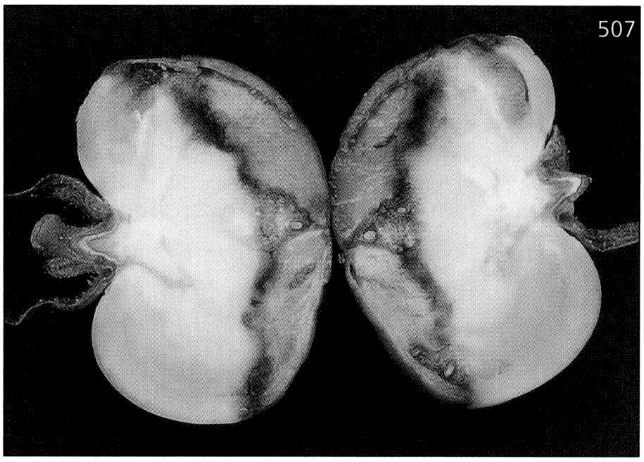

507 Internal fruit symptom of blossom end rot.

Spinacia oleracea Spinach

SPINACH (*Spinacia oleracea*) is in the Amaranthaceae (amaranth family) and has become an important leafy vegetable crop in many parts of the world. The plant probably originated in Iran (formerly Persia). Leaves of this plant are high in antioxidants and nutrients, and hence spinach is gaining popularity as a fresh salad and cooked vegetable commodity. In the USA spinach is now commonly grown on broad beds (2 meters wide) with extremely high seed density; such plantings can be harvested early for 'baby leaf' spinach or held longer and cut for standard fresh market spinach. Spinach in the USA is often harvested mechanically.

Pseudomonas syringae pv. *spinaciae*

BACTERIAL LEAF SPOT

Introduction and significance

Bacterial leaf spot was first reported on spinach in Europe (Italy) in 1988. Since that time the disease has also been detected in Japan and in the USA (California). The disease thus far is a minor problem.

Symptoms and diagnostic features

Initial symptoms consist of water-soaked, irregularly shaped spots that measure 2–5 mm in diameter (508). As the disease develops, these small spots enlarge to as much as 10–15 mm in diameter, are angular in shape due to delimiting from leaf veins, and turn dark brown with streaks of black (509). On occasion, faint yellow halos surround the spots. On leaves having numerous spots, the spots sometimes merge together, resulting in the death of large areas of the leaf (510). Spots are visible from both top and bottom sides of leaves. The disease occurs on both newly expanded and mature foliage.

Causal agent

Bacterial leaf spot is caused by the bacterium *Pseudomonas syringae* pv. *spinaciae*. This pathogen is an aerobic, Gram-negative bacterium. The pathogen can be isolated on standard microbiological media and produces cream-colored colonies typical of most pseudomonads. Strains are non-fluorescent when cultured on Kings medium B. The pathogen appears to be host specific to spinach.

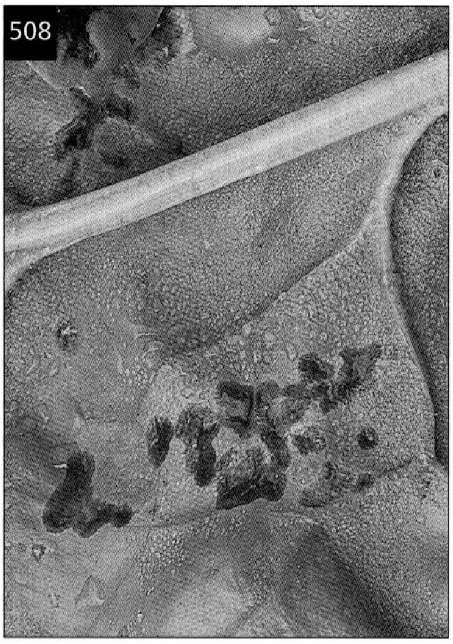

508 Initial water-soaked lesions due to bacterial leaf spot of spinach.

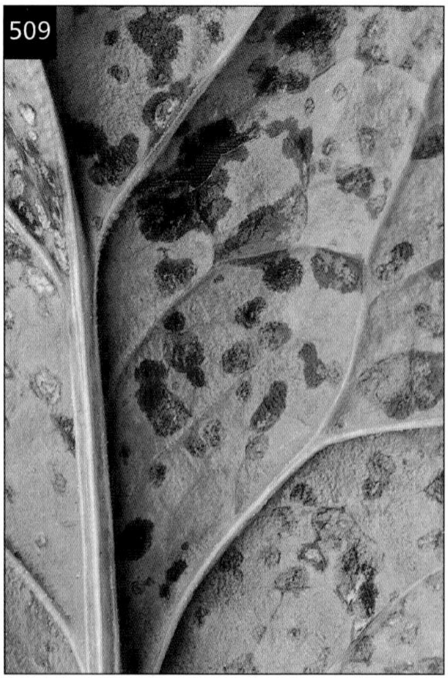

509 Advanced water soaked symptom of bacterial leaf spot of spinach.

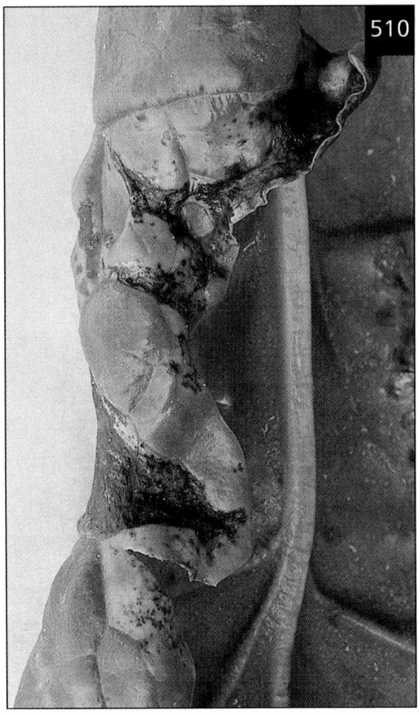

510 Advanced leaf spots due to bacterial leaf spot of spinach.

Disease cycle

Disease epidemiology is still being determined. *Pseudomonas syringae* pv. *spinaciae* may possibly be seedborne; if this is the case, then seedborne inoculum may be the primary means of pathogen introduction to spinach fields. Weed or other reservoir hosts have not been identified.

Control

Until the source of the pathogen is identified, control strategies will be incomplete. Avoid using overhead sprinkler irrigation because splashing water will spread the pathogen and encourage infection and disease development.

References

Bazzi, C., Gozzi, R., Stead, D., and Sellwood, J. 1988. A bacterial leaf spot of spinach caused by a non-fluorescent *Pseudomonas syringae*. *Phytopathologia Mediterranea* 27:103–107.

Koike, S. T., Azad, H. R., and Cooksey, D. C. 2002. First report of bacterial leaf spot of spinach, caused by a *Pseudomonas syringae* pathovar, in California. *Plant Disease* 86:921.

Ozaki, K., Kimura, T., and Matsumoto, K. 1998. *Pseudomonas syringae* pv. *spinaciae*, the causal agent of bacterial leaf spot of spinach in Japan. *Annals Phytopathological Society of Japan* 64:264–269.

Albugo occidentalis

WHITE RUST

Introduction and significance

White rust is a very damaging disease of spinach that occurs only in the USA. Thus far the disease is found in states that are east of the Rocky Mountains.

Symptoms and diagnostic features

Initial symptoms consist of small chlorotic spots and lesions. As the disease develops, raised white pustules or blisters appear in and around these chlorotic areas and can be seen on both leaf surfaces (**511**). These pustules often develop in concentric circles or in rings that surround the yellow lesion. In advanced stages of the disease, the white pustules coalesce, lesions appear grainy due to the presence of oospores, and the leaf tissues can become necrotic.

Causal agent

White rust is caused by the obligate pathogen *Albugo occidentalis*, an oomycete organism. *Albugo occidentalis* can infect spinach and some *Chenopodium* species. Distinct physiological races have not been identified. The pathogen produces hyaline, globose to oval sporangia that measure 10–14 µm in diameter when they are dry and 10–19 x 20–22 µm when hydrated. Sporangia are borne in chains. Yellowish oospores have a finely reticulate surface and are 44–62 µm in diameter.

Disease cycle

Sporangia are produced within the white rust pustules (sori) and are released when the leaf epidermis that

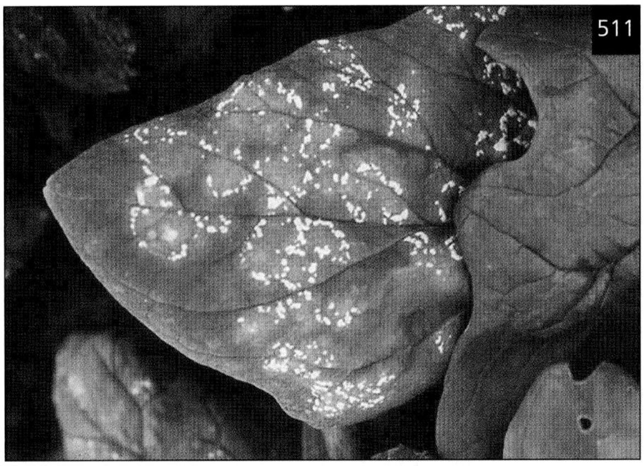

511 Spinach leaf showing white rust signs and symptoms.

encloses the sori ruptures. They are then blown onto leaf surfaces via winds and splashing water and usually germinate indirectly by releasing six to nine motile zoospores. The zoospores encyst and produce germ tubes that penetrate the surface and infect the leaf. Optimum temperatures for sporangia production and germination are 22° C and 12–16° C, respectively. At least 12 hours of leaf wetness is required for infection, and disease severity increases with wetness duration up to 84 hours. Oospores are produced extensively in infected tissues. Due to their resilient nature, oospores are an important survival structure for *A. occidentalis* and probably are a major source of primary inoculum. However, their precise role in white rust epidemiology is uncertain.

Control

Plant resistant cultivars. Practice crop rotation to help reduce soil inoculum levels. Apply preplant and foliar fungicides such as metalaxyl. Weather data are being used to attempt to schedule fungicide applications.

References

Brandenberger, L. P., Correll, J. C., Morelock, T. E., and McNew, R. W. 1994. Characterization of resistance of spinach to white rust (*Albugo occidentalis*) and downy mildew (*Peronospora farinosa* f. sp. *spinaciae*). *Phytopathology* 84:431–437.

Dainello, F. J., Black, M. C., and Kunkel, T. E. 1990. Control of white rust of spinach with partial resistance and multiple soil applications of metalaxyl granules. *Plant Disease* 74:913–916.

Sullivan, M. J. and Damicone, J. P. 2003. Development of a weather-based advisory program for scheduling fungicide applications for control of white rust of spinach. *Plant Disease* 87: 923–928.

Aphanomyces cochlioides, Fusarium oxysporum, Pythium aphanidermatum, P. irregulare, Rhizoctonia solani

DAMPING-OFF, ROOT ROTS

Introduction and significance

Spinach is very susceptible to damping-off and root rot diseases and overwatering situations. Significant plant stand loss can be the result of damping-off that occurs early in the crop cycle.

Symptoms and diagnostic features

With pre-emergence damping-off, spinach seed and newly germinated seedlings are attacked and rotted prior to the above-ground emergence of the seedling. Symptoms of post-emergence damping-off consist of stunted plants, yellowed lower leaves, general poor growth (**512**), wilting, and eventual collapse and death of plants. Roots of infected plants appear water-soaked or brown to black in color (**513**). Areas of the taproot may be girdled or damaged by a necrotic lesion. In severe cases, nearly all roots may be girdled or rotted off. Care must be taken when diagnosing these soil-borne diseases because symptoms resulting from abiotic problems, caused by factors such as overwatering and poor planting technique, can look identical to symptoms caused by damping-off pathogens. As a plant, spinach is very sensitive to excess soil moisture.

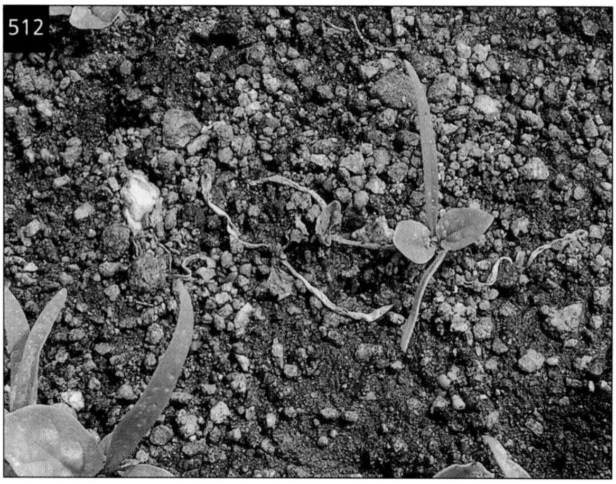

512 *Pythium* damping-off of spinach.

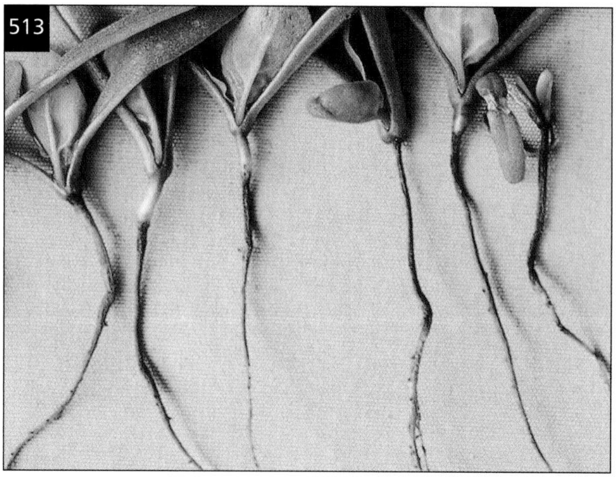

513 Root lesions of *Pythium* damping-off of spinach.

Causal agents

Damping-off diseases are caused by several pathogens, including the following: *Fusarium oxysporum*, *Pythium aphanidermatum*, *P. irregulare*, *Aphanomyces cochlioides*, and *Rhizoctonia solani*. In some areas, the Fusarium wilt pathogen (*F. oxysporum* f. sp. *spinaciae*) can cause damping-off. In Europe, *Phytophthora cryptogea* causes spinach root rot disease. In California, there are fields where *Fusarium*, *Pythium*, and *Rhizoctonia* all are present and contribute to root rot problems.

Disease cycle

These organisms are soil inhabitants and are therefore permanent residents in infested soils. Damping-off can result from the activity of an individual pathogen or from a complex of several of these agents working simultaneously. Severity of damping-off is influenced by cultivar, soil texture and profile, irrigation management, production technique, and pathogen populations. Severe damping-off is often associated with warmer temperatures, heavy textured or poorly draining soils, low areas in the field, and fields that have a history of frequent spinach production.

Control

Plant spinach in soils that drain well. Do not plant spinach seeds too deeply into the soil. Prepare seed beds so that even, rapid germination is enhanced. Carefully manage the irrigation schedule to prevent flooding and saturated soil conditions. Plant seed that is treated with fungicides. Applying metalaxyl to spinach seed lines immediately after planting can help control *Pythium* species.

References

Larsson, M. 1994. Prevalence and pathogenicity of spinach root pathogens of the genus *Pythium* in Sweden. *Plant Pathology* 43: 261–268.

Larsson, M. and Gerhardson, B. 1992. Disease progression and yield losses from root diseases caused by soilborne pathogens of spinach. *Phytopathology* 82:403–406.

Larsson, M. and Olofsson, J. 1994. Prevalence and pathogenicity of spinach root pathogens of the genera *Aphanomyces, Phytophthora, Fusarium, Cylindrocarpon*, and *Rhizoctonia* in Sweden. *Plant Pathology* 43: 251–260.

Sumner, D. R., Kays, S. J., and Johnson, A. W. 1976. Etiology and control of root diseases of spinach. *Phytopathology* 66:1267–1273.

Cladosporium variabile

CLADOSPORIUM LEAF SPOT

Introduction and significance

Cladosporium leaf spot is regularly found in regions such as California, but the disease is rarely severe on production spinach unless there are significant rains during the growing season. However, the disease can significantly damage spinach seed crops.

Symptoms and diagnostic features

This disease is characterized by mostly round, tan leaf spots that rarely exceed 1 cm in diameter (**514, 515**). Dark green spores and mycelium later develop in the

514 Leaf spots due to Cladosporium leaf spot of spinach.

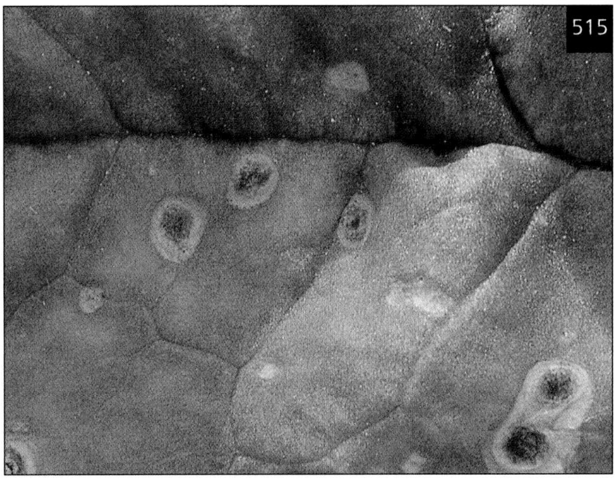

515 Close-up of Cladosporium leaf spots.

centers of these spots and are characteristic signs of this pathogen. The presence of dark green sporulation distinguishes Cladosporium leaf spot from anthracnose and Stemphylium leaf spot diseases, both of which also form circular lesions. The disease is commonly seen on spinach seed crops grown in the northwest USA, especially if weather is cool and moist. For seed crops, disease severity is greater if pollen from spinach flowers falls on leaves prior to infection.

Causal agent

Cladosporium leaf spot is caused by the fungus *Cladosporium variabile* (formerly named *Heterosporium variabile*). Conidiophores of *C. variabile* emerge from leaf tissue and appear in clusters. Conidiophores are straight, unbranched, geniculate, and are light brown in color. Conidia are borne in short chains (three to five spores) and are ellipsoidal to subglobose with rounded ends, brown to green-brown, and roughly textured (verrucose). Conidia have zero to three septa and measure 5–30 x 3–13 µm. In culture *C. variabile* forms aerial hyphae that are characteristically twisted and coiled (torulose).

Disease cycle

The complete epidemiology of this disease has not been documented. The pathogen is seedborne and has been detected on spinach seed produced in both Europe and the USA. Conidia are dispersed by winds and splashing water from rain and sprinkler irrigation. In California the disease is always most severe in early spring (February through April) if there is rainy weather.

Control

No control recommendations are available for production spinach in the field. Disease severity may increase if overhead sprinkler irrigation is used. Use seed that does not have significant levels of the pathogen. Hot water or 1.2% bleach seed treatments can significantly reduce seedborne inoculum.

References

Correll, J. C., Morelock, T. E., Black, M. C., Koike, S. T., Brandenberger, L. P., and Dainello, F. J. 1994. Economically important diseases of spinach. *Plant Disease* 78:653–660.

David, J. C. 1995. *Cladosporium variabile. Descriptions of Pathogenic Fungi and Bacteria No. 1229*. International Mycological Institute, Surrey, England.

du Toit, L. J., Derie, M. L., and Hernandez-Perez, P. 2005. Evaluation of fungicides for control of leaf spot in spinach seed crops, 2004. *Fungicide & Nematicide Tests* 60:V044.

du Toit, L. J., Derie, M. L., and Hernandez-Perez, P. 2005. Evaluation of yield loss caused by leaf spot fungi in spinach seed crops, 2004. *Fungicide & Nematicide Tests* 60:V047.

du Toit, L. J. and Hernandez-Perez, P. 2005. Efficacy of hot water and chlorine for eradication of *Cladosporium variabile, Stemphylium botryosum*, and *Verticillium dahliae* from spinach seed. *Plant Disease* 89:1305–1312.

Hernandez-Perez, P. and du Toit, L. J. 2006. Seedborne *Cladosporium variabile* and *Stemphylium botryosum* in spinach. *Plant Disease* 90:137–145.

Inglis, D. A., Derie, M. L., and Gabrielson, R. L. 1997. Cladosporium leaf spot on spinach seed crops and control measures. *Washington State University Extension Bulletin* No. 865.

Colletotrichum dematium f. sp. *spinaciae*

ANTHRACNOSE

Introduction and significance

Anthracnose outbreaks occur sporadically in spinach growing areas. In Europe, Asia, and North America it is usually only a minor problem.

Symptoms and diagnostic features

Initial symptoms are small, circular, water-soaked lesions on both young and old leaves (**516**). Lesions later enlarge, first turning chlorotic and then later brown to tan in color (**517**). The brown lesions can become dry, thin, and papery in texture. In severe cases, lesions coalesce and result in severe blighting of foliage. Tiny, black fruiting bodies (acervuli) form profusely in diseased tissue and are a characteristic sign of the pathogen. The presence of acervuli distinguishes anthracnose from Cladosporium and Stemphylium leaf spot diseases, both of which also form circular lesions.

Causal agent

Anthracnose is caused by the fungus *Colletotrichum dematium* f. sp. *spinaciae*. The fungus produces a cup-shaped acervulus fruiting body that usually contains long, dark brown setae along with the hyaline, fusiform, slightly curved conidia having acute apices and measuring 20–24 x 2–3 µm. Conidia are extruded in a light pink gelatinous matrix that is primarily spread by splashing water. When *C. dematium* isolates from spinach, tomato, and onion were inoculated onto spinach, the non-spinach isolates were less virulent than the spinach isolates. Such research indicates that host specificity for this pathogen, indicated by the forma specialis (f. sp.) designation, is demonstrated.

Disease cycle

The fungus survives as dormant mycelium in infected plant debris. Seedborne inoculum is also possible, though this source of the pathogen is not well documented. Conidia are moved from plant to plant by splashing water from rain or sprinklers. Wet conditions, along with dense leaf canopies, limited air movement, and low plant fertility favor infection and disease development. This fungus also acts as a secondary decay organism on spinach and readily colonizes tissue damaged by pathogens such as the white rust pathogen (*Albugo occidentalis*).

Control

Reduce leaf wetness by eliminating overhead sprinkler irrigations, or by irrigating by sprinklers early in the day. Provide adequate fertilizer for the spinach crop. Effective fungicides are not registered, and resistant cultivars are still under development.

References

Correll, J. C., Morelock, T. E., and Guerber, J. C. 1993. Vegetative compatibility and virulence of the spinach anthracnose pathogen, *Colletotrichum dematium*. *Plant Disease* 77:688–691.

Koike, S. T. and Correll, J. C. 1993. First report of spinach anthracnose caused by *Colletotrichum dematium* in California. *Plant Disease* 77:318.

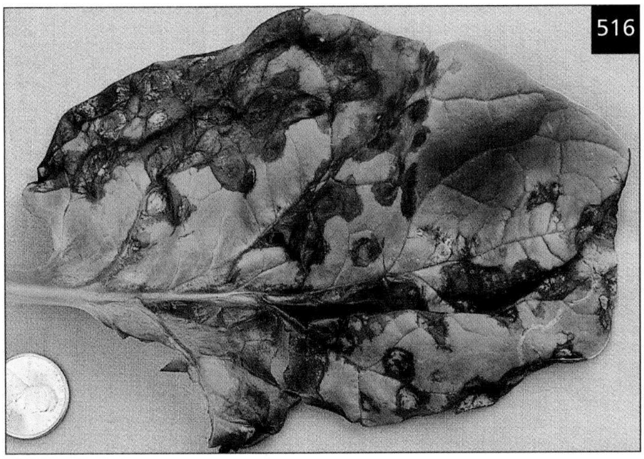

516 Water soaked lesions caused by anthracnose of spinach.

517 Advanced lesions and acervuli caused by anthracnose of spinach.

Fusarium oxysporum f. sp. *spinaciae*

FUSARIUM WILT

Introduction and significance

Fusarium wilt is an important disease of spinach worldwide. However, disease incidence varies from region to region. In the extensive spinach industry in California, this disease occurs only sporadically.

Symptoms and diagnostic features

Typical symptoms include plant stunting, chlorosis and necrosis of older leaves, wilting of the older leaves, and plant collapse (**518**). The vascular tissue of the diseased plant is often discolored and can be brown, dark brown, or black. In particular, the vascular tissue of the taproot tip may show the greatest degree of discoloration and the surrounding root cortex may be necrotic and black (**519**).

Causal agent

Fusarium wilt is caused by the soilborne fungus *Fusarium oxysporum* f. sp. *spinaciae*. The pathogen morphology and colony characteristics are similar to other *F. oxysporum* fungi. The fungus forms one- or two-celled, oval to kidney-shaped microconidia on monophialides, and three- to five-celled, fusiform, curved macroconidia. Macroconidia are usually produced in cushion-shaped structures called sporodochia. Chlamydospores are also formed. The pathogen is usually readily isolated from symptomatic vascular tissue. The pathogen is host specific to spinach and may be seedborne. In some spinach producing regions this pathogen can also contribute to damping-off disease.

Disease cycle

The pathogen produces chlamydospores that enable it to survive long periods of time in the soil. If spinach is planted in the field, pathogen propagules germinate and invade the spinach plants via the roots. From the roots, the pathogen can become systemic and grow into spinach vascular tissue.

Control

Plant resistant cultivars as they become available. Practice crop rotation so that soilborne inoculum does not build up to high levels. Avoid planting in fields with history of severe Fusarium wilt.

518 Chlorosis and decline due to Fusarium wilt of spinach.

519 Darkened roots due to Fusarium wilt of spinach.

References

Bassi, A., Jr. and Bode, M. J. 1978. *Fusarium oxysporum* f. sp. *spinaciae* seedborne in spinach. *Plant Disease Reporter* 62:203–205.

Fiely, M. B., Correll, J. C., and Morelock, T. E. 1995. Vegetative compatibility, pathogenicity, and virulence diversity of *Fusarium oxysporum* recovered from spinach. *Plant Disease* 79:990–993.

O'Brien, M. J. and Winters, H. F. 1977. Evaluation of spinach accessions and cultivars for resistance to Fusarium wilt: I. Greenhouse method. *Journal of the American Society for Horticultural Science* 102:424–426.

Reyes, A. A. 1979. Populations of the spinach wilt pathogen, *Fusarium oxysporum* f. sp. *spinaciae*, in the root tissues, rhizosphere, and soil in the field. *Canadian Journal of Microbiology* 25:227–229.

Takehara, T., Kuniyasu, K., Mori, M., and Hagiwara, H. 2003. Use of a nitrate-nonutilizing mutant and selective media to examine population dynamics of *Fusarium oxysporum* f. sp. *spinaciae* in soil. *Phytopathology* 93:1173–1181.

Peronospora farinosa f. sp. *spinaciae*

DOWNY MILDEW, BLUE MOLD

Introduction and significance

Downy mildew, or blue mold, is one of the most widespread and destructive diseases of spinach. In most areas, downy mildew is generally considered the most important disease of spinach.

Symptoms and diagnostic features

Early symptoms consist of initially light green to dull yellow, irregularly shaped lesions on cotyledons and true leaves (**520**). These lesions later turn bright yellow. With time, the lesions can enlarge and become tan and dry, or if wet conditions persist the tissue can become soft and necrotic. Examination of the underside of the leaf, opposite the yellow area, usually reveals the purple growth of the pathogen (**521**). Such sporulation can at times occur on the top leaf surfaces. If disease development is extensive, leaves can appear curled and distorted and may take on a blighted effect due to the numerous infection sites.

Causal agent

Downy mildew is caused by *Peronospora farinosa* f. sp. *spinaciae*, which belongs in the oomycete group of organisms. Older literature names the pathogen as *P. effusa*. The conidiophores of this pathogen emerge from leaf stomates, branch extensively, and have tapered tips on which blue-gray conidia are borne. Conidia are ovoid to ellipsoid and measure 21–27 x 16–19 µm. *Peronospora farinosa* f. sp. *spinaciae* is heterothallic and requires two mating types for the production of the sexual oospore. Oospores are spherical, thick walled, and measure 20–38 µm. *Peronospora farinosa* f. sp. *spinaciae* is a complex pathogen and consists of multiple races. Races are identified by testing their virulence on differential sets of spinach cultivars having various resistance genes. In the early 2000s, races 5 and 6 were prevalent in the USA, while races 5 and 7 occurred in Europe. However, additional new races are being detected in both the USA (race 10) and Europe (race 8). This pathogen infects only spinach and a few *Chenopodium* weed species, such as *C. album* (lamb's quarters or fat hen).

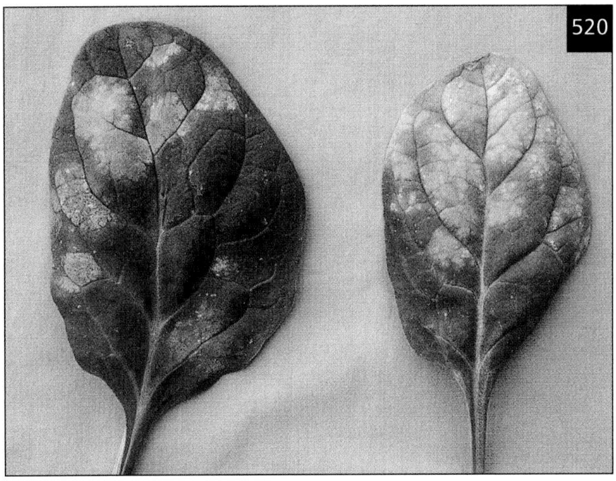

520 Leaf chlorosis caused by downy mildew of spinach.

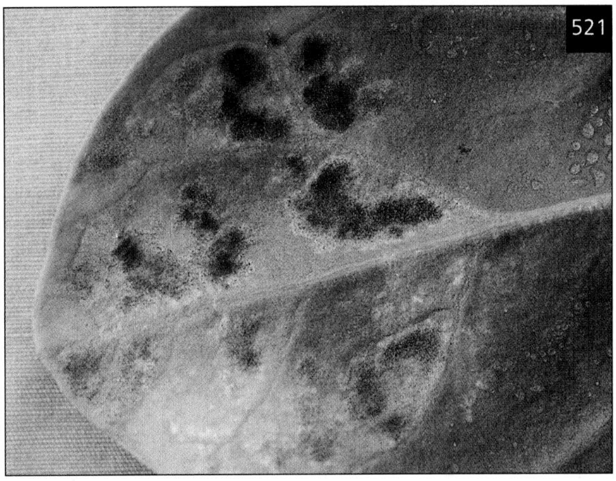

521 Sporulation of downy mildew on underside of spinach leaf.

Disease cycle

Like most downy mildews, this pathogen requires cool, wet conditions for infection and disease development. The heavy canopy of densely planted spinach retains much moisture and creates ideal conditions for infection and disease development. Spores are dispersed in the air from plant to plant and field to field by winds and to a lesser extent splashing water. Lesion development is enhanced at temperatures between 15–25° C. If favorable temperatures and leaf wetness are present, downy mildew can progress rapidly and result in significant disease. Resilient oospores have been detected on spinach seed and presumably can occur in leaf tissue. However, the role of oospores in disease epidemiology is uncertain.

Control

Use resistant cultivars, as this is the most satisfactory means of controlling spinach downy mildew. Historically, single gene resistance has been relied on to breed resistant cultivars. When new races of the pathogen develop, however, cultivars with this type of resistance are readily infected and new sources of resistance must be found. To prevent downy mildew on young seedlings, treat seed with metalaxyl, or apply metalaxyl post-plant to the seed lines. Apply foliar fungicides to protect emerged crops.

References

Brandenberger, L. P., Correll, J. C., and Morelock, T. E. 1991. Identification of and cultivar reactions to a new race (race 4) of *Peronospora farinosa* f. sp. *spinaciae* on spinach in the United States. *Plant Disease* 75:630–634.

Brandenberger, L. P., Correll, J. C., Morelock, T. E., and McNew, R. W. 1994. Characterization of resistance of spinach to white rust (*Albugo occidentalis*) and downy mildew (*Peronospora farinosa* f. sp. *spinaciae*). *Phytopathology* 84:431–437.

Brandenberger, L. P., Morelock, T. E., and Correll, J. C. 1992. Evaluation of spinach germplasm for resistance to a new race (Race 4) of *Peronospora farinosa* f. sp. *spinaciae*. *HortScience* 27:1118–1119.

Brandenberger, L. P., Correll, J. C., and Morelock, T. E. 1991. Nomenclature of the downy mildew fungus on spinach. *Mycotaxon* 41:157–160.

Byford, W. J. 1967. Host specialization of *Peronospora farinosa* on *Beta, Spinacia* and *Chenopodium*. *Transactions of the British Mycological Society* 50: 603–607.

Correll, J. C., Morelock, T. E., Black, M. C., Koike, S. T., Brandenberger, L. P., and Dainello, F. J. 1994. Economically important diseases of spinach. *Plant Disease* 78:653–660.

Frinking, H. D., Harrewijn, J. L., and Geerds, C. F. 1985. Factors governing oospore production by *Peronospora farinosa* f. sp. *spinaciae* in cotyledons of spinach. *Netherlands Journal of Plant Pathology* 91:215–223.

Inaba, T. and Morinaka, T. 1984. Heterothallism in *Peronospora effusa*. *Phytopathology* 74:214–216.

Inaba, T. and Morinaka, T. 1985. The relationship between conidium and oospore production in spinach leaves infected with *Peronospora effusa*. *Annals Phytopathological Society of Japan* 51:443–449.

Inaba, T., Takahashi, K., and Morinaka, T. 1983. Seed transmission of spinach downy mildew. *Plant Disease* 67:1139–1141.

Irish, B. M., Correll, J. C., Koike, S. T., Schafer, J., and Morelock, T. E. 2003. Identification and cultivar reaction to three new races of the spinach downy mildew pathogen from the United States and Europe. *Plant Disease* 87:567–572.

Koike, S. T., Smith, R. F., and Schulbach, K. F. 1992. Resistant cultivars, fungicides combat downy mildew of spinach. *California Agriculture* 46:29–31.

Stemphylium botryosum

STEMPHYLIUM LEAF SPOT

Introduction and significance

This new disease of spinach was first documented in 2001 in the Salinas Valley of California, though the problem may have been present for many years before. Since that time Stemphylium leaf spot has been found elsewhere in the USA (Arizona, Delaware, Florida, Maryland, and Washington) and in Europe. This disease occurs both on production spinach and spinach seed crops.

522 Initial gray spots of Stemphylium leaf spot of spinach.

523 Advanced tan spots of Stemphylium leaf spot of spinach.

Symptoms and diagnostic features

Initial symptoms consist of small, 2–6 mm diameter, circular to oval, gray-green leaf spots on leaves (**522**). As the disease progresses, leaf spots enlarge, remain circular to oval in shape, and turn tan in color. Older spots coalesce, dry up, and become papery in texture (**523**). Visual signs of fungal growth are generally absent from the spots; hence this problem is readily differentiated from foliar diseases in which purple growth (downy mildew), green spores (Cladosporium leaf spot), or acervuli (anthracnose) develop within lesions. Overall, symptoms can closely resemble the tan, circular spots caused by pesticide or fertilizer damage. This disease can cause significant leaf spots to develop on seed spinach crops, especially if pollen from spinach flowers falls on leaves prior to infection.

Causal agent

The pathogen is presently named *Stemphylium botryosum*, though this fungus may be a new, different species of *Stemphylium*. *Stemphylium botryosum* isolates from spinach are host specific to spinach. On V-8 juice agar incubated under lights, the pathogen produces dark green-brown mycelium and abundant conidia after approximately 10 days. Conidiophores are mostly unbranched, 5 μm wide with distinctly swollen apical cells (7 μm wide) having darkly pigmented bands. The multicelled conidia are brown-colored, broadly ellipsoidal to ovoid, and borne singly. A conspicuous constriction is present at the median transverse septum. Outer conidial walls are roughly textured (verrucose). Conidia dimensions are mostly 19–28 x 14–19 μm, and the mean length/width ratio is 1.46. Pseudothecia from an ascomycete perfect stage, *Pleospora herbarum*, have been found on spinach seed and senescent leaf infections. The role of *P. herbarum* in the disease cycle has not been clearly established.

Disease cycle

Details on disease development are lacking. The pathogen is seedborne, and has been detected on spinach seed from the USA, Europe, and New Zealand. In production fields the disease spreads slowly and economic damage is usually limited. In spinach seed production fields, however, the epidemiology of the disease is much more complex. Spinach pollen that falls onto leaves can provide a food base for the pathogen and therefore favor spore germination and infection. There are also indications that *Cladosporium* and *Stemphylium* spinach pathogens can form a complex and result in severe disease in seed fields.

Control

For production spinach, this disease appears to be of minor importance and no controls are currently recommended. Overhead sprinkler irrigation may cause the disease to be worse. Use seed that does not have significant levels of the pathogen. Hot water or 1.2% bleach seed treatments can significantly reduce seedborne inoculum.

References

du Toit, L. J. and Derie, M. L. 2001. *Stemphylium botryosum* pathogenic on spinach seed crops in Washington. *Plant Disease* 85:920.

du Toit, L. J., Derie, M. L., and Hernandez-Perez, P. 2005. Evaluation of fungicides for control of leaf spot in spinach seed crops, 2004. *Fungicide & Nematicide Tests* 60:V044.

du Toit, L. J., Derie, M. L., and Hernandez-Perez, P. 2005. Evaluation of yield loss caused by leaf spot fungi in spinach seed crops, 2004. *Fungicide & Nematicide Tests* 60:V047.

du Toit, L. J. and Hernandez-Perez, P. 2005. Efficacy of hot water and chlorine for eradication of *Cladosporium variabile*, *Stemphylium botryosum*, and *Verticillium dahliae* from spinach seed. *Plant Disease* 89:1305–1312.

Everts, K. L. and Armentrout, D. K. 2001. Report of leaf spot of spinach caused by *Stemphylium botryosum* in Maryland and Delaware. *Plant Disease* 85:1209.

Hernandez-Perez, P. and du Toit, L. J. 2006. Seedborne *Cladosporium variabile* and *Stemphylium botryosum* in spinach. *Plant Disease* 90:137–145.

Koike, S. T., Henderson, D. M., and Butler, E. E. 2001. Leaf spot disease of spinach in California caused by *Stemphylium botryosum*. *Plant Disease* 85:126–130.

Koike, S. T., Matheron, M. E., and du Toit, L. J. 2005. First report of leaf spot of spinach caused by *Stemphylium botryosum* in Arizona. *Plant Disease* 89:1359.

Verticillium dahliae

VERTICILLIUM WILT

Introduction and significance

Verticillium wilt has not yet been detected as a problem of production spinach in the field, but significant disease can be observed in spinach seed fields in the Pacific northwest USA. The ability of the pathogen to be seedborne in spinach may create opportunities for this pathogen to cause spinach problems in the future. In addition, infested spinach seed could introduce the pathogen to fields that later are planted to susceptible rotation crops.

Symptoms and diagnostic features

Symptoms typically begin to show when seed crop plants begin to bolt and enter their reproductive phase. Interveinal tissue of lower leaves turns yellow while veins remain green (**524**). Such leaves later wilt. The outside lower stems can be red in color. Vascular tissue of roots and crowns is light tan to brown. In severe cases plants can be significantly stunted and die (**525**). Vascular discoloration caused by Verticillium wilt is noticeably lighter in color than the vascular discoloration seen with Fusarium wilt (**526**).

Causal agent

Verticillium wilt is caused by the fungus *Verticillium dahliae*. The pathogen can be isolated on standard

524 Chlorosis of lower spinach leaves caused by Verticillium wilt.

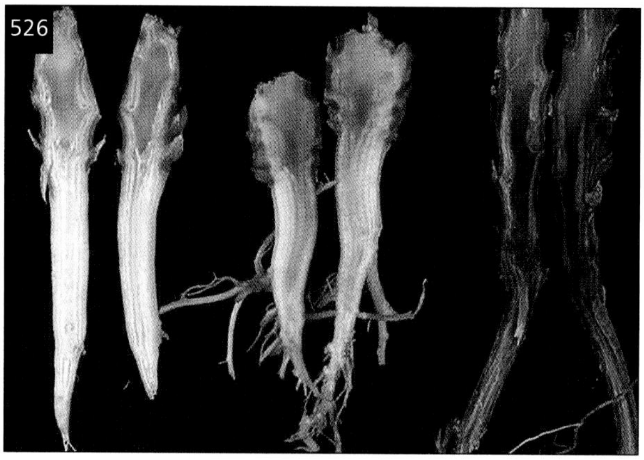

526 Light discoloration of spinach vascular tissue caused by Verticillium wilt (center) versus dark discoloration from Fusarium wilt (right). Healthy plant is on the left.

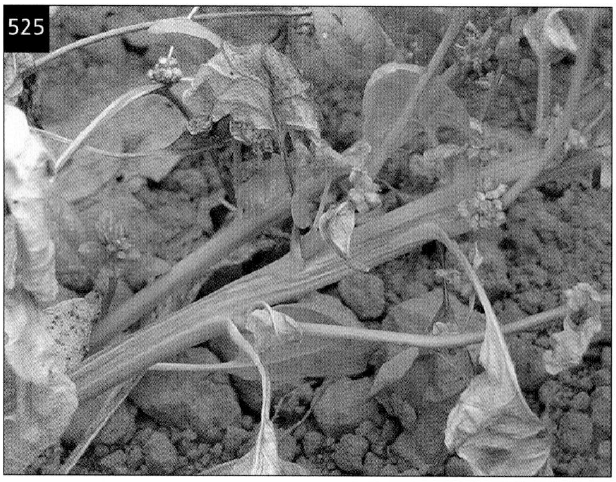

525 Foliage dieback and stem necrosis caused by Verticillium wilt.

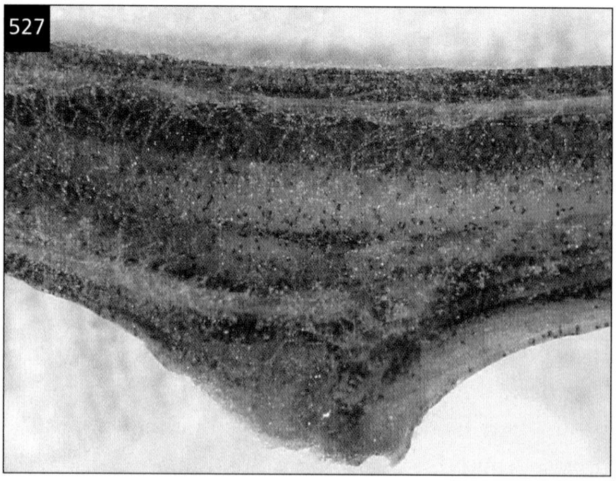

527 Dissected spinach stem showing small, black microsclerotia of *Verticillium dahliae*.

microbiological media, though semi-selective media such as NP-10 can be useful for isolation. On general purpose media, the pathogen forms the characteristic hyaline, verticillate conidiophores bearing three to four phialides at each node, and hyaline, single-celled, ellipsoidal conidia. Older cultures form dark brown to black torulose microsclerotia that consist of groups of swollen cells formed by repeated budding. Microsclerotia size varies greatly and is in the range of 15–100 μm in diameter. This pathogen is seedborne in spinach, and has been detected in spinach seed from both the USA and Europe. Therefore, infested spinach seed could be a means of spreading the pathogen to previously uninfested fields.

Disease cycle

The pathogen survives in the soil as dormant microsclerotia for extended periods of time (up to 8 to 10 years) but can also persist as an epiphyte on non-host roots. Diseased tissue can also harbor microsclerotia (**527**). Cool to moderate weather conditions favor the pathogen, and disease is enhanced at temperatures between 20–24° C. The microsclerotia germinate, and infective hyphae enter host roots through wounds. On spinach seed plants, disease severity becomes greater as plants mature and develop seed stalks (bolting).

Control

For production spinach, no control measures are yet recommended, though growers should be aware that infested spinach seed is a source of the pathogen. For spinach seed crops, use seed that has been treated or found to be uninfested. Hot water or 1.2% bleach seed treatments can significantly reduce seedborne inoculum. Avoid planting spinach seed crops in fields having a history of Verticillium wilt.

References

du Toit, L. J., Derie, M. L., and Hernandez-Perez, P. 2005. Verticillium wilt in spinach seed production. *Plant Disease* 89:4–11.

du Toit, L. J. and Hernandez-Perez, P. 2005. Efficacy of hot water and chlorine for eradication of *Cladosporium variabile*, *Stemphylium botryosum*, and *Verticillium dahliae* from spinach seed. *Plant Disease* 89:1305–1312.

Snyder, W. C. and Wilhelm, S. 1962. Seed transmission of Verticillium wilt of spinach. *Phytopathology* 52:365.

van der Spek, J. 1973. Seed transmission of *Verticillium dahliae*. *Mededelingen van de Faculteit Landbouwwetenschappen, Rijksuniversiteit Gent* 38(3):1427–1434.

Beet curly top virus

BEET CURLY TOP

Introduction and significance

Beet curly top virus (BCTV) occurs in North and South America, Asia, the Middle East, and the Mediterranean region. This virus is an important pathogen of many crops such as pepper, tomato, and *Chenopodium* plants. Occurrence on spinach varies between regions.

Symptoms and diagnostic features

BCTV infections on spinach initially appear as leaf stunting and chlorosis (**528**). Affected plants appear compact and yellow. Younger leaves in the center of the plant are often very pale yellow, curled, and brittle in texture. Plants can die a few weeks after symptoms appear.

Causal agent

BCTV is a geminivirus with isometric particles that measure 18–22 nm in diameter and which occur singly or in pairs. The BCTV genome is a single-stranded circular DNA. BCTV is vectored in a persistent manner by the beet leafhopper (*Circulifer tenellus*). *Circulifer opacipennis* is a vector in the Mediterranean region. In the plant, BCTV is restricted to the phloem tissue. On a molecular level, researchers have compared strains of BCTV from North America and the Middle East and found them to be similar, providing evidence that these various BCTV strains share a common origin.

528 Stunting and leaf chlorosis due to *Beet curly top virus*.

Disease cycle

This virus infects many weed and crop hosts. Recent research on curly top disease as it occurs in Chenopodiaceae, Fabaceae, Solanaceae, and other crops indicates that the viral agent may differ depending upon the host being considered. Beet curly top as a disease may actually be caused by one of four different curly top virus species: *Beet curly top virus* (BCTV), *Beet mild curly top virus* (BMCTV), *Beet severe curly top virus* (BSCTV), and *Spinach curly top virus* (SCTV). Research is on-going to further determine the relationships of these various viruses.

Control

Follow general suggestions in Part 1.

References

Baliji, S., Black, M. C., French, R., Stenger, D. C., and Sunter, G. 2004. *Spinach curly top virus*: a newly described curtovirus species from southwest Texas with incongruent gene phylogenies. *Phytopathology* 94:772–779.

Briddon, R. W., Stenger, D. C., Bedford, I. D., Stanley, J., Izadpanah, K., Markham, P. G. 1998. Comparison of a *beet curly top virus* isolate originating from the old world with those from the new world. *European Journal of Plant Pathology* 104:77–84.

Sams, D. W. and Bienz, D. R. 1974. Relative susceptibility of spinach plant introduction accessions to curly top. *HortScience* 9:600–601.

Soto, M. J. and Gilbertson, R. L. 2003. Distribution and rate of movement of the curtovirus *Beet mild curly top virus* (family Geminiviridae) in the beet leafhopper. *Phytopathology* 93:478–484.

Beet western yellows virus

BEET WESTERN YELLOWS

Introduction and significance

The *Polerovirus* genus of viruses has three different species that affect chenopodium crops. *Beet western yellows virus* (BWYV) is prevalent in the USA and elsewhere and infects many other plants in addition to beet, Swiss chard, and spinach. *Beet mild yellowing virus* (BMYV) is the most important yellows virus on Chenopodiaceae hosts in the UK and Western Europe. A third *Polerovirus* pathogen is *Beet chlorosis virus* (BCHV).

Symptoms and diagnostic features

BWYV infections on spinach appear as interveinal and leaf margin chlorosis on older leaves (**529, 530**). Leaf veins and areas adjacent to the veins usually remain green. This yellows symptom may be confused with various nutrient deficiency symptoms, notably those caused by nitrogen, iron, or magnesium deficiency. As the disease progresses, the yellowing increases and the older, lower leaves become completely chlorotic. Older symptomatic leaves are sometimes colonized by secondary fungi which contribute further to the decline in quality.

Causal agent and disease cycle

BWYV is a polerovirus and has isometric particles with diameters of 25 nm and single-stranded, linear RNA genomes. BWYV may be found in numerous crop and weed plants, and the host list is extensive, including over 150 plant species. Some isolates or strains of this virus have different abilities to infect certain plants; thus, not all strains of BWYV may be able to infect all known hosts, greatly complicating the etiology of this disease. The role of different strains for spinach has not been clarified. BWYV is vectored by several aphid vectors, especially the green peach aphid (*Myzus persicae*), in a persistent manner.

Control

Follow general suggestions for managing virus diseases (see Part 1).

529

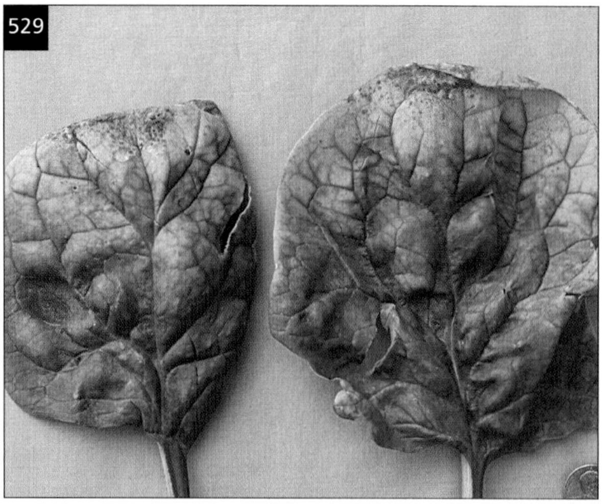

529 Chlorosis and green vein symptoms of BWYV on spinach.

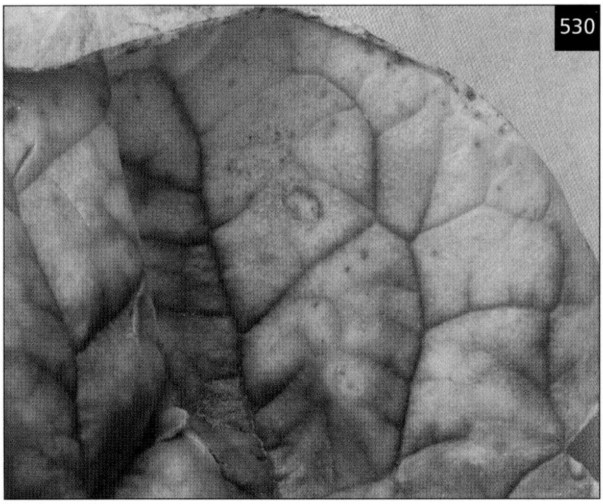

530 **Close-up of chlorosis and green vein symptoms of BWYV on spinach.**

Cucumber mosaic virus

CUCUMBER MOSAIC

Introduction and significance

Cucumber mosaic virus (CMV) is commonly found throughout the world and can cause disease on over 800 plant (crop and weed) hosts, including plants in the Chenopodiaceae. CMV is most prevalent in temperate regions.

Symptoms and diagnostic features

Symptoms of CMV infection on spinach include a slight to moderate chlorosis (**531**) and a narrowing of young leaves. Leaves also can have an inward rolling of margins and be extremely wrinkled and distorted. In advanced stages of disease, the plants usually appear stunted and the immature leaves in the center of the plant may become completely blighted, killing the growing point (**532**). In some reports this disease is called spinach blight.

Causal agent and disease cycle

CMV is a cucumovirus with virions that are isometric (29 nm in diameter) and contain three single-stranded RNAs. Many CMV strains have been documented. CMV is transmitted by several aphid vectors in a non-persistent manner. CMV occurs in many weed hosts and can also be seedborne in spinach

Control

CMV is difficult to control because of its extremely wide host range. Follow general suggestions for managing virus diseases (see Part 1). Some resistant spinach cultivars are available, though such resistance may not be complete. Use spinach seed that does not have the virus.

References

Yang, Y., Kim, K. S., and Anderson E. J. 1997. Seed transmission of cucumber mosaic virus in spinach. *Phytopathology* 87:924931.

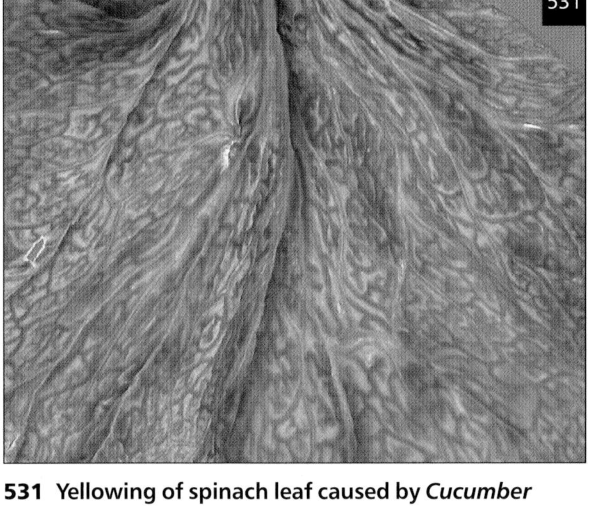

531 **Yellowing of spinach leaf caused by *Cucumber mosaic virus*.**

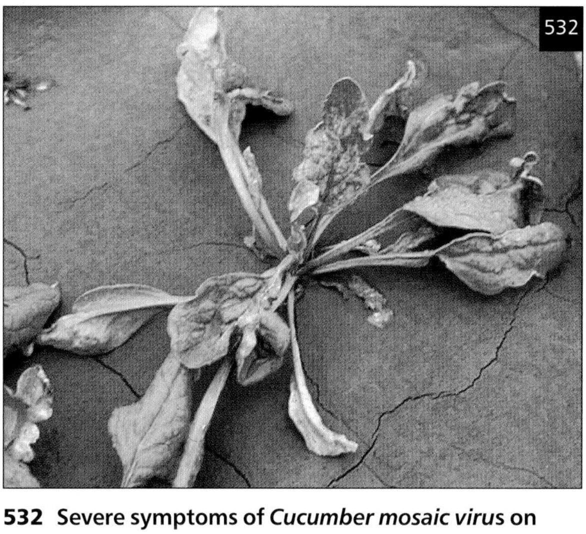

532 **Severe symptoms of *Cucumber mosaic virus* on spinach.**

Specialty and herb crops

SPECIALTY CROPS is a category that usually refers to agricultural or horticultural commodities that are grown in a very limited volume, comprise small production areas, are new crops, tend to be produced or consumed by certain ethnic groups, or are used for specialized cooking purposes. Both organic and conventionally grown specialty items find their way to farmers' markets, specialty grocery stores, and other specialized outlets. This chapter outlines some disease problems found on various specialty vegetables, gourmet commodities, and herbs and spices. In many cases these specialty vegetables are susceptible to the same range of pathogens that affect more familiar, widely grown vegetables that are in the same plant families as the specialty items. However, in contrast to the more well-known vegetables, the specialty crops face particular challenges with regard to disease management. Research projects and funding are often not available for dealing with specialty crop problems. Therefore, precise information on specialty crop diseases is usually very limited. Commercial plant breeding efforts can rarely be justified for specialty crops that are planted on such a small scale. Specialty crops are often omitted from fungicide registrations, or such registrations are not approved in all regions or countries. For example, a fungicide product such as azoxystrobin may be registered in the Eastern USA on some minor crops, but the California label may not include those same crops; therefore, chemical control options may not be available for use on specialty commodities.

Controlling diseases of specialty crops follows the general integrated disease management strategies used for other vegetable crops. Choose and plant resistant cultivars, if such are available. Rotate specialty crops so that the same commodity is not planted in the same location for back-to-back plantings or in the same season; this strategy is particularly important for soilborne pathogens. Select the area to be planted so as to favor plant development and discourage pathogens; for example, selection of well-draining, lighter textured soils will help reduce disease pressure caused by *Pythium* and other damping-off pathogens. Avoid fields and locations that have a history of the diseases of concern. As mentioned, fungicides are often not registered for these crops, though some mineral products (sulfur, copper) are generally approved. When possible, use appropriate cultural practices to create conditions unfavorable for disease development: avoid using overhead sprinkler irrigation, do not plant until the previous crop residues have dissipated, use seed that does not have significant levels of the pathogen, use transplants that do not show disease symptoms, practice good weed and insect management. Refer to the appropriate sections in this book that discuss the same pathogens and diseases as found on the crops in this chapter.

Amoracia rusticana

HORSERADISH

Horseradish is a crucifer plant (family Brassicaceae) cultivated for its leaves and fleshy, white, pungent roots that are processed into the horseradish condiment. Horseradish is propagated from root cuttings. This plant is probably native to southeast Europe and western Asia. Horseradish has escaped from cultivation and become widely naturalized in Europe, North America, and New Zealand. Fungicide recommendations exist for some of these diseases in some countries.

533 Cercospora leaf spot of horseradish.

534 Close-up of Cercospora leaf spot of horseradish.

Albugo candida

White rust

White rust or white blister is a common and sometimes severe foliar disease on horseradish. The undersides of affected leaves bear small, white or pale brown raised blisters. These blisters break open to release powdery white spores. The pathogen is *Albugo candida*, the same pathogen that affects other crucifers. Perennial plantings of horseradish facilitate *A. candida* survival and re-infection, and the pathogen can infect the fleshy roots of horseradish. Hot water treatment of root cuttings (44° C for 10–15 minutes) may control root infection. For more information see the white rust section in the crucifer chapter

Cercospora amoraciae

Cercospora leaf spot

This disease produces round to angular, pale brown leaf spots (**533**, **534**). Some spots have a darker margin and chlorotic halo. Lesions are often numerous on older leaves and coalesce to produce large pale blotches or severe blighting of foliage. Cercospora leaf spot is very prevalent on wild horseradish in the UK. The pathogen, *Cercospora amoraciae*, forms hyaline spores that measure 50–140 x 3-5 µm, have 7 to 18 septa, and are broader at the attachment point and tapered at the distal end. Young lesions caused by *C. amoraciae* may be difficult to distinguish from *Alternaria brassicae* lesions.

Peronospora parasitica
(= *Hyaloperonospora parasitica*)

Downy mildew

This disease is occasionally reported on horseradish and causes irregular, angular-shaped leaf lesions. The pathogen is *Peronospora parasitica*. The precise relationship between the horseradish downy mildew pathogen and *P. parasitica* isolates from other brassica hosts requires additional research. For more information see the downy mildew section in the crucifer chapter.

Ramularia amoraciae

Ramularia leaf spot

This disease is common in Europe and North America and sometimes causes significant losses. Ramularia leaf spot produces circular to irregular-shaped, pale green or yellowish spots that measure up to 10 mm in diameter (535). Spots later turn white and develop dark margins. These lesions coalesce to form larger blotches. Severe infection results in death of the leaves; affected leaves can have a tattered appearance as the dead areas tear away. The pathogen is *Ramularia amoraciae*. Spores are visible on the undersurfaces of lesions and have one septum or none and measure 8–30 x 3–4 µm. The pathogen overwinters as black stromata on fallen leaves; stromata produce new conidia in the spring. This pathogen apparently infects only horseradish; the reported occurrence of *R. amoraciae* on oilseed rape (*Brassica napus*) in France is now considered to involve the white leaf spot pathogen.

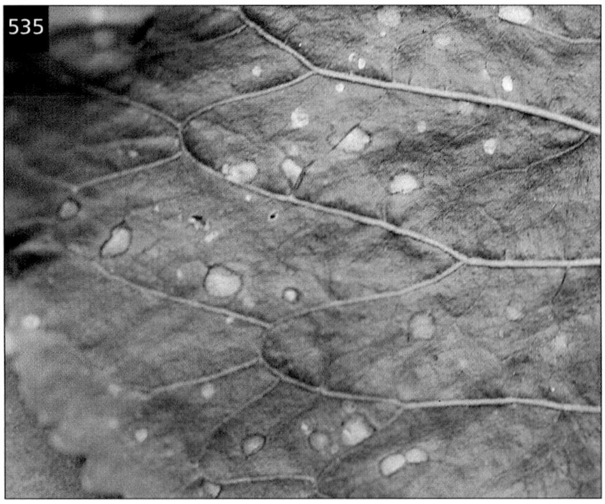

535 Ramularia leaf spot of horseradish.

Turnip mosaic virus (TuMV)

Turnip mosaic

This virus disease causes a chlorotic, blotchy mottle, sometimes with vein clearing. As the leaves age, a more general chlorosis develops followed by necrotic rings at the edge of the chlorotic patches. Affected plants have stunted growth. Flecking and streaking occur on the petioles and leaf veins and this may extend into the roots. TuMV is a potyvirus.

Verticillium dahliae

Verticillium wilt

Horseradish plants affected by Verticillium wilt may not show foliar symptoms. If symptoms of the foliage occur, these consist of wilting, chlorosis, necrosis, and overall stunting of growth; lower leaves are affected first. The vascular tissue of the roots (and also plant petioles) shows dark brown to black discoloration and streaking. This discoloration later spreads to the root cortex and core, making the harvested root unusable for processing. The main causal agent is *Verticillium dahliae*. Microsclerotia enable the pathogen to survive in the soil for extended periods of time (up to 8 to 10 years). Because horseradish is propagated vegetatively, cuttings and plant divisions probably spread this pathogen to new plantings. For more information see the Verticillium wilt section in the brassica chapter (page 194). Resistant or tolerant cultivars are being developed. This disease of horseradish may be caused by a complex of soilborne pathogens; researchers recently have also implicated *V. longisporum* and *Fusarium solani* with vascular discoloration.

Xanthomonas campestris pv. *amoraciae*

Black rot

Black rot causes localized leaf spots that tend to be angular in shape and resemble those of bacterial leaf spot disease caused by *Pseudomonas syringae* pv. *maculicola*. Black rot of horseradish is caused by *Xanthomonas campestris* pv. *amoraciae*. This pathogen is closely related to *X. campestris* pv. *campestris*, causal agent of the systemic black rot disease that affects many brassica plants. *Xanthomonas campestris* pv. *amoraciae* from horseradish is only weakly pathogenic to cabbage and cauliflower. For more information see the black rot section in the brassica chapter (page 159).

References

Atibalentja, N. and Chang, R. J. 1998. *Verticillium dahliae* resistance in horseradish germ plasm from the University of Illinois collection. *Plant Disease* 82:176–180.

Babadoost, M., Chen, W., Bratsch, A. D., and Eastman, C. E. 2004. *Verticillium longisporum* and *Fusarium solani*: two new species in the complex of internal discoloration of horseradish roots. *Plant Pathology* 53:669–676.

Brun H., Renard M., Jouan B., Tanguy, X., and Lamarque, C. 1979. Observations préliminaires sur quelques maladies du colza en France: *Sclerotinia sclerotiorum, Cylindrosporium concentricum, Ramularia amoraciae*. *Sciences Agronomiques Rennes* (1979), 7–77.

Eastburn, D. M. and Chang, R. J. 1994. *Verticillium dahliae*: a causal agent of root discoloration of horseradish in Illinois. *Plant Disease* 78:496–498.

Percich, J. A. and Johnson, D. R. 1990. A root rot complex of horseradish. *Plant Disease* 74:391–393.

Anethum graveolens

DILL

Dill is produced for its foliage and is used as a culinary herb and for seasonings. The plant is in the Apiaceae family.

Erysiphe heraclei

Powdery mildew

On dill, powdery mildew forms white mycelium and conidia on leaves, stems, and flower stalks. The disease causes slight twisting of foliage and early senescence. Conidia form singly, are cylindric, and measure 30–40 x 12–15 um. Conidia lack fibrosin bodies and germinate at the ends with either a very short or a long germ tube forming a lobed appressorium (polygoni-type). Spherical cleistothecia have been observed. The pathogen is *Erysiphe heraclei.*

Itersonilia perplexans

Itersonilia blight

This is a minor disease of dill that has been reported only occasionally. Initial symptoms consist of a gray-green discoloration and wilting of the tips of dill leaves. As disease develops, many of the leaves discolor and collapse, which gives the foliage a blighted appearance and makes the leaves unsuitable for harvest. The pathogen is the fungus *Itersonilia perplexans*. In culture the fungus produces slow-growing colonies that are cream-colored, velutinous, and flat with minimal aerial mycelium. Mycelium is hyaline and has clamp connections. Ovoid to subglobose sporogenous cells are produced and measure 11–14 x 12–16 μm. Spores are ballistospores that are bilaterally symmetrical, lunate, 16–18 x 12–13 μm, and germinate with hyphae or secondary ballistospores. Studies indicate *I. perplexans* consists of pathogenic forms that are host-specific to either chrysanthemum and closely related Compositae, or to Apiaceae plants such as dill, carrot, parsley, and parsnip.

References

Koike, S. T. and Tjosvold, S. A. 2001. A blight disease of dill in California caused by *Itersonilia perplexans*. *Plant Disease* 85:802.

Soylu, E. M. and Soylu, S. 2003. First report of powdery mildew caused by *Erysiphe heraclei* on dill (*Anethum graveolens*) in Turkey. *Plant Pathology* 52:423.

Anthriscus cerefolium

CHERVIL

Chervil, or salad chervil, another member of the Apiaceae, is produced for its foliage and is used as a culinary herb, garnish, salad ingredient, and tea.

Erysiphe heraclei

Powdery mildew

On chervil, powdery mildew forms white mycelium and conidia on both sides of leaves (**536**), causes slight twisting of foliage, and results in quality loss of the harvested product. Severely affected plants are not harvested. Conidia form singly, are cylindric, and measure 40–50 x 15–17 um. Conidia lack fibrosin bodies and germinate at the ends with either a very short or a long germ tube forming a lobed appressorium (polygoni-type). No cleistothecia have been observed. The pathogen is *Erysiphe heraclei.*

536 White fungal growth of powdery mildew on chervil.

Sclerotinia sclerotiorum

White mold

White mold infections cause tan to gray lesions to develop on crowns and lower sections of stems (**537**). These lesions are surrounded by pink to orange-colored tissue, which is similar in color to celery petioles infected by this same fungus. Affected stems wilt and plants eventually collapse and rot. White mycelium and large, irregular, black sclerotia (5–10 mm in diameter) are observed on infected stems and crowns. The pathogen is *Sclerotinia sclerotiorum*. For more information see the white mold on bean section in the legume chapter (page 262).

References

Koike, S. T. and Saenz, G. S. 2004. First report of powdery mildew caused by *Erysiphe heraclei* on chervil in California. *Plant Disease* 88:1163.

Koike, S. T. 1999. Stem and crown rot of chervil, caused by *Sclerotinia sclerotiorum*. *Plant Disease* 83:1177.

Beta vulgaris subsp. *cicla*

SWISS CHARD

Swiss chard is in the Amaranthaceae family and is closely related to beet. However, Swiss chard is grown exclusively for its leaves and is most often used as a cooked vegetable. In California, Swiss chard is also grown for only a few weeks and then harvested for a small, 'baby leaf' fresh vegetable used in salads. Two common chard types are the green leaf varieties that have white petioles and the red-green to purple-green leaf types that have red petioles. Table beet (*Beta vulgaris* subsp. *vulgaris*), or beetroot, is in the same plant family as Swiss chard and is grown for its swollen, fleshy taproot that is eaten as a cooked or pickled commodity.

Beet curly top virus (BCTV)

Beet curly top

Beet curly top virus can result in severely stunted, chlorotic Swiss chard plants (**538**). Such plants have older leaves that turn dull to bright yellow and are brittle, thickened, and rolled. BCTV is vectored by the beet leafhopper (*Circulifer tenellus*) and other leafhopper species. For more information see the beet curly top section in the beet chapter (page 148).

537 Chervil crown infected with *Sclerotinia sclerotiorum*.

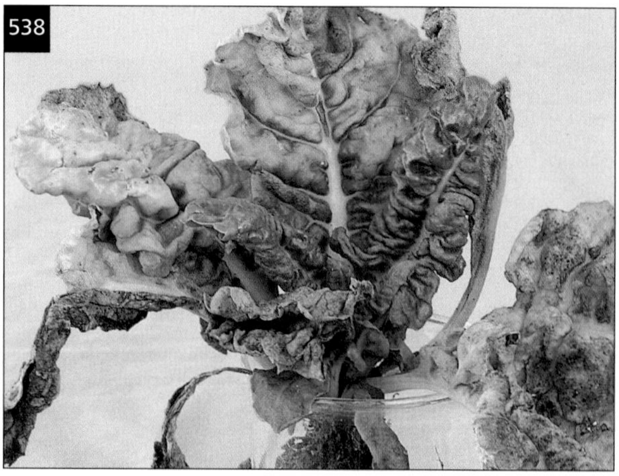

538 Chlorosis and leaf deformity of Swiss chard infected with *Beet curly top virus*.

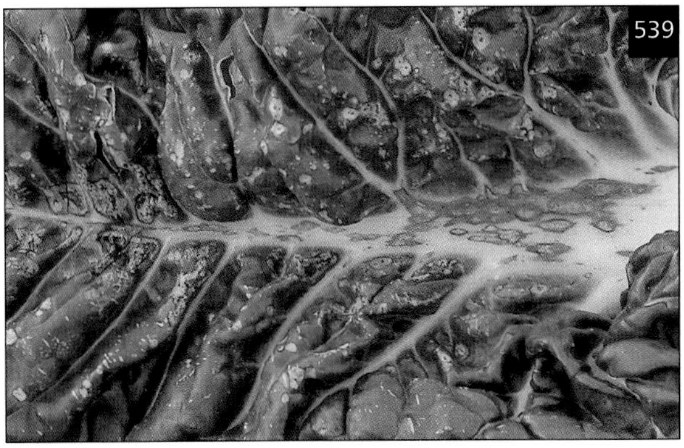

539 Cercospora leaf spot of Swiss chard.

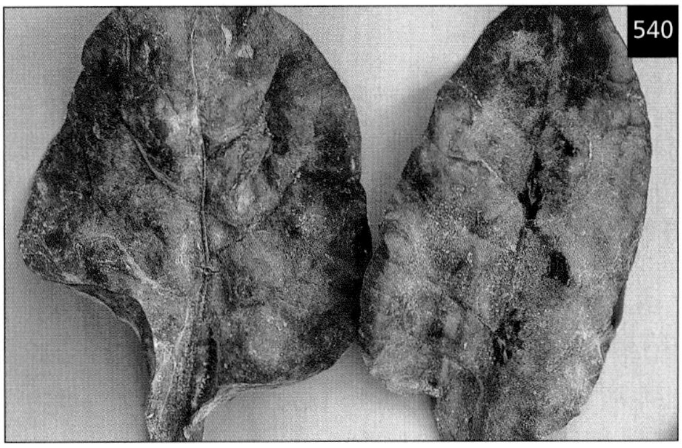

540 Powdery mildew on Swiss chard.

Cercospora beticola

Cercospora leaf spot

Initial symptoms are small, angular leaf spots that are brown or tan with purple to red borders (**539**). As spots enlarge, the tissue becomes tan to gray in color and may or may not retain the purple border. In severe cases the spots can coalesce and cause large sections of the leaf to die. The pathogen is *Cercospora beticola,* and it can infect both Swiss chard and table beet. The fungus is seedborne. For more information see the Cercospora leaf spot section in the beet chapter.

Erysiphe polygoni

Powdery mildew

Swiss chard and table beet are susceptible to powdery mildew caused by *Erysiphe polygoni.* The pathogen appears as typical white to gray mycelium and conidia growing on both top and bottom sides of leaves (**540**). If disease is severe, the leaves may be twisted and distorted. See also the powdery mildew section in the beet chapter.

Peronospora farinosa f. sp. *betae*

Downy mildew

On both Swiss chard and table beet, downy mildew causes leaves to twist, curl, and grow in abnormal ways. The newest leaves and growing point can be extremely deformed (**541**). If conditions are favorable, affected tissue can be covered with the purple-gray growth of the fungus (**542**). This pathogen (*Peronospora farinosa* f. sp. *betae*) does not infect spinach; spinach downy mildew is caused by the closely related *P. farinosa* f. sp. *spinaciae*. See also the downy mildew section in the beet chapter (page 142).

541 Infected crown and emerging leaves of Swiss chard caused by downy mildew.

542 Infected leaves of Swiss chard caused by downy mildew.

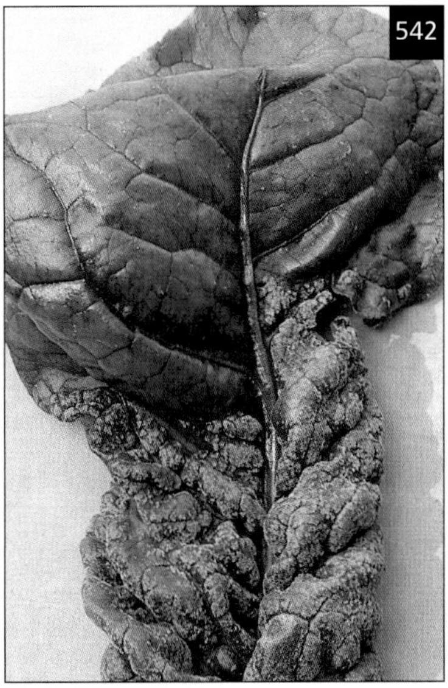

Pseudomonas syringae pv. *aptata*

Bacterial leaf spot

Initial symptoms of bacterial leaf spot on Swiss chard and table beet consist of water-soaked leaf spots that measure 2–3 mm in diameter (543). As bacterial leaf spot develops, spots become circular to ellipsoid in shape, measure 3–8 mm in diameter, and are tan with distinct brown to black borders (544). Spots are visible from both top and bottom sides of leaves. The pathogen is *Pseudomonas syringae* pv. *aptata*. This pathogen also infects sugar beet.

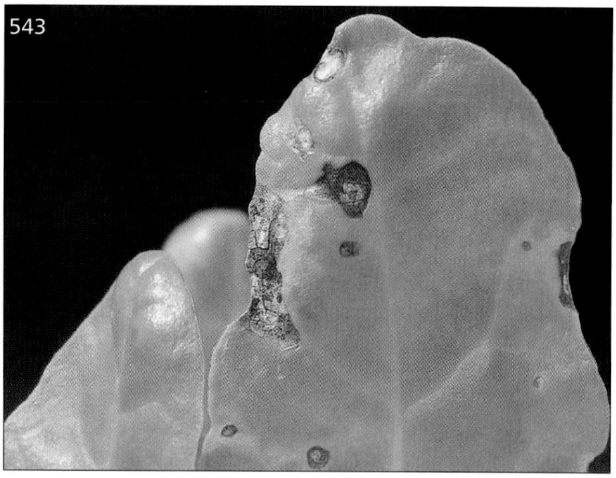

543 Early symptoms of bacterial leaf spot of Swiss chard.

Rhizoctonia solani, Pythium spp.

Damping-off

Symptoms of damping-off of Swiss chard occur on newly emerging plants and consist of wilting, brown necrosis of crown tissue, and eventual death of seedlings (545). If Swiss chard is planted in high plant densities, disease incidence can increase rapidly and affect large numbers of seedlings. Damping-off pathogens are *Rhizoctonia solani* and species of *Pythium*.

544 Advanced leaf spots of bacterial leaf spot of Swiss chard.

Uromyces betae

Rust

On both Swiss chard and table beet, the most obvious indication of rust is the presence of discrete, brown to orange-brown pustules on both sides of the infected leaf (546, 547). The brown spores emerge in large numbers from these pustules. If infected severely, the leaf can turn yellow and begin to decline. The pathogen is *Uromyces betae*. For more information see the rust section in the beet chapter.

545 Damping-off of Swiss chard caused by *Rhizoctonia solani*. Healthy plants are on the right.

References

Koike, S. T., Henderson, D. M., Bull, C. T., Goldman, P. H., and Lewellen, R. T. 2003. First report of bacterial leaf spot of Swiss chard caused by *Pseudomonas syringae* pv. *aptata* in California. *Plant Disease* 87:1397.

Koike, S. T. and Subbarao, K. V. 1999. Damping-off of Swiss chard, caused by *Rhizoctonia solani*, in California. *Plant Disease* 83:695.

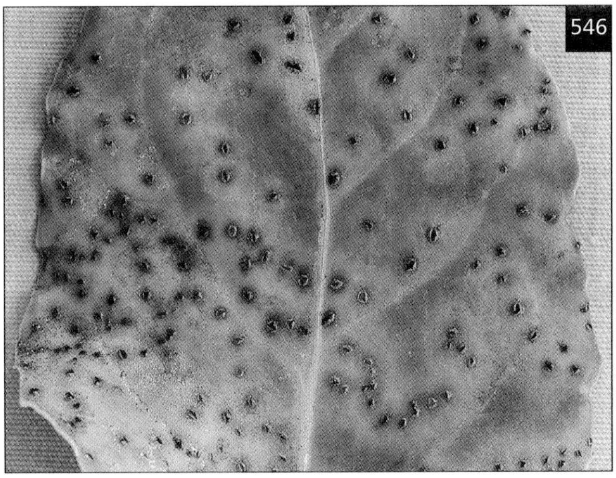

546 Rust pustules on Swiss chard.

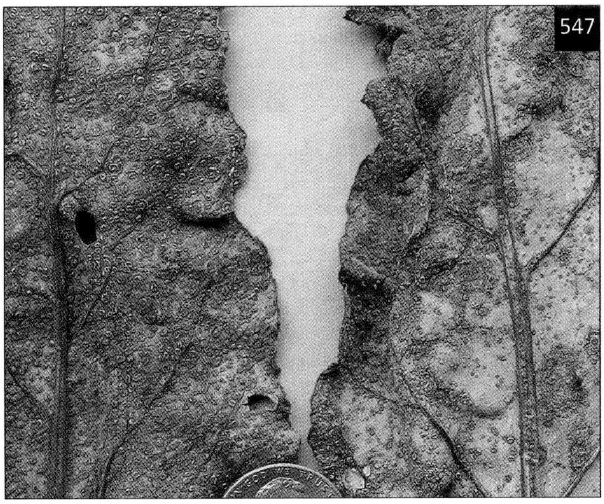

547 Advanced rust on Swiss chard.

Brassica rapa subsp. *rapa*

BROCCOLI RAAB

Broccoli raab, also known as rapini, broccoletto di rapa, or cima de rapa, is a crucifer vegetable (family Brassicaceae) grown for its succulent young shoots, leaves, and immature flower buds. In addition to bacterial blight, broccoli raab is susceptible to Alternaria leaf spot and powdery mildew (see the brassica chapter).

Pseudomonas syringae pv. *alisalensis*

Bacterial blight

Initial symptoms consist of small (2–4 mm), angular, water-soaked flecks on lower foliage that are visible on both sides of the leaves (**548**). These flecks expand and become surrounded by bright yellow borders. As the disease progresses, multiple leaf spots coalesce and result in large, irregular necrotic areas and extensive leaf yellowing. Affected leaf tissue later dries, turns tan, and becomes papery in texture (**549**). Disease progresses upwards from these lower leaves to leaves higher in the plant canopy. If symptoms develop on the uppermost leaves attached to the flower stalks, the shoot loses market quality and will not be harvested. Under conditions favorable for disease development, leaf infections can be extensive and cause significant reduction in plant growth. Bacterial blight is caused by *Pseudomonas syringae* pv. *alisalensis*. This pathogen also causes bacterial blight on broccoli. See also the bacterial blight section in the brassica chapter (page 157).

548 Water-soaked spots of early bacterial blight infection of broccoli raab.

549 Leaf lesions of advanced bacterial blight infection of broccoli raab.

References

Cintas, N. A., Koike, S. T., and Bull, C. T. 2002. A new pathovar, *Pseudomonas syringae* pv. *alisalensis* pv. *nov.*, proposed for the causal agent of bacterial blight of broccoli and broccoli raab. *Plant Disease* 86:992–998.

Koike, S. T., Cintas, N. A., and Bull, C. T. 2000. Bacterial blight, a new disease of broccoli caused by *Pseudomonas syringae* in California. *Plant Disease* 84:370.

Koike, S. T., Henderson, D. M., Azad, H. R., Cooksey, D. A., and Little, E. L. 1998. Bacterial blight of broccoli raab: a new disease caused by a pathovar of *Pseudomonas syringae*. *Plant Disease* 82:727–731.

Brassica spp.

MUSTARDS

Specialty leafy mustard crops are grown for use as fresh vegetables for salads. Japanese or mizuna mustard (*Brassica campestris* ssp. *nipposinica*) is a dark green leafy mustard that usually has finely dissected, feathery leaves. Red Asian mustard (*B. juncea* ssp. *rugosa*), known also as purple or giant-leafed mustard, is a dark red to red green leafy mustard that has spatulate leaves. Tatsoi or tah tsai (*B. campestris* ssp. *narinosa*) is a leafy mustard that has oval, dark green leaves and almost white petioles.

Albugo candida

White rust

On Japanese mustard and tah tsai, white rust causes typical white raised pustules to develop profusely on the lower sides of leaves (550). Leaf hypertrophy is observed only on infected Japanese mustard. The pathogen is *Albugo candida*. Spherical sporangia diameters range from 18–19 µm. Red mustard is apparently not a host. For more information see the white rust section in the brassica chapter (page 162).

Alternaria brassicae

Alternaria leaf spot

Alternaria leaf spot symptoms consist of small (2–6 mm in diameter), circular to oblong brown spots that contain concentric rings (551 and also 553). On tah tsai, yellow borders may develop around the spots. Elongated, brown spots also develop on leaf petioles. The pathogen is *Alternaria brassicae*. For more information see the Alternaria leaf spot section in the brassica chapter (page 164).

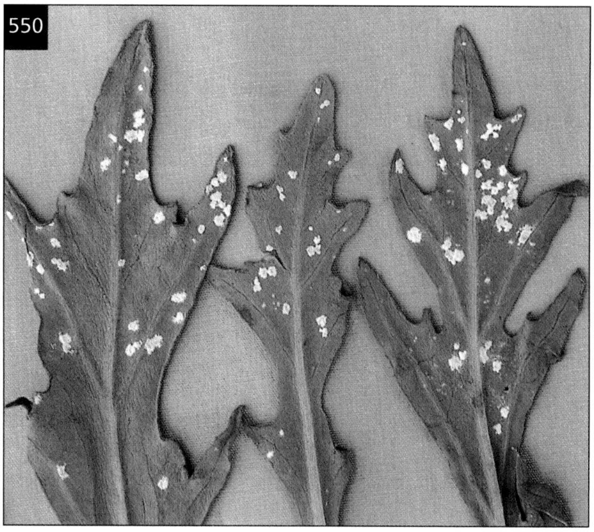

550 White rust pustules of Japanese mustard.

551 Alternaria leaf spot of Japanese mustard.

Pseudocercosporella capsellae

White leaf spot

On leafy mustards, white leaf spot causes round to oval, light tan to off-white leaf spots measuring 2–8 mm in diameter (552). White sporulation is common on leaf spot surfaces, and the hyaline, cylindrical conidia measure 60–79 µm x 2–3 µm. Isolations onto acidified potato-dextrose agar (2 ml 25% lactic acid/liter) yield slow-growing, raised, black, stromatic colonies that produce few conidia. On 2% water agar, colony morphology is similar to that on PDA, but colonies also release a purple-pink pigment into the media. The pathogen is *Pseudocercosporella capsellae*. For more information see the white leaf spot section in the brassica chapter (page 176).

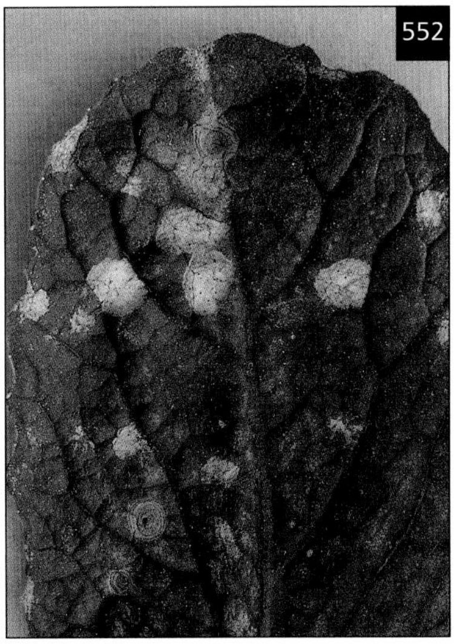

552 White leaf spot of red mustard.

553 White leaf spot (left) and Alternaria leaf spot (right) of tatsoi mustard.

References

Koike, S. T. 1996. Outbreak of white rust, caused by *Albugo candida*, on Japanese mustard and tah tsai in California. *Plant Disease* 80:1302.

Koike, S. T. 1996. Japanese mustard, tah tsai, and red mustard as hosts of *Alternaria brassicae*. *Plant Disease* 80:822.

Koike, S. T. 1996. Red mustard, tah tsai, and Japanese mustard as hosts of *Pseudocercosporella capsellae* in California. *Plant Disease* 80:960.

Cichorium endivia, C. intybus

ENDIVE/ESCAROLE, RADICCHIO

There are several head-forming leafy vegetables that are in the chicory plant group (family Asteraceae). These crops are grown for leaves that are used fresh in salads or cooked in other foods. Endive (*Cichorium endivia*) and escarole (the broad-leaf type of *C. endivia*) are green, open headed types that form dense rosettes that have a growth habit similar to low, prostrate leaf lettuces. Radicchio or red chicory (*C. intybus*) typically is a dark red to red-green chicory that is field grown (in contrast to witloof chicory that is forced in dark, indoor incubation conditions) and forms rosettes that later develop into enclosed heads. Some *C. intybus* types grow more like romaine lettuce and have an open architecture.

Alternaria cichorii

Alternaria leaf spot

Symptoms of Alternaria leaf spot on endive and escarole consist of small (1–2 mm diameter), necrotic, circular leaf spots (554). The pathogen is *Alternaria cichorii*. The fungus sporulates on the leaf spots. Conidia from leaves are obclavate in shape (larger at the base) with slender, unbranched beaks extending from the narrow end of the spore body. Spore body dimensions measure 56–78 x 14–20 µm, and beaks measure 36–81 x 1–2 µm. Spore bodies have seven to nine transverse septa. Longitudinal septa are usually not present.

554 Alternaria leaf spot of endive.

555 Early leaf spots of Alternaria leaf spot of radicchio.

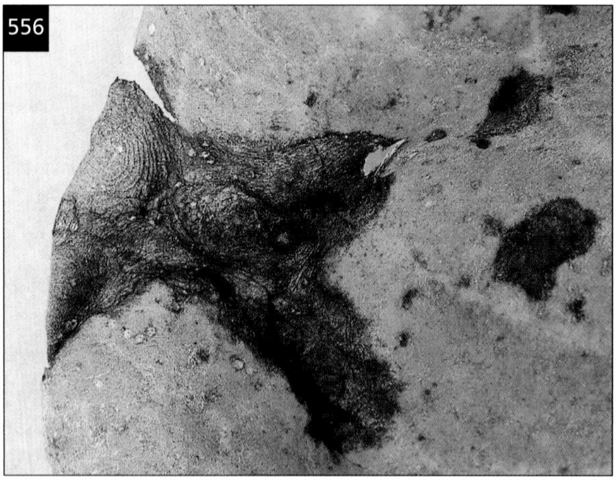

556 Advanced leaf spots of Alternaria leaf spot of radicchio.

557 Endive infected with *Beet western yellows virus*.

On radicchio the pathogen causes circular to oblong, necrotic spots ranging in diameter from 3–20 mm and having concentric zones of darker tissue (555, 556).

Beet western yellows virus (BWYV)

Beet western yellows

Symptoms of this virus disease in endive and escarole resemble those of lettuce infected with the same virus. The leaves develop yellow blotches in interveinal tissue. This yellowing continues until the oldest, lower leaves are bright yellow to sometimes almost white in color (557), with the main leaf veins remaining green. It appears that this virus does not cause significant stunting or reduction in plant size. Overall symptoms of this yellows disease may resemble nutrient deficiencies, such as iron chlorosis. BWYV is spread by aphid vectors. For more information see the beet western yellows section in the lettuce chapter (page 317).

Erwinia spp., *Pseusdomonas* spp.

Bacterial soft rot

Bacterial soft rot is a problem on many vegetables, including endive and escarole. On endive and escarole, symptoms consist of a brown to black, soft mushy rot of lower leaves as well as leaves within the tightly whorled heads (558). Such leaves are usually slimy and wet. Bacterial soft rot occurs when plant tissue is damaged from tipburn, disease infections, and other factors. This problem is usually considered to be a secondary decay and is caused by a complex of *Erwinia, Pseudomonas*, and other bacteria.

Golovinomyces cichoracearum

Powdery mildew

On endive and radicchio, powdery mildew forms white mycelium and conidia on both sides of leaves (559), causes slight twisting of foliage, and results in quality loss of the harvested product. Infected leaves must be trimmed off the plant at harvest. Conidia are produced in chains. The pathogen is *Golovinomyces cichoracearum* (previously named *Erysiphe cichoracearum*).

558 Leaf and petiole decay of bacterial soft rot of endive.

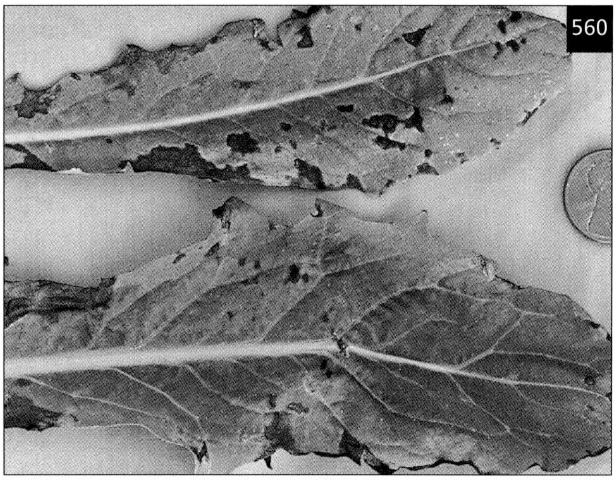

560 Bacterial leaf spot of Italian dandelion (=*C. intybus*).

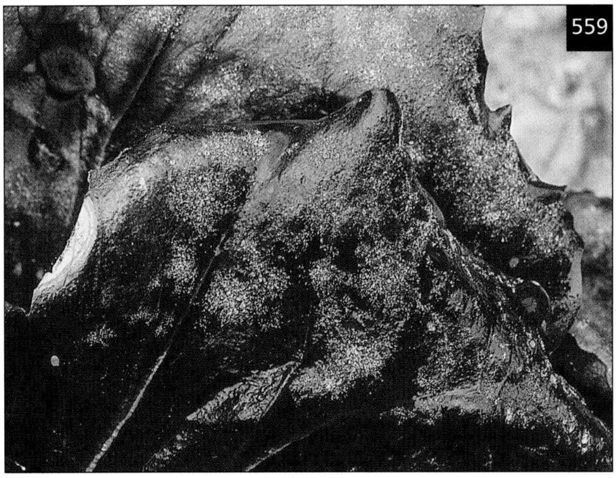

559 Sporulation of powdery mildew of radicchio.

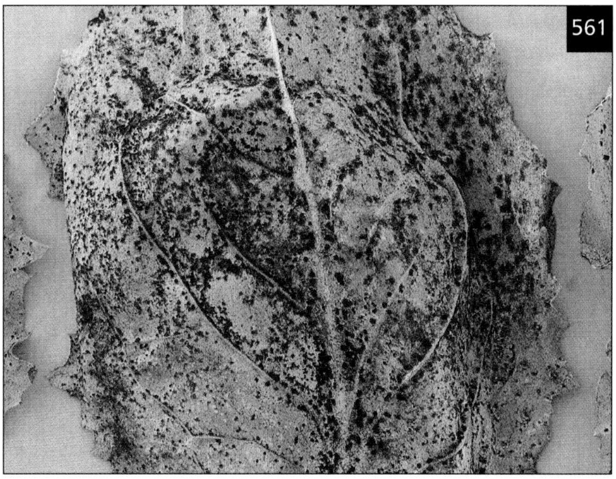

561 Rust pustules on endive.

Pseudomonas syringae

Bacterial leaf spot

Italian dandelion (*Cichorium intybus*) is a chicory plant related to radicchio. Italian dandelion does not form a head, but rather grows long, upright, loose foliage that superficially looks similar to true dandelion (*Taraxacum officinale*). Early symptoms of bacterial leaf spot are angular, vein delimited, dark, water-soaked leaf spots that measure 2–7 mm in diameter. As disease develops, spots retain the angular edges but exhibit various irregular shapes (**560**). Spots commonly form along the edges of the leaves; in some cases these spots develop into long lesions that measure between 10 and 30 mm in length. Spots are visible on both the top and bottom sides of the leaf and at maturity are a dull black color. The pathogen is *Pseudomonas syringae*.

Puccinia hieracii

Rust

Endive and escarole are susceptible to a rust disease. The most obvious indication of this problem is the presence of the rust pathogen itself. The fungus forms discrete, dark brown to red-brown pustules on both sides of the diseased leaf (**561**). The red brown spores emerge in large numbers from these pustules. If infected severely, the leaf can turn yellow and begin to decline. The pathogen is *Puccinia hieracii*.

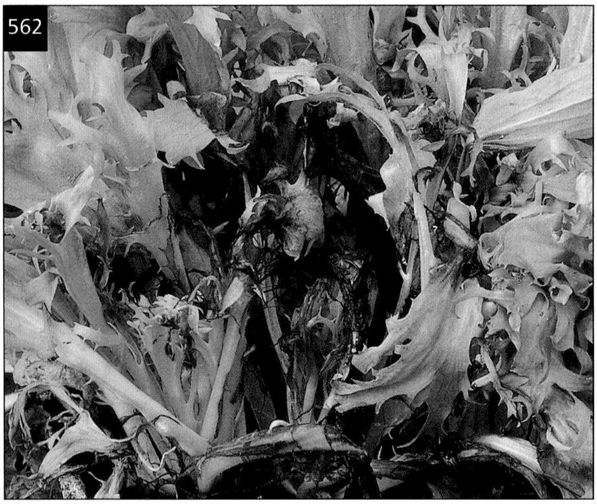

562 Leaf and petiole decay of Rhizoctonia blight of endive.

563 Collapsed radicchio infected with *Sclerotinia minor.*

564 Mycelium and sclerotia of *Sclerotinia minor* on radicchio.

Rhizoctonia solani
(teleomorph = *Thanatephorus cucumis*)

Rhizoctonia blight

On endive and escarole, Rhizoctonia blight causes a soft, watery, brown decay of the leaves (**562**). In the tightly appressed heads of these plants, the decay generally spreads in a concentric circle, resulting in circular whorls of brown, rotted leaves within diseased heads. Such infections make the heads unmarketable. The pathogen is *Rhizoctonia solani*. The perfect stage of this fungus, *Thanatephorus cucumeris*, is not commonly observed on these hosts.

Sclerotinia minor

White mold

In coastal California, *Sclerotinia minor* is the main *Sclerotinia* species on these crops. White mold affects endive, escarole, and radicchio. Plants nearing maturity wilt and collapse (**563**). Crown tissues become necrotic and develop a soft rot. White mycelium and small (0.5–3.0 mm) black sclerotia form on infected tissues (**564**). Isolates of *S. minor* are able to infect both lettuce and these chicory species. *Sclerotinia sclerotiorum* can also infect these plants. For more information see the white mold on bean section in the legume chapter (page 262).

Tomato spotted wilt virus (TSWV)

Tomato spotted wilt

On radicchio, TSWV causes chlorotic spots, streaks, mottles, and other symptoms on leaves (**565**). If radicchio is infected when young, the plants can be stunted and leaves deformed and twisted. TSWV is vectored by thrips. For more information see the tomato spotted wilt section in the tomato chapter (page 364).

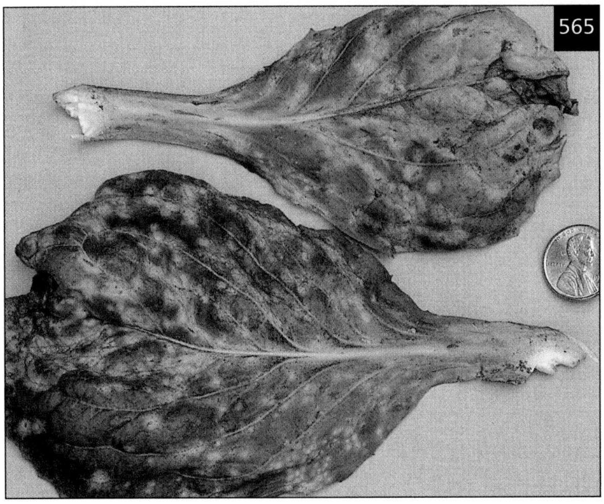

565 Radicchio leaves infected with *Tomato spotted wilt virus*.

566 Tipburn disorder on radicchio.

Tipburn

Tipburn is a physiological disorder of leafy vegetables caused by an imbalance of calcium in leaf tissue. Symptoms occur on the margins of developing leaves and consist of light to dark brown speckling, lesions, and necrosis (**566**, **567**). In severe cases tipburn can result in extensive damage to these leaf margins. Symptomatic leaves are usually found within the inner whorls of chicory heads. Calcium deficiency occurs when conditions cause plants to grow so rapidly that the plant cannot supply sufficient calcium to leaf margins. This disorder is similar to tipburn in lettuce and spinach, blackheart of celery, and blossom end rot of tomato. For more information see the tipburn section in the lettuce chapter (page 326).

567 Tipburn disorder on endive.

References

di Mario, E. 1968. *Alternaria porri* f. sp. *cichorii* on endive and chicory in Apulia (Italy). *Phytopathologia Mediterranea* 7:7–14.

Koike, S. T. and Butler, E. E. 1998. Leaf spot of radicchio caused by *Alternaria cichorii* in California. *Plant Disease* 82:448.

Koike, S. T. and Bull, C. T. 2006. First report of bacterial leaf spot of Italian dandelion (*Cichorium intybus*) caused by a *Pseudomonas syringae* pathovar in California. *Plant Disease* 90:245.

Koike, S. T and Saenz, G. S. 1996. Occurrence of powdery mildew, caused by *Erysiphe cichoracearum*, on endive and radicchio in California. *Plant Disease* 80:1080.

Koike, S. T. and Subbarao, K. V. 1999. Leaf blight of endive and escarole, caused by *Rhizoctonia solani*, in California. *Plant Disease* 83:1070.

Koike, S. T. and Subbarao, K. V. 1995. First report of endive and escarole as hosts of *Sclerotinia minor*. *Plant Disease* 79:642.

Koike, S. T. and Subbarao, K. V. 1995. First report of radicchio as a host of *Sclerotinia minor*. *Plant Disease* 79:966.

Schober, B. M. and Vermeulen, T. 1999. Enzymatic maceration of witloof chicory by the soft rot bacteria *Erwinia carotovora* subsp. *carotovora*: the effect of nitrogen and calcium treatments of the plant on pectic enzyme production and disease development. *European Journal of Plant Pathology* 105:341–349.

Vakalounakis, D. J. and Christias, C. 1985. Light intensity, temperature and conidial morphology in *Alternaria cichorii*. *Transactions of the British Mycological Society* 85:425–430.

Coriandrum sativum

CILANTRO

Also known as Chinese parsley or Mexican parsley, Cilantro is a leafy vegetable in the Apiaceae family and is grown for use as a fresh herb, salad ingredient, and seasonings in cooking. The same plant is also grown for its seed, which is used in seasoning and cooking. The seed crop is usually called coriander.

Cilantro yellow blotch virus (CYBV)

Cilantro yellow blotch

Symptoms on leaves consist of chlorotic blotches with diffuse margins, yellowed veins, and slight leaf deformities (**568**). Some cultivars can be mildly stunted. This disease reduces the quality of the harvested product. When symptomatic leaves are examined with an electron microscope, unusual virus-like particles measuring approximately 2 μm in length can be observed. Particle morphology resembles that of closteroviruses, but with a broadly twisted symmetry. In thin-sectioned leaf material, aggregates of virus-like particles are seen in phloem tissue. The pathogen is not yet fully characterized and is tentatively named *Cilantro yellow blotch virus*.

Fusarium oxysporum f. sp. *corianderii*

Fusarium wilt

This disease has been documented in Argentina, India, and the USA (California). Affected plants show stunting and poor growth, yellowing and reddening of foliage, and eventual collapse (**569**). Vascular tissues in crowns and large roots have a light tan to orange discoloration. Infected cilantro roots become translucent and later rot. The pathogen is the soilborne fungus *Fusarium oxysporum* f. sp. *corianderii*. In preliminary experiments this pathogen appears to be host specific to cilantro.

568 Foliar symptoms of *Cilantro yellow blotch virus* of cilantro.

569 Yellowing and reddening of cilantro infected by *Fusarium oxysporum*. Healthy plant on right.

570 Bacterial leaf spot of cilantro.

Pseudomonas syringae pv. *coriandricola*

Bacterial leaf spot

Bacterial leaf spot initially causes water-soaked, vein delimited lesions on leaves. These spots rapidly turn dark brown in color, remain angular in shape, and can be seen from both top and bottom sides of leaves (**570**). If disease is severe, the foliage can take on a blighted appearance when leaf spots coalesce. The pathogen is *Pseudomonas syringae* pv. *coriandricola*. This pseudomonad has both fluroescent and nonfluorescent strains and is seedborne on cilantro.

A new foliar disease of cilantro was found in 1999 in Australia. A *Microdochium* species caused light gray, irregular, sunken stem lesions on seed crops. Initial studies indicate this fungus infects cilantro but not other Apiaceae plants.

References

Cooksey, D. A., Azad, H. R., Paulus, A. O., and Koike, S. T. 1991. Leaf spot of cilantro in California caused by a nonfluorescent *Pseudomonas syringae*. *Plant Disease* 75:101.

Dennis, J. I. 2003. New disease of coriander in Australia associated with a *Microdochium* species. *Plant Pathology* 52:408.

Koike, S. T. and Gordon, T. R. 2005. First report of Fusarium wilt of cilantro caused by *Fusarium oxysporum* in California. *Plant Disease* 89:1130.

Madia, M., Gaetan, S., and Reyna, S. 1999. Wilt and crown rot of coriander by a complex of *Fusarium* species in Argentina. *Fitopatologia* 34:155–159.

Mayhew, D. E. 2002. Cilantro yellow blotch. *Compendium of Umbelliferous Crop Diseases*, p.55. R. M. Davis and R. N. Raid, editors. American Phytopathological Society Press. St. Paul, Minnesota.

Taylor, J. D. and Dudley, C. L. 1980. Bacterial disease of coriander. *Plant Pathology* 29:117–121.

Toben, H. M. and Rudolph, K. 1996. *Pseudomonas syringae* pv. *coriandricola*, incitant of bacterial umbel blight and seed decay of coriander in Germany. *Journal of Phytopathology* 144:169–178.

Srivastava, U. S. 1972. Effect of interaction of factors on wilt of coriander caused by *Fusarium oxysporum* f. sp. *coriánderii*. *Indian Journal of Agricultural Sciences* 42:618–621.

Cymbopogon citratus

LEMONGRASS

Lemongrass is a perennial, aromatic grass species (family Poaceae) that has a lemon-like scent to its foliage. The crop is grown for its edible stem and oil.

Puccinia nakanishikii/P. cymbopogonis

Rust

Symptoms of rust consist of elongated, stripe-like, dark brown lesions that develop on both sides of leaf surfaces (**571**). Only lesions on bottom leaf surfaces erupt and produce dark, cinnamon brown pustules (uredinia). Lesion development can be substantial, and coalescing lesions can result in significant foliage drying and death. Several rust pathogens, such as *Puccinia nakanishikii* and *P. cymbopogonis*, are known to infect lemongrass. It appears possible that more than one rust pathogen may occur on the crop in any particular area.

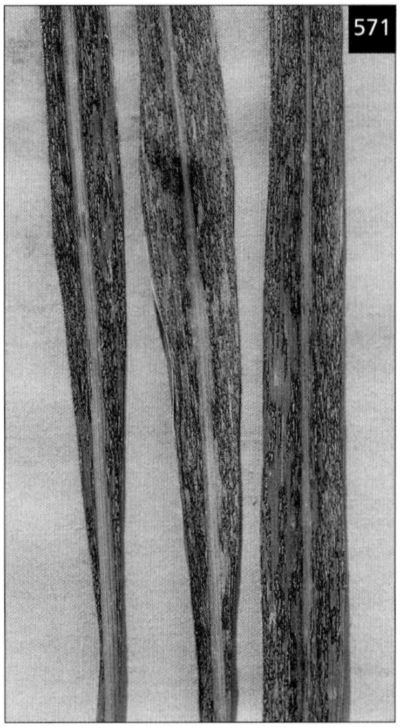

571 Lemongrass leaves infected with rust.

References

Bandara, J. M. R. S. 1981. Puccinia rust of citronella and lemongrass in Sri Lanka. *Plant Disease* 65:164.

Gardner, D. E. 1985. Lemongrass rust caused by *Puccinia nakanishikii* in Hawaii. *Plant Disease* 69:1100.

Koike, S. T. and Molinar, R. H. 1999. Rust disease on lemongrass in California. *Plant Disease* 83:304.

Cynara scolymus

ARTICHOKE

Artichoke is in the Asteraceae family and is an herbaceous thistle. These plants are grown for the edible bracts and receptacles of immature flower buds. Artichokes traditionally have been grown as large, perennial shrubs that were propagated by crown divisions. In recent years artichokes have also been grown as annuals that are started as transplants in greenhouses.

Artichoke curly dwarf virus (ACDV)

Artichoke curly dwarf

Artichoke curly dwarf disease causes significant reduction in growth and vigor. Diseased plants later become severely stunted. Leaves are distorted and have dark, necrotic spots and sections. Infected plants are less productive, with up to 40% less yield than healthy plants. Buds that are produced are often deformed and hence unmarketable. Severely affected plants may die. The virus particle is filamentous, measures 582 nm in length, and is possibly a member of the potexvirus group. The virus has no known vector. Artichokes infected with ACDV are apparently always co-infected with *Artichoke latent virus*. However, *Artichoke latent virus* by itself apparently causes no disease symptoms in artichoke. Therefore, the exact etiology of artichoke curly dwarf disease has not been determined. ACDV is spread to new plantings when diseased plants are divided for propagation material. In the field, only artichoke has been found to be a natural host of ACDV. Under experimental conditions, ACDV can also infect other plants in the Asteraceae family such as cardoon, sunflower, and zinnia.

Artichoke Italian latent virus (AILV)/
Artichoke yellow ringspot virus (AYRSV)

Artichoke Italian latent/Artichoke yellow ringspot

Both of these nepoviruses occur in Europe and are thought to be transmitted by the needle nematode (*Longidorus fasciatus*). AYRSV is also carried in seed, pollen, and plant material and produces bright yellow blotches, ringspots, and line patterns. AILV may be symptomless in some cultivars but can cause yellowing and stunting in others. *Artichoke mosaic, Bean yellow mosaic, Broad bean wilt, Tobacco streak*, and *Artichoke mottle crinkle viruses* are some of the other agents infecting artichoke. AILV has also been reported from Greece.

Ascochyta hortorum

Ascochyta rot or black rot

Ascochyta rot, or black rot, is an important disease in Mediterranean Europe. Symptoms usually develop first on the tips of the outermost bracts of the flower bud. Under dry conditions there is little further development of the disease, and infections remain as superficial blemishes. However, in wet weather the pathogen spreads externally and internally within the artichoke flower bud. The entire bud may be affected by a dark wet rot; secondary soft rot often follows. Cultivars with compact buds are more prone to Ascochyta rot than types with more open bud structure. The pathogen is *Ascochyta hortorum*. Older lesions have numerous black pycnidia that produce one-septate conidia. Brown lesions also occur on stems and foliage. Ascochyta rot may sometimes develop on old downy mildew (*Bremia lactucae*) lesions. The presence of pycnidia distinguishes Ascochyta rot from Itersonilia rot caused by *Itersonilia perplexans*.

572 Collapsed artichoke plants due to bacterial crown rot.

Erwinia chrysanthemi

Bacterial crown rot

Plants with bacterial crown rot may be stunted and show poor growth. The leaves wilt during the day when temperatures are warm. In advanced stages of the disease, plants may collapse entirely (572). New leaves in the center of plants may not expand and instead turn brown and dry. Crown and tap root tissues are soft, rotted, and turn brown or black (573). Blackened crown tissues can be readily observed when perennial plantings are pruned back at ground level, exposing the black, discolored cross sections of the crowns and stems.

After cutting, infected plants may fail to sprout, or will re-grow more slowly than healthy plants. The cause of bacterial crown rot is *Erwinia chrysanthemi*. This bacterium is a Gram-negative, non-spore producing rod and is in the Erwinia soft rot group. On CVP medium, *E. chrysanthemi* forms iridescent, translucent colonies in pits or depressions in the agar surface. Little information is available on disease development. It is likely that the pathogen is spread to other plants by the machines used to cut the plants. The digging and dividing of infected crowns for propagation results in diseased new plantings. The bacterium probably survives on both plant tissue and on dead organic matter. Other bacterial pathogens of artichokes include a *Xanthomonas* species that is usually associated with frost injury, and *Ralstonia solanacearum* that causes bacterial wilt.

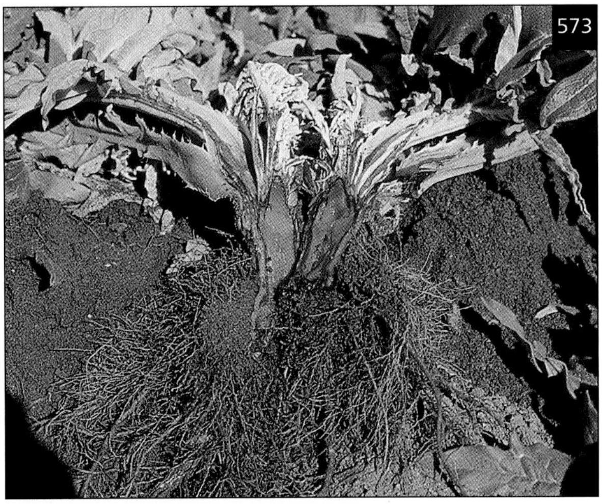

573 Discolored crown symptom of bacterial crown rot of artichoke.

574 Bract infection of gray mold of artichoke.

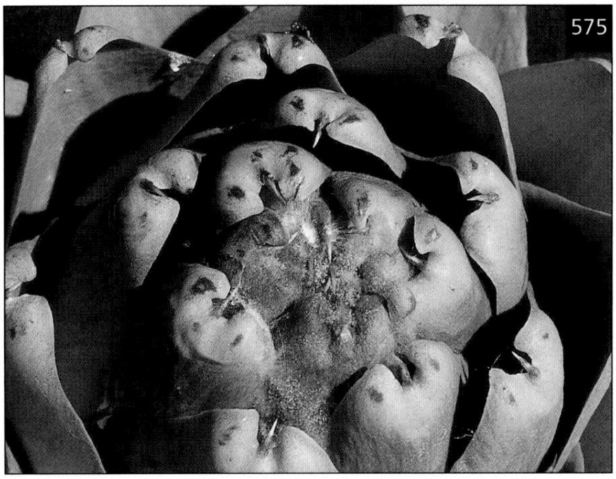

575 Infected flower bud and mycelium of gray mold of artichoke.

Botrytis cinerea
(teleomorph = *Botryotinia fuckeliana*)

Gray mold

In the field, individual flower bracts turn brown and later dry up (574). The top of the artichoke flower bud can also turn brown if infected. Characteristic gray to gray-brown, fuzzy growth of the fungus develops on the inner surfaces of the bracts and in the center of the bud apex (575). In the field, artichoke foliage generally is not affected by this pathogen. Gray mold can be a postharvest problem, with boxed and stored artichoke buds showing the same brown infections and fungal growth. The cause of gray mold is *Botrytis cinerea*. *Botrytis cinerea* survives in and around fields as a saprophyte on senescent leaves and crop debris, as a pathogen on numerous crops and weed plants, and as sclerotia in the soil. Conidia develop from these sources and become windborne. When conidia land on senescent or damaged flower bracts, they will germinate if free moisture is available and rapidly colonize this food source. The fungus usually invades bracts that have been damaged by insects, snails and slugs, frost, or other factors. Cool temperatures, free moisture, and high humidity favor the development of the disease. Gray mold is typically more severe following periods of wet weather.

Calcium deficiency

Calcium deficiency manifests itself in artichokes by causing the young, developing flower buds to develop black lesions on the tips of the bracts (576). Such bracts later shrivel and curl inwards. Sunken, brown sections develop on inner bracts within the flower bud. The young leaves that encircle the artichoke flower buds can also develop black lesions and spots. Calcium deficiency occurs when conditions cause plants to grow rapidly or are stressed. This disorder is similar to tipburn in lettuce and spinach, blackheart of celery, and blossom end rot of tomato.

Golovinomyces cichoracearum
*Leveillula taurica (*anamorph = *Oidiopsis taurica)*

Powdery mildew

There are two powdery mildew fungi that infect artichokes. *Leveillula taurica* (anamorph = *Oidiopsis taurica*) is found primarily colonizing the undersides of older leaves. Careful examination of the undersides of these leaves reveals the white powdery growth of this pathogen (577); however, the profuse white hairs of the leaf can obscure this sign. Severely infected leaves will first turn yellow, then brown (578). The brown leaves dry up and become papery and tattered. *Leveillula taurica* infects only the older leaves; the younger leaves escape infection until they mature (579). For *L. taurica*, all conidiophores develop from endophytic mycelium and emerge through stomata. Conidiophores can be branched and carry one or sometimes two conidia. Research suggests that *L. taurica* isolates from artichoke are distinct from the ones occurring on tomato and other vegetable hosts.

Another powdery mildew on artichoke, *Golovinomyces cichoracearum* (previously named *Erysiphe cichoracearum*), usually results in less severe symptoms. The white to gray growth of this fungus develops on the outside of flower bracts and on upper surfaces of both young and old leaves. Underlying tissue can turn purple to brown. *Erysiphe cichoracearum* produces epiphytic mycelium that grows superficially on host surfaces.

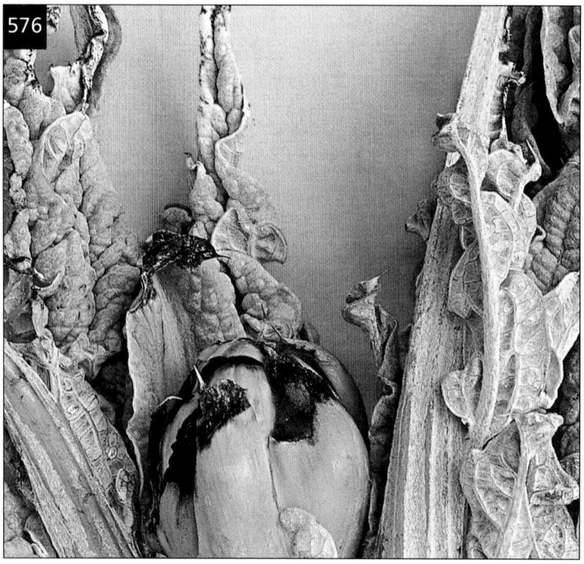

576 Blackened tips of bracts and leaves caused by calcium deficiency.

577 Sporulation of *Leveillula taurica* on an artichoke leaf.

578 **Severe (first two leaves on the left) and light infection (third leaf from the left) of *Leveillula taurica* of artichoke. Leaf on the right is healthy.**

579 ***Leveillula taurica*** **on older foliage of artichoke.**

Conidiophores are borne on this surface mycelium. Conidia are produced in long chains, are hyaline, ellipsoid to barrel-shaped, and measure 25–45 x 14–25 µm. Like *L. taurica, E. cichoracearum* likely exists as different physiological strains that have slightly different host ranges.

580 **Collapse of articoke plants caused by Pythium root rot.**

Pythium spp.

Pythium root rot

Newly planted seedling artichokes, either direct seeded or transplanted, wilt and collapse. Foliage becomes dull green and later brown and dry. Root tissues are discolored, decayed, and soft. Severely infected seedlings rarely recover and will die (**580**). Seedlings that do live can be stunted and delayed in their development. In California, artichokes were traditionally propagated vegetatively by digging up crowns of older plantings, dividing them into large pieces, then planting them into new fields. Recent production practices now use seed or transplants to initiate new plantings. Therefore this root rot problem, which occurs mainly on seedling artichokes, has only recently occurred in California. Root rot is caused by various species of the soilborne organism *Pythium*, including *P. aphanidermatum*. *Pythium aphanidermatum* is an oomycete that produces coenocytic mycelium, swollen, toruloid sporangia, and spherical, thick-walled oospores. Sporangia develop and release zoospores in soil water. The pathogen has a broad host range and can infect the seed and seedlings of many plant species. *Pythium* is favored by wet soil conditions that enable it to grow, produce swimming zoospores, and infect host roots. The disease has most often been observed on crops grown in sandy soil.

Ramularia cynarae

Ramularia leaf spot

Initial symptoms of Ramularia leaf spot consist of small (2–5 mm in diameter) pale to yellow-green circular spots. With time the spots expand up to 20 mm in diameter and turn brown (**581**). Spots are visible from both upper and lower surfaces of leaves. If disease is severe, lesions will coalesce and the entire leaf can turn brown and dry up. White growth of the fungus (**582**) will usually develop in the center of leaf lesions. This pathogen can also cause elongated lesions on stems (**583**). Ramularia leaf spot is economically important when the pathogen moves from the leaves to the flower bud bracts. On bracts, brown, irregularly shaped, patchy lesions will form (**584**), causing the bracts to curl and dry out. Symptoms on flower bracts usually make the artichoke bud unmarketable.

The pathogen is *Ramularia cynarae*. Conidiophores develop in the center of the leaf spots and grow in clustered groups (fascicles) that arise from stromata. Conidiophores are simple, rarely branch, geniculate, and measure 5–60 µm long. Conidia are catenulate, ellipsoid–ovoid to fusiform in shape, have from zero to three septa, are hyaline with thickened and darkened hila, and measure 10–5 x 2–5 µm. Ramularia leaf spot is managed by applying fungicides when leaf disease is significant and prior to flower bract infection.

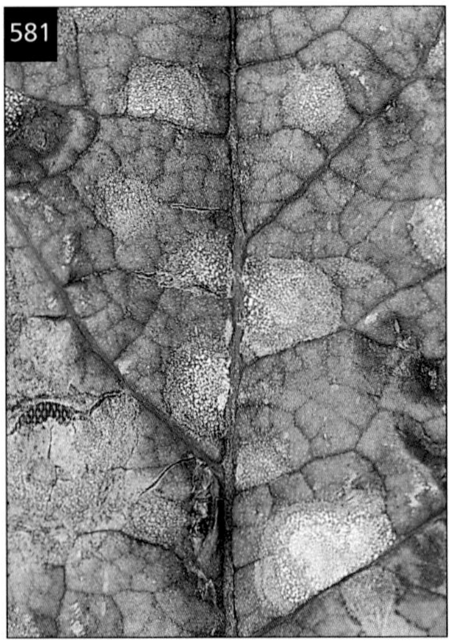

581 Leaf spots due to Ramularia leaf spot of artichoke.

583 Stem lesions due to Ramularia leaf spot of artichoke.

582 Sporulation of *Ramularia cynarae* on an artichoke leaf.

584 Bract lesions of Ramularia leaf spot on artichoke flower buds.

Verticillium dahliae

Verticillium wilt

Verticillium wilt symptoms initially consist of wilting of foliage, often on only one side of the plant. Such foliage turns chlorotic and later brown and dry. Infected plants grow poorly and are stunted. In severe cases an artichoke can lose all leaves and result in the plant consisting of only the bare, tall, thick green branches (585, 586). The vascular tissue of the lower stems, crowns, and roots shows dark brown to black discoloration and streaking (587, 588). Diseased plants produce smaller flower buds, and in severe cases the buds are discolored and dried. The causal agent is *Verticillium dahliae*. Microsclerotia enable the pathogen to survive in the soil for extended periods of time (up to 8–10 years). Infected artichoke plants may not always exhibit symptoms of disease; it is possible that only stressed plants show indications of infection. Annual artichoke varieties appear to be more susceptible to *V. dahliae* than perennial cultivars such as Green Globe.

References

Aly, M. M. and Abd El Ghafar, N. Y. 2000 Bacterial wilt of artichoke caused by *Ralstonia solanacearum* in Egypt. *Plant Pathology* 49: 807.

Bhat, R. G., Subbarao, K. V., and Bari, M. A. 1999. First report of *Verticillium dahliae* causing artichoke wilt in California. *Plant Disease* 83:782.

585 Collapsed artichoke plant infected with Verticillium wilt.

586 Collapsed plants due to Verticillium wilt of artichoke. Healthy plant is on the right.

587 Crown discoloration due to Verticillium wilt of artichoke.

588 Close-up of vascular discoloration in an artichoke stem infected with Verticillium wilt.

Braun, U. 1998. *A Monograph of Cercosporella, Ramularia, and Allied Genera (Phytopathogenic Hyphomycetes)* Vol. 2. IHW-Verlag, Eching, Germany.

Brown, D. J. F., Kyriakopoulou, P. E., and Robertson, W. M. 1997. Frequency of transmission of artichoke Italian latent nepovirus by *Longidorus fasciatus* (Nematoda: Longidoridae) from artichoke fields in the Iria and Kandia areas of Argolis in northeast Peloponnesus, Greece. *European Journal of Plant Pathology* 103:501–506.

Cirulli, M., Ciccarese, F., and Amenduni, M. 1994. Evaluation of Italian clones of artichoke for resistance to *Verticillium dahliae*. *Plant Disease* 78:680–682.

Correll, J. C., Gordon, T. R., and Elliott, V. J. 1987. Host range, specificity, and biometrical measurements of *Leveillula taurica* in California. *Plant Disease* 71:248–251.

Cragg, I. A. 1966. New or uncommon plant diseases and pests. *Itersonila perplexans* on globe artichoke. *Plant Pathology* 15:47.

Francois, L. E., Donovan, T. J., and Maas, E. V. 1991. Calcium deficiency of artichoke buds in relation to salinity. *HortScience* 26:549–553.

Lipton, W. J. and Harvey, J. M. 1960. Decay of artichoke bracts inoculated with spores of *Botrytis cinerea* Fr. at various constant temperatures. *Plant Disease Reporter* 44:837–839.

Malençon, G. 1936. Une grave maladie des artichauts au Maroc. *Revue de Mycologie* NS 1:165–175.

Morton, D. J. 1961. Host range and properties of the globe artichoke curly dwarf virus. *Phytopathology* 51:731–734.

Palti, J. 1971. Biological characteristics, distribution, and control of *Leveillula taurica*. *Phytopathologia Mediterranea* 10:139–153.

Rana, G. L., Russo, M., Gallitelli, D., and Martelli, G. P. 1980. Artichoke latent virus: characterization, ultrastructure and geographical distribution. *Annals of Applied Biology* 101:279–289.

Rana, G. L., Kyriakopoulou, P. E., Gallitelli, D., Russo, M., and Martelli, G. P. 1980. Host range and properties of artichoke yellow ringspot virus. *Annals of Applied Biology* 96:177–185.

Ride, M. 1956. Sur une maladie nouvelle de l'architaut (*Cynara scolymus* L.). *Comptes Rendus hebdomadaires des Séances de l'Académie des Sciences, Paris* No. 243: 174–177.

Russo, M. and Rana, G. L. 1978. Occurrence of two legume viruses in artichoke. *Phytopathologia Mediterranea* 17:212–216.

Stanghellini, M. E., Vilchez, M., Kim, D. H., Aguiar, J. L., and Armendariz, J. 2000. First report of root rot caused by *Pythium aphanidermatum* on artichoke. *Plant Disease* 84:811.

Eruca sativa

ARUGULA

Arugula is a leafy crucifer (family Brassicaceae) grown for salads or as a cooked vegetable. The plant is also called roquette, rucchetta, garden rocket, and rocket salad.

Albugo candida

White rust

Signs of white rust consist of white blister-like pustules (sori) that develop beneath the raised host epidermis on the underside of leaves (**589**). The corresponding upper leaf surface turns yellow. If disease is severe, leaves can turn chlorotic and then necrotic. Flower heads can also become infected and cause tissues to become swollen and twisted. This symptom is called a staghead. For more information see the white rust section in the brassica chapter (page 162).

Peronospora parasitica
(= *Hyaloperonospora parasitica*)

Downy mildew

Symptoms of downy mildew consist of small (1–4 mm diameter), irregular, dark brown to black speckling on the top and bottom sides of leaves (**590, 591**). Specks expand into tan spots 3–8 mm in diameter. The corresponding lower leaf surfaces supports the white fungal growth of the downy mildew (**592**). If sporulation is not present, downy mildew symptoms may be mistaken for

589 Pustules of white rust of arugula.

ARTICHOKE

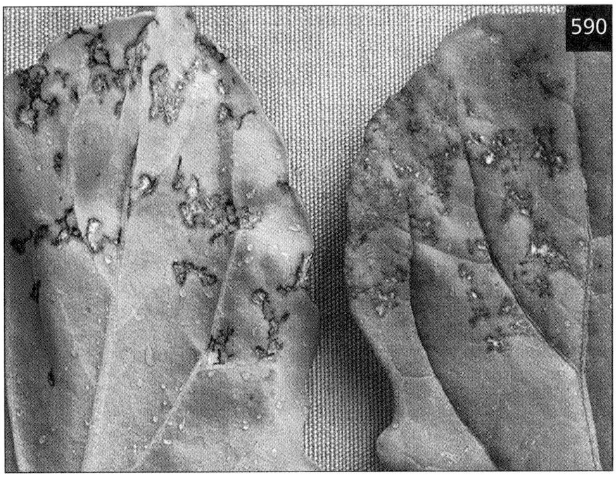

590 **Downy mildew lesions on arugula leaves.**

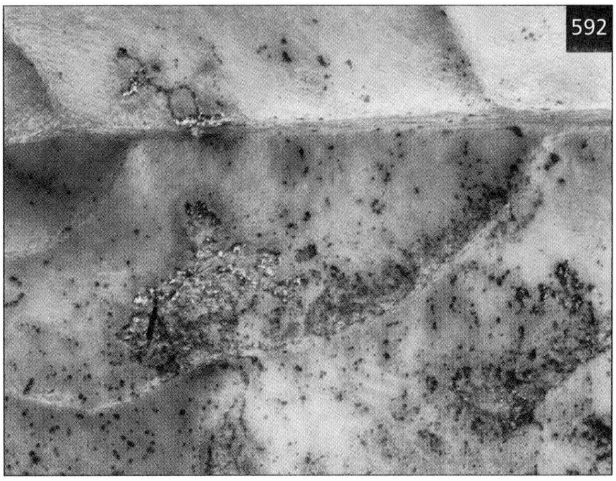

592 **Close-up of sporulation of downy mildew on arugula leaf.**

591 **Leaf lesions of downy mildew of arugula.**

593 **Leaf spots of bacterial blight of arugula.**

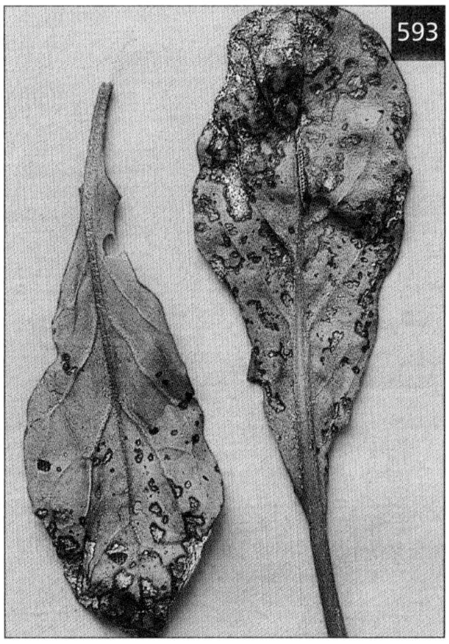

bacterial blight disease. The pathogen is *Peronospora parasitica*. For more information see the downy mildew section in the brassica chapter (page 178).

Pseudomonas syringae pv. *alisalensis*

Bacterial blight

Initial symptoms of bacterial blight are small (less than 2 mm in diameter), water-soaked spots on both fully expanded and younger leaves. As disease develops, lesions enlarge and become angular in shape and tan in color (593). Older lesions become desiccated and have purple margins. Bacterial blight symptoms may be similar to those of downy mildew. The pathogen is *Pseudomonas syringae* pv. *alisalensis*. It appears that this is the same bacterium that infects broccoli raab and occasionally broccoli. For more information see the bacterial blight section in the brassica chapter (page 157).

References

Bull, C. T., Goldman, P., and Koike, S. T. 2004. Bacterial blight on arugula, a new disease caused by *Pseudomonas syringae* pv. *alisalensis* in California. *Plant Disease* 88:1384.

Koike, S. T. 1998. Downy mildew of arugula, caused by *Peronospora parasitica*, in California. *Plant Disease* 82:1063.

Koike, S. T., Smith, R. F., Van Buren, A. M., and Maddox, D. A. 1996. A new bacterial disease of arugula in California. *Plant Disease* 80:464.

Scheck, H. J. and Koike, S. T. 1999. First occurrence of white rust of arugula, caused by *Albugo candida*. *Plant Disease* 83:877.

594 Blighted foliage due to Cercosporidium blight of fennel. Healthy foliage on the right.

595 Close-up of fennel leaves infected with *Cercosporidium punctum*.

596 Close-up of fennel petiole infected with *Cercosporidium punctum*.

Foeniculum vulgare dulce

FENNEL

Fennel, Florence fennel, or finochio is in the Apiaceae family and is used as both a vegetable and herb commodity. Fennel is eaten as a fresh and cooked vegetable. The foliage, wide basal portions of the petioles, and swollen above-ground stem are used. Fennel is also produced for its seed. Fennel is distinct from the closely related anise (*Pimpinella anisum*).

Cercosporidium punctum

Cercosporidium blight

This leaf blight disease affects the older foliage of fennel but does not infect new leaves. Affected leaf tips and stems turn brown, wither, and dry up (**594**). Close examination of the stems and the threadlike leaves will reveal the presence of tiny, discrete, dark brown to black pustules (**595, 596**). These pustules consist of clusters of the conidiophores and conidia of the pathogen.

Sclerotinia minor, S. sclerotiorum

White mold

Crown and lower petiole tissues in contact with soil develop a brown rot (**597, 598**). This necrotic tissue rapidly turns into a soft rot. White mycelium and small (0.5–3.0 mm) black sclerotia form on infected tissues.The foliage of affected stems declines and turns bright yellow, then brown (**599**). Isolates of this pathogen are able to infect both lettuce and fennel. The other main white mold species, *S. sclerotiorum,* also infects fennel and causes foliar blight and lower petiole decay similar to that seen on celery.

References

Cirulli, M. 1981. *Cercosporidium punctum* (Lacroix) Deighton su prezzemolo in Italia. *Informatore Fitopatologico* 3:33-37.

Koike, S. T. 1994. First report of stem rot of fennel in the United States caused by *Sclerotinia minor. Plant Disease* 78:754.

Koike, S. T., Butler, E. E., and Greathead, A. S. 1992. Occurrence of *Cercosporidium punctum* on fennel in California. *Plant Disease* 76:539.

Sisto, D. 1983. *Cercosporidium punctum* (Lacroix) Deighton su finocchio (*Foeniculum vulgare* Mill. var. *Azoricum* Thell.) in Italia meridionale. *Informatore Fitopatologico* 7-8:55-58.

597 Diseased petioles of fennel infected with *Sclerotinia minor*.

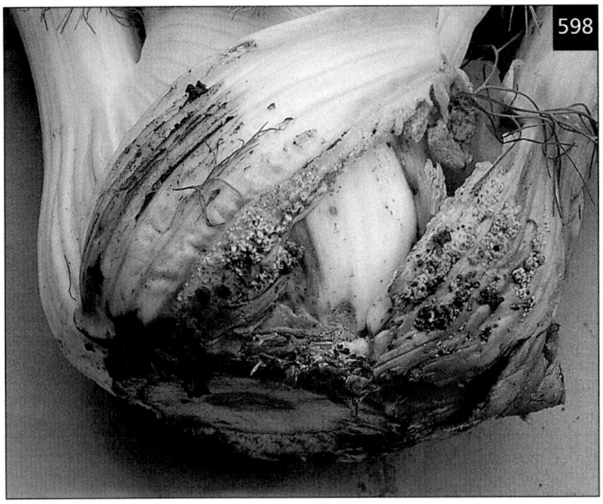

598 Fennel crown with sclerotia of *Sclerotinia minor*.

Helianthus tuberosus

JERUSALEM ARTICHOKE

Jerusalem artichoke or sunchoke (family Asteraceae) is a native of North America and was introduced into Europe in the 17th century. It is grown for its tubers and is eaten as a fresh vegetable in salads, a cooked vegetable ingredient, and as a pickled relish. In the UK, Jerusalem artichoke is also commonly grown to provide cover for game birds around field margins. Harvested tubers are prone to postharvest fungal rots.

Golovinomyces orontii

Powdery mildew

Powdery mildew is a common foliar disease of Jerusalem artichoke. Typical white powdery growth develops on leaves and stems (**600**). Severe disease results in extensive fungal growth on the foliage and both upper and lower leaf surfaces are colonized. Powdery mildew can cause yellowing and early death of leaves. The pathogen is *Golovinomyces orontii* (previously named *Erysiphe orontii*). This organism consists of various strains that differ in their respective host ranges. Conidia are borne in long chains and are elliptical to barrel shaped (25–45 x14–26 µm). Dark brown cleistothecia may be produced in autumn and have unbranched flexuous appendages up to 500 µm long. The cleistothecia contain asci that produce two ascospores each.

599 Chlorotic foliage of fennel infected with *Sclerotinia minor*.

600 Powdery mildew of Jerusalem artichoke.

Sclerotinia minor, S. sclerotiorum

Sclerotinia rot/watery soft rot

Sclerotinia rot or watery soft rot is the most important disease of Jerusalem artichoke. Both *Sclerotinia minor* and *S. sclerotiorum* have been recorded on this host, with the latter being more frequently reported. Ascospores of *S. sclerotiorum* infect aerial parts of the plant and cause brown petiole and stem lesions. Subsequently, there is spread down the stem to the tubers and formation of numerous black sclerotia on the surface of the stem and within the pith cavity. Affected plants wilt and die. In the case of *S. minor*, soilborne sclerotia germinate to produce mycelium that invades the stem base directly.

This disease is often associated with heavy clay soils and poorly draining areas. Infected Jerusalem artichoke may act as an inoculum source of *S. sclerotiorum* for other crops. In the UK, these plants are maintained as game covers. If infected, these plantings harbor apothecia that release ascospores that are carried by winds into oilseed rape (*Brassica napus*) fields. For more information see the white mold on bean section in the legume chapter (page 262).

Sclerotium rolfsii

Southern blight

Infected Jerusalem artichoke plants have foliage and branches that turn chlorotic, wilt, and eventually collapse. Lower stem tissue in contact with soil is the site of infection, and the fungus causes a dry rot of these tissues. The characteristic white mycelium and small (1-2 mm in diameter), spherical, tan to light brown sclerotia form on the stems and adjacent soil (**601, 602**). The pathogen is *Sclerotium rolfsii*. For more information see the southern blight on carrot section in the Apiaceae chapter (page 101).

601 Mycelium and sclerotia of *Sclerotium rolfsii* growing from infected Jerusalem artichoke and onto soil surface.

602 Mycelium and spherical sclerotia of *Sclerotium rolfsii* on Jerusalem artichoke crowns, resulting in plant dieback.

References

Laberge, C. and Sackston, W. E. 1987. Adaptability and diseases of Jerusalem artichoke (*Helianthus tuberosus*) in Quebec. *Canadian Journal of Plant Science* 67:349–352.

Koike, S. T. 2004. Southern blight of Jerusalem artichoke caused by *Sclerotium rolfsii* in California. *Plant Disease* 88:769.

McCarter, S. M. 1993. Reactions of Jerusalem artichoke genotypes to two rusts and powdery mildew. *Plant Disease* 77:242–245.

McCarter, S. M. and Kays, S. J. 1984. Diseases limiting production of Jerusalem artichokes in Georgia. *Plant Disease* 68:299–302.

Mentha spp.

MINT

Mint (family Lamiaceae) is grown as a leafy herb and vegetable and is used in various ways in cooking. It is also valuable as a processed commodity for oil and mint flavoring. There are several mint types; two of the more commonly grown species are spearmint (*Mentha spicata*) and peppermint (*M. x piperita*).

Erysiphe biocellata, E. biocellata, E. orontii

Powdery mildew

Powdery mildew is a problem in some commercial production areas in the USA, being more severe on spearmint than peppermint. The disease results in the growth of white mycelium that is epiphytic and occurs on both leaf surfaces (**603**). Conidia are produced in chains but lack fibrosin bodies. The pathogen is *Erysiphe orontii*, though the new name *Golovinomyces orontii* has been suggested. In the UK, the powdery mildew pathogen is reported as *E. biocellata*, which forms cleistothecia and few conidia on *M. aquatica* and *M. arvensis*. Older records indicate *E. cichoracearum* is the pathogen.

603 Mint leaf infected with powdery mildew.

Puccinia menthae

Rust

Rust is probably the most important foliar disease of mint worldwide, capable of causing reductions in both yield and quality. The most obvious indication of this problem is the presence of the rust pathogen itself. The fungus forms discrete, dark brown uredinia pustules on both sides of the infected leaf (**604**). The brown urediniospores emerge in large numbers from these pustules. If infected severely, the leaf can turn yellow and defoliate. From late summer onwards, dark brown teliospores are produced on leaves and stems, with telia visible as black spots. A systemic infection of shoots in the spring can occur if overwintering rhizomes are infected by teliospores in soil. The systemically affected shoots are swollen and distorted, with elongated chlorotic internodes and chlorotic leaves. Like many rusts, the life cycle is complex and there is even a third spore stage (aeciospores) that occurs (**605**).

604 Rust pustules on mint.

605 Rust aecia on eau de cologne mint.

The pathogen is *Puccinia menthae*, which is an autoecious and macrocyclic rust fungus. *P. menthae* is regarded as a collective species and there are numerous strains or races, which show specialization to different host species. For example, the *P. menthae* isolates that infect oregano are distinct from those that infect mint. *Puccinia angustata* is the only other rust recorded on *Mentha*, on which it produces aecia.

606 One-sided leaf chlorosis caused by *Verticillium dahliae* on mint.

Control mint rust by using resistant cultivars, treating stolons or cuttings with hot water dips prior to planting, and applying fungicides. In some cases, growers use propane flamers to destroy diseased crop residues in early spring.

Verticillium dahliae

Verticillium wilt

Verticillium wilt is a serious disease of peppermint and other mint species. Foliage becomes chlorotic and shows unevenness in growth (**606**). The plant can be stunted, internodes are shortened, and the plant eventually will die. Wilt infection also increases susceptibility to winter kill. The pathogen is *Verticillium dahliae*. If plants are colonized by the lesion nematode (*Pratylenchus penetrans*) the Verticillium wilt problem can be even more severe. Spearmint is apparently tolerant to this disease. For more information see the Verticillium wilt section in the tomato chapter (page 360).

References

Hagan, R. A. and Walters, D. R. 1994. Chemical control of rust on mint. *Proceedings of the Brighton Crop Protection Conference – Pests & Diseases* 3:791–796.

Horner, C. E. 1963. Field disease cycle of peppermint rust. *Phytopathology* 53:1063–1067.

Johnson, D. A. and Santo, G. S. 2001. Development of wilt in mint in response to infection by two pathotypes of *Verticillium dahliae* and co-infection by *Pratylenchus penetrans*. *Plant Disease* 85:1189–1192.

Koike, S. T. and Saenz, G. S. 1999. Powdery mildew of spearmint caused by *Erysiphe orontii* in California. *Plant Disease* 83:399.

Nepeta cataria

CATNIP

Catnip is a perennial shrub in the mint family (Lamiaceae) and is probably best known as a plant that is attractive to cats. This plant is also grown commercially and used as an herb, seasoning, and tea.

Xanthomonas campestris

Xanthomonas leaf spot

Initial symptoms of Xanthomonas leaf spot consist of small brown flecks (1–2 mm in diameter) that are visible from both top and bottom sides of the leaves. These flecks expand into larger leaf spots (2–8 mm diameter) that are dark brown to black and angular in shape (**607**). The pathogen is *Xanthomonas campestris*. Catnip strains are not able to infect carrot, cauliflower, pepper, tomato, or the ornamental flower stock (*Matthiola incana*). Therefore, this pathogen may be a new pathovar in the *X. campestris* group.

References

Koike, S. T., Azad, H. R., and Cooksey, D. A. 2001. Xanthomonas leaf spot of catnip: a new disease caused by a pathovar of *Xanthomonas campestris*. *Plant Disease* 85:1157–1159.

607 Foliar symptoms of Xanthomonas leaf spot of catnip.

Ocimum basilicum

BASIL

Basil is a widely produced herb in the mint family (Lamiaceae). The plant has extensive uses as a fresh leafy vegetable, cooked vegetable, spice and seasoning in foods, and for teas and medicinal purposes. There is a broad range of cultivars that have green, purple, or maroon-colored leaves, flat or crinkled leaves, and diverse flavors and fragrances.

Botrytis cinerea
(teleomorph = *Botryotinia fuckeliana*)

Gray mold

Gray mold disease affects all above-ground parts of basil. If conditions of high humidity are present, the pathogen can cause brown leaf and stem lesions to form (**608**, **609**). Infected tissues become soft and rotted. Gray sporulation and black, oblong sclerotia can sometimes form. This pathogen is particularly aggressive on damaged foliage and on basil stems remaining after a harvest. Gray mold can be a significant post-harvest problem, as well.

Fusarium oxysporum f. sp. *basilicum*

Fusarium wilt

Infected basil plants show stunting, twisted and asymmetrical growth, and foliar wilting (**610**). Plants can eventually collapse and die. Stem and crown xylem tissues show a red-brown to almost black discoloration (**611**). The pathogen can survive in soil for extended periods, spread via airborne conidia, and be seedborne. Worldwide, this disease is an important concern for basil production.

608 Leaf infection due to gray mold of basil.

610 Wilting and collapsing of basil infected with Fusarium wilt.

609 Gray mold stem infection of basil.

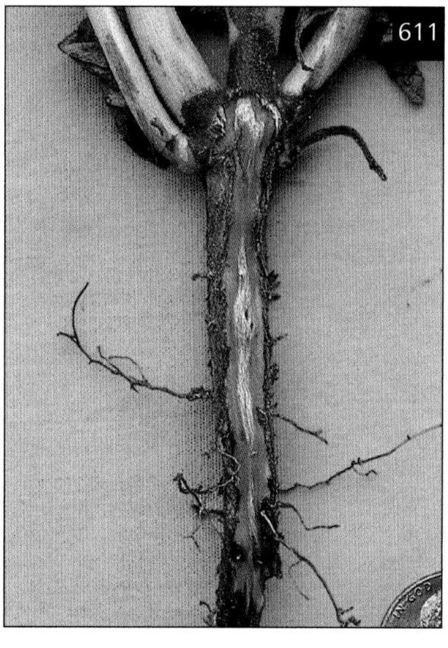

611 Discolored vascular tissue of Fusarium wilt of basil.

612

621 Leaf lesions caused by *Pseudomonas viridiflava* on basil.

613

613 Basil leaf on left is infected with *Tomato spotted wilt virus*. Leaf on the right is healthy.

Pseudomonas viridiflava

Bacterial leaf spot

Bacterial leaf spot symptoms initially consist of angular water-soaked spots. These spots rapidly turn black in color, but remain vein-delimited (**612**). Leaf spots often develop along the margins of the leaves.

Sclerotinia minor, S. sclerotiorum

White mold

White mold disease is caused by two pathogens. *Sclerotinia minor* causes a brown soft rot to develop on the lower sections of stems that are in contact with soil. Such stems wilt and plants eventually collapse and rot. White mycelium and small, irregular, black sclerotia (3–5 mm in diameter) are observed on diseased stems. The other species, *S. sclerotiorum*, can cause identical symptoms if the large, soilborne sclerotia directly infect basil stems. However, *S. sclerotiorum* forms apothecia and airborne ascospores. These can land on the upper foliage of basil and cause a brown, soft decay of leaves and upper stems. Infected foliage will also develop white mycelium and large (5–10 mm) black sclerotia. For more information see the white mold on bean section in the legume chapter (page 262).

Tomato spotted wilt virus (TSWV)

Tomato spotted wilt

TSWV causes basil leaves to develop a light green to yellow mottle, mosaic, and blotches (**613**). Symptomatic leaves can be buckled. TSWV is vectored by thrips. See also the tomato spotted wilt section in the tomato chapter (page 364).

References

Elmer, W. H., Wick, R. L., and Haviland, P. 1994. Vegetative compatibility among *Fusarium oxysporum* f. sp. *basilicum* isolates recovered from basil seed and infected plants. *Plant Disease* 78:789–791.

Gamliel, A., Katan, T., Yunis, H., and Katan, J. 1996. Fusarium wilt and crown rot of sweet basil: involvement of soilborne and airborne inoculum. *Phytopathology* 86:56–62.

Garibaldi, A., Lodovica Gullino, M., and Minuto, G. 1997. Diseases of basil and their management. *Plant Disease* 81:124–132.

Katan, T., Gamliel, A., and Katan, J. 1996. Vegetative compatibility of *Fusarium oxysporum* from sweet basil in Israel. *Plant Pathology* 45:656–661.

Keinath, A. P. 1994. Pathogenicity and host range of *Fusarium oxysporum* from sweet basil and evaluation of disease control methods. *Plant Disease* 78:1211–1215.

Koike, S. T. 2000. Occurrence of stem rot of basil, caused by *Sclerotinia sclerotiorum*, in coastal California. *Plant Disease* 84:1342.

Koike, S. T. and O'Brien, R. D. 1995. Basil as a host of *Sclerotinia minor. Plant Disease* 79:859.

Little, E. L., Gilbertson, R. L., and Koike, S. T. 1994. First report of *Pseudomonas viridiflava* causing a leaf necrosis on basil. *Plant Disease* 78:831.

Reuveni. R., Dudai, N., Putievsky, E., Elmer, W. H., and Wick, R. L. 1997. Evaluation and identification of basil germ plasm for resistance to *Fusarium oxysporum* f. sp. *basilicum*. *Plant Disease* 81:1077–1081.

Sharabani, G., Shtienberg, D., Elad, Y., and Dinoor, A. 1999. Epidemiology of *Botrytis cinerea* in sweet basil and implications for disease management. *Plant Disease* 83:554–560.

Origanum majorana/O. vulgare

MARJORAM, OREGANO

Oregano (*Origanum vulgare*) and sweet marjoram (*Origanum majorana*) are two closely related herbs in the mint family (Lamiaceae). Both plants are used for a variety of purposes in cooking. The two species have similar flavor, with marjoram tending to be milder than oregano.

Puccinia menthae

Rust

Both herbs are susceptible to the same rust pathogen. Symptoms on both plants are similar and mostly consist of small (2–5 mm diameter), circular, brown, necrotic leaf spots that develop cinnamon-brown pustules in the center of the spot or in concentric groups around the spot periphery (**614**). Pustules sometimes develop without spots. On sweet marjoram, leaf spots can be surrounded by a chlorotic halo. Ellipsoidal urediniospores measure 22–25 µm x 19–22 µm and contain two to three germ pores in an equatorial configuration. The pathogen, *Puccinia menthae,* is regarded as a collective species, and there are numerous strains or races which show specialization to different host species. For example, the *P. menthae* isolates that infect mint are distinct from those that infect oregano and marjoram.

References

Koike, S. T., Subbarao, K. V., Roelfs, A. P., Hennen, J. F., and Tjosvold, S. A. 1998. Rust disease of oregano and sweet marjoram in California. *Plant Disease* 82:1172.

614 Rust pustules on oregano.

Physalis ixocarpa

TOMATILLO

Tomatillo, or tomate verde, is native to Mexico and is a member of the Solanaceae family. This annual plant is grown for its relatively small (3–6 cm in diameter) spherical, light green to yellow fruit that resembles a small immature tomato. However, the tomatillo fruit is noticeably sticky and is encased in an enlarged, papery sheath or husk (which is a modified calyx). The fruit is used in sauces, salsa, and many other foods.

Fusarium oxysporum

Fusarium wilt

Plants affected with Fusarium wilt grow poorly. Foliage turns chlorotic, and later wilts (**615**). In advanced stages the foliage turns brown and falls off the plant. The vascular tissue in lower stems turns brown (**616**).

615 Declining tomatillo infected with Fusarium wilt.

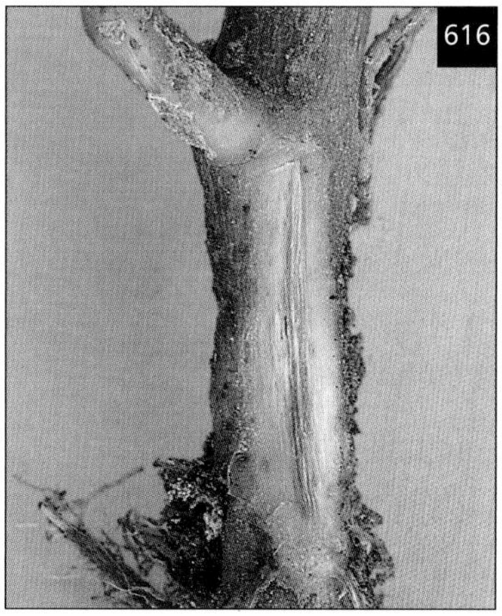

616 Discolored vascular tissue of Fusarium wilt of tomatillo.

617 Tomatillo foliage infected with *Turnip mosaic virus*. Healthy foliage on the right.

628 Tomatillo fruit infected with *Turnip mosaic virus*.

Turnip mosaic virus (TuMV)

Turnip mosaic

The foliage of affected plants becomes extremely deformed, twisted, and chlorotic (**617**). Stem internodes are shortened, giving the leaves a bunched up look. Fruit appear blotchy with white and green patterns (**618**). TuMV is vectored by aphids.

Rheum rhubarbarum

RHUBARB

Rhubarb is a perennial plant in the Polygonaceae family that is grown mostly for its thick, edible stalks or stems. The roots of some rhubarb selections are used for medicinal purposes. However, the rhubarb leaves are toxic.

Ascochyta rhei

Ascochyta leaf spot

This disease is common in North America. Symptoms are small, green-yellow spots that turn brown within a few days. Spots are variable in size and often have a pale center, a red margin, and are surrounded by a gray-green zone. The centers of individual or coalesced lesions tend to fall out and leave ragged holes. Dark pycnidia are produced but can be difficult to observe in the lesions. The pathogen is *Ascochyta rhei.*

Botrytis cinerea
(teleomorph = *Botryotinia fuckeliana*)

Gray mold

Gray mold is common in forcing sheds where the humidity is high and soft new growth is easily damaged. On outdoor crops gray mold can develop after frost damage on new growth, and after wind damage to mature foliage. The pathogen is *Botrytis cinerea*. The perfect stage, *Botryotinia fuckeliana*, is rarely observed on the crop.

Colletotrichum erumpens

Anthracnose

Anthracnose is caused by *Colletotrichum erumpens*. The disease produces small, oval watery spots on petioles. These spots increase in size and number particularly under warm, humid conditions. Foliage later collapses and there can be postharvest rots of the petioles and stems. The diagnostic dark acervuli are produced abundantly in lesions.

Erwinia carotovora, Pseudomonas marginalis

Bacterial soft rot

Two bacterial pathogens cause water-soaked, slimy rots on rhubarb stalks. Severely affected stems can collapse. This disease occurs both in the field and in postharvest storage situations. Bacterial soft rot is caused by *Erwinia carotovora* and *Pseudomonas marginalis*.

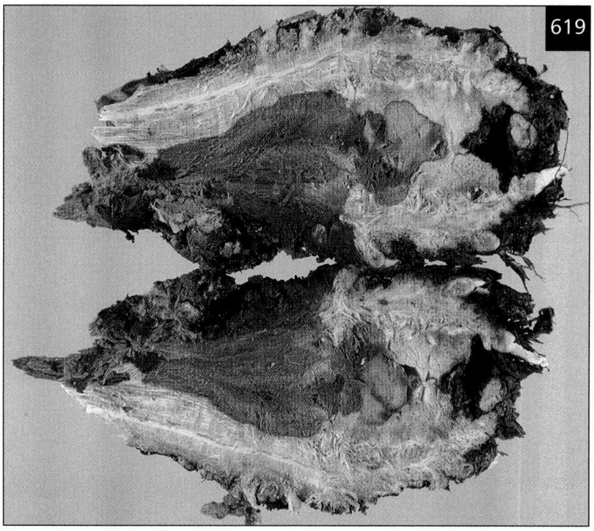

619 Discolored rhubarb crown tissue caused by *Erwinia* crown rot.

Erwinia rhapontici

Crown rot

Crown rot is caused by the bacterium *Erwinia rhapontici*, which results in a rot of the terminal bud. The rot may extend into the central pith (**619**), where a dark brown soft rot develops and then breaks down to form cavities. Side shoots can grow from infected plants, but these shoots often wilt and rot. In wet weather, older leaves can develop rot. Infection is thought to be associated with wounds, and the stem and bulb nematode (*Ditylenchus dipsaci*) may spread the pathogen.

Peronospora jaapiana

Downy mildew

Severe attacks of downy mildew, caused by *Peronospora jaapiana*, have occurred in eastern England in recent years. Downy mildew is a destructive disease on seedlings but also affects crops at any stage. Symptoms are large, brown leaf lesions that bear a purple-white fungal growth on the undersurface. Small spots often disintegrate and cause foliage to have a tattered appearance. Severe infection results in leaf death, reducing yield in the following year.

Puccinia phragmitis

Rust

Rust is caused by the fungus *Puccinia phragmitis* and is occasionally reported on rhubarb, causing red or deep purple leaf spots that measure 5–15 mm in diameter (**620, 621**). Dense clusters of the aecidial stage are produced at the center of the leaf spots; aeciospores measure 16–26 µm and appear white on the leaf. The urediniospores and teliospores are formed on the alternate host, water reed (*Phragmites communis*).

Pythium spp., *Phytophthora* spp.

Root rot diseases

Various root rots have been recorded on rhubarb, notably those caused by *Pythium* and *Phytophthora* species when sites are poorly drained. Violet root rot (*Helicobasidium brebissonii*), Armillaria root rot (*Armillaria mellea*), and southern blight (*Sclerotium rolfsii*) are occasionally detected.

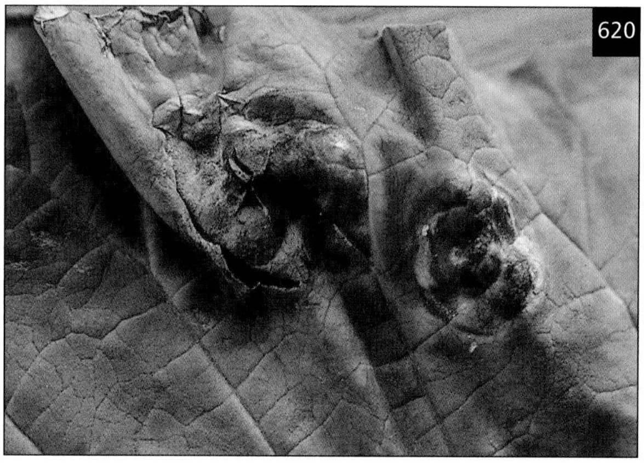

620 Rust disease on rhubarb leaf.

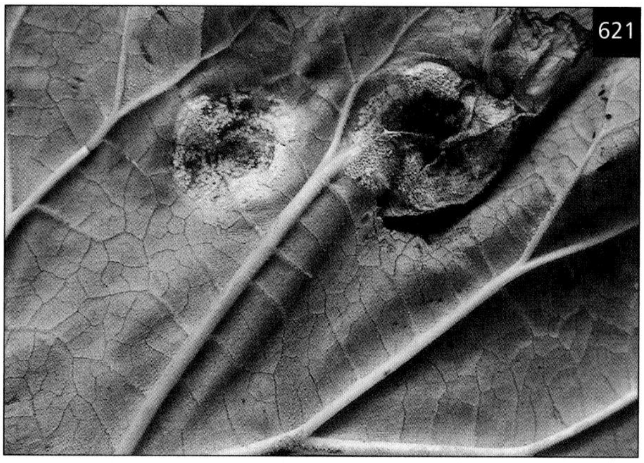

621 Underside of rhubarb leaf infected by rust.

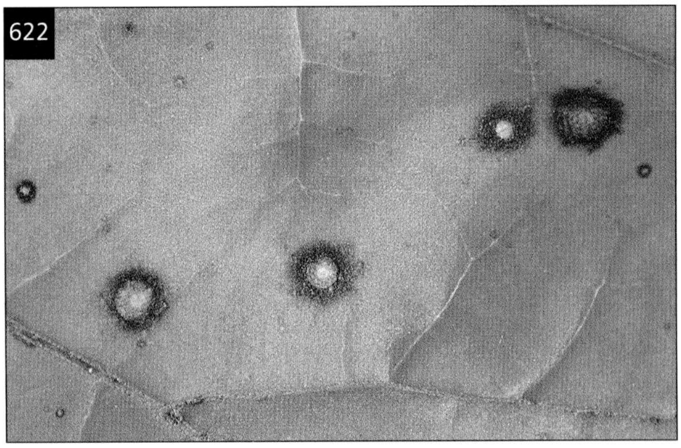

622 Circular to oval tan leaf spots on rhubarb caused by *Ramularia rhei*.

Ramularia rhei

Ramularia rot

Ramularia rot occurs in Europe, North America, and New Zealand. The disease causes oval sunken lesions on leaf stalks and leaves (622). It can cause crop loss and hence justifies a program of fungicide sprays when infection occurs early in the season. The pathogen is *Ramularia rhei*.

Virus diseases

Rhubarb is affected by a number of viruses. *Turnip mosaic virus* (TuMV) is the most important, but *Arabis mosaic virus* (ArMV), *Cucumber mosaic virus* (CMV), *Cherry leaf roll virus* (CLRV), and *Strawberry latent ringspot virus* (SLRSV) have been reported. Aphids are the vectors for TuMV and CMV. ArMV, CLRV, and SLRSV have nematode vectors while ArMV and CLRV may be seedborne. Combinations of different viruses may occur in the same plant, producing leaf mosaic and ring spot symptoms. Foliar growth may be stunted.

References

Ormrod, D. J., Sweeney, M. E., and MacDonald, L. S. 1985. Effect of fungicides on Ramularia leaf and stalk spot of rhubarb in coastal British Columbia. *Canadian Plant Disease Survey* 65:29–30.

Sellwood, J. E. and Lelliot, R. A. 1978. Internal browning of hyacinth caused by *Erwinia rhapontici*. *Plant Pathology* 27:120–124.

Tomlinson, J. A. and Walkey, D. G. A. 1967. The isolation and identification of rhubarb viruses in Britain. *Annals of Applied Biology* 59:415–427.

Walkey, D. G. A., Creed, C., Delaney, H., and Whitwell, J. D. 1982. Studies on re-infection and yield of virus-tested and commercial stocks of rhubarb cv. Timperley Early. *Plant Pathology* 31:253–260

Rorippa nasturtium-aquaticum

WATERCRESS

Watercress is a crucifer plant (family Brassicaceae) that is usually started as a transplant and then later placed in specially constructed beds with shallow water (10–15 cm deep). There is slow movement of water through the bed (usually from an underground source and at a constant temperature of 10–12° C), so waterborne problems are a significant threat to production.

Peronospora parasitica (= *Hyaloperonospora parasitica*), *Septoria sisymbrii*

Downy mildew, Septoria leaf spot

There are several minor foliar diseases of watercress including downy mildew (*Peronospora parasitica*) and Septoria leaf spot (*Septoria sisymbrii*). Downy mildew causes yellow blotches on leaves (**623**) with white sporulation on the leaf underside. It is most often found in autumn and spring when humidity is high and there is poor air circulation. See also the downy mildew section in the brassica chapter (page 178).

Septoria sisymbrii develops on the lower leaves as small (5–10 mm) pale brown or yellowish spots containing prominent dark pycnidia (**624**). Spread occurs by splashing water. *Septoria* may originate from infected debris, wild hosts, or be seedborne.

Spongospora nasturtii

Crook root

Crook root was first recorded in Wiltshire, England, in 1947. It is a serious threat to commercial production in parts of Europe and the USA and also occurs on wild watercress. It is caused by *Spongospora nasturtii*, which is now classified as a protozoan (like the clubroot pathogen of crucifers) rather than as a fungus. This pathogen causes stunting and chlorosis of the foliage. Diseased roots are swollen and brittle with a characteristic curved or crook-like appearance. The roots decay and plants may float away. Symptoms are most pronounced when temperatures are low, appearing from October through to April under UK conditions. Symptoms often appear first near the water outlet and spread back up the bed towards the water inlet. Plants show some recovery as temperatures rise in the spring. In addition, *S. nasturtii* is the vector of two viruses affecting watercress: *Watercress yellow spot virus* and *Watercress chlorotic leaf spot viroid*.

Virus diseases

Watercress is susceptible to several virus pathogens. *Turnip mosaic virus* (TuMV) is synonymous with *Watercress mosaic virus*. The symptoms include green and yellow mottling, yellowish spotting, leaf distortion, and mosaics. Affected plants are usually stunted. It is common in Europe, where it is spread by aphids, particularly *Myzus persicae* and *Brevicoryne brassicae*. *Turnip yellow mosaic virus* is also common in watercress and is transmitted by the mustard beetle (*Phaedon cochleariae*) and flea beetles (*Phyllotreta* spp.).

Watercress chlorotic leaf spot viroid (WCLVd) initially causes small (3–5 mm), bright yellow or golden spots to appear on leaves (**625**), followed by more numerous spots, many of which are chlorotic and measure 3–9 mm in diameter. Badly affected plants are unmarketable. Symptoms appear when temperatures are low (less than 10° C) and fade when plants are grown at 10–15° C. The affected plants often have crook root symptoms because the crook root pathogen (*Spongospora nasturtii*) is the vector of WCLVd. Initially thought to be a virus, WCLVd is a viriod and consists of an RNA genome that lacks a protein coat.

Watercress yellow spot virus (WYSV) has been reported in France since 1962 and southern England since 1983. It causes bright yellow spots, irregularly shaped yellow blotches, and sometimes ringspots. The vector is the crook root fungus *Spongospora nasturtii*. The virus has isometric particles that are reported to be 27 nm in diameter in French isolates and 37–38 nm in isolates from England.

623 Downy mildew of watercress.

624 Septoria leaf spot of watercress.

625 *Watercress chlorotic leaf spot viroid* on watercress.

References

Claxton, J. R., Arnold, D. L., Blakeley, D., and Clarkson, J. M. 1995. The effects of temperature on zoospores of the crook root fungus *Spongospora subterranea* f. sp. *nasturtii*. *Plant Pathology* 44:765–771.

Down, G. J. and Clarkson, J. M. 2002. Development of a PCR-based diagnostic test for *Spongospora subterranea* f. sp. *nasturtii*, the causal agent of crook root of watercress (*Rorippa nasturtium-aquaticum*). *Plant Pathology* 51:275–280.

Gungoosingh, A., Beni Madhu, S. P., and Dumur, D. 2001. First report of *turnip mosaic virus* in watercress in Mauritius. *Plant Disease* 85:919.

Hashimoto, T., Kimura, T., and Katsuki, S .1987. [Leaf spot disease of watercress (*Rorippa nasturtium-aquaticum* Hayek) caused by *Cercospora nasturtii Passerini*.] *Scientific Report of the Miyagi Agricultural College* No. 35:77–79.

Stevens, C. P. 1983. *ADAS/MAFF Reference Book 136. Watercress: Production of the cultivated crop*. Grower Books, London, 56 pp.

Strandberg, J. O. and Tucker, C. A. 1968. Diseases of watercress in Florida. *Proceedings of Florida State Horticultural Society* 81:194–196.

Tomlinson, J. A. 1958. Crook root of watercress. I. Field assessment of the disease and the role of calcium bicarbonate. *Annals of Applied Biology* 46:593–607.

Tomlinson, J. A. 1958. Crook root of watercress. III. The causal organism *Spongospora subterranea* (Wallr.) Lagerh. f. sp. *nasturtii* f. sp. *nov. Transactions of the British Mycological Society* 41:491–498.

Tomlinson, J. A. and Hunt, J. 1987. Studies on *watercress chlorotic leaf spot virus* and on the control of the fungus vector (*Spongospora subterranea*) with zinc. *Annals of Applied Biology* 110:75–88.

Tomlinson, J. A. and Hunt, J. 1987. Studies on *watercress chlorotic leaf spot virus* and control of the fungal vector (*Spongospora subterranea*) with zinc. *Annals of Applied Biology* 110:75–88.

Walsh, J. A, Clay, C. M., and Miller, A. 1989. A new virus disease of watercress in England. *EPPO Bulletin* 19:463-470.

Walsh, J. A. and Phelps, K. 1991. Development and evaluation of a technique for screening watercress (*Rorippa nasturtium-aquaticum*) for resistance to *watercress yellow spot virus* and crook-root fungus (*Spongospora subterranea* f. sp. *nasturtii*). *Plant Pathology* 40:212–220.

626 Sage crown and roots infected with *Phytophthora cryptogea*.

627 Seedling sage infected with *Phytophthora cryptogea*. Healthy plant on right.

Salvia officinalis

SAGE

Sage is a commonly grown herb from the mint family (Lamiaceae). It is widely used as a seasoning and ingredient in cooking. There are numerous varieties having different leaf shapes and colors.

Golovinomyces cichoracearum

Powdery mildew

Typical white growth of powdery mildew develops on the mature foliage of sage. Under favorable warm, dry conditions, there can be active development on young leaves. As plants are often cropped for several years, powdery mildew can survive by overwintering on the lower stems. The pathogen is *Golovinomyces cichoracearum* (previously named *Erysiphe cichoracearum*).

Phytophthora cryptogea

Phytophthora root rot

Phytophthora root rot causes roots to turn necrotic and black (**626**). Crowns and lower stems also turn black. Affected plants can wither and die (**627**). The pathogen is *Phytophthora cryptogea.*

References

Koike, S. T., Henderson, D. M., MacDonald, J. D., and Ali-Shtayeh, M. S. 1997. Phytophthora root and crown rot of sage caused by *Phytophthora cryptogea* in California. *Plant Disease* 81:959.

Tragopogon porrifolius, Scorzonera hispanica

SALSIFY, SCORZONERA

Salsify (*Tragopogon porrifolius*) and black salsify or scorzonera (*Scorzonera hispanica*) are members of the Asteraceae and are grown for their edible taproots. Both commodities are steamed or otherwise cooked prior to eating.

Albugo tragopogonis

White rust

White rust is the most common and important disease on salsify and black salsify. The pathogen affects closely related wild and ornamental species including *Gerbera* and sunflower. The characteristic small white blisters occur on leaves (**628**) and stems (**629**) and develop after initial chlorotic spotting. Severe infection can cause death of plants.

Puccinia hysterium, P. jackyana

Rust

There are occasional problems with the autoecious rust *Puccinia hysterium* infecting salisfy. In the spring this rust produces aecia that deform the new growth. Aecidiospores are produced and measure 20–30 x 18–24 µm; however, this species does not have urediniospores. Brown teliospores are uniseptate, variable in size, and mostly measure 30–44 x 22–30 µm. Teliospores are produced on the stems. On black salsify, the rust pathogen is *P. jackyana*, which produces aecia and uredinia. *Puccinia scorzonerae* is a form of *P. jackyana* which does not produce aecia.

628 **White rust on salsify leaf.**

629 **White rust on salsify stem.**

Valerianella locusta, V. olitoria

CORN SALAD

Corn salad is a leafy vegetable that is also known as lamb's lettuce, field salad, and fetticus. The plant is in the Valerianaceae family. The term 'mache' is a market term applied to corn salad. The plant is grown for its succulent leaves that are used in salads or as cooked greens.

Golovinomyces orontii

Powdery mildew

On corn salad, powdery mildew forms white mycelium and conidia on both sides on leaves (**630**), causes slight twisting of foliage, and results in quality loss of the harvested product. Severely affected leaves turn necrotic and collapse, and such plants are not harvested. Conidia are cylindric to doliform, measure 27–32 x 15–17 µm, and lack fibrosin bodies. Conidial length-to-width ratios are usually less than, or equal to, 2.0. Conidia germinate at the ends (cichoracearum-type). No cleistothecia have been observed. The pathogen is *Golovinomyces orontii* (previously named *Erysiphe orontii*).

Sclerotinia minor

White mold

White mold infections cause a gray-brown soft rot to develop on crowns and lower sections of stems. Affected stems wilt and plants eventually collapse and rot (**631**). White mycelium and small, irregular, black sclerotia (3–5 mm in diameter) are observed on infected stems and crowns. The pathogen is *Sclerotinia minor*. For more information see the white mold on bean section in the legume chapter (page 262).

630 White fungal growth of powdery mildew on corn salad.

631 Diseased corn salad crowns infected with *Sclerotinia minor*.

References

Koike, S. T. and Saenz, G. S. 2005. First report of powdery mildew caused by *Golovinomyces orontii* on corn-salad in California. *Plant Disease* 89:686.

Koike, S. T. 2003. Sclerotinia stem and crown rot of corn-salad caused by *Sclerotinia minor* in California. *Plant Disease* 87:1264.

Zea mays var. *saccharata*

SWEETCORN

Sweetcorn or sugar corn is a variety of maize, *Zea mays* var. *saccharata* (family Poaceae). Baby corn refers to standard corn cultivars that are harvested when the ears are very small and before they develop the silk.

Fusarium spp.

Stalk rot

Root, stem, and ear diseases of sweetcorn are caused by various *Fusarium* species. These *Fusarium* diseases are of particular concern because some of these pathogens produce mycotoxins such as fumonisin, zearalenone, and trichothene. Stalk rot produces a gray coloration in the foliage of mature plants. This discoloration is often followed by wilting and yellowing as the plant senesces prematurely. Plants break off at the basal internode (**632**), resulting in considerable crop lodging. The stem base has red-brown internal discoloration and rotting of the internodes and pith. There are several species of *Fusarium* involved, and laboratory isolation is required to identify them: *F. avenaceum, F. culmorum, F. graminearum, F. poae, F. verticillioides (= F. moniliforme)*.

Ustilago maydis

Maize smut

This disease occurs worldwide and is both seedborne and soilborne. Affected cobs are unmarketable, but losses are usually limited except where sweetcorn is grown continuously or in short rotations. Symptoms occur at any growth stage but are often only noticed close to harvest when cobs are examined. The most obvious symptoms consist of conspicuously swollen, white to gray galls on the cobs or on other aerial parts of the plant (**633**). The galls vary in size from a few millimeters to several centimeters in diameter. At maturity the galls break open and release a dark brown or black mass of chlamydospores (also known as brand spores or smut spores). These spores are globose to elliptical, have blunt spines, and measure 8–12 µm in diameter. Spores are dispersed by wind or insects such as frit fly (*Oscinella frit*). Infection often occurs in young meristematic tissues or through wounds. The pathogen (*Ustilago maydis*) remains localized and does not become systemic. In Mexico and the USA, the young galls are considered a delicacy and are eaten before the spores form and mature.

632 Stalk rot of sweetcorn caused by *Fusarium verticillioides* (syn. *F. moniliforme*).

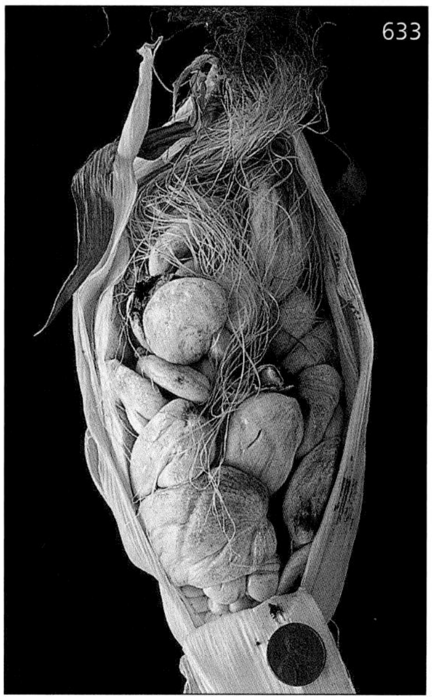

633 Smut galls on cob of sweetcorn.

References

Banuett, F. 1992. *Ustilago maydis*, the delightful blight. *Trends in Genetics* 8:174–180.

Magan, N., Marin, S., Ramos, A. J., Sanchis, V. 1997. Impact of ecological factors on germination, growth, fumonisin production by *F. moniliforme* and *F. proliferatum* and their interactions with other common maize fungi. *Cereal Research Communications* 25:643–646.

Pataky, J. K. and Chandler, M. A. 2003. Production of huitlacoche, *Ustilago maydis*: timing inoculation and controlling pollination. *Mycologia* 95:1261–1270.

Glossary

Abiotic Pertaining to nonliving factors. Abiotic problems of plants are often caused by environmental factors.

Abscission The physiological process in which leaves, petioles, flowers, and fruits fall off the plant, either as a natural part of plant senescence or as the result of infection and disease.

Acervulus A small, saucer-shaped, fungal fruiting body that produces asexual conidia. Acervuli are embedded within plant host tissue. Some dark acervuli can be seen as small dark specks or dots within symptomatic tissue, while others are colorless. An example of a fungus that produces dark acervuli is *Colletotrichum*.

Actinomycete A microorganism that forms branching filaments which later break into fragments that look like individual bacterial cells. Actinomycetes are classified as a type of Gram-positive bacterium and are often found in soil. An example of an actinomycete is *Streptomyces*.

Aeciospore One of the several different types of rust fungi spores. Aeciospores are borne in and on aecia and germinate to form dikaryotic mycelium. Some references call these spores aecidiospores. *See also* urediniospore, teliospore, and basidiospore.

Aecium The fruiting structure of rust fungi that produces aeciospores.

Aerobic Referring to environmental conditions, usually in the soil, in which oxygen levels are sufficient for normal physiological growth of most organisms, including plants.

Agar A gelatinous substance that is extracted and purified from algae. Agar is an ingredient in semi-solid culture media that are used for culturing bacteria and fungi.

Agonomycetes One of three subgroups of mitosporic fungi (the other two groups are hyphomycetes and coelomycetes). Mitosporic fungi (also known as deuteromycetes or fungi imperfecti) do not have sexually produced spores. Agonomycetes are asexual fungi that are sterile and do not make any spores at all. An example of an agonomycete is *Rhizoctonia solani*.

Alga A primitive, non-vascular plant ranging from microscopic single-celled species to large, multicelled species such as seaweed. Most algae are aquatic. A few algae (an example is *Cephaleuros virescens*, a green alga) are pathogens of higher vascular plants. Recent taxonomic research indicates that the oomycete group, an important group of pathogens, is closely related to algae.

Allele One of two or more alternate forms of a gene. Corresponding alleles are located at the same place on a particular chromosome.

Alternate host One of two distinct kinds of plants that serve as hosts for some pathogenic fungi. Some rust species, for example, require alternate hosts to complete its life cycle.

Anaerobic Referring to environmental conditions, usually in the soil, in which oxygen levels are not sufficient for normal physiological growth of most organisms. Under such conditions, only specialized organisms such as anaerobic bacteria can survive.

Anamorph The asexual stage of a fungus. Spores and propagules produced in this stage do not involve sexual recombination or processes. This form of fungi is also called the imperfect stage. Anamorphs are placed in a category that uses one of the following synonymous names: mitosporic fungi, deuteromycetes, fungi imperfecti.

Anastomosis The fusion of one fungal hypha with another. This fusion usually results in combination and exchange of genetic material. For some fungi, such as *Rhizoctonia* species, anastomosis groups are used to classify and differentiate isolates (the method is called vegetative compatibility pairing).

Antagonism The interaction between microorganisms in which one species inhibits, competes with, or kills a second species. The antagonist accomplishes this inhibition through competition for resources, parasitism, predation, or the production of toxic substances.

Antheridium Specialized male structure for fungi that have sexual phases. Antheridia form male sexual cells (gametes).

Anthocyanin Water-soluble, flavonoid plant pigments that are blue, purple, or red in color. In some plant species, the development of anthocyanins in plant tissue indicates that the plant is damaged, diseased, or under stress.

Anthracnose The common name used for some foliar plant diseases that are caused by fungi (acervular coelomycetes) that produce acervuli. An example of this kind of fungus is *Colletotrichum*.

Antibiotic A chemical that is produced by one microorganism and which can inhibit or kill other microorganisms.

Antibody A protein that is produced by the lymphatic system of warm-blooded animals in response to the presence of a foreign protein or substance. The foreign substance is called an antigen. The antibody reacts to the antigen, binds with it, and causes the antigen to become inactive. This antibody–antigen interaction is the basis for serology and pathogen detection techniques such as ELISA.

Antigen A substance, usually a protein, that is foreign to the body of warm-blooded animals. Scientists intentionally inject plant pathogen antigens into animals so that antibodies are produced. The resulting antibodies can be used in serological tests to detect the presence of the plant pathogen.

Antiserum The serum part of a warm-blooded animal's blood that contains antibodies to introduced antigens of plant pathogens. Various antisera are used in serological tests such as ELISA.

Aphid Small, soft bodied insect in the order Homoptera. Aphids cause problems by direct feeding on plant tissues, producing honeydew that causes plant surfaces to be sticky and discolored, and vectoring virus pathogens.

Apical Referring to the end or upper part of a structure.

Apothecium Fruiting body of ascomycete fungi that is a cup or vase shaped structure that usually forms on top of a stalk. The fertile layer of the apothecium is lined with ascospore-producing asci.

Appressorium A fungal cell that is often enlarged and which attaches the fungus to the plant host surface.

Ascocarp The fruiting body of ascomycete fungi. The different shapes (apothecium, cleistothecium, perithecium, pseudothecium) all have asci that contain ascospores.

Ascomycete One of the major divisions of fungi having a perfect or sexual stage. All ascomycetes have sexual spores (ascospores) that are produced within a sac-like or club-shaped structure (ascus).

Ascospore The sexually produced spore that is formed within the ascus of ascomycete fungi.

Ascus The sac-like or club-shaped structure in which ascospores are formed. Asci are contained within various ascocarp forms. Asci-producing fungi are placed in the ascomycete group.

Aseptate Referring to fungal hyphae that do not have cross walls. A synonym for aseptate is coenocytic. The aseptate nature of certain fungi is used in classification systems.

Aseptic Free of living microorganisms.

Asexual Reproduction that does not involve gamete union and meiosis. Fungi that do not reproduce sexually are called imperfect or anamorphic fungi.

Autoecious fungus A pathogenic fungus that completes its entire life cycle on one host plant. Rust fungi are good examples of this life cycle. Compare with heteroecious.

Avirulent Referring to an organism that is nonpathogenic and unable to cause disease.

Bacillus
A bacterium that is rod-shaped.

Bactericide
A chemical or substance that kills bacteria.

Bacterium A microscopic, single-celled organism, usually but not always enclosed in a rigid cell wall. Bacteria do not have membrane-bound nuclei and are therefore classified as prokaryotes. The science and study of bacteria is bacteriology.

Bar Unit of pressure that expresses water potential, with 1.013 bar = 1 atmosphere.

Basal Referring to the base or bottom part of a structure.

Basidiomycete One of the major divisions of fungi having a perfect or sexual stage. All basidiomycetes have sexual spores (basidiospores) that are produced and borne on pegs (sterigma) that extend from a specialized cell called the basidium.

Basidiospore The sexually produced spore that is produced on the club shaped structure, the basidium, of basidiomycete fungi. Basidiospores are usually haploid.

Basidium The specialized cell, usually club shaped, that produces and bears externally formed basidiospores. Fungi that produce basidia are placed in the basidiomycete group.

Binucleate Referring to a cell that contains two nuclei.

Bioassay A test that uses a living organism to measure that organism's sensitivity to a chemical substance or susceptibility to a pathogen.

Biological control The control or management of a pathogen or pest by means of another living organism. The biological control agent can act in various ways: predator, parasite, competitor.

Biotechnology The study, design, and use of genetically modified organisms (GMOs). Also, the modern techniques (such as sequencing and inserting genes) used in manipulating DNA and other genetic material.

Biotic Pertaining to living organisms. Biotic problems of plants are usually called diseases and are caused by fungi, bacteria, viruses, and other microorganisms.

Biotype A subgroup or subdivision of a species or race. The individual organisms making up the biotype share common features and genetics. Similar to physiologic race and strain.

Bitunicate ascus An ascus having a double wall. Bitunicate asci are characteristic of certain groups of ascomycetes.

Blast The plant disease symptom of death of buds, flowers, or young fruit.

Bleached The plant disease symptom of light- colored to white tissue, particularly of the foliage.

Blight The plant disease symptom of extensive withering and dying of leaves, flowers, shoots, fruit, or the entire above-ground portion of the plant.

Blotch The plant disease symptom of large, irregular, necrotic spots or areas on foliage.

Bordeaux mixture One of the first fungicides, consisting of copper sulfate and lime.

Bract A modified leaf that is associated with a flower structure.

Broadcast The act of spreading seed, fertilizers, pesticides, or other amendments evenly over the soil surface and not in lines.

Canker A diseased area that has well defined edges, is localized, and occurs on stems, twigs, branches, or trunks of plants. Cankers are usually necrotic and sunken in appearance.

Capsid The protein coat of viruses that forms the outer shell of the virus genome.

Carotenoid A plant compound that is usually yellow or red in color. Examples of carotenoids are carotenes and xanthophylls.

Catenulate Referring to the development of objects in chains or attached end-to-end. Powdery mildew spores often grow in sequential chains and are therefore catenulate.

Causal agent The organism that is responsible for causing the disease in question. Causal agents include bacteria, fungi, phytoplasmas, nematodes, viruses, and viroids.

Certification A registration procedure in which an official agency declares plant material to be acceptably free of pathogenic organisms, weeds, and pests. Plant material can also be certified to be true-to-type and therefore representative of the cultivar name. Certification procedures are based upon tests and inspections of the plant material.

Chlamydospore Usually a single-celled, thick-walled asexual spore produced within a hypha by a fungus. Chlamydospores enable the fungus to survive unfavorable conditions.

Chlorophyll The green, light-sensitive pigment found in the chloroplasts of leaves and other green organs of plants. Chlorophyll uses light energy to convert water and carbon dioxide into food via photosynthesis. Plant diseases can interfere with this process. Some pathogens prevent the normal formation of chlorophyll.

Chlorosis The plant disease symptom of fading, yellowing, and whitening of green leaves, stems, and other foliage. Chlorosis occurs when plant chlorophyll fades, is degraded, or does not form. Yellowing is generally considered a synonym of chlorosis.

Chromista The name of the taxonomic kingdom that some taxonomists believe should contain the important oomycete pathogen group. Kingdom Chromista is distinct from fungi and contains certain types of primitive golden algae, diatoms, and giant kelp. However, other researchers believe the appropriate name for this group should be Kingdom Straminipila.

Circulative virus Referring to a plant pathogenic virus that is acquired by an insect vector, accumulates internally in the insect, then is transmitted to plants.

Cirrus A small mass of spores exuded from fungal fruiting bodies called pycnidia. Cirri are curled, tendril-like, and held together by a slimy matrix material. Also called a tendril.

Clamp connection A hyphal outgrowth that occurs on the hyphae of some basidiomycete fungi. During cell division the clamp connection makes a link between the two resulting cells.

Clavate A club-shaped structure. The base of the structure is narrower than the apex. Some fungal structures are described as being clavate in shape.

Cleistothecium Fruiting body of ascomycete fungi that is an entirely enclosed, spherical structure. The inner cavity of cleistothecia contains ascospore producing asci.

Clone A population of organisms or cells that is genetically identical and was produced asexually from one such organism or cell.

Coalesce Merging or growing together through growth and expansion. For some disease symptoms, such as leaf spots, continued growth of adjacent spots can result in a merging together, or coalescence.

Coelomycetes One of three subgroups of mitosporic fungi (the other two groups are hyphomycetes and agonomycetes). Mitosporic fungi (also known as deuteromycetes or fungi imperfecti) do not have sexually produced spores. Coelomycetes are asexual fungi that produce spores in pycnidia, acervuli, or other similar structures.

Coenocytic Referring to fungal hyphae that do not have cross walls. A synonym for coenocytic is aseptate. The coenocytic nature of certain fungi is used in classification systems.

Colonize To grow and become established on either a host plant or in culture.

Colony Growth of a microscopic organism, usually on culture media, to the degree that the growth is visible with the unaided eye.

Complex (as in disease complex) Referring to the situation in which a plant disease is caused by more than one pathogen or factor.

Compost A mixture of plant residues and other organic materials that has been incubated and managed to allow for decomposition and aging. Composts are used as soil amendments for plants, and in some circumstances may have possible disease suppressing effects.

Concentric spot The plant disease symptom where multiple circles (which usually are alternating in appearance between dark and light colors) form within one larger circle or lesion. All the circles share the same center. This symptom is also called ring spot.

Conical A cone-shaped structure. Some fungal structures are conical in shape.

Conidiogenous cell Any fungus cell that directly produces one or more conidia

Conidiophore The specialized hypha that bears one or more conidiogenous cell.

Conidium An asexually produced, non-motile spore, primarily from deuteromycete fungi (also called mitosporic fungi or fungi imperfecti). Conidia form from, and externally to, a conidiogenous cell. Conidia do not develop as a result of the division of cytoplasm within a specialized structure (*see* sporangiospore). Conidia are released from the conidiogenous cells and serve to disperse the organism.

Contaminant A substance or organism that makes an otherwise pure culture or system impure, unfit, and unclean.

Coremium A reproductive structure of asexual fungi. *See* synnema.

Cotyledon The first leaf (monocot plants) or a leaf of the first leaf pair (dicot plants) developed by the embryo of a seed plant.

Cover crop Plants that are seeded, grown, and used to add organic matter to fields, improve soil conditions, reduce erosion, retain nutrients in soil profiles, and provide a rotation for agronomic cash crops.

Crop rotation In a given field, the practice of not growing the same crop consecutively or too frequently in sequence. Crop rotation strategies use diverse crops from different plant groups and families. Poor crop rotation can increase problems with soilborne pathogens.

Crown Typically the lower base of a plant from which both the main root and above ground stems or trunk arise. The crown is the tissue that serves as the root-stem junction. Note that for trees, the crown refers to the upper part, from which the foliage emerges.

Cull The unusable or unmarketable portion of the harvest that is left in the field or is discarded after harvest.

Cultivar (abbreviation: cv.) A cultivated variety of plant. All individuals of a particular cultivar share characteristic features. Cultivar is generally used synonymously with variety.

Culture (V.) to grow microorganisms or plant cells on prepared media and under controlled, artificial conditions. (N.) the growth and colonies of microorganisms or plant cells on prepared media and under controlled conditions.

Cylindrical Round in cross section and of equal width throughout the length. Some fungal structures are cylindrical in shape.

Cyst In fungi and oomycetes, a resting spore or propagule that usually is enclosed in a protective, thick cell wall. Such encysted cells can survive during unfavorable conditions. In nematodes, the resilient, hardened outer body of non-viable females in the genera *Heterodera* and *Globodera*. These cysts enclose and protect eggs of these nematodes.

Damping-off A disease affecting seeds and young seedlings, caused by soilborne pathogens. The disease has different aspects. Planted seeds can be infected and die. Seeds may germinate, but the resulting seedling may become infected and die prior to emerging above ground (pre-emergence damping-off). Seedlings may emerge above ground but later become infected and die (post-emergence damping-off).

Decay The plant disease symptom of decomposition and rotting of plant tissue.

Decline The plant disease symptom of reduced plant vigor and deterioration of a plant's condition and growth.

Decomposition Breakdown and rotting of plant tissue, usually caused by microorganisms.

Defoliate The plant disease symptom of the dropping of leaves from a plant. For perennial deciduous plants, defoliation is a normal process of senescence.

Dematiaceous Referring to fungal spores and structures that are darkly colored.

Deuteromycetes Fungi (also known as mitosporic fungi or fungi imperfecti) that do not have sexually produced spores. This is an artificially created group and is not an official taxonomic category. Deuteromycetes can be further organized based on the way that spores are made (*see* hyphomycetes, agonomycetes, coelomycetes). While a deuteromycete fungus does not produce any sexual spore, some deuteromycetes can enter a phase of development in which it becomes associated with a sexual form (an ascomycete or a basidiomycete teleomorph). In such situations, the deuteromycete is called the anamorph form of that teleomorph.

Diagnosis The systematic, problem-solving process of identifying the cause of a plant problem or disease.

Dieback The plant disease symptom of a progressive decline and death of plant parts. Dieback usually begins at the tip of the roots or branches, then progresses towards the main part of the host.

Differential hosts A series of different cultivars or lines of a particular plant species, each line of which varies in susceptibility to a given pathogen. Inoculating such a series of differential hosts can help identify pathogen races.

Diploid The genetic condition in which an organism has twice the number of chromosomes (2n) as the haploid (1n) condition.

Disease Any abnormality in a plant's growth, development, functioning, value, or appearance due to the activity of a pathogenic, biotic agent. A disease results in symptoms caused by the pathogens. Injury or effect from an abiotic agent is usually considered a disorder.

Disease cycle The sequence of events that takes place during disease development, appearance of disease symptoms, and growth and development of the pathogen.

Disease triangle The pathology concept in which all plant diseases can be analyzed and understood in terms of three components: host, pathogen, and environment.

Disinfectant A chemical or treatment (such as steam or heat) that kills or inactivates microorganisms once such agents have infected a plant.

Disinfestant A chemical or treatment (such as steam or heat) that kills or removes microorganisms from the surfaces of plants, seeds, tools, working surfaces, and other objects.

Disorder Any abnormality in a plant's growth, development, functioning, value, or appearance due to the effects of an abiotic factor. A disorder results in symptoms caused by these nonliving factors and is not infectious. Injury or effect from a biotic agent is usually considered a disease.

DNA Deoxyribonucleic acid. DNA is any of various nucleic acids that occur in nuclei (for eukaryotes) or cell contents (for prokaryotes) of living organisms. DNA molecules are composed of repeating sub-units of deoxyribose (a five-carbon sugar), phosphoric acid, and four nitrogenous bases (adenine, cytosine, guanine, and thymine).

Dolipore septum A complex septum found in hyphae of some basidiomycete fungi. This septum has a barrel-shaped structure in the middle of the septum.

Dominant gene A gene that is fully expressed and completely dominates the phenotype of the organism. When paired with a corresponding gene or allele, the dominant gene will be the one that is expressed.

Echinulate Having sharply pointed spines. This term is used to describe spiny textures on fungal spores, conidiophores, and other structures.
A synonym is echinate.

Ecology The study of the interrelationships between organisms and their environments.

Electron microscope A specialized microscope that uses a focused beam of electrons to create images of tiny, subcellular structures and entities. For example, electron microscopy is the only means of seeing virus particles.

ELISA Enzyme-linked immunosorbent assay. A serological method that uses antibodies to recognize and detect target antigens of pathogens and other organisms. The ELISA method uses polystyrene microtiter plates and registers a positive test by a color change in the reaction liquid.

Ellipsoidal Elliptical or oval in shape. Some fungal spores are ellipsoidal.

Emergence Appearance of plant structures, primarily shoots, above the soil surface following germination of seeds and below-ground growth of the seedling.

Enation Typically small, malformed tissue and swellings that occur on leaves and other organs of plants.

Encyst When fungi or oomycete spores form a cyst.

Endemic Native or naturally occurring to a region or area.

Endophyte An organism, particularly a fungus, that lives inside a plant but produces no external signs of infection.

Environment The complete surroundings, conditions, and forces that influence living organisms. For a susceptible plant host to become infected by a virulent pathogen, environmental conditions favorable for disease development must occur.

Enzyme A complex protein that catalyzes a specific biochemical reaction but which is not changed or degraded during the reaction. Some plant pathogens release enzymes that break down plant tissues.

Epidemic Widespread and rapidly developing outbreak of an infectious disease. This term is commonly used for diseases of all organisms, including plants. Such an outbreak in plants can also be called an epiphytotic.

Epidemiology The science and study of the factors influencing disease development, outbreak, and spread.

Epidermis The outer layer of cells that line the surfaces of plant tissues.

Epinasty The plant disease symptom of abnormal downward twisting, curling, or bending of leaves and other above-ground parts.

Epiphyte An organism that is present and growing on the surface of a plant, but which has not infected the plant.

Epiphytotic Widespread and rapidly developing outbreak of an infectious disease of plants. Such an outbreak in plants can also be called an epidemic.

Eradicant A chemical substance that can stop the growth of a pathogen even if the pathogen has already infected the host plant.

Eradicate To control plant pathogens by completely eliminating them from plant tissue or planted area. Complete eradication is rarely possible.

Erumpent The act of breaking through to the outside of a surface. This term is often used for pathogenic fungi that colonize plant tissues and later form erumpent fruiting bodies that break through the plant epidermis.

Escape The situation that exists when a susceptible plant avoids infection and subsequent disease due some feature of the plant or because the plant is located where there are no pathogens.

Etiology The science of the causes of diseases, including the study of the responsible agents and their interactions with the host.

Eukaryote An organism that has a membrane-bound nucleus.

Exclusion A disease control step in which pathogens, pathogen-contaminated materials, and infected plants are not allowed into a particular field or area.

Exudate A liquid or slimy substance that is released by diseased or damaged plant tissue. In culture, some fungi also produce exudates.

Facultative parasite An organism that is normally saprophytic (non-pathogenic) but under certain conditions can act as a parasite or pathogen.

Facultative saprophyte or saprobe An organism that is normally pathogenic but under certain conditions can act as a saprophyte (non-pathogen).

Fallow Referring to land that is not planted with crops and is uncultivated. Leaving land fallow is one aspect of controlling soilborne pathogens.

Family A taxonomic group of organisms that is one category above a genus and below an order.

Fasciation The plant disease symptom of striking distortion in which multiple stems are flattened and appear fused together.

Fatty acid An organic fat compound consisting of carbon, hydrogen, oxygen, and glycerol. The fatty acid profile of bacteria and other organisms is used to identify and classify these organisms.

Filamentous Thread-like in shape. A synonym is filiform. Some pathogenic fungi have filamentous spores.

Flagellate Having one or more flagella or whiplike structures. Bacteria and zoospores that move in water are flagellates.

Flagellum The flexible whip-like or hair-like filament that is found on the cells of zoospores and some bacteria. The twisting of the flagella enables these cells to swim through water and be motile.

Fleck The plant disease symptom of small, white to tan lesions.

Flexuous Referring to the edge or shape of an object that is serpentine, wavy, or sinuate.

Floret One individual flower that is a part of a flower cluster or compact group of florets.

Fluorescent The visible reaction, in the form of a glow of light, when a substance or organism is placed under ultraviolet radiation.

Foliar Referring to leaves and other parts of the above-ground portion of plants.

Foot cell A structure found in fungi that is used to identify and classify fungi to species. For powdery mildews, the foot cell is the bottom cell of the conidiophore that connects the conidiophore to supporting hyphae. For *Fusarium* species, the foot cell is the bent basal cell of the conidium.

Foot rot The plant disease symptom of a decay of the lower stem, hypocotyl, or upper root.

Forma specialis Abbreviated f. sp., refers to a 'special form' or biotype of a pathogen species. A particular forma specialis differs from other forms in its ability to infect certain plant species. For example, the pathogen *Fusarium oxysporum* f. sp. *apii* infects only celery. This term is used primarily for fungi. A similar designation for bacteria is pathovar.

Formulation The nature and form of a pesticide or fertilizer preparation. Formulations influence how chemicals function and are handled. Examples of formulations are wettable powders, granules, dry flowable granules, emulsifiable concentrates, and liquids.

Free-living Referring to an organism that is able to survive in the environment without dependence on a host organism.

Fruiting body A multicellular structure of fungi that produces and contains reproductive spores.

Fumigant A volatile gas that is used to kill pathogens and other pests in soil, agricultural products, and structures. The practice of applying a fumigant is fumigation.

Fungicide A chemical substance used to kill or inhibit fungi.

Fungi imperfecti Fungi (also known as mitosporic fungi or deuteromycetes) that do not have sexually produced spores. This is an artificially created group and is not an official taxonomic category. Fungi imperfecti can be further organized based on the way that spores are made (*see* hyphomycetes, agonomycetes, coelomycetes). While a fungus in the fungi imperfecti does not produce any sexual spore, some fungi imperfecti can enter a phase of development in which it becomes associated with a sexual form (an ascomycete or a basidiomycete teleomorph). In such situations, the fungi imperfecti species is called the anamorph form of that teleomorph.

Fungus An organism that lacks chlorophyll, is usually multicelled, and usually has a vegetative body consisting of microscopic tube-like structures (hyphae) that grow and become organized into branched systems (mycelium). Fungi usually reproduce via spores.

Fusiform Spindle-shaped, with the ends narrower than the middle. Some fungal spores are fusiform.

Gall The plant disease symptom of abnormal plant growths that are localized, swollen, and usually spherical. Galls are caused by a number of pathogens and other pests.

Gametangium A fungal cell that contains gametes. Gametangia fuse together and enable sexual reproduction and meiosis to take place.

Gamete A male or female reproductive cell on fungal gametangia.

Gene A discrete, linear segment of a chromosome which codes for one or more genetic traits. A gene is the smallest functioning unit of genetic material in an organism's genome.

Geniculate Referring to structures that are bent.

Genome The genetic material of an organism.

Genotype The genetic constitution of an organism that dictates its physical characteristics (phenotype).

Genus A taxonomic group of organisms that is one category above a species and below a family.

Germ theory The idea that contagious diseases are caused by infectious microorganisms.

Germ tube The initial growth of mycelium made when a fungal spore germinates.

Globose Spherical or globe-shaped. Some fungal spores are globose.

Gram stain A microbiological technique in which bacteria are stained with various solutions. The resulting violet (Gram-positive) or red (Gram-negative) coloration helps distinguish bacterial groups based on differing cell wall features.

Guttation The collection and exudation of water from plant leaves, especially from glands (hydathodes) in the edges of leaves.

Halo A plant disease symptom in which a ring of discolored, necrotic, or otherwise symptomatic tissue surrounds a lesion or spot.

Haploid The genetic condition in which an organism has half the number of chromosomes (1n) as the diploid (2n) condition.

Haustorium A specialized hyphal branch that absorbs nutrients for the fungus. Haustoria sometimes penetrate and grow into host cells.

Heteroecious Pertaining to a pathogenic fungus that needs two distinct kinds of hosts to complete its life cycle. Many rust fungi are heteroecious.

Heterothallic Referring to fungi that produce male or female organs and gametes on different individuals. Sexual reproduction between isolates is therefore possible only if each mating type is present.

Holomorph The mycological concept of the 'whole fungus.' The holomorph includes the whole fungus in both its teleomorph (sexual or perfect) and anamorph (asexual or imperfect) stages.

Homothallic Referring to fungi that produce male and female organs and gametes on the same individual. Sexual reproduction therefore does not require the presence of two distinct mating types or individuals.

Horizontal resistance Partial resistance in plants that is effective against most or all distinct races of a pathogen. In contrast to vertical resistance, this type of resistance is usually determined by multiple genes.

Host The susceptible plant that is subject to infection and colonization by a virulent agent, the pathogen. For disease to develop on the host, conditions favorable for the pathogen–host interaction must be present.

Host range A documented range of plant species or cultivars that is susceptible to a particular pathogen.

Hyaline Clear or colorless. This term is often used to describe fungal structures (spores, mycelium, conidiophores) that are clear in appearance.

Hybrid The resulting offspring from a cross between two species or genotypes.

Hydathodes Glands in the edges of leaves, from which water collects and is discharged via guttation.

Hydroponics The growing of plants in an aerated water solution instead of soil or solid rooting media. The water solutions include nutrients necessary for plant growth.

Hyperplasia An abnormal increase in plant cell division due to the influence of a pathogen. Galls, enations, and witches' broom symptoms are caused by hyperplasia.

Hypersensitive reaction (HR) Ultrasensitivity of a plant and its cells to invasion by pathogens. When cells are infected, the cells rapidly die and prevent the pathogen from spreading beyond the dead cells. Synonymous with hypersensitivity.

Hypertrophy An abnormal increase in plant cell size due to the influence of a pathogen.

Hypha The microscopic tube-like filament or thread that comprises the basic vegetative body of a fungus. Hyphae grow and are organized into extensive branched systems (mycelium).

Hyphomycetes One of three subgroups of mitosporic fungi (the other two groups are coelomycetes and agonomycetes). Mitosporic fungi (also known as deuteromycetes or fungi imperfecti) do not have sexually produced spores. Hyphomycetes are asexual fungi that produce spores on individual or multiple hyphae. Such conidia-bearing hyphae, however, are not contained within other structures or fruiting bodies.

Hypocotyl The part of a plant embryo or seedling that is below the point of cotyledon attachment. The tissue between the stem and the root. In some plants, such as radish, the hypocotyl is extremely swollen.

Imbibition The taking up and absorption of water by solid materials, such as plant seeds.

Immune Not subject to infection by a particular pathogen. An immune plant does not become infected. Immunity is an extreme form of resistance.

Imperfect fungus A fungus that does not produce sexual spores. *See* fungi imperfecti or anamorph.

Incidence The frequency of a disease within a plant population. Incidence is often expressed as the percentage of plants that are affected.

Incipient Referring to the beginning, initial phase of infection and disease development.

Inclusion bodies Various structures, crystals, and bodies that form in plant cells as a result of virus infections. Inclusion bodies are visible with the compound microscope and can be used to diagnose some virus diseases.

Incubate To maintain in a controlled environment with a given set of conditions (temperature, humidity, light, etc.).

Incubation period The time between initial infection by the pathogen and expression of disease symptoms by the host. Incubation period can also refer to the length of time that plants are kept at controlled environments (*see* incubate).

Indexing Examining and testing plant materials for the presence of pathogens.

Indicator plant A plant that reacts to virus infections by developing diagnostic and characteristic symptoms. Indicator plants can therefore be used to detect and confirm some virus pathogens.

Infection The invasion, penetration, and establishment of a pathogen within a host. Distinguished from infestation, in which the pathogen is present on the host but has not yet infected it.

Infestation Occurs when a pathogen or pest is present on and contaminates surfaces, fields, soil, plant material, and other objects. If a plant is infested, however, it may not be actually infected.

Inoculation The intentional act of introducing a pathogen to a host plant so that infection will take place followed by the development of disease. Inoculations are important steps in proving pathogenicity of an organism (*see* Koch's postulates).

Inoculum The propagules or parts of a pathogen that initiate disease in a host. Inocula are often the parts of pathogens that are dispersed and spread in the environment. Examples of inocula are the spores of fungi, cells of bacteria, and particles of viruses.

Insensitive Often referring to pathogens that have adapted and changed so that the insensitive isolate is no longer killed by and susceptible to a pesticide or other control chemical. For example, an insensitive fungal pathogen is not controlled by a particular fungicide, and an insensitive bacterial pathogen is not controlled by a particular antibiotic. Some pathologists consider insensitive to be synonymous with resistant.

Integrated pest management (IPM) The use of all available methods (biological, chemical, genetic, physical, and cultural) to manage plant diseases and pests for maximum control, while minimizing damage to the environment. Examples of IPM components are the following: resistant plants, exclusion, pesticides, crop rotations, sanitation, environmental manipulation, and other cultural practices.

In vitro Referring to the growing and culturing of organisms in artificial environments or culture. A pathogen being grown *in vitro* (Latin 'in glass') is not living inside its natural host.

In vivo Referring to the growing of organisms in natural environments. A pathogen being grown *in vivo* (Latin 'in living') is living inside its natural host.

Isolate (V.) to recover and separate a pathogen from the infected tissue of its host plant. Once recovered, the isolation procedure usually includes purifying the pathogen so it is a pure culture. (N.) a pure, viable culture of a pathogen. Such a culture is obtained by isolating the pathogen from host tissue, and is usually followed by obtaining one distinct individual by growing out a single spore or hypal tip.

Koch's postulates The underlying principle that is used to identify the causal agent of a disease. The suspected pathogen must be present in all disease cases, be isolated and grown in culture, cause disease when inoculated into healthy plants, and be re-isolated from the test plant.

Latent To be hidden. To be present but not visible.

Latent infection An infection in a plant that does not yet show visual symptoms.

Latent period In virology, referring to the period of time between when an insect vector acquires a plant pathogenic virus and when it can transmit the virus to a plant host. For fungi, the latent period is the period of time between infection and the initial appearance of fungal signs (mycelium or spores).

Leaf spot The plant disease symptom of a well defined, localized, diseased lesion on a leaf.

Lesion The plant disease symptom of a well defined, localized area of diseased tissue.

Localized Related to the restricted movement of a chemical, pathogen, or infection that in plants only spreads in a small area of cells and tissues. In contrast to systemic.

Lunate Crescent-shaped like a new moon. Some fungal spores are lunate.

Macronutrients Major chemical elements, needed in relatively large amounts, that allow for normal plant growth. There are three macronutrients: nitrogen (N), phosphorus (P), and potassium (K).

Mechanical injury Physical damage to a plant organ by abrasion, crushing, or other wounding factor.

Mechanical transmission Referring to the successful introduction and establishment of a pathogen into a host by means of a physical action. In the field, the rubbing of an infected leaf against a healthy leaf can result in mechanical transmission of the pathogen. Mechanical transmission is used in virus research by taking virus laden plant sap and rubbing it onto leaves so that the virus can become established in the test plant.

Medium The artificial food or substrate that is prepared and used to grow and culture microorganisms.

Meristem Layer or zone of undifferentiated plant cells that functions primarily in cell division and is thus responsible for the first phase of plant growth.

Micronutrients Minor chemical elements, needed in relatively small amounts, that allow for normal plant growth. There are many micronutrients, such as boron, chlorine, copper, iron, manganese, molybdenum, and zinc.

Microorganism An organism that usually is so small that it is not visible without microscopes. A synonym is microbe. Microorganisms include actinomycetes, algae, bacteria, fungi, protozoans, and viruses. The science and study of microorganisms is microbiology.

Midrib The main, central, thickened vein of a leaf.

Mitosporic fungi Fungi (also known as fungi imperfecti or deuteromycetes) that do not have sexually produced spores. This is an artificially created group and is not an official taxonomic category. Mitosporic fungi can be further organized based on the way that spores are made (*see* hyphomycetes, agonomycetes, coelomycetes). While a fungus in the mitosporic group does not produce any sexual spore, some mitosporic fungi can enter a phase of development in which it becomes associated with a sexual form (an ascomycete or a basidiomycete teleomorph). In such situations, the mitosporic species is called the anamorph form of that teleomorph.

Mollicute Prokaryotic microorganisms that are single-celled, lack a cell wall, but are bounded by flexible cell membranes. In plant pathology, phytoplasmas and spiroplasmas are mollicutes.

Monoclonal antibody A homogeneous antibody made from a single clone of antibody producing cells. Such antibodies react with only one of the antigens of a pathogen or protein. Monoclonal antibodies are important components of serological detection methods such as ELISA.

Monocyclic Having only one cycle per season or host crop.

Morphology The science and study of the form and structure of organisms.

Mosaic The plant disease symptom of foliage, particularly leaves, showing abnormal, irregular patterns and patches of light green, dark green, yellow, and other colors. Many virus pathogens cause leaves to show mosaic patterns. Similar to mottle.

Motile Able to move about and be mobile. Bacteria and zoospores with flagella are motile and can swim in water.

Mottle The plant disease symptom of irregular patterns of indistinct light and dark areas. Similar to mosaic.

Mucilaginous A viscous, slimy, or sticky substance or surface.

Mulch In agriculture, a substance or material that is placed in a layer on the soil surface. Mulches are used to catch water, reduce soil moisture loss, prevent weed growth, regulate soil temperatures, keep plants clean from splashing soil, and other purposes. Mulch material include straw or hay, dried leaves, compost, plastic films, and other materials.

Multinucleate Having more than one nucleus per cell.

Mutant An individual organism that has a novel inheritable feature or trait as a result of undergoing a mutation.

Mutation A permanent change in the genetic material of an organism. The resulting organism, or mutant, will have some new trait or character.

Mycelium The collection or mass of hyphae of a fungus.

Mycology The science and study of fungi.

Mycoplasma A bacterium-like organism that is prokaryotic, usually single- celled, and not having a rigid cell wall. Mycoplasmas belong in the class mollicutes.

Mycoplasma-like organism (MLO) Obsolete term for a microorganism that has the characteristics of mycoplasmas, but which has not been proven to be a mycoplasma. MLOs live in the phloem tissue of their plant hosts. These pathogens are now called phytoplasmas (non-helical species) and spiroplasmas (helical species).

Mycorrhiza A fungus that survives in the soil and lives in association with plant roots. This type of fungus has a symbiotic relationship with the plant. Mycorrhizae probably assist plants by helping in the uptake of nutrients.

Mycotoxins Poisonous, toxic substances produced by certain fungi. If toxigenic fungi are present on materials such as animal feed, grains, and other foods, the mycotoxins can cause disease and death to the animal or person eating the contaminated food.

Myxomycetes True slime molds, which are not pathogenic to plants.

Necrosis The plant disease symptom of darkened, brown to black tissue, usually consisting of dead cells.

Nematicide A chemical pesticide used to kill or inhibit nematodes.

Nematode An unsegmented, usually microscopic, roundworm in the phylum Nematoda. Different nematode species are parasitic on plants or animals or are free-living in soil, soil organic matter, or water. Some plant parasitic nematodes form disease complexes with pathogenic fungi. The science and study of nematodes is nematology.

Node On a plant the enlarged joint on a stem, usually located where leaves are attached.

Non-circulative virus Referring to a plant pathogenic virus that is borne on the stylet of its insect vector and which does not accumulate internally in the insect.

Nonpathogenic Not able to cause disease.

Nonpersistent A feature of the relationship between a plant virus and its arthropod vector in which the virus does not remain on or in the vector indefinitely. The vector eventually loses the virus, often within a few hours.

Nucleic acid Genetic material that consists of DNA or RNA.

Nucleus The cellular organelle of eukaryotic organisms that contains the chromosomes. Nuclei are bounded by a membrane.

Obligate parasite A parasite that can grow and multiply only on or in living tissue and that cannot be cultured on an artificial medium.

Oogonium The female gametangium of oomycete organisms. Oogonia each contain one or more female gametes.

Oomycete An important group of plant pathogens that includes the downy mildews, *Aphanomyces*, *Phytophthora*, *Pythium*, and others. These organisms lack chitin in their cell walls, usually have zoospores with heterokont flagella (one whiplash, one tinsel type), and possess other features that separate them from the true fungi and align them more closely with certain types of primitive golden algae, diatoms, and giant kelp in Kingdom Chromista or Kingdom Straminipila.

Oospore A sexual spore produced by the union of two morphologically different gametangia (oogonium and antheridium). Oospores are resting spores that can survive during unfavorable conditions.

Ostiole A pore-like opening in some fungal fruiting bodies (perithecia, pycnidia) through which spores leave the fruiting body.

Paraphysis A sterile hypha that is present in some fungal fruiting bodies. The presence or absence of paraphyses, and their morphology if present, are used to help identify fungal species.

Parasite An organism that lives in or on another living organism (the host). The parasite obtains its food and sustenance from the host.

Parasitism The association where one organism (parasite) grows at the expense of another organism (the host).

Pascal (Pa) A metric unit of pressure equivalent to 0.00014504 lb/in^2; 1 kPa = 0.14504 lb/in^2; 100 kPa = 1 bar.

Pathogen A virulent agent that is able to infect plants. For disease to develop on the susceptible host, conditions favorable for the pathogen–host interaction must be present.

Pathogenicity The ability of a pathogen to cause disease. The state of being pathogenic.

Pathovar Abbreviated pv., refers to a 'special form' or biotype of a bacterial pathogen species. A particular pathovar differs from other pathogens in its ability to infect certain plant species. For example, the pathogen *Xanthomonas campestris* pv. *campestris* infects only *Brassica* plants. A similar designation for fungi is forma specialis.

Pedicel The plant stalk that supports and bears flowers.

Perfect fungus A fungus that produces sexual spores. *See* teleomorph.

Perithecium Fruiting body of ascomycete fungi that is a spherical to flask-shaped structure. The inner cavity of perithecia contains ascospore producing asci. Ascospores are released through ostioles in the perithecia.

Persistent A feature of the relationship between a plant virus and its arthropod vector in which the virus remains on or in the vector indefinitely. The vector can transmit the virus throughout its life.

Phenotype The physical characteristics of an organism as dictated by its genetic constitution (genotype).

Phycomycete An obsolete term previously used to describe the 'lower fungi' such as zygomycetes and oomycetes.

Phyllody A plant disease symptom in which flower petals and other floral organs are transformed into green, leaf-like structures. Phyllody can be caused by certain pathogens, insects, mites, and genetic mutations.

Phylloplane The surface of plant leaves.

Phylogenetic tree A diagram that consists of branches that indicate evolutionary relationships between members of the tree.

Phylogeny The science and study of changes and evolution of organisms over time. The evolutionary history of an organism.

Phylum A taxonomic group of organisms that is one category above a class and below a kingdom.

Physiologic race A subgroup or subdivision of a species. The individual organisms making up a physiologic race share common features and genetics. Similar to biotype and strain.

Physiology The science and study of the functions of living organisms or their parts, and organic processes that take place in organisms.

Phytoplasma A plant pathogenic microorganism that has the characteristics of mycoplasmas, but which has not been proven to be a mycoplasma. Phytoplasmas are nonhelical in shape and live in the phloem tissue of their plant hosts. These pathogens were formerly called mycoplasma-like organisms. The helical species are called spiroplasmas.

Phytotoxicity The damage or death that plants sustain due to exposure to injurious factors, primarily chemicals such as pesticides and fertilizers. Symptoms of phytotoxicity include burning or scorching, chlorosis, deformities, inhibited growth and development, and plant death.

Pith The soft, spongy, thin-walled cellular tissue found most often in the central core of plant stems and roots.

Plasmid A small piece of circular DNA that is not part of a cell's chromosomes, is self-replicating, and is found in certain bacteria and fungi.

Pleomorphic Referring to an organism that exists in various shapes and sizes. For example, phytoplasmas have no rigid cell wall, so their shapes vary and are therefore pleomorphic.

Polyclonal antibodies The natural mix of diverse antibodies present in the blood serum of an animal that had been injected with a pathogen or protein. Polyclonal antibodies are components of serological detection methods.

Polycyclic Having more than one cycle per season or host crop. In fungi, this occurs when the pathogen produces additional generations of spores that spread within the same crop and cause new infections.

Polygenic A character or trait that is controlled by many genes.

Polymerase chain reaction (PCR) A technique in molecular biology in which a specific DNA fragment is greatly amplified. The technique uses specially designed primers that hybridize with the target DNA. Temperature changes in the PCR protocol allow for the rapid multiplication of the DNA fragment.

Post-emergence The period of time after the appearance of plant structures above the soil surface following germination of seeds and below-ground growth of the seedling.

Predispose To make susceptible to infection, or to increase susceptibility to infection. Some environmental stresses and factors can predispose plants to infection by pathogens.

Pre-emergence The period of time prior to the appearance of plant structures above the soil surface following germination of seeds and below-ground growth of the seedling.

Primary inoculum The inoculum, or pathogen propagule, that first infects the host and causes initial disease. Infection caused by primary inoculum usually takes place following an overwintering period.

Prokaryote An organism that lacks a membrane-bound nucleus.

Propagule A discrete, separate unit of an organism that is able to grow and propagate the organism.

Protectant A chemical substance that protects the plant against infection by a pathogen only if the pathogen has not yet infected and established itself in the plant. Compare eradicants.

Protozoan An organism in the phylum Protozoa. Protozoans are single-celled, mostly motile, and include a few plant pathogens such as the plasmodiophoromycetes (example clubroot of crucifers) and flagellate protozoans.

Pseudothecium Fruiting body of ascomycete fungi that consists of a mass of hyphae (stroma) and hollow cavities that develop within the stroma. The ascospore-bearing asci are formed within these cavities. Ascomycetes that make pseudothecia are called loculoascomycetes.

Pustule Small, raised, blister-like structure that forms on a leaf infected by fungi such as rusts. Pustules initially are covered by the leaf epidermis, but the plant tissue later ruptures and allows release of spores.

Pycnidium A small, flask-shaped to spherical, fungal fruiting body that produces asexual conidia. Pycnidia are either embedded within or form on the outside of host tissue. Conidia develop within pycnidia and are later released through pores in the structure (ostiole) or when the pycnidia rupture.

Pyriform Pear-shaped. Some fungal spores are pyriform.

Quarantine Temporary holding of imported plants or plant materials in isolation so that the materials can be determined to be free from diseases and pests. Quarantines are regulatory requirements designed to exclude pathogens.

Race A subgroup or subdivision of a species. The individual organisms making up a physiologic race share common features and genetics. Similar to biotype and strain.

Recessive gene A gene that is not fully expressed when paired with a corresponding dominant gene or allele.

Relative humidity (RH) A standard weather measurement of the ratio of the amount of water vapor actually in the air to the greatest amount of water vapor that air can hold at that same temperature. RH is often important in the germination and survival of spores of pathogenic fungi

Replication The process in which a virus pathogen reproduces by inducing the host cell to make more virus entities. In molecular biology, the process in which a nucleic acid (DNA or RNA) is copied.

Resistant (regarding host plants) The genetically based ability of a plant to not become infected, or when a plant minimizes the effect of a pathogen. Related to immune.

Resistant (regarding pesticides) Often referring to pathogens that have adapted and changed so that the resistant isolate is no longer killed by and susceptible to a pesticide or other control chemical. For example, a resistant fungal pathogen is not controlled by a particular fungicide, and a resistant bacterial pathogen is not controlled by a particular antibiotic. Some pathologists consider resistant to be synonymous with insensitive.

Resting spore Any spore that is able to survive despite unfavorable conditions such as drying, high temperatures, and lack of a host. An example of a resting spore is the thick walled, sexual oospore of oomycetes.

Rhizoid Fungal filaments and hyphae that grow root-like towards nutrients and substrates.

Rhizomorph An aggregation or bundle of hyphal strands that appears root-like. Rhizomorphs function as a means of spreading the fungus and as an aid to survival.

Rhizosphere The soil environment near and around the roots of plants.

Ring spot The plant disease symptom where multiple circles (which usually are alternating in appearance between dark and light colors) form within one larger circle or lesion. All the circles share the same center. This symptom is also called a concentric spot.

RNA Ribonucleic acid. RNA is any of various nucleic acids that occur in nuclei (for eukaryotic organisms) and cell contents (eukaryotic and prokaryotic organisms) of living organisms. RNA molecules are composed of repeating sub-units of ribose (a five-carbon sugar), phosphoric acid, and four nitrogenous bases (adenine, cytosine, guanine, and uracil). RNA is involved with protein synthesis. RNA is the only nucleic acid in some viruses.

Rogue The act of removing undesirable plants from a field or planting. Plants are rogued because they may show symptoms of disease, exhibit abnormal growth, or not conform (off-types) to the desired cultivar.

Rosette The plant disease symptom where plant growth is short and stunted, and leaves and stems are bunched together and compressed.

Rot The plant disease symptom of soft, mushy, and discolored tissue. Rots usually involve the disintegration of plant tissues due to release of enzymes and other factors by the pathogen.

Sanitation The removal and disposal of diseased plants and other contaminated material. Also, the cleansing or decontamination of production equipment, tools, hands, and other objects that might be sources of pathogen inoculum.

Saprobe An organism, usually a microbe, that colonizes, feeds on, and decomposes dead organic matter. A true saprobe is not pathogenic to plants. However, some pathogens can function as saprobes. A synonym is saprophyte.

Scab The plant disease symptom where affected plant surfaces are rough, crusty, and dry.

Sclerotium A compact mass of fungal growth that enables the organism to survive unfavorable conditions. Sclerotia are mostly dark-colored, hard, with smooth and rounded surfaces, and having a distinct outer rind (as for *Sclerotinia* and *Sclerotium* species). For some fungi, such as *Rhizoctonia solani*, sclerotia are made up of closely interwoven hyphae and lack the rind layer.

Scorch The plant disease or damage symptom that resembles the effects of burning on a leaf or other plant part. Scorch can be caused by pathogens, nutrient deficiencies, or various abiotic factors such as chemical damage, high salt content in soil and water, or weather extremes.

Secondary inoculum The inoculum, or pathogen propagule, that develops following initial disease development. Secondary inoculum is produced after the primary inoculum already incited disease on the host.

Secondary organism An organism that colonizes and multiplies in already diseased tissue. Secondary organisms do not cause disease on their own.

Seed transmission Passage of inoculum from an infected or infested seed to the germinating plant that grows from that seed.

Seed treatment Chemical or other treatment used to reduce or eliminate the presence of pathogens on seeds. Also, the application of fungicides to seeds so that when seeds are planted they are protected against damping-off pathogens.

Seedborne The situation in which a disease, pathogen, or pathogen inoculum is present on or in seed.

Selection In agriculture, the process of choosing certain plants, having desired traits, for use in growing and breeding crops. Also, selection is the biological process where certain microorganisms, having survival traits such as resistance to pesticides or the ability to overcome plant host resistance, are able to survive in greater numbers than individuals that do not possess the same trait.

Senescence The natural aging process of plant tissues as tissues mature, then decline. Senescence can also be induced by diseases.

Septate Referring to the presence of cross walls (septa) between cells.

Septum The cross wall that divides fungal hyphae or spores into separate cells. The presence or absence of septa characterizes some fungal groups.

Serology The science and study of antigen and antibody interactions. Serological methods are critical for the identification of protein antigens of plant pathogens, and are the basis for detection methods such as ELISA.

Seta In fungi, a sterile, hair-like structure.

Sexual Reproduction that involves the union of gametes and meiosis. Fungi that reproduce sexually are called perfect or teleomorphic fungi.

Shot-hole The plant disease symptom of small, often circular or oval leaf spots that later become necrotic. The necrotic tissue will dry up and fall out of the leaf, leaving a hole.

Sign The visible presence of the pathogen on host tissue. Signs usually are limited to fungal growth, spores, and other structures, and bacterial ooze and accumulation. Signs are distinct from symptoms.

Sinuate Referring to the edge or shape of an object that is serpentine, wavy, or flexuous.

Soil inhabitant A microorganism that is able to survive in the soil for many years or indefinitely. Soil inhabitants that are pathogenic to plants usually have a saprophytic ability as well.

Soil invader A microorganism that is not able to survive and persist in the soil for more than a relatively short time of one or two years. A synonym is soil transient.

Soilborne pathogen A pathogen that is present in soil, is usually a soil inhabitant, and which survives by inoculum that is present in soil.

Solarization The use of solar radiation to treat soil and reduce populations of soilborne pathogens. Solarization usually involves the placement of clear plastic tarps on the soil surface, then allowing the sun to heat the soil for a number of days or weeks.

Sorus A compact mass of spores in a fruiting structure. This term is used especially for rust and smut fungi.

Species The smallest formal classification unit in taxonomy. Species is one category below genus.

Spiroplasma A plant pathogenic microorganism that has the characteristics of mycoplasmas, but which has not been proven to be a mycoplasma. Spiroplasmas are helical in shape and live in the phloem tissue of their plant hosts. These pathogens were formerly called mycoplasma-like organisms. The nonhelical species are called phytoplasmas.

Sporangiophore The specialized hypha that bears one or more sporangia.

Sporangiospore The asexual spore that differentiates within and is released from a sporangium. If the sporangiospores are motile and swim in water, they can also be called zoospores.

Sporangium A fungal reproductive structure that encloses cytoplasm and cell contents that later differentiate into individual spores (sporangiospores). The cell walls of the spores are distinct and not derived from the sporangium. In some cases the cytoplasm does not differentiate, and the sporangium itself can function as a spore (hence can be called a conidium in this situation). If the sporangiospores are motile and swim in water, the sporangium can also be called a zoosporangium.

Spore The reproductive unit of fungi, some bacteria, and other microorganisms.

Sporodochium A reproductive structure of asexual fungi that consists of a raised cushion-shaped mass of hyphae. The sporodochia then support short conidiophores.

Sporulate The action of forming and releasing spores.

Sterile For an organism, the state of being infertile. Also, the condition of being free from living microorganisms and contaminants.

Strain A subgroup or subdivision of a species. The individual organisms making up a strain share common features and genetics. Similar to biotype and race. Also, in bacteriology and virology a group or collection of clonally related individuals derived from one isolation and maintained as a pure culture.

Straminipila The name of the taxonomic kingdom that some taxonomists believe should contain the important oomycete pathogen group. Kingdom Straminipila is distinct from fungi and contains certain types of primitive golden algae, diatoms, and giant kelp. However, other researchers believe the appropriate name for this group should be Kingdom Chromista.

Stroma A mass of hyphae of fungi. The stromata themselves do not produce spores and are therefore vegetative in nature. Stromata can contain or support the reproductive structures.

Stylet The long, slender, hollow mouth structure of piercing and sucking insects such as aphids and leafhoppers. Viruliferous aphids insert their stylets into plant tissues and transmit virus particles. Also, the spear-like structure of plant parasitic nematodes that enables these nematodes to feed on plants.

Stylet-borne virus Referring to a non-circulative virus that is borne on the stylet of its insect vector.

Substrate The surface or medium upon which a microorganism grows.

Suppressive soil A soil in which some soilborne diseases are reduced in severity or incidence. The mechanisms of suppressive soils are not well understood, but probably involve the activity of antagonistic and beneficial microorganisms.

Susceptible Subject to infection.

Symptom The visual manifestation of disease (or abiotic disorder). The reaction of the host to infection. Symptoms do not include the actual physical presence of the pathogen (see sign).

Synnema A reproductive structure of asexual fungi that consists of a column of compacted, erect, fused conidiophores. On synnemata, conidia are produced only on the outer surfaces of these columns. The term coremium is also used.

Systemic Related to the movement of a chemical, pathogen, or infection that spreads internally through the plant's vascular system (primarily the xylem). In contrast to localized.

Taproot The main, central root of a plant from which secondary or lateral roots arise.

Taxonomy The science and study of systematic naming, ordering, and classifying of organisms on the basis of their natural relationships, characteristics, and genetics.

Teleomorph The sexual stage of a fungus. Spores and propagules produced in this stage involve sexual recombination or processes. This form of fungi is also called the perfect stage. Teleomorph fungi include ascomycetes, basidiomycetes, and zygomycetes.

Teliospore One of the several different types of rust fungi spores. Teliospores are borne in and on telia and often are the overwintering spore stage. Teliospores germinate to form basidia and basidiospores. *See also* aeciospore, urediniospore, and basidiospore.

Telium The fruiting structure of rust fungi that produces teliospores.

Tendril (of fungi) A small, linear mass of spores exuded from fungal fruiting bodies called pycnidia. Tendrils are curled and held together by a slimy matrix material. Also called a cirrus.

Tolerance Ability of a plant to endure a disease without sustaining serious damage or yield loss. Related to and sometimes considered synonymous with resistance.

Toxin A non-enzymic metabolite produced by one organism that is damaging or inhibitory to another organism. Mycotoxins are toxins that are produced by fungi, which are therefore considered to be mycotoxigenic species.

Transformation The change of a cell through uptake and expression of additional genetic material.

Transgenic Referring to organisms that have been altered by the insertion of genes via modern molecular techniques.

Transmission The transfer or spread of pathogens, especially viruses, from plant to plant, seed to plant, or from vector to plant.

Transverse section Viewing plant tissue that has been cut or dissected crosswise, at right angles to the longitudinal axis.

Trichome A hair-like cell projection on the epidermis of plants.

Tylosis Balloon-like growth that develops and blocks vascular cells in the xylem of plants. Tyloses sometimes form in reaction to infection by vascular pathogens such as *Verticillium dahliae*.

Umbel A type of flower structure in which flowers are borne on pedicels that arise from a common attachment. The result is a cluster of flowers having a rounded or flat top. Umbels are characteristic of plants in the Apiaceae family.

Urediniospore One of the several different types of rust fungi spores. Urediniospores are borne in and on uredinia, windborne, and responsible for widespread distribution of inoculum. *See also* aeciospore, teliospore, and basidiospore.

Uredinium The fruiting structure of rust fungi that produces urediniospores.

Variability The phenomenon in which some organisms have slightly different traits even though they are of the same species.

Variety In agriculture and horticulture, a cultivated variety of plant. All individuals of a particular variety share characteristic features. Variety is generally used synonymously with cultivar.

Vascular pathogen A pathogen that is able to colonize and move about in the vascular tissue of plants.

Vascular tissue The conductive tissue (xylem and phloem) of plants.

Vector An organism that can acquire and subsequently transmit a pathogen (mostly viruses but also fungi and bacteria) to plants. Primary vectors include insects, mites, nematodes, and primitive soil fungi.

Vegetative compatibility The ability of vegetative (non-reproductive) hyphae to fuse together (anastomose). This fusion usually results in combination and exchange of genetic material. For some fungi, pairing of vegetatively compatible isolates assists in the classification and identification of groups within a species.

Verrucose Having a surface with small, rounded structures or warts.

Verruculose Delicately or minutely verrucose.

Vertical resistance Complete resistance in plants that provides resistance to some races of a pathogen but not to others. In contrast to horizontal resistance, this type of resistance is often determined by a single gene.

Verticillate In fungi, having phialides or other structures that arise from a common point. Structures occurring in whorls.

Vesicle A balloon-like structure that expands from sporangia of *Pythium* organisms and in which the zoospores differentiate and are released. Vesicles also refer to other mycology structures, such as swellings in *Aspergillus* conidiophores.

Virescence The plant disease symptom in which normally white or non-green tissue becomes green.

Virion The complete virus particle that consists of a nucleic acid (RNA or DNA) and a surrounding protein coat.

Viroid A type of pathogen that consists only of a single-stranded RNA. Viroids do not have an enclosing protein coat, as is the case with viruses.

Virulence The degree or severity of pathogenicity.

Virulent Referring to something that is pathogenic.

Viruliferous Referring to a vector that has acquired a virus and is able to transmit it.

Virus A submicroscopic pathogen that consists of a nucleic acid (RNA or DNA) and an enclosing protein coat. The science and study of virus and virus-like organisms is virology.

Water-soaked The plant disease symptom where succulent tissue appears wet, greasy or dark due to the activity of a pathogen. Water-soaked symptoms are often the initial symptoms of bacterial or fungal infections of leaves and shoots.

Wilt or wilting The plant disease symptom where leaves, shoots, and foliage in general will droop and sag. Wilting is often caused by insufficient water in the plant as a result of damaged roots or plugged vascular systems.

Witches'-broom The plant disease symptom where shoots are abnormally clustered into a tight brush-like mass. These witches'-broom shoots result from a proliferation of buds from one section of the plant.

Yellows The plant disease category characterized by yellowing of plant foliage and sometimes stunting of the diseased plant. The yellowing symptom can also be called chlorosis.

Zonation The plant disease symptom where multiple circles (usually alternating in appearance between dark and light colors) form within one larger circle or lesion. All the circles share the same center. This symptom is also called a concentric spot or ring spot.

Zoosporangium A sporangium which produces and releases swimming zoospores.

Zoospore An asexual sporangiospore that is motile and can swim in water.

Zygomycete A group of fungi that produce sexual zygospores. Zygomycetes lack cross walls (septa) in their vegetative hyphae and produce asexual sporangiospores.

Zygospore The sexual or resting spore of zygomycetes. Zygospores are produced by the fusion of two morphologically similar gametangia.

Index

Note: Page numbers in *italic* refer to tables in the text

B

P

R

T

U

V